P R O T E I N S

PROTEINS

Structures and Molecular Principles

THOMAS E. CREIGHTON

Medical Research Council
Laboratory of Molecular Biology
Cambridge, England

W. H. FREEMAN AND COMPANY   NEW YORK

Cover and text design: Edward A. Butler

Library of Congress Cataloging in Publication Data
Creighton, Thomas E., 1940 -
    Proteins.

    Includes bibliographies and index.
    1. Proteins. I. Title. [DNLM: 1. Proteins. QU 55
C914p]
QP551.C737   1984   574. 19'245   83-19541
ISBN 0-7167-1566-X

CONTENTS

PREFACE

While working with proteins during the past 20 years, I often felt the need for a comprehensive text on their physical and chemical properties. Such a text was simply not available. A search of the literature for information about any of the many diverse aspects of proteins required countless hours. No less daunting was the task of reconciling the different, often conflicting, explanations given for these properties. This book attempts to provide such information within a coherent framework.

Proteins have been studied for so long by so many different people that there is no shortage of material in the literature. However, because it is scattered very widely, from journals of physical chemistry to those of microbiology and genetics, the information needed to be gathered together and then sifted, sorted, and rationalized. This process revealed that most of the information is incomplete; in many cases, research of a particular problem apparently ended when only the vague outline of the answer became evident. Perhaps those working directly on such a problem see the answer clearly but usually do not bother to write a definitive account, presumably moving on to newer, more exciting fields. Much of the material is also contradictory. When a choice between conflicting views became necessary, I have attempted to write as definitive an account as possible and have presented the view that is both more plausible and more compatible with all other knowledge about proteins.

Upon some aspects, such as protein stability, this volume presents a very personal point of view, which does not represent a consensus. I have been driven to this only when there was no plausible treatment of the problem in the literature. Otherwise, I have attempted to convey the conventional wisdom that has evolved, while trying to analyze its flaws, point out its deficiencies, and present it in the most convincing manner. Although this volume is as comprehensive as possible within the limits of the space and information available, it cannot claim to be complete, for there are still many aspects of protein structure and function that are not understood.

The reader is expected to have primarily a background in biology, with some knowledge of biochemistry, genetics, and cell biology. Accordingly, this book addresses at the lowest level the physical and chemical properties of proteins that are least likely to be intuitively obvious to such a reader; the treatment is not rigorous but is often empirical, presented in a manner meant to convey principles without obscuring them in mathematical derivations. (The reader is directed to textbooks of biophysical chemistry for such rigorous derivations.) To a chemist or physicist with no biology background, many of the discussions of physical chemistry may seem trivial, but the significance of many of the biological properties should not be overlooked. (A complete explanation of all the biological phenomena would have required a volume many times as large; and there are very many biologic textbooks that present such information clearly and comprehensively.)

The text is organized on a combined basis of the chronological events in the life of a protein and of structural and functional hierarchy. Biosynthesis is discussed early (Chapter 2), although first it is necessary to introduce the parts that make up proteins (Chapter 1); the volume ends with a discussion of degradation (Chapter 10). The first three chapters deal with properties of linear polypeptide chains, where only the primary structure is immediately relevant, as in biosynthesis, and includes a discussion of the origins of contemporary primary structures (Chapter 3). The next three chapters describe the three-dimensional aspects of proteins, starting with a brief description of the forces controlling this level of structure (Chapter 4), continuing with their effects on conformational properties of the polypeptide chain (Chapter 5), and ending with a description of the folded structures of globular proteins (Chapter 6). The subsequent three chapters describe the consequences of the folded structures of proteins, including their dynamic properties in solution (Chapter 7), their interactions with other molecules (Chapter 8), and their ability to catalyze chemical reactions and interconvert different forms of energy (Chapter 9).

Documentation of each statement in the text with a reference is not necessary for a volume at this level. Instead, a few of the most pertinent references are listed at the end of each section, for convenience and effeciency; the more general references are listed after the introduction to a chapter or to a group of sections.

I have endeavoured to select references on the basis of their usefulness to the reader, rather than the gratification of scientists' need for recognition. Consequently, the choice of references should not be taken as evidence of their contribution to advancing knowledge but simply reflects that they are useful compendiums of data, present coherent discussions, or simply have nice illustrations. Only very exceptionally important references more than ten years old are cited; some of the most recent references were included simply to provide a recent entry into the literature of a particular field, as a starting point for tracing back earlier references, and as an indication of the current status of the field.

This seems an appropriate opportunity to thank those who contributed both directly and indirectly to this book. First mention must go to my colleagues and the many visitors at the Medical Research Council Laboratory of Molecular Biology, who provide an unsurpassed scientific environment, especially those with whom I have spent many hours in the canteen discussing proteins and other topics. In particular, David Goldenberg provided numerous perceptive comments and questions in such conversations; he also contributed directly to this volume by reading the entire manuscript and pointing out errors, inconsistencies and shortcomings. More remote, but no less important, contributions were made by those who provided the very important support and encouragement at the start of my career: Robert L. Sinsheimer and George Guthrie, who introduced me to molecular biology and to research while I was an undergraduate at the California Institute of Technology; and Charles Yanofsky, who patiently guided me to the Ph.D. degree at Stanford, and who provided an exemplary model of how being a great scientist need not prevent one from also being a kind and generous human being.

1

CHEMICAL NATURE
OF POLYPEPTIDES

Confronting the full three-dimensional model of a typical protein, with its multitude of H, C, O, N, and S atoms, can be an awesome experience. To the uninitiated eye, it is simply an incomprehensible myriad of atoms and bonds. Nevertheless, it can be comprehended, even admired, once a few principles of protein structure are grasped. Then, many of the biological properties of the protein can be rationalized. The first step is to divide the protein up into its various components and to become acquainted with their individual properties.

Proteins fall into the general class of polymers, which are simply linear molecules built up from simple repeating units, the monomers. In the case of proteins, the monomers are the amino acids; 20 different amino acids are used.

In one sense, proteins are more complex than most polymers in that 20 different monomers are used in their construction, whereas many polymers have only a single type of monomer. In another sense, however, proteins are structurally less complex: Most chemical polymers are created by polymerizing a mixture of the monomers, thereby producing a distribution of chain lengths and an approximately random sequence if more than one type of monomer is present. In contrast, proteins have precise lengths of polypeptide chain and exact sequences of the amino acids. They also are linear and unbranched. Most important, they have the property of acquiring very specific folded three-dimensional conformations, which will be described in later chapters.

Of the 20 amino acids normally used to build proteins, 19 have the general structure

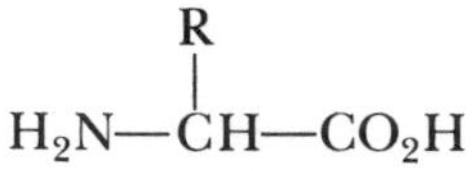

$$R$$
$$H_2N-CH-CO_2H \qquad\qquad (1-1)$$

and differ only in the chemical structure of the side chain, R. Proline, the 20th natural amino acid, is similar but has the side chain bonded to the nitrogen atom, to give the imino acid:

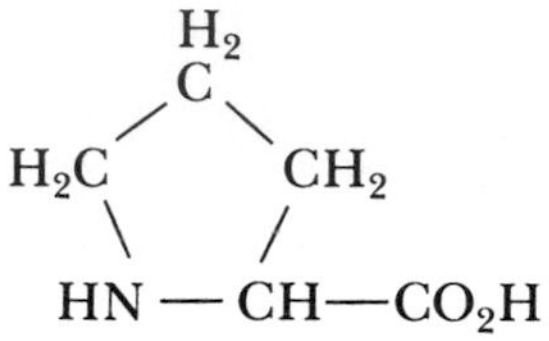

$$
\begin{array}{c}
H_2 \\
C \\
H_2C \qquad CH_2 \\
HN - CH-CO_2H
\end{array}
\qquad (1-2)
$$

Except in glycine, where the side chain is simply a hydrogen atom, the α-carbon atom is asymmetric and always of the L isomer:

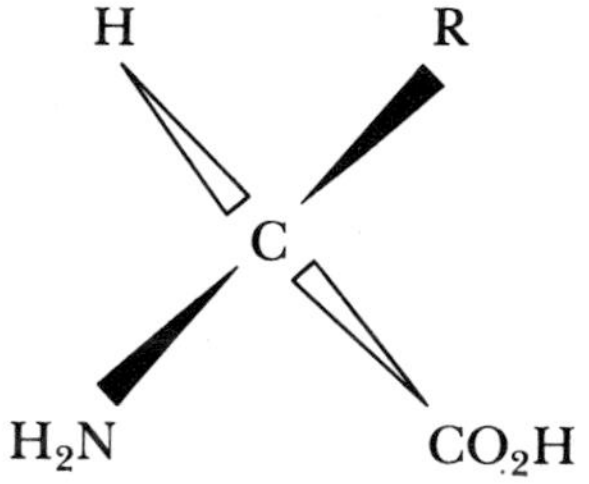

$$
\begin{array}{c}
H \qquad\qquad R \\
C \\
H_2N \qquad\qquad CO_2H
\end{array}
\qquad (1-3)
$$

The structures of the side chains of the 20 amino acids are illustrated in Figure 1–1. Also given are their three- and one-letter abbreviations. Because the former are obvious and used widely, they will be employed here.

The 20 amino acids are assembled into proteins by linking them together via peptide bonds, as illustrated here for the condensation of two amino acids:

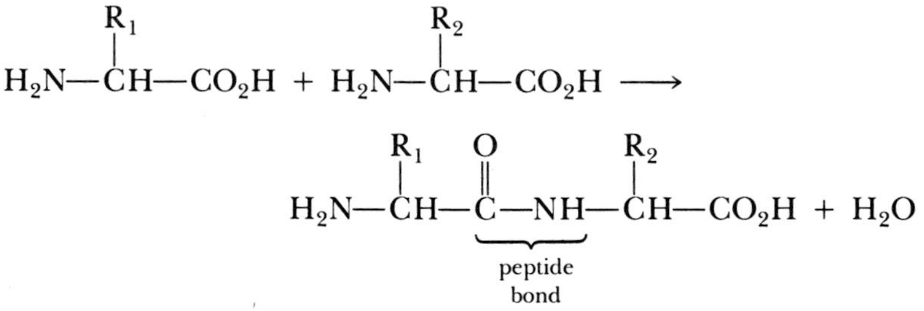

$$
\begin{array}{cc}
R_1 & R_2 \\
H_2N-CH-CO_2H & + \quad H_2N-CH-CO_2H \longrightarrow
\end{array}
$$

$$
\begin{array}{ccc}
R_1 & O & R_2 \\
H_2N-CH-C-NH-CH-CO_2H & + & H_2O
\end{array}
\qquad (1-4)
$$

peptide
bond

Many such amino acids, generally from 50 to 1000, are linked together in this way to form a linear polypeptide chain. The polypeptide backbone is simply a repetition of the basic amino acid unit, which is described as an amino acid **residue** when incorporated into a polypeptide chain:

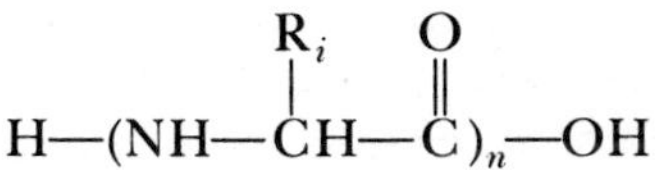

$$
\begin{array}{cc}
R_i & O \\
H-(NH-CH-C)_n-OH
\end{array}
\qquad (1-5)
$$

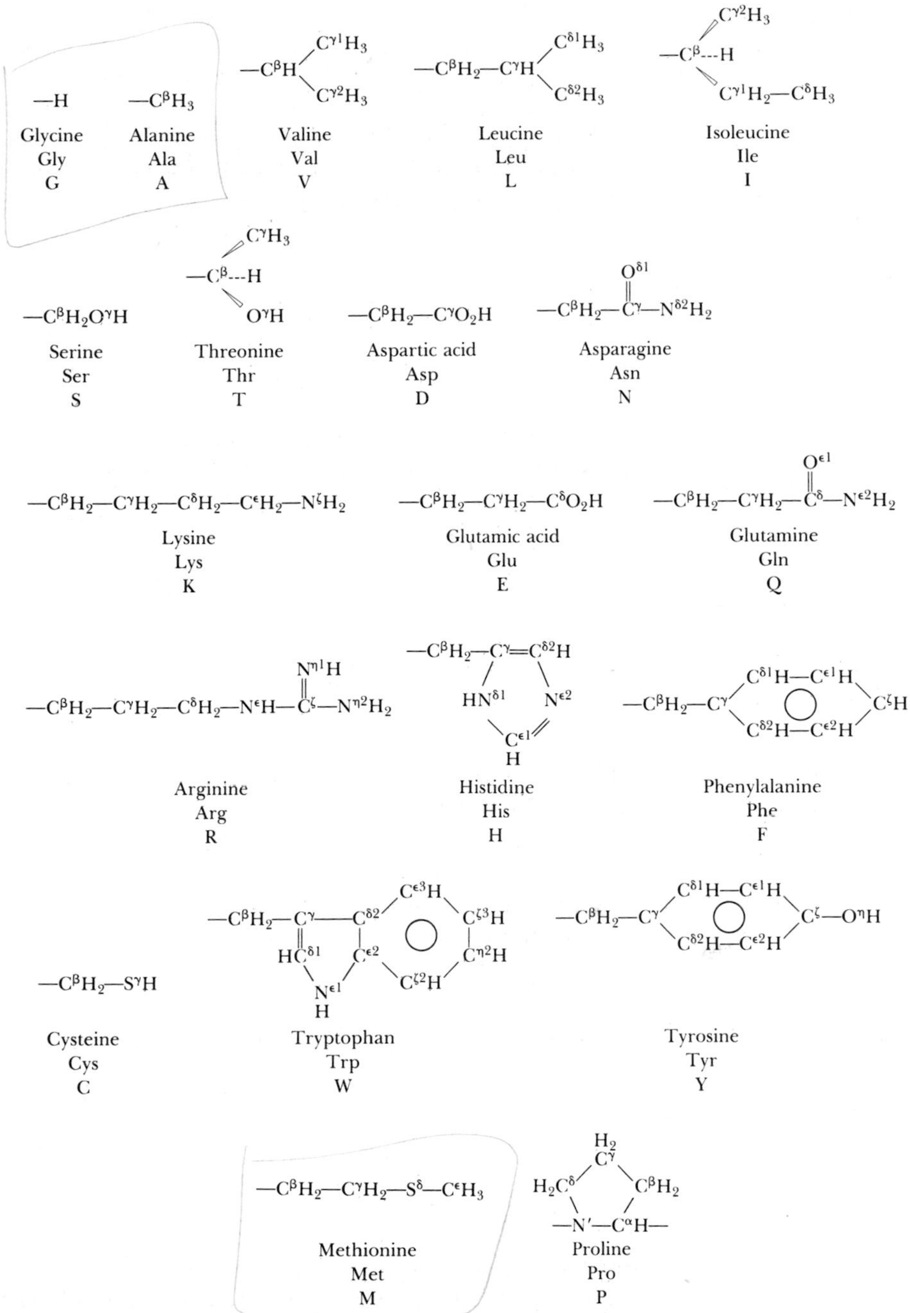

Figure 1–1

Side chains of the 20 amino acids used to synthesize proteins. The full names and the three- and one-letter abbreviations are given below the structures. The C and N atoms of the backbone are also included in the unique case of Pro. The designations of the atoms are those recommended by the IUPAC–IUB Commission on Biochemical Nomenclature. (In J. Mol. Biol. 52:1–17, 1970.)

All proteins and polypeptides have this simple basic structure and differ only in the number of amino acid residues linked together in the chain (n in Eq. 1–5) and in the sequence in which the various amino acids occur in the polypeptide chain.

It may be useful at this stage to define the various terms that are used. A **peptide** generally refers to only a small number of amino acid residues linked together, usually with a defined sequence. No particular maximum number of residues may be specified, but the term peptide is appropriate if the physical properties are generally those expected from the total of the constituent amino acids. A **polypeptide** generally refers to longer chains, but with either the sequence or the length not defined. Such polymers are often prepared by chemical polymerization of one or a few amino acids into random sequences of varying lengths. They usually have no defined conformation, or they acquire simple repetitive structures such as helices or sheets (see Chapter 4). Proteins of specific sequence are often referred to as polypeptides if they are not in a defined conformation.

The term **protein** is usually reserved for those chains with a specific sequence, length, and folded conformation. It is this class that is the subject of this volume, although useful reference will also be made to peptides and polypeptides.

Proteins, Amino Acids, and Peptides. E. J. Cohn and J. T. Edsall. Princeton, N.J., Van Nostrand-Reinhold, 1943.

X-ray studies of amino acids and peptides. R. B. Corey. Adv. Protein Chem. 4:385–406, 1948.

Crystal structure studies of amino acids and peptides. R. E. Marsh and J. Donohue. Adv. Protein Chem. 22:235–256, 1967.

The Structure and Action of Proteins. R. E. Dickerson and I. Geis. New York, Harper & Row, 1969.

Nomenclature of α-amino acids. IUPAC Commission on the Nomenclature of Organic Chemistry and IUPAC–IUB Commission on Biochemical Nomenclature. Eur. J. Biochem. 53:1–14, 1975.

THE POLYPEPTIDE BACKBONE

The peptide backbone consists of a repeated sequence of three atoms: the amide N, the alpha C, and the carbonyl C:

$$\underset{\begin{array}{c}\text{H} \\ | \\ \end{array}}{\text{—N—}}\underset{}{\text{CH—}}\overset{\begin{array}{c}\text{O} \\ \| \\ \end{array}}{\text{C—}} \qquad (1\text{–}6)$$

which are generally represented as N_i, C_i^α, and C_i', respectively, where i is the number of the residue, starting from the amino end.

The dimensions of the peptide group derived from three-dimensional crystal-structure analyses of small peptides are given in Figure 1–2. The maximum distance between repeating atoms in the polypeptide backbone is 3.80 Å, when the peptide bond is *trans*. In a fully extended chain consisting of many residues, the repeating units are staggered, so that the maximum linear dimension of a polypeptide with n residues is $n \times 3.63$ Å.

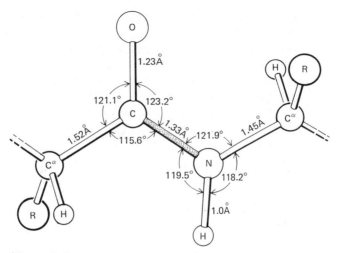

Figure 1–2
The geometry of the peptide backbone, with a trans peptide bond, showing all the atoms between two C^α atoms of adjacent residues. The peptide bond is stippled. The dimensions given are the averages observed crystallographically in amino acids and small peptides. (Adapted from G. N. Ramachandran, et al., Biochim. Biophys. Acta 359:298–302, 1974.)

The presence of an asymmetric center at the C^α carbon atom, with only L-amino acid residues, results in an inherent asymmetry of the polypeptide chain. This will be shown to be important for its spectral properties and conformation. In principle, rotation may occur about any of the three bonds of the polypeptide backbone. However, the peptide bond appears to have partial double-bonded character, presumably owing to resonance:

$$O \qquad C^\alpha \qquad O^- \qquad C^\alpha$$
$$C-N \qquad \longleftrightarrow \qquad C=N$$
$$C^\alpha \qquad H \qquad C^\alpha \qquad H^+ \qquad\qquad (1\text{–}7)$$

so that the six atoms depicted above have a strong tendency to be coplanar. Consequently, the C'—N bond length is observed crystallographically to be 1.33 Å, shorter than a normal bond length of 1.45 Å, as in the C^α—N bond, but also longer than the value of 1.25 A for the average C=N bond length in model compounds. The peptide bond appears to have approximately 40 percent double-bonded character. Rotation of this bond is then markedly restricted, but two configurations of the planar peptide bond are possible, one in which the C^α atoms are *trans,* the other with them *cis:*

$$O \qquad C^\alpha \qquad O \qquad H$$
$$C-N \qquad \longleftrightarrow \qquad C-N$$
$$C^\alpha \qquad H \qquad C^\alpha \qquad C^\alpha \qquad\qquad (1\text{–}8)$$
$$\text{trans} \qquad\qquad\qquad \text{cis}$$

The *trans* form is favored energetically, probably owing to fewer repulsions between nonbonded atoms.

As a result of this resonance, the peptide backbone is relatively polar, with the electrons tending to be redistributed to give the permanent dipole indicated in the resonance state depicted in Equation 1–7. The H and N atoms appear to have positive and negative, respectively, equivalent charges of 0.28 electron, while C and O have respective positive and negative equivalent charges of 0.39 electron.

The other two types of bonds of the peptide backbone appear to behave as normal C—C and C—N bonds.

The peptide backbone is chemically rather unreactive. A proton is added or lost only at extremes of pH. The apparent pK_a value for deprotonation of the amide NH of the polypeptide backbone is between 15 and 18, whereas it is about -8 to -12 for protonation. The oxygen atom is protonated more readily, with an apparent pK_a value of about -1. These properties are extremely useful for making possible measurement of hydrogen exchange from the backbone, which is important for studying the conformation in solution (see Chapter 7). At pH values where protonation or deprotonation becomes significant, the polypeptide chain is generally hydrolyzed to yield the substituent amino acids. For example, the standard method for determining the amino acid composition of proteins or peptides is to heat them at 105° C for about 24 hours in 6 M HCl. Other chemical alterations of the peptide chain require either drastic conditions or the relatively close proximity of certain reactive groups attached to the amino acid side chains, which may be used profitably for selective cleavage of the polypeptide chain at certain amino acid residues (see later section, Fragmentation of a Protein into Peptides).

The Nature of the Chemical Bond, 3rd ed. L. Pauling. Ithaca, N.Y., Cornell University Press, 1960.

Conformational energy estimates for statistically coiling polypeptide chains. D. A. Brant, et al. J. Mol. Biol. 23:47–65, 1967.

Abbreviations and symbols for the description of the conformation of polypeptide chains. IUPAC–IUB Commission on Biochemical Nomenclature. J. Mol. Biol. 52:1–17, 1970.

Hydrogen exchange. S. W. Englander, et al. Ann. Rev. Biochem. 41:903–924, 1972.

Coordinating properties of the amide bond. Stability and structure of metal ion complexes of peptides and related ligands. H. Sigel and R. B. Martin. Chem. Rev. 82:385–426, 1982.

AMINO ACID SIDE CHAINS

The 20 different amino acid side chains possess a variety of chemical properties which, when combined on a single molecule, give a protein properties far beyond those possible with organic molecules. This chemical diversity is vital for the unique functions of biological proteins. As will become apparent, the chemical properties of the whole protein are far greater than the sum of its constituent amino acids, but the intrinsic chemical properties of these 20 different side chains are important as a starting point for considering their roles in proteins.

Some pertinent chemical and physical properties of the 20 amino acid residues are summarized in Table 1–1. References to their detailed geometries determined crystallographically are given in Table 1–2. Each of the side chains will be discussed briefly, describing their normal chemical properties. However, specific residues in folded proteins may have very different properties, as will be discussed in Chapter 7.

The reactions by which the various side chains can be modified will be those that are moderately selective and specific, so that they may be used on proteins and peptides.

Chemistry of the Amino Acids. J. P. Greenstein and M. Winitz. New York, John Wiley & Sons, 1961.

Chemical Modification of Proteins. G. E. Means and R. E. Feeney. San Francisco, Holden-Day, 1971.

Table 1–1 *Properties of Amino Acid Residues*

	ONE-LETTER ABBREVIATION	MASS[a] (daltons)	VOLUME[b] (Å^3)	ACCESSIBLE SURFACE AREA[c] (Å^2)	PARTIAL SPECIFIC VOLUME[b] (ml/gm)	pK_a OF IONIZING SIDE CHAIN[d]	OCCURRENCE IN PROTEINS[e] (%)
Ala	A	71.08	88.6	115	0.748		9.0
Arg	R	156.20	173.4	225	0.666	~12	4.7
Asn	N	114.11	117.7	160	0.619		4.4
Asp	D	115.09	111.1	150	0.579	4.5	5.5
Cys	C	103.14	108.5	135	0.631	9.1 to 9.5	2.8
Gln	Q	128.14	143.9	180	0.674		3.9
Glu	E	129.12	138.4	190	0.643	4.6	6.2
Gly	G	57.06	60.1	75	0.632		7.5
His	H	137.15	153.2	195	0.670	6.2	2.1
Ile	I	113.17	166.7	175	0.884		4.6
Leu	L	113.17	166.7	170	0.884		7.5
Lys	K	128.18	168.6	200	0.789	10.4	7.0
Met	M	131.21	162.9	185	0.745		1.7
Phe	F	147.18	189.9	210	0.774		3.5
Pro	P	97.12	122.7	145	0.758		4.6
Ser	S	87.08	89.0	115	0.613		7.1
Thr	T	101.11	116.1	140	0.689		6.0
Trp	W	186.21	227.8	255	0.734		1.1
Tyr	Y	163.18	193.6	230	0.712	9.7	3.5
Val	V	99.14	140.0	155	0.847		6.9
						α-Amino 6.8 to 7.9	
						α-Carboxyl 3.5 to 4.3	

[a]Molecular weight of amino acid minus that of water. Values from Handbook of Chemistry and Physics, 43rd ed. Cleveland, Chemical Rubber Publishing Co., 1961.

[b]Values from A. A. Zamyatnin, Prog. Biophys. Mol. Biol. 24:107–123, 1972.

[c]Values from C. Chothia, J. Mol. Biol. 105:1–14, 1975. The accessible surface area is defined in Figure 6–20.

[d]Estimated from the pK_a values of small model compounds, from C. Tanford, Adv. Protein Chem. 17:69–165, 1962.

[e]Frequency of occurrence of each amino acid residue in the primary structures of 207 unrelated proteins of known sequence. Values from M. H. Klapper, Biochem. Biophys. Res. Commun. 78:1018–1024, 1977.

Table 1–2   Reference Sources for Detailed Geometries of the Amino Acids

Alanine	M. S. Lehmann, et al., J. Am. Chem. Soc. 94:2657–2663, 1972
Arginine	M. S. Lehmann, et al., J. Chem. Soc. [Perkin II] 2:133–135, 1973
Asparagine	J. J. Verbist, et al., Acta Cryst. B28:3006–3013, 1972
Aspartic acid	D. S. Eggleston, et al., Acta Cryst. B37:1428–1430, 1981
Cysteine	K. A. Kerr and J. P. Ashmore, Acta Cryst. B29:2124–2127, 1973
Cystine	D. D. Jones, et al., Acta Cryst. B30:1220–1228, 1974
Glutamic acid	M. S. Lehmann, et al., J. Cryst. Mol. Struct. 2:225–233, 1972
Glutamine	T. F. Koetzle, et al., Acta Cryst. B29:2571–2575, 1973
Glycine	D. S. Eggleston, et al., Acta Cryst. B37:1427–1430, 1981
Histidine	T. J. Kristenmacher, et al., Acta Cryst. B28:3352–3361, 1972
Isoleucine	K. Torii and Y. Iitaka, Acta Cryst. B27:2237–2246, 1971
Leucine	L. Golic and W. C. Hamilton, Acta Cryst. B28:1265–1271, 1972
Lysine	T. F. Koctzle, et al., Acta Cryst. B28:3207–3214, 1972
Methionine	C. Chen and R. Parthasarathy, Acta Cryst. B33:3332–3336, 1977
Phenylalanine	A. R. Al-KaraGhouli and T. F. Koetzle, Acta Cryst. B31:2461–2465, 1975
Proline	I. Tanaka, et al., Acta Cryst. B33:116–119, 1977
Serine	M. N. Frey, et al., Acta Cryst. B29:876–884, 1973
Threonine	V. S. Yadava and V. M. Padmanabhan, Acta Cryst. B29:854–858, 1973
Tryptophan	L. C. Andrews, et al., Amer. Cryst. Assoc. Abstracts G6, 1974
Tyrosine	M. N. Frey, et al., J. Chem. Phys. 56:2547–2556, 1973
Valine	T. F. Koetzle, et al., J. Chem. Phys. 60:4690–4696, 1974

The chemical modification of proteins by group-specific and site-specific reagents. A. N. Glazer. In The Proteins, 3rd ed. H. Neurath and R. L. Hill (eds.). Vol. 2, pp. 1–103. New York, Academic Press, 1976.

Electronic structure and bonding of the amino acids containing first row atoms. D. A. Dixon and W. N. Lipscomb. J. Biol. Chem. 251:5992–6000, 1976.

Gly

Gly is the simplest amino acid, with no side chain. Because there are two H atoms on C^α, it is not asymmetric and is the only amino acid that cannot exist as D or L isomers. The absence of a side chain gives the polypeptide backbone at Gly residues much greater conformational flexibility than that possible otherwise, as will be discussed further in Chapter 5.

The Aliphatic Residues: Ala, Val, Leu, Ile

With solely aliphatic side chains, containing no polar or functional chemical groups, Ala, Val, Leu, and Ile comprise a rather homogeneous class. Their inert side chains have the chemical property of being hydrophobic; thus they have the potential for being structural units, since they provide a variety of molecular surfaces and shapes that seem to be well-suited for such a purpose. It might be noted that there occurs among the normal 20 amino acids used in proteins none with a single ethyl side chain, corresponding to that of the amino acid α-amino butyric acid. The side chain of isoleucine has an extra center of asymmetry; only the one isomer

$$-C \hspace{-0.5em}\begin{array}{c} CH_3 \\[0.5em] H \\[0.5em] CH_2-CH_3 \end{array}$$

(1–9)

occurs naturally and is incorporated into proteins.

The Cyclic Imino Acid: Pro

The side chain of proline is aliphatic, with no functional groups, but it is unique in that it is bonded covalently to the nitrogen atom of the peptide group. Therefore the backbone at Pro residues within a polypeptide chain has no amide hydrogen for participation in hydrogen bonding or in resonance stabilization of the peptide bond of which it is part. The cyclic five-membered ring also imposes rigid constraints on rotation about the N—C$^\alpha$ bond of the backbone, and the adjacent peptide bond is more likely to adopt the *cis* configuration; Pro residues, therefore, have very significant effects on the conformation of the polypeptide backbone, as will be discussed in Chapter 5. The five-membered pyrrolidine ring is invariably puckered, with the C$^\alpha$, C$^\beta$, C$^\delta$, and N atoms approximately coplanar, but with the C$^\gamma$ atom displaced from the plane by about 0.5 Å.

The Hydroxyl Residues: Ser and Thr

The side chains of Ser and Thr are small and aliphatic, except for the presence of a hydroxyl group on each, so they can be either hydrophobic or hydrophilic. The hydroxyl groups are normally no more reactive than that of ethanol, but they are somewhat polar, being able to function as both hydrogen donors and acceptors in hydrogen bonding. The hydroxyl groups may be acetylated by reaction with acetylchloride in aqueous trifluoroacetic acid; there are no other, milder procedures for modifying such groups.

It should be noted that the side chain of Thr, like that of Ile, has a center of asymmetry, and only the one isomer occurs naturally:

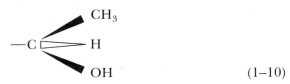

$$(1\text{--}10)$$

The Acidic Residues: Asp and Glu

The side chains of Asp and Glu differ only in having one and two —CH$_2$— groups, respectively, each with a terminal carboxyl group. The carboxyl groups are generally ionized above pH 5 and thus are extremely polar. The ionized forms may be effective chelators of metal ions when held in appropriate proximity. The nonionized forms are capable of serving as both hydrogen donors and acceptors in hydrogen bonds, but they are encountered in proteins at neutral pH only in exceptional circumstances. The carboxyl groups of Asp and Glu are normally as reactive as those of corresponding organic molecules, such as acetic acid; they may be chemically esterified, coupled with amino or other nucleophiles, and reduced to alcohols. The side-chain carboxyls differ only marginally from the terminal carboxyl group of the polypeptide chain.

Susceptibility to modification by different reagents can be used to determine the ionization state of carboxyls. Diazo compounds, such as diazoacetate esters and amides, react with the nonionized form:

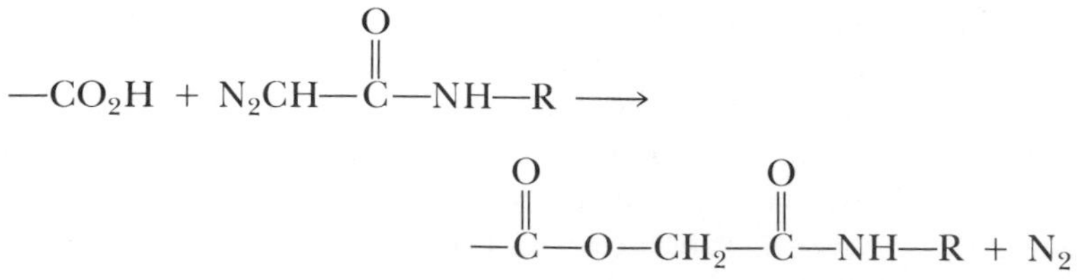

$$-CO_2H + N_2CH-\overset{\overset{\displaystyle O}{\|}}{C}-NH-R \longrightarrow$$

$$-\overset{\overset{\displaystyle O}{\|}}{C}-O-CH_2-\overset{\overset{\displaystyle O}{\|}}{C}-NH-R + N_2 \qquad (1-11)$$

In contrast, epoxides react with the ionized form:

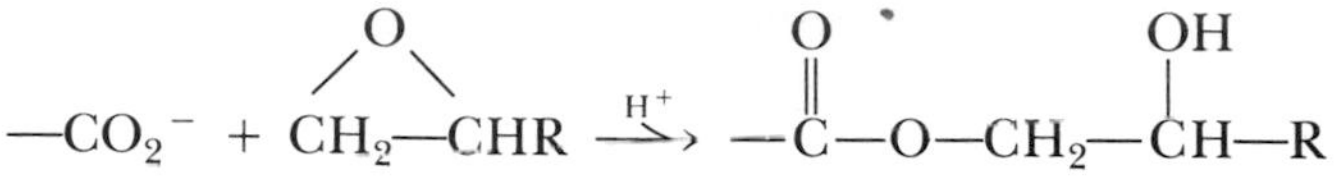

$$-CO_2{}^- + CH_2-CHR \xrightarrow{\text{H}^+} -\overset{\overset{\displaystyle O}{\|}}{C}-O-CH_2-\overset{\overset{\displaystyle OH}{|}}{CH}-R \qquad (1-12)$$

Esterification. P. E. Wilcox. Methods Enzymol. 25:596–615, 1972.
Carbodiimide modification of proteins. K. L. Carraway and D. E. Koshland. Methods Enzymol. 25:616–623, 1972.
The carboxylate ion in the active center of pepsin. J. A. Hartsuck and J. Tang. J. Biol. Chem. 247:2575–2580, 1972.

The Amide Residues: Asn and Gln

The amide forms of Asp and Glu—Asn and Gln, respectively—occur normally as amino acids and are incorporated directly into proteins. The side chains do not ionize but are relatively polar, being capable of both donating and accepting hydrogen atoms in hydrogen bonds. The amide groups are somewhat labile at extremes of pH, and the residues may be converted spontaneously by hydrolysis to Asp and Glu. When it is the amino-terminal group in a peptide, the Gln residue tends spontaneously to cyclize to pyrrolidone carboxylic acid:

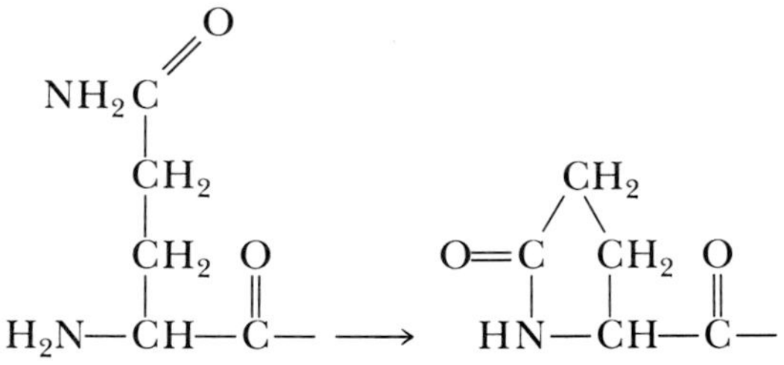

$$(1-13)$$

which occasionally is found as the amino-terminal residue of protein. This cyclized form makes the amino terminal group refractory to the procedures normally used for amino acid sequence determination; thus, it must be removed with a specific enzyme.

Rates of nonenzymatic deamidation of glutaminyl and asparaginyl residues in pentapeptides. A. B. Robinson, et al. J. Am. Chem. Soc. 95:8156–8159, 1973.

The Basic Residues: Lys and Arg

The side chain of Lys is a hydrophobic chain of four methylene groups ($-CH_2-$) capped by an amino group. At pH values less than 10, the amino group is ionized and thus is a very polar charged group. However, a fraction of the molecules are always nonionized, this fraction decreasing by a factor of 10 for each decrease in pH of one unit. The nonionized amino group is a potent nucleophile, and it may readily undergo a variety of acylation, alkylation, arylation, and deamination reactions; the rates of such reactions generally increase with increasing pH, as expected from the proportion of nonionized amino group present.

Arylation with 2,4,6-trinitrobenzene sulfonate (TNBS)

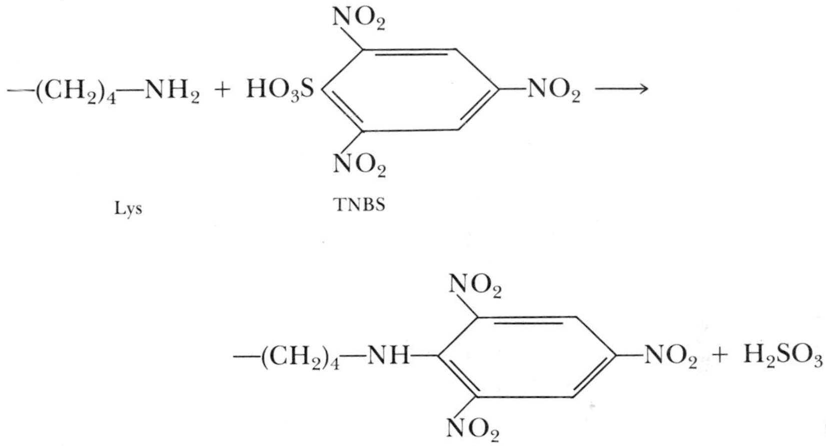

$$-(CH_2)_4-NH_2 \; + \; HO_3S \;\text{(benzene ring with } NO_2, NO_2, NO_2\text{)}-NO_2 \longrightarrow$$

Lys   TNBS

$$-(CH_2)_4-NH\;\text{(benzene ring with } NO_2, NO_2\text{)}-NO_2 \; + \; H_2SO_3 \qquad (1-14)$$

may be used to quantitate easily the concentration of amino groups, by the spectrophotometric appearance of the new derivative; it absorbs strongly at 367 nm. Acetylation with acetic, succinic, maleic, and citraconic (i.e., methylmaleic) anhydrides occurs readily:

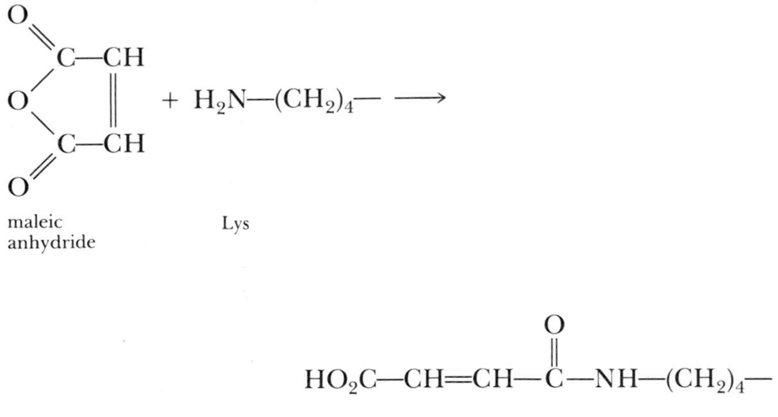

$$\text{(maleic anhydride structure)} \; + \; H_2N-(CH_2)_4- \longrightarrow$$

maleic
anhydride   Lys

$$HO_2C-CH{=}CH-\overset{\overset{\textstyle O}{\|}}{C}-NH-(CH_2)_4- \qquad (1-15)$$

The maleic and citraconic anhydrides are especially useful, as the modification can be reversed under acidic conditions.

The amino group may also be guanylated with O-methyl isourea to convert the side chain to that of homoarginine:

$$-(CH_2)_4-NH_2 + CH_3O-C\underset{NH_2}{\overset{NH}{\diagup}} \longrightarrow$$

Lys   O-methyl isourea

$$-(CH_2)_4-NH-C\underset{NH_2}{\overset{NH}{\diagup}} + CH_3OH \qquad (1\text{--}16)$$

A related reaction is amidination:

$$-(CH_2)_4-NH_2 + CH_3CH_2-O-\overset{\overset{\displaystyle NH}{\|}}{C}-R \longrightarrow$$

Lys

$$-(CH_2)_4-NH-\overset{\overset{\displaystyle NH}{\|}}{C}-R + CH_3CH_2OH \qquad (1\text{--}17)$$

In both of these instances, the side chains remain very basic and are normally positively charged.

Lys residues also form reversibly Schiff bases with aldehydes, such as that of the natural cofactor pyridoxal phosphate, which may be stabilized by reduction, e.g., by sodium borohydride ($NaBH_4$):

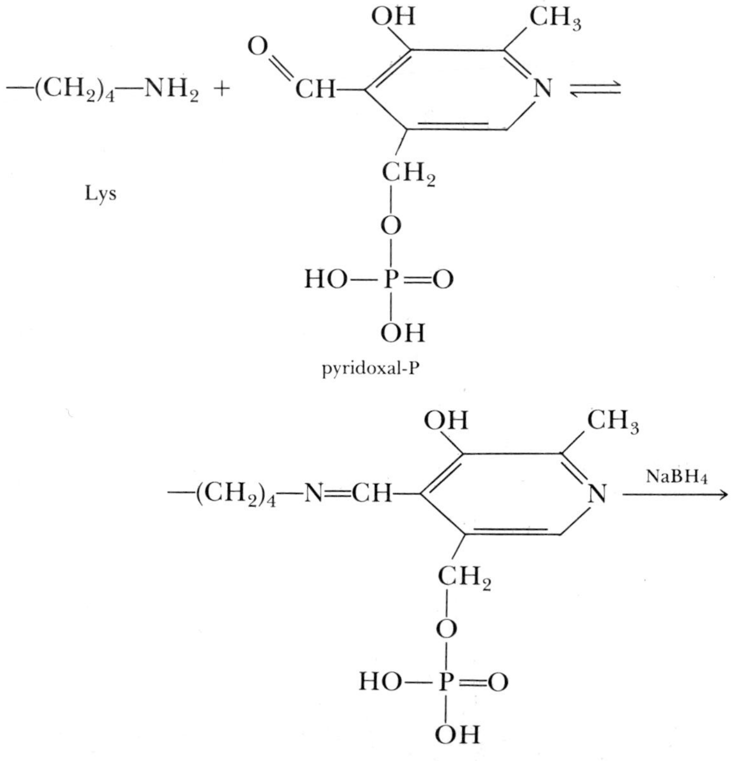

pyridoxal-P

Schiff base

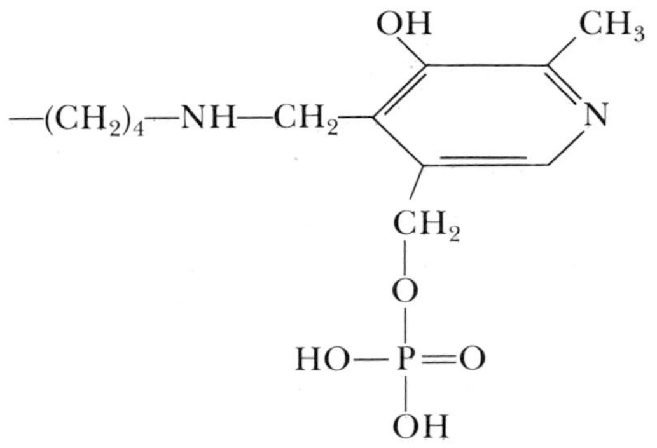

$$—(CH_2)_4—NH—CH_2- \text{(pyridine ring with OH, CH}_3\text{, N, CH}_2—O—\overset{O}{\underset{OH}{\overset{\|}{P}}}—OH\text{)} \qquad (1\text{–}18)$$

A widely used modification is that of carbamylation by cyanate, to form a homocitrulline residue:

$$—(CH_2)_4—NH_2 + HN{=}C{=}O \longrightarrow —(CH_2)_4—NH—\overset{O}{\overset{\|}{C}}—NH_2 \qquad (1\text{–}19)$$

$$\underset{\text{Lys}}{} \qquad \underset{\text{cyanate}}{} \qquad \underset{\text{homocitrulline}}{}$$

The amino groups of Lys residues participate in a multitude of reactions; this reactivity makes it possible to convert these side chains to a variety of analogues that have positive, negative, or no charge under normal conditions. However, the α-amino group of the polypeptide chain will also tend to participate in the same reaction; the ϵ-amino and α-amino groups inherently differ only slightly in their intrinsic pK_a values (Table 1–1).

Maleylation of amino groups. P. J. G. Butler and B. S. Hartley. Methods Enzymol. 25:191–199, 1972.

Acetylation. J. F. Riordan and B. L. Vallee. Methods Enzymol. 25:494–499, 1972.

Amidination. M. J. Hunter and M. L. Ludwig. Methods Enzymol. 25:585–596, 1972.

Carbonyl-amine reactions in protein chemistry. R. E. Feeney, et al. Adv. Protein Chem. 29:135–203, 1975.

Tritium labelling of proteins to high specific radioactivity by reductive methylation. B. F. Tack, et al. J. Biol. Chem. 255:8842–8847, 1980.

The Arg side chain consists of three hydrophobic methylene groups and the strongly basic δ-guanido group, which is ionized over the entire pH range in which proteins are normally studied. The ionized guanido group is planar owing to resonance:

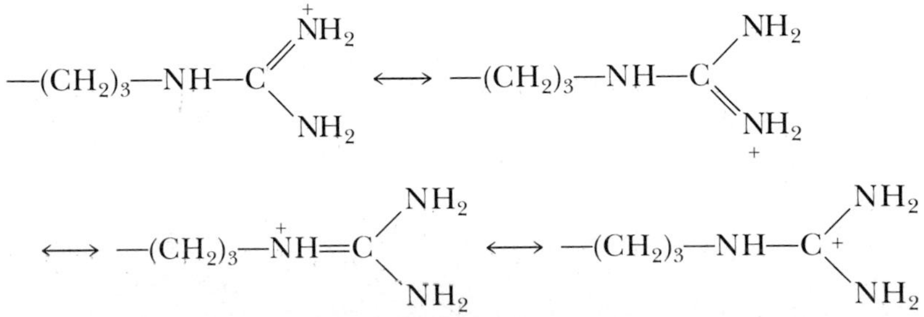

$$—(CH_2)_3—NH—C\overset{\overset{+}{N}H_2}{\underset{NH_2}{}} \longleftrightarrow —(CH_2)_3—NH—C\overset{NH_2}{\underset{\overset{NH_2}{+}}{}}$$

$$\longleftrightarrow —(CH_2)_3—\overset{+}{N}H{=}C\overset{NH_2}{\underset{NH_2}{}} \longleftrightarrow —(CH_2)_3—NH—C\overset{NH_2}{\underset{NH_2}{+}} \qquad (1\text{–}20)$$

and the positive charge is effectively distributed over the entire group. In the protonated form, the guanido group is virtually unreactive, and only very small fractions of the nonionized form are present at normal pH values. However, 1,2- and 1,3-dicarbonyl compounds (e.g., phenylglyoxal, 2,3-butanedione, 1,2-cyclohexanedione) readily form heterocyclic condensation products with the guanido group because the distance of the two carbonyl groups closely matches that of the two unsubstituted nitrogen atoms of the guanido group:

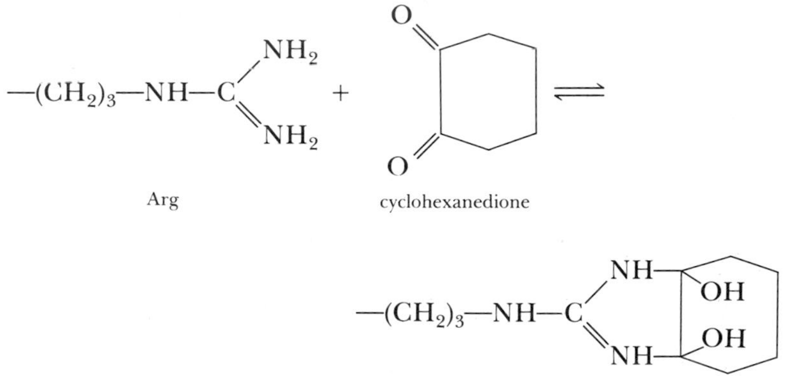

$$\text{—(CH}_2)_3\text{—NH—C} \Big\langle \substack{\text{NH}_2 \\ \text{NH}_2} \quad + \quad \text{cyclohexanedione} \quad \rightleftharpoons$$

Arg   cyclohexanedione

$$\text{—(CH}_2)_3\text{—NH—C} \Big\langle \substack{\text{NH} \\ \text{NH}} \Big\rangle \substack{\text{OH} \\ \text{OH}} \tag{1–21}$$

The adduct may be stabilized by the presence of borate, which complexes with the vicinal OH groups.

The guanido group may be cleaved by hydrazine to produce the amino side chain of ornithine:

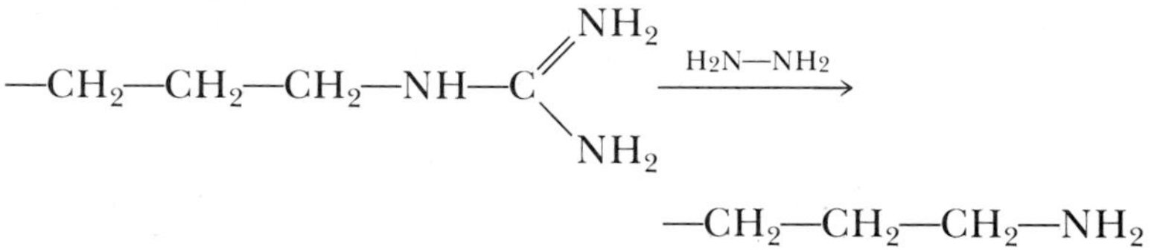

$$\text{—CH}_2\text{—CH}_2\text{—CH}_2\text{—NH—C} \Big\langle \substack{\text{NH}_2 \\ \text{NH}_2} \quad \xrightarrow{\text{H}_2\text{N—NH}_2}$$

$$\text{—CH}_2\text{—CH}_2\text{—CH}_2\text{—NH}_2 \tag{1–22}$$

but this reaction is often accompanied by cleavage of the polypeptide backbone.

Reversible blocking at arginine by cyclohexanedione. E. L. Smith. Methods Enzymol. 47:156–161, 1977.

Chemical modification of peptides by hydrazine. A. Honegger, et al. Biochem. J. 199:53–59, 1981.

His

The imidazole side chain of His possesses several special properties that make it effective as a nucleophilic catalyst. It is an amine, and amines are much more reactive than hydroxide ion in terms of their basicity. Furthermore, it is a tertiary amine; this form generally has a higher intrinsic nucleophilic reactivity than that of primary or secondary amines. The enhanced reactivity of tertiary amines is usually canceled out by the greater steric hindrance in such amines, but in imidazole the substituents on the nitrogen atom are held back in an aromatic five-membered ring and cause relatively little steric hindrance. The amine has a pK near 7, so that it is one of the strongest bases that can exist at neutral pH; a weaker base would have a lower nucleophilic reactivity, whereas a stronger base would be protonated to a greater extent at neutral pH, so that it would be less reactive with a given total concentration of amine. In the nonionized form, one N atom is an electrophile and donor for hydrogen bonding, while the other is a nucleophile and acceptor for hydrogen bonding; this one side chain is thus extremely versatile—almost the chemical equivalent of being ambidextrous.

The nonionized imidazole ring may exist as two tautomers, with the H atom on either the $\delta 1$ or $\epsilon 2$ N atom:

$$-CH_2 \quad HN^{\delta 1} \quad N \qquad\qquad -CH_2 \quad N \quad NH^{\epsilon 2} \qquad (1\text{--}23)$$

These two N atoms are also often designated as N-1 and N-3, or as π and τ, respectively. ^{13}C-NMR studies have concluded that the hydrogen atom normally is predominantly on the $N^{\epsilon 2}$ atom in model peptides. However, this can vary with different local environments.

The nonprotonated N atom is the potent nucleophile, which is capable of interacting with a variety of substances, including metal ions, and of catalyzing a variety of reactions; this has been demonstrated with imidazole in model chemical reactions, and His side chains will be shown to be involved in the catalytic activities of numerous enzymes.

Although the imidazole side chain is, in principle, capable of undergoing numerous reactions (e.g., acylation, alkylation, and electrophilic substitution), there are very few reactions suitable for specifically modifying His residues. The classical approach has been oxidation by light in the presence of dye sensitizers, such as methylene blue or rose bengal; the reaction products appear to be aspartic acid and urea, but the reaction is not well-characterized. The imidazole N atoms will also react with many of the reagents used to modify amino and sulfhydryl groups; this is usually an undesirable side reaction.

The imidazole side chain is readily protonated, with a pK_a of about 7, at the second N atom, which destroys its nucleophilicity. The positive charge is generally shared between the two N atoms by resonance:

$$-CH_2 \quad ^+HN \quad NH \qquad \longleftrightarrow \qquad -CH_2 \quad HN \quad NH^+ \qquad (1\text{--}24)$$

There is also a finite, but small, probability of the deprotonation of the $C^{\epsilon 1}$ atom (also often designated as C-2):

$$
\text{(1-25)}
$$

As a result of this rare occurrence, this H atom is observed to exchange slowly with the solvent, which is a useful probe of the environments of histidine residues in proteins (see Chapter 7). The imidazole N atoms can also be deprotonated, with an apparent pK_a of about 14.4, to give the aromatic anion:

$$
\text{(1-26)}
$$

This anion possesses two equivalent coordination sites and would be a potent bond-forming ligand, but it is rarely present.

Histidine side chains are especially useful in ^{1}H-NMR studies of proteins, as the H atom on the $C^{\epsilon 1}$ (or C-2) atom is usually well-resolved from the multitude of resonances of the other atoms of proteins. The resonance is also usually shifted by about 1 ppm to a lower field strength upon protonation of the side chain, thereby often making it possible to determine pK_a values of individual His residues.

The roles of imidazole in biological systems. E. A. Barnard and W. D. Stein. Adv. Enzymol. 20:51–110, 1959.

Titration behaviour and tautomeric states of individual histidine residues of myoglobin. D. J. Wilbur and A. Allerhand. J. Biol. Chem. 252:4968–4975, 1977.

Tautomeric states of the histidine residues of bovine pancreatic ribonuclease A. D. E. Walters and A. Allerhand. J. Biol. Chem. 255:6200–6204, 1980.

^{1}H-NMR study on the tautomerism of the imidazole ring of histidine residues. I. Microscopic pK values and molar ratios of tautomers in histidine-containing peptides. M. Tanokura. Biochim. Biophys. Acta 742:576–585, 1983.

The Aromatic Residues: Phe, Tyr, and Trp

The aromatic side chains are responsible for most of the ultraviolet absorbance and fluorescence properties of proteins, parameters that are useful probes of protein structure (Table 1–3 and Figure 1–3). (Histidine might also be considered aromatic but does not have such spectral properties; therefore, it has been treated separately.)

The aromatic ring of Phe is comparable to that of benzene or toluene. Consequently, it is very hydrophobic and is chemically reactive only under rather extreme conditions.

The hydroxyl group of the phenolic ring of Tyr makes it relatively reactive in electrophilic substitution reactions at the positions designated $\epsilon 1$ and $\epsilon 2$, but which are often numbered as 3 and 5. Tyr side chains may be readily nitrated and iodinated:

Table 1–3   Spectroscopic Properties of the Aromatic Amino Acids at Neutral pH

	ABSORBANCE[a]		FLUORESCENCE[b]	
	max (nm)	Molar absorbance ($M^{-1}cm^{-1}$)	max (nm)	Quantum yield
Phenylalanine	257.4	197	282	0.04
Tyrosine	274.6	1420	303	0.21
Tryptophan	279.8	5600	348	0.20

[a]From J. E. Bailey, Ph.D. thesis, London University, 1966.
[b]From F. W. J. Teale and G. Weber, Biochem. J. 65:476–482, 1957.

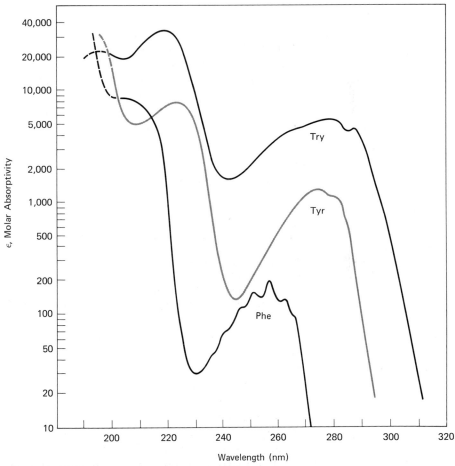

Figure 1–3
Absorption spectra of the aromatic amino acids at pH 6. Incorporation of these amino acids into peptides has little direct effect on the absorbance of their side chains. (From D. B. Wetlaufer, Adv. Prot. Chem. 17:303–390, 1962; © Academic Press.)

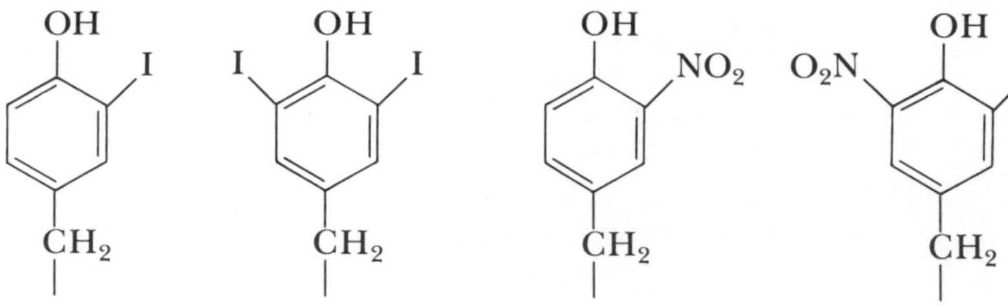

$$(1\text{--}27)$$

with consequent alterations of their spectrophotometric properties and in ionization of the hydroxyl group (Table 1–4).

The hydroxyl group of the Tyr side chain is able to participate in hydrogen bonding. It may ionize at alkaline pH values, when the spectroscopic properties are also altered. It may also be acetylated by acetic anhydride:

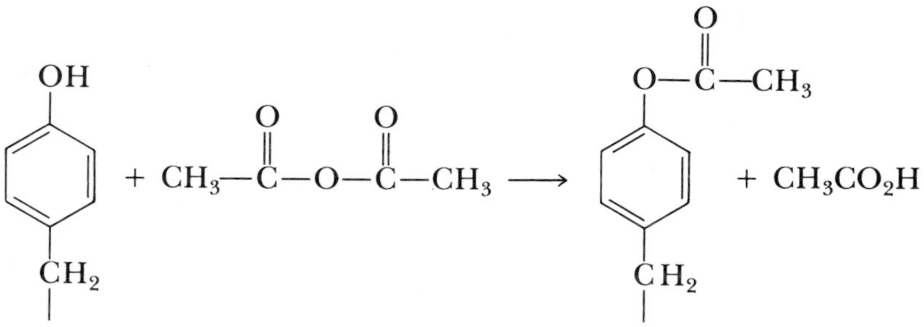

$$(1\text{--}28)$$

The spectral properties in particular make Tyr residues very useful in studying protein structure, as they are sensitive to the environment of the side chain.

The indole side chain of Trp is the largest and most fluorescent of those that occur in proteins. This amino acid also occurs least frequently, so that proteins often have only one or a few residues. The spectral properties of such residues are often useful probes of protein structure (Chap-

Table 1–4   Spectroscopic Properties of Tyrosine and Several Derivatives

	pK$_{app}$ of —OH	NONIONIZED OH		IONIZED OH	
		max (nm)	Molar absorbance (M^{-1}cm^{-1})	max (nm)	Molar absorbance (M^{-1}cm^{-1})
Tyrosine	10.1	274.5	1400	293	2400
ϵ-Iodotyrosine	8.2	283	2750	305	4100
ϵ1,ϵ2-Diiodotyrosine	6.5	287	2750	311	6250
ϵ-Nitrotyrosine	7.2	360	2790	428	4200
ϵ-Aminotyrosine	10.0[a]	275	1600	320	4200
O-Acetyltyrosine	—	262	262	—	—

[a]The pK$_{app}$ of the aromatic amino group is approximately 4.8.

From A. N. Glazer, in H. Neurath and R. L. Hill (eds.), The Proteins, 3rd ed. Vol. 2, pp. 1–103. New York, Academic Press, 1976.

ter 7). The fluorescence properties in particular are sensitive to the environment of the side chain.

However, the indole ring is susceptible to irreversible oxidation, which can hinder such studies. These oxidation reactions may also be controlled to provide controlled modifications. Iodine or *N*-bromosuccinimide may be used to oxidize the indole ring to oxindolealanine:

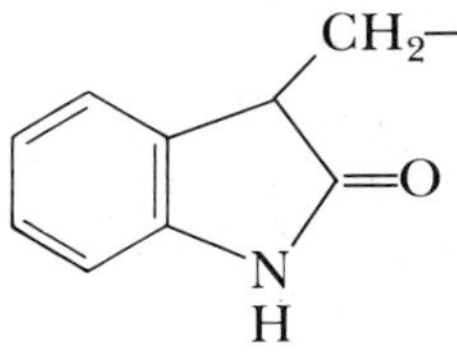

$$(1\text{--}29)$$

Ozone will open the indole ring to give *N*-formylkynurenine:

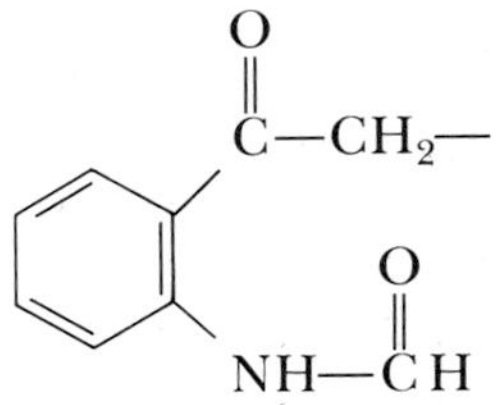

$$(1\text{--}30)$$

The nitrogen atom of Trp may also be reversibly formylated by anhydrous formic acid containing HCl:

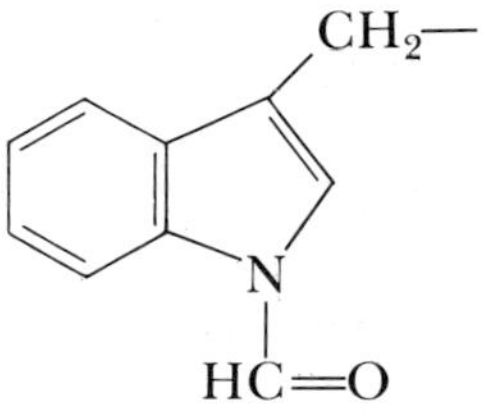

$$(1\text{--}31)$$

Certain reactive benzyl halides and sulfenyl halides have been designed that will alkylate the $C^{\delta 1}$ atom of the indole ring:

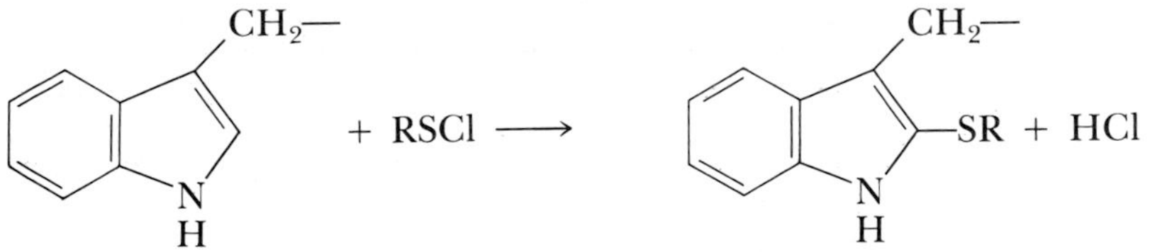

$$+ \text{RCl} \longrightarrow \qquad (1\text{--}32)$$

A variety of groups with special spectral properties may be introduced in this way.

The Trp side chain is the only one capable of entering into charge-transfer interactions as a donor; indoles are excellent donors for formation of such charge-transfer complexes with pyridinium compounds and other electrophilic structures. The pyrrole nitrogen of the indole ring may also be a hydrogen donor in hydrogen bond interactions.

Ultraviolet fluorescence of the aromatic amino acids. F. J. W. Teale and G. Weber. Biochem. J. 65:476–482, 1957.

Ultraviolet spectra of proteins and amino acids. D. B. Wetlaufer. Adv. Protein Chem. 17:303–390, 1962.

Modification of proteins with active benzyl halides. H. R. Horton and D. W. Koshland. Methods Enzymol. 25:468–482, 1972.

Sulfenyl halides as modifying reagents for polypeptides and proteins. A. Fontana and E. Scoffone. Methods Enzymol. 25:482–494, 1972.

Nitration with tetranitromethane. J. F. Riordan and B. L. Vallee. Methods Enzymol. 25:515–521, 1972.

Fluorescence of aminotyrosyl residues in peptides and helical proteins. R. L. Seagle and R. W. Cowgill. Biochim. Biophys. Acta 439:461–469, 1976.

The Sulfur-containing Residues: Met and Cys

The long alkyl side chain of Met renders it rather hydrophobic, but the sulfur atom of the thioether group is a potent nucleophile. It readily forms sulfonium salts with alkylating agents, such as methyl iodide:

$$-(CH_2)_2-S-CH_3 + CH_3I \longrightarrow -(CH_2)_2-\overset{CH_3}{\underset{+}{S}}-CH_3 + I^- \qquad (1-33)$$

This reaction is reversed by thiol reagents; since the methyl group removed is equally likely to be the original one or that introduced by the methyl iodide, this offers the possibility of introducing an isotopic label in 50 per cent of the residues by treating with, for example, ^{13}C-labeled methyl iodide. The sulfur atom also has a tendency to bind noncovalently the platinum derivative $PtCl_4^{2-}$, which is a useful heavy atom derivative in protein crystallography (see Chapter 6).

The sulfur atom is also susceptible to oxidation, by air or more potent oxidants such as peroxide, to the sulfoxide and then to the sulfone:

$$-(CH_2)_2-S-CH_3 \xrightarrow{[O]} -(CH_2)_2-\overset{O}{\underset{}{S}}-CH_3 \xrightarrow{[O]}$$

$$-(CH_2)-\overset{O}{\underset{O}{S}}-CH_3 \qquad (1-34)$$

The first step, but not the latter, may be reversed readily by thiols or other reducing agents. Such oxidation makes the residue unreactive with alkylating reagents.

The thiol group of Cys is the most reactive of any amino acid side chain. It ionizes at slightly alkaline pH; the thiolate anion is the reactive form in most instances. It reacts very rapidly with alkyl halides, such as iodoacetate, iodoacetamide, methyl iodide, and so on, to give the corresponding stable alkyl derivatives:

$$-CH_2-S^- + ICH_2CO_2^- \longrightarrow -CH_2-S-CH_2-CO_2^- + I^- \quad (1-35)$$

$$\text{Cys} \qquad\qquad \text{iodoacetate} \qquad\qquad \text{carboxymethyl-Cys}$$

The thiol group can also add across double bonds, such as those of *N*-ethylmaleimide or of maleic anhydride:

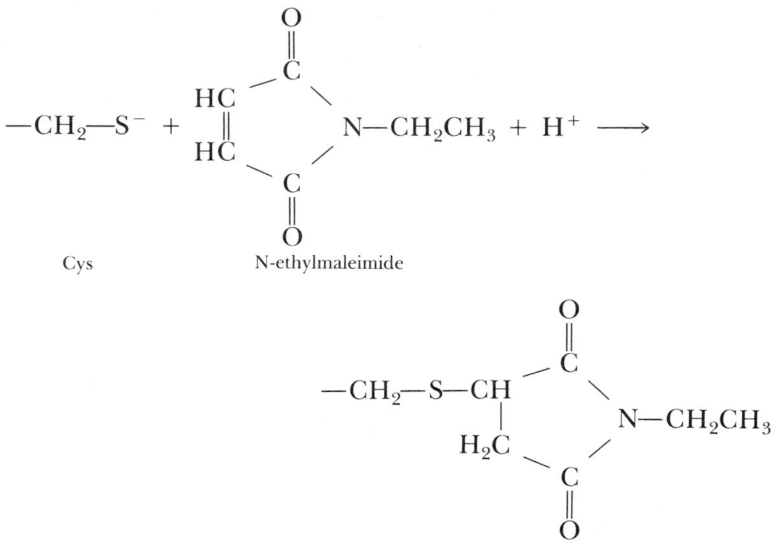

Cys   N-ethylmaleimide

$$(1-36)$$

and also is able to open the ring of ethyleneimine:

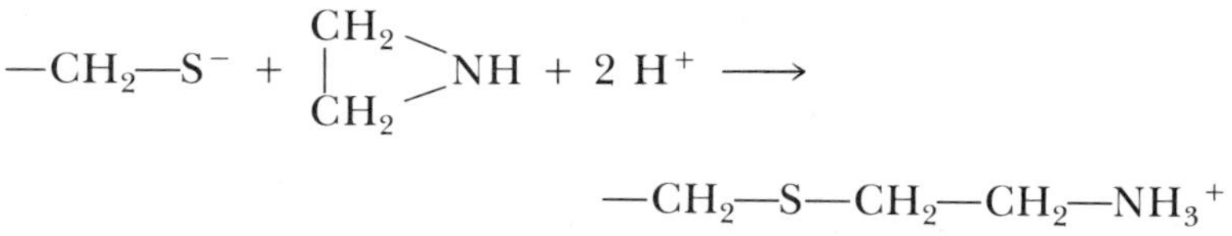

$$-CH_2-S-CH_2-CH_2-NH_3^+ \quad (1-37)$$

The resulting positively charged side chain provides a new site for cleavage of polypeptides by trypsin, which is useful for amino acid sequence determination (see below).

Thiols form complexes of varying stability with a wide variety of metal ions. The most stable are those with divalent mercury, Hg^{++}; complexes with a variety of stoichiometries are formed. Because of this variability, the univalent organic mercurials of the type $R-Hg^+$ tend to be used, since they form more reproducible one-to-one complexes with thiols. The best-known is ρ-mercuribenzoic acid,

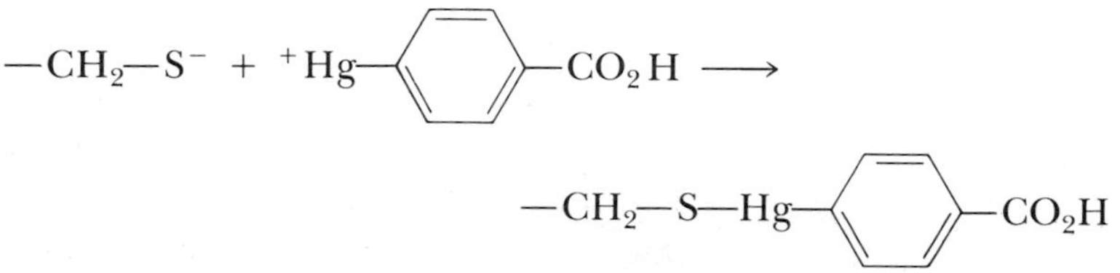

$$(1-38)$$

often employed to titrate thiol groups using the spectral change of the reagent that occurs upon binding. Reactions forming such mercurial complexes of cysteine thiol groups are one of the most obvious ways of making heavy atom derivatives for protein crystallography (see Chapter 6).

Thiol complexes with silver ion are less stable than those with mercury, but because Ag^+ is univalent, it reacts stoichiometrically and may be used to titrate thiols. The reaction of thiols with arsenic compounds is of considerable biological interest. Copper, iron, zinc, cobalt, molybdenum, manganese, and cadmium ions all form various complexes with thiols.

The sulfur atom can exist in a variety of oxidation states, and thiols are readily oxidized by the oxygen of air or other agents. Air oxidation is markedly enhanced by trace levels of metal ions, such as Cu^{2+}, Fe^{2+}, Co^{2+}, and Mn^{2+}, and it is possible that such catalysts are required; the metal complexes of thiols are likely to be the actual reactants with oxygen. Many of the oxidation states of sulfur are unstable, and only two oxidation derivatives of thiols are generally encountered, the disulfide and the sulfonic acid. The disulfide is generally the end-product of air oxidation:

$$2 -CH_2SH + \tfrac{1}{2}O_2 \longrightarrow -CH_2S-SCH_2- + H_2O \qquad (1-39)$$

The sulfonic acids are produced by the more potent oxidizing agents. For example, performic acid will oxidize both thiol and disulfide forms of Cys residues to cysteic acid residues (Cys O_3H):

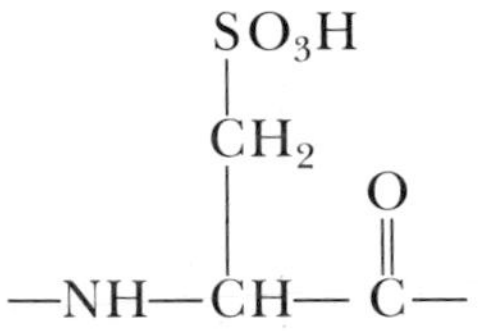

$$
\begin{array}{c}
SO_3H \\
| \\
CH_2 \qquad\quad O \\
| \qquad\qquad \parallel \\
-NH-CH-\ C- \\
\text{Cys } O_3H
\end{array}
\qquad (1-40)
$$

The number of Cys residues in a protein, whether present as thiols or disulfides, is most accurately measured after oxidation with performic acid.

Disulfide bonds between Cys residues occur in some proteins. Because such bonds are covalent, they may be kept intact under appropriate conditions and isolated after disruption of the polypeptide chain. Indeed, Cys residues present in the thiol form in a protein are recovered after acid hydrolysis primarily as the amino acids linked by disulfides, owing to air oxidation. Its isolation as such led to the designation of the amino acid **cystine:**

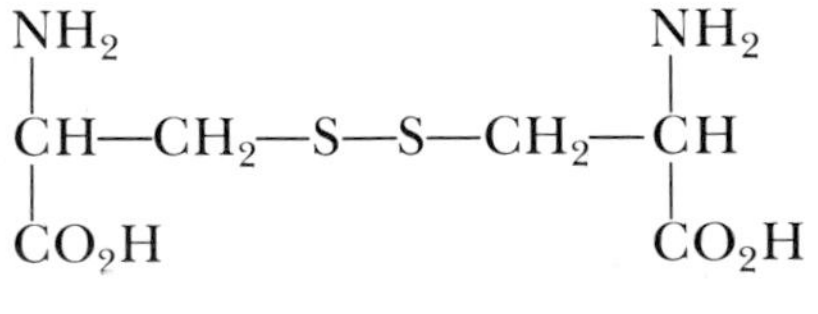

$$
\begin{array}{ccc}
NH_2 & \qquad & NH_2 \\
| & & | \\
CH-CH_2-S-S-CH_2-CH \\
| & & | \\
CO_2H & & CO_2H
\end{array}
\qquad (1-41)
$$

Cys residues are often referred to as "$\tfrac{1}{2}$ cystine" residues. However, it is now clear that cystine is not incorporated into proteins as such; instead, the thiol form, Cys, is used in protein biosynthesis, and disulfide bonds between them are added later. The designation of cystine as an amino acid of proteins is thus somewhat misleading, but references to it in the literature are still common.

The disulfide group appears to be very hydrophobic relative to thiol groups. This is reflected in the very low solubility of cystine in water (0.01 g/100 ml), less than 10^{-3} that of cysteine. The disulfide group has a preferred geometry in which the angle of rotation about the disulfide bond has a value close to either ± 90 degrees. Other angles are less stable by up to 10 kcal/mole; interconversion of the two favorable isomers has an activation barrier of about 7 kcal/mole.

Thiols and disulfides undergo exchange very rapidly at neutral to alkaline pH values:

$$R_1{-}S^- + R_2{-}S{-}S{-}R_2 \rightleftharpoons R_2{-}S^- + R_1{-}S{-}S{-}R_2 \qquad (1\text{--}42)$$

The ionized thiol group is the reactive species, so that the rate decreases at lower pH values. With simple aliphatic thiols, the equilibrium constant is close to unity.

Thiol–disulfide exchange with aromatic disulfides, such as dithionitrobenzoic acid (DTNB), or with Ellman's reagent provides the most convenient method of assaying for thiol groups, because the liberated aromatic thiol is brightly colored:

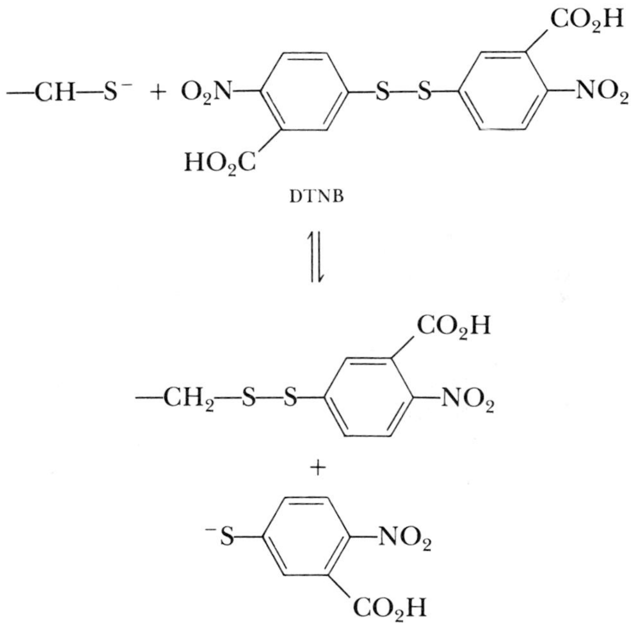

$$(1\text{--}43)$$

The equilibrium with such aromatic disulfides lies far to the right, so that addition of excess DTNB results in formation of the yellow aromatic thiol to an extent stoichiometric with original thiol groups.

Disulfides in proteins may be effectively reduced by thiol–disulfide exchange with thiol compounds (RSH), such as mercaptoethanol:

$$-CH_2-S-S-CH_2- + RS^- \rightleftharpoons -CH_2-S-S-R + {}^-S-CH_2-$$
protein disulfide $\qquad\qquad$ mixed disulfide

$$-CH_2-S-S-R + RS^- \rightleftharpoons R-S-S-R + {}^-S-CH_2- \qquad (1-44)$$

Two sequential thiol–disulfide exchanges are involved, proceeding via the mixed disulfide intermediate. Protein disulfides are generally stabilized by the protein conformation, so that a large excess of thiol reagent is required to break them completely. The stabilities of protein disulfide bonds may be easily varied by controlling the relative concentrations of the thiol and disulfide forms of the reagent.

A most useful reagent for reducing protein disulfides, introduced by Cleland, is dithiothreitol (and its isomer dithioerythritol):

$$
\begin{array}{l}
CH_2SH \\
| \\
HOCH \\
| \\
HCOH \quad + \ R-S-S-R \ \rightleftharpoons \\
| \\
CH_2SH
\end{array}
\qquad\text{[ring structure]} \quad + \quad 2\ RSH
$$

dithiothreitol $\qquad\qquad\qquad\qquad\qquad\qquad\qquad\qquad (1-45)$

It forms an intramolecular disulfide bond within a stable six-membered ring, so that the equilibrium constant lies far to the right. The reaction proceeds in two steps through the mixed disulfide, but this is very unstable.

Disulfide bonds may also be reduced directly by electrodes or chemically with reagents such as sodium borohydride. They may also be cleaved with nucleophiles such as cyanide, sulfite, or hydroxide ion:

$$RS^- + RSCN \overset{CN^-}{\rightleftharpoons} R-S-S-R \overset{OH^-}{\rightleftharpoons} RSOH + RS^-$$

$$\updownarrow SO_3^=$$

$$R-S-SO_3^-$$

$$+$$

$$RS^- \qquad (1-46)$$

The equilibria for these reactions are such that it is difficult to drive them to completion.

Tissue sulfhydryl groups. G. L. Ellman. Arch. Biochem. Biophys. 82:70–77, 1959.
The role of sulfur in proteins. R. Cecil. In The Proteins, 2nd ed. H. Neurath (ed.). Vol. 1, pp. 379–476. New York, Academic Press, 1963.
Dithiothreitol, a new protective reagent for SH groups. W. W. Cleland. Biochemistry 3:480–482, 1964.
Biochemistry of the SH group. P. C. Jocelyn. New York, Academic Press, 1972.
Formation and reactions of sulfenic acids in proteins. W. S. Allison. Acc. Chem. Res. 9:293, 1976.
Biochemistry and physiological role of methionine sulfoxide residues in proteins. N. Brot and H. Weissbach. Arch. Biochem. Biophys. 223:271–281, 1983.

DETECTION OF AMINO ACIDS, PEPTIDES, AND PROTEINS

This volume does not describe the methods used in *purifying* proteins, because there are so many that a discussion of these technical aspects would require an entire volume in itself. Most such methods rely upon the specific properties of individual proteins, which will be the main topic of the remainder of this text. A discussion now would be premature. However, it seems worthwhile to describe briefly the most popular methods of *detecting* proteins, as they are encountered in virtually every paper on the subject of proteins.

There are very many ways in which a few amino acids, such as His, Cys, Arg, Trp, or Tyr, may be detected specifically, either in peptides or as the free amino acid. These methods are too numerous to describe here, but they rely upon the chemical or spectral properties of particular side chains.

One of the most widely used reagents that will detect most amino acids or peptides is **ninhydrin.** It reacts with the amino groups that are almost invariably present. The reaction is complex, but a major component appears to be first the oxidation deamination of the amino acid to CO_2, NH_3, and an aldehyde containing one less carbon atom than the original amino acid; the ninhydrin is simultaneously reduced to the hydrindantin:

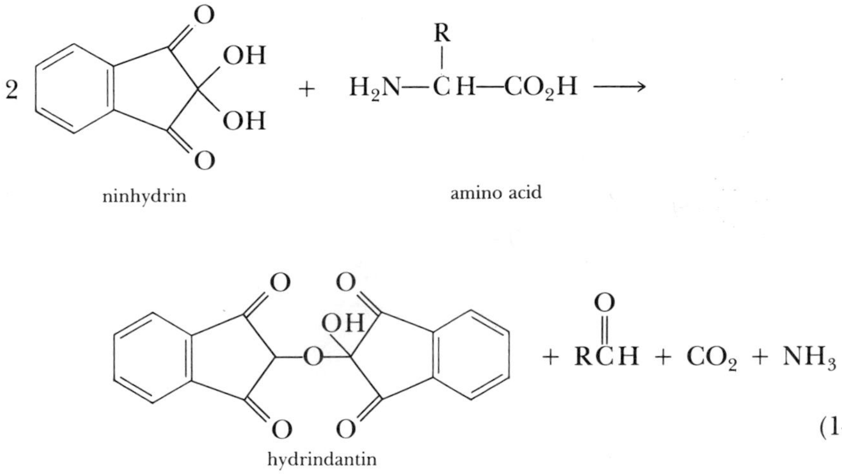

$$(1-47)$$

The hydrindantin then reacts with the liberated NH_3 to give the purple product

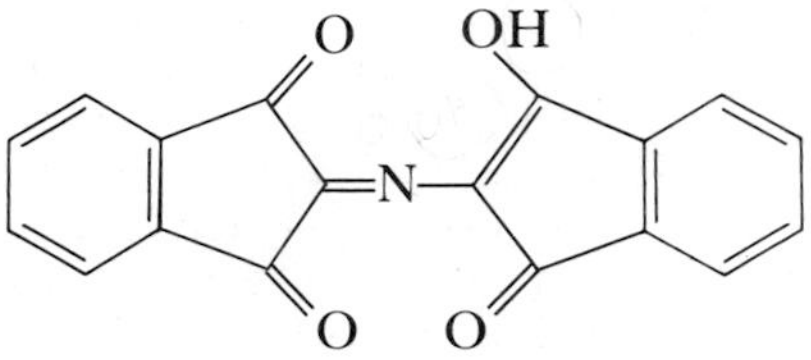

$$(1-48)$$

This product is intensely colored, with a maximum absorbance at 570 nm; the original ninhydrin is virtually colorless. The imino amino acid Pro does not react in this way but usually forms a yellow product of uncertain structure, with maximum absorbance at 440 nm.

Ninhydrin is usually used to detect amino acids and peptides, either in solution (as in the normal amino acid analyzers) or on paper or thin-layer plates. It has the disadvantage of reacting irreversibly, so that the reacted groups are not recoverable and cannot be identified. Also, excess reagent will react with amino acids liberated upon hydrolysis of a treated peptide.

A similar reagent that is more sensitive and does not have these disadvantages is **fluorescamine.** This reagent also reacts with amino groups:

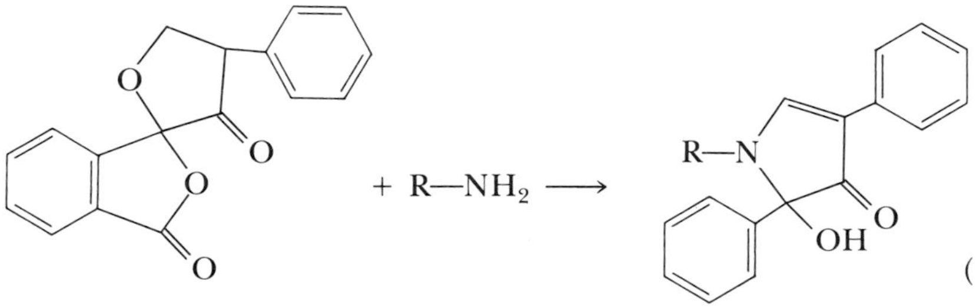

fluorescamine

$$+ \ R\text{---}NH_2 \ \longrightarrow \qquad\qquad\qquad (1\text{--}49)$$

but in this case the derivative is fluorescent, whereas the original reagent is not. Reacted amino acids appear to be recovered in high yield upon acid hydrolysis. The original reagent is also unstable in the presence of H_2O, yielding nonfluorescent hydrolysis products; therefore, any excess reagent is destroyed. It does not then react further upon acid hydrolysis of peptides, so that peptides detected on paper with this reagent may be eluted directly from the paper, hydrolyzed, and analyzed.

Both ninhydrin and fluorescamine react with most amines; they are not specific and are most useful with purified components.

The most widely used method of specifically detecting and measuring quantitatively proteins in solution is the **Folin phenol method** of Lowry, Rosebrough, Farr, and Randall. The active constituents of the Folin phenol reagent are the phosphomolybdic-tungstic mixed acids:

$$3 \ H_2O \cdot P_2O_5 \cdot 13 \ WO_3 \cdot 5 \ MoO_3 \cdot 10 \ H_2O$$

$$3 \ H_2O \cdot P_2O_5 \cdot 14 \ WO_3 \cdot 4 \ MoO_3 \cdot 10 \ H_2O \qquad\qquad (1\text{--}50)$$

Proteins reduce the mixed acids to cause loss of one, two, or three oxygen atoms from the tungstate and molybdates, to produce a number of reduced species with a characteristic blue color ($\lambda_{max} = 745$–750 nm). Copper ions are generally included, as they facilitate the reduction process. The principal chromogenic groups are the peptide linkages, the aromatic amino acids Tyr and Trp, and the polar side chains. Therefore, the color yield depends upon the amino acid sequence and composition of the protein, but this has not been observed in practice to vary greatly with normal proteins. The greatest disadvantage with this assay procedure has been the wide variety of nonprotein substances that interfere, either producing the blue color by themselves or interfering with color development by proteins.

A more specific but less sensitive method of quantitatively assaying proteins is the **biuret reaction.** A dilute solution of cupric sulfate in strongly alkaline tartrate is added to the protein solution. A purplish-violet color is formed, with maximum absorbance at 540 nm. The nature of the colored compound is uncertain, but the color is probably due to the formation of a tetracoordinated cupric ion complex with adjacent peptide groups, e.g.:

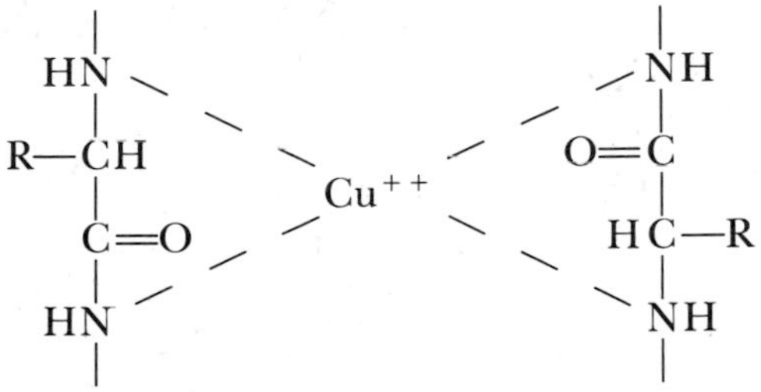

$$
\begin{array}{ccc}
\overset{|}{HN} & & \overset{|}{NH} \\
R-\overset{|}{CH} & Cu^{++} & O=\overset{|}{C} \\
\overset{|}{C}=O & & H\overset{|}{C}-R \\
\overset{|}{HN} & & \overset{|}{NH} \\
| & & |
\end{array}
\qquad (1\text{--}51)
$$

Proteins in gels, e.g., after electrophoresis, are generally stained with **Coomassie Brilliant Blue R250:**

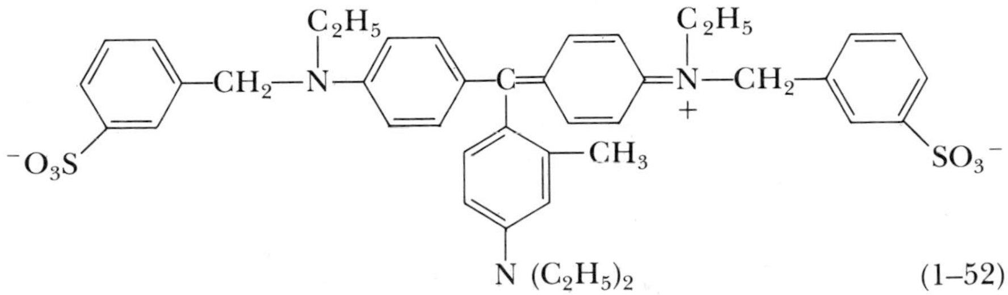

$$ (1\text{--}52) $$

This dye forms rather strong, but noncovalent, complexes with proteins, possibly by electrostatic interactions with NH_3^+ groups and by van der Waals forces. The protein must then be rendered insoluble in the gel and the excess dye removed after staining to give a clear background.

A staining procedure 100 times more sensitive using **silver** has recently been developed empirically. The reactions involved are not at all understood but seem to have similarities with the reduction of silver from ionic to metallic forms that occurs in the photographic development process. Nevertheless, the unsurpassed sensitivity of the staining procedure has resulted in its very rapid widespread adoption.

The most convenient and accurate measures of proteins use their ultraviolet absorbance, either around 280 nm, where the aromatic side chains absorb, or at lower wavelengths, such as 210 nm, where the peptide bond also absorbs. Proteins vary widely in their spectral properties; therefore, the molar absorbance value of each protein must be determined for quantitative results.

Protein measurement with the folin phenol reagent. O. H. Lowry, et al. J. Biol. Chem. 193:265–275, 1951.

Two new staining procedures for quantitative estimation of proteins on electrophoretic strips. S. Fazekas de St. Groth, et al. Biochim. Biophys. Acta 71:377–391, 1963.

Fluorescamine: a reagent for assay of amino acids, peptides, proteins, and primary amines in the picomole range. S. Udenfriend, et al. Science 178:871–872, 1972.

Fluorometric assay of proteins in the nanogram range. P. Böhlen, et al. Arch. Biochem. Biophys. 155:213–220, 1973.

Detection of fluorescamine-labelled amino acids, peptides, and other primary amines on thin-layer chromatograms. K. Imai, et al. Arch. Biochem. Biophys. 161:161–163, 1974.

Review of the Folin phenol protein quantitation method of Lowry, Rosebrough, Farr, and Randall. G. L. Peterson. Anal. Biochem. 100:201–220, 1979.

Sequence analysis of fluorescamine-stained peptides and proteins purified on a nanomole scale. J. Vandekerckhove and M. van Montagu. Eur. J. Biochem. 44:279–288, 1974.

A highly sensitive silver stain for detecting proteins and peptides in polyacrylamide gels. R. C. Switzer, et al. Anal. Biochem. 98:231–237, 1979.

A rapid sensitive stain for polypeptides in polyacrylamide gels. C. R. Merril, et al. Anal. Biochem. 110:201–207, 1981.

Silver stain for proteins in polyacrylamide gels: a modified procedure with enhanced uniform sensitivity. J. H. Morrissey. Anal. Biochem. 117:307–310, 1981.

DETERMINATION OF THE COVALENT STRUCTURES OF PROTEINS

A typical protein of between 100 and 1000 amino acid residues will generally consist of from 1500 to 15,000 atoms, making it orders of magnitude more complex than an average organic molecule. To determine directly the covalent structure of such a molecule would be an impossible task were it not known that the protein is a linear polymer of the 20 amino acids. The covalent structure of such a regular polymer is defined by merely the sequence of the amino acid residues within the one-dimensional polypeptide; the detailed covalent structure is defined by the known structures of the amino acid side chains and by the nature of the regular polypeptide backbone.

Length of the Polypeptide Chain

Proteins vary tremendously in length of the polypeptide chain, which may be composed of as many as 2000 residues; therefore, the most pertinent property of a new protein to be determined initially is its size. In many cases, this item of information may determine whether a protein is studied further; the prospect of determining the covalent structure or the three-dimensional structure of a protein of 50 or 100 residues is much more appealing than in the case of 1000 or more residues. The much greater sizes of proteins relative to normal organic molecules require special techniques for their accurate measurement. Often, proteins aggregate into an even larger complex; therefore, if the individual polypeptide chains are to be studied, denaturing solvents must be used.

Sedimentation Analysis

The method most widely used in the recent past for measurement of the molecular weight of proteins was analytic centrifugation in the ultracentrifuge, since the rate at which any particle sediments in a gravitational field is proportional to its size, as well as to other parameters. The rate of sedimentation in a centrifugal field, dr/dt, where r is the radius at which the protein is situated and t is time, is given by

$$\frac{dr}{dt} = \frac{M(1 - \bar{v}\rho)}{N_A f}\, \omega^2 r \qquad (1\text{–}53)$$

where the pertinent parameters of the protein are M, the molecular weight; $\bar{v}$, the partial specific volume—that is, essentially the inverse of the density; and f, the translational frictional coefficient (see p. 178), which depends upon the shape, size, and hydration of the protein. The density of the solvent is ρ, ω is the radial velocity of the rotor in radians per second, and N_A is Avogadro's number. This equation simply states that the rate of sedimentation of a protein, or of any molecule, is proportional to its molecular weight, M; the centrifugal force, $\omega^2 r$; and the density difference between it and the solvent, $1 - \bar{v}\rho$ (if a molecule has the same density as the solvent, it will not sediment irrespective of the centrifugal force; if it is lighter i.e., $\bar{v} > 1/\rho$, it will move in the opposite direction). On the other hand, the greater the frictional coefficient, the lower the rate of sedimentation. The sedimentation coefficient, s, of a molecule is defined as $(dr/dt)(1/\omega^2 r)$, to yield the well-known Svedberg equation

$$s = \frac{M(1 - \bar{v}\rho)}{N_A f} \qquad (1\text{–}54)$$

The value of s is usually expressed in Svedberg units, S ($\equiv 10^{-13}$ second).

The rate of sedimentation of a protein thus depends upon three parameters of the protein: M, $\bar{v}$, and f. The value of $\bar{v}$ may be measured from the density of aqueous solutions of the protein of known concentrations, but this requires large quantities of protein. More frequently, it is estimated from the amino acid composition of the protein by calculating the weighted average of the $\bar{v}$ values of the constituent amino acid residues (Table 1–1); proteins in this case appear to be the sum of their parts. Unfortunately, the precise value of $\bar{v}$ is quite important: Values for proteins generally lie in the range 0.70–0.75; with an aqueous solvent with ρ of about 1, for example, a 1 per cent error in $\bar{v}$ leads to an error of 3 to 4 per cent in the term $(1 - \bar{v}\rho)$, and consequently in the value of M. The partial specific volume also depends upon interactions with the solvent, so that special corrections must be made in solvents other than water.

The parameter f introduces even more uncertainty into determinations of molecular weight from only the rate of sedimentation, because proteins have been found to vary widely in shape and hydration. One solution is to place the protein in a denaturing solvent so that all proteins become disordered linear polypeptide chains, with similar values of f. Alternatively, this factor can be determined from the rate of diffusion of the protein, since the two are related by the Einstein-Sutherland equation:

$$f = \frac{k_B T}{D} \tag{1--55}$$

where D is the diffusion constant (see p. 183), k_B is Boltzmann's constant, and T the temperature. The Svedberg equation may then be expressed as

$$s = \frac{M(1 - \bar{v}\rho)}{D\,RT} \tag{1--56}$$

where R is the gas constant ($\equiv k_B N_A$). Consequently, calculation of the molecular weight of a protein from hydrodynamic data requires knowledge of two parameters, s and D (or f).

The best solution is to use sedimentation equilibrium, where the rate of sedimentation is balanced by the reverse process of diffusion. The frictional coefficient f enters equally into these two processes and thus cancels out; thus, its exact value does not affect the equilibrium. In an ideal situation with a homogeneous protein species, the concentration gradient of the protein at equilibrium is given by

$$\frac{d \ln c}{dr^2} = \frac{M(1 - \bar{v}\rho)\omega^2}{2\,RT} \tag{1--57}$$

where c is the protein concentration; the other symbols are as defined earlier. Consequently, calculation of the molecular weight of a protein by such a procedure is limited only by uncertainty in its value of $\bar{v}$.

The use of the ultracentrifuge is becoming less important for the measurement of protein molecular weights owing to the development of alternative techniques that are less demanding in equipment, purity of protein, and expertise in physical chemistry. However, the ultracentrifuge has been of immense importance in recent research on protein structure and still provides the method of choice for study of proteins that interact reversibly with each other.

Ultracentrifugation in Biochemistry. H. K. Schachman. New York, Academic Press, 1959.

Ultracentrifugation of Macromolecules. J. W. Williams. New York, Academic Press, 1972.

Protein volume in solution. A. A. Zamyatnin. Prog. Biophys. Mol. Biol. 24:109–123, 1972.

Characterization of proteins by sedimentation equilibrium in the analytical ultracentrifuge. D. C. Teller. Methods Enzymol. 27:346–441, 1973.

The calculation of partial specific volumes of proteins in guanidine hydrochloride. J. C. Lee and S. N. Timasheff. Arch. Biochem. Biophys. 165:268–273, 1974.

The calculation of partial specific volumes of proteins in 8 M urea solution. V. Prakash and S. N. Timasheff. Anal. Biochem. 117:330–335, 1981.

Gel Filtration

The newer methods of determining protein size depend upon the physical effect of passing the proteins through molecular sieves. One technique is **gel filtration,** in which the protein is passed through a chromatographic column containing particles with porous networks, made of a cross-linked dextran (i.e., Sephadex), polyacrylamide (Bio-Gel P), agarose (Sepharose or Bio-Gel A), or porous glass (Bio-Glass). The gel matrix has a network of pores into which molecules of less than a maximum size may penetrate; the size of the pores may vary, ranging from a few up to several hundred angstroms. The protein molecules in solution then partition between the internal volume of the pores of the gel particles and the volume of the liquid external to the gel particles. The smaller the protein, the greater the probability that it will enter the internal volume of the gel particles. Consequently, proteins too large to enter the pores are eluted first, in the void volume of the column, and then the smaller proteins as a function of the fraction of the internal volume of the gel matrix they could enter. There is no satisfactory method of calculating directly the size of a protein only from its elution position with a given column material, but a column may be calibrated by passing through a set of proteins of known molecular weights. A plot of the logarithms of the molecular weights of the known proteins versus their elution positions usually gives a smooth curve, with asymptotes for particles above and below the fractionation range of the gel matrix used. Within the fractionation range, linear plots may be devised (Figure 1–4). By determining the elution volume of an unknown protein on an appropriately calibrated column, its molecular weight may be estimated from such a standard curve.

The elution position is determined by both the molecular weight and the shape of the protein, just as in sedimentation analysis, unless the gel filtration is run under denaturing conditions, when the calibration and unknown proteins are disordered polypeptide chains, with similar shapes (Chapter 5). Other complications may arise, such as affinity of a protein for the gel matrix, but the method has the advantage that the proteins need not be purified for such an analysis if they can be detected specifically, e.g., by their biological activity. The procedure can also be used to determine the shape of a protein resulting from its conformational properties (Chapters 5 through 10).

Determination of molecular weights of proteins by gel filtration on Sephadex. J. R. Whitaker. Anal. Chem. 35:1950–1953, 1963.

The gel filtration behaviour of proteins related to their molecular weights over a wide range. P. Andrews. Biochem. J. 96:595–606, 1965.

Estimation of molecular size and molecular weights of biological compounds by gel filtration. P. Andrews. Methods Biochem. Anal. 18:1–53, 1970.

Protein polypeptide chain molecular weights by gel chromatography in guanidinium chloride. K. G. Mann and W. W. Fish. Methods Enzymol. 26:28–42, 1972.

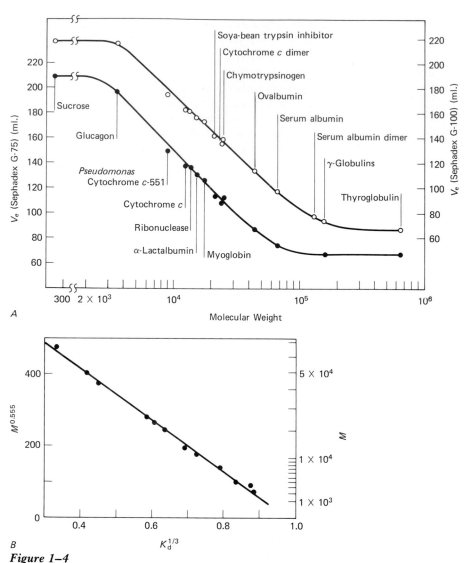

Figure 1–4
Determination of the size of a protein by gel filtration. A, Plot of elution volume, V_e, of native proteins of known molecular weight, M, on Sephadex G-75 (●) and G-100 (○) versus log M. (From P. Andrews, Biochem. J. 91:222–233, 1964.) B, Empirical linear plot with unfolded proteins in 6 M guanidinium chloride (From N. Ui, Anal. Biochem. 97:65–71, 1979; © Academic Press.)

The parameter K_d is determined as in A from V_e, the void volume V_0, and the total volume V_t,

$$K_d = \frac{V_e - V_0}{V_t - V_0}$$

V_0 and V_t are, respectively, the lower and upper asymptotes in A. The value of K_d is believed to be proportional to the effective volume of the molecule, i.e., to the radius cubed. The effective radius of an unfolded polypeptide chain is observed to be proportional to $M^{0.555}$, so a plot of $K_d^{1/3}$ versus $M^{0.555}$ is usually found to be linear with unfolded proteins.

SDS Polyacrylamide Gel Electrophoresis

A related technique is **polyacrylamide gel electrophoresis,** wherein the rate of migration through a gel with small pores is dependent upon the size of the protein. As with the other techniques discussed, the shape of the protein will also determine the rate of migration. An additional factor in this case is the net charge of the protein, since electrophoresis is the driving force. These two factors may be determined by varying the sieving effect of the gel, achieved by altering the concentration of acrylamide and the cross-linking agent, methylene-*bis*-acrylamide; all other parameters are kept constant so that the shape and net charge do not vary.

A much more satisfactory approach is to abolish all shape and charge differences of proteins by adding the denaturant sodium dodecyl sulfate (SDS):

$$CH_3—(CH_2)_{11}—SO_4^-Na^+ \tag{1–58}$$

This approach has arisen from a series of empirical observations, so that its detailed physical basis is not yet fully understood. Binding studies of a variety of different proteins have shown that above an SDS monomer concentration of 8×10^{-4} M, a constant of 1.4 grams of SDS is bound per gram of protein, i.e., one molecule of SDS for every two amino acid residues of the chain. This high level of binding of the charged detergent and the constant binding ratio will generally "swamp out" the intrinsic charge contribution of most proteins, so that an approximately constant negative charge per unit mass will be obtained. All polypeptides also appear to have a similar shape when SDS is bound, generally considered as elongated particles, with a constant diameter and a length proportional to the number of amino acid residues in the polypeptide chain. The exact nature of the protein–SDS complex is not known; none of the models proposed is entirely consistent with the many experimental observations.

Nevertheless, the constant net charge and shape lead to SDS–protein complexes having electrophoretic mobilities in polyacrylamide gels that are generally directly proportional to the logarithm of the length of the polypeptide chain (Figure 1–5). By comparing the mobility of an unknown protein with that of a set of standard marker proteins, the molecular weight of the unknown polypeptide chain may usually be determined within 10 per cent of the true value. In certain cases, abnormalities in SDS binding or protein conformation, large differences in intrinsic protein charge, or covalently attached nonprotein moieties may lead to increased or decreased electrophoretic mobilities; therefore, caution is advisable in use of this technique. Nevertheless, the remarkable resolution and ease of SDS electrophoresis has made it the most widely used method for determining the absolute values of polypeptide chain sizes. It is especially useful in studies of processes in which the molecular weights of proteins are altered.

Molecular weight estimation of polypeptide chains by electrophoresis in SDS–polyacrylamide gels. A. L. Shapiro, et al. Biochem. Biophys. Res. Commun. 28:815–820, 1967.

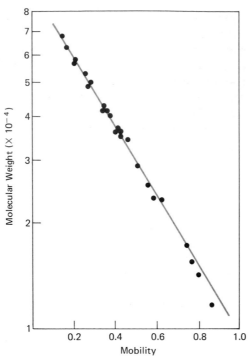

Figure 1–5

Correlation between mobility in SDS polyacrylamide gel electrophoresis and molecular weight of the polypeptide chain. Proteins of known molecular weight within the range 14,000 to 70,000 were used. (From K. Weber and M. Osborne, in H. Neurath and R. L. Hill (eds.), The Proteins, 3rd ed. Vol. 1, pp. 179–223. New York, Academic Press, © 1975.)

Molecular weight determinations and the influence of gel density, protein charges, and protein shape in polyacrylamide gel electrophoresis. D. B. Blattler and F. J. Reithel. J. Chromatogr. 46:286–292, 1970.

Molecular characterization of proteins in detergent solutions. C. Tanford, et al. Biochemistry 13:2369–2376, 1974.

Proteins and sodium dodecyl sulfate: molecular weight determination on polyacrylamide gels and related procedures. K. Weber and M. Osborne. In The Proteins, 3rd ed. H. Neurath and R. L. Hill (eds.). Vol. 1, pp. 179–223. New York, Academic Press, 1975.

Behavior of glycopolypeptides with empirical molecular weight estimation methods. 1. In sodium dodecyl sulfate. B. S. Leach, et al. Biochemistry 19:5734–5741, 1980.

Gels. T. Tanaka. Sci. Amer. 244:110–123, 1981.

Amino Acid Composition

The amino acid residues that make up a protein are now determined routinely by standard procedures of amino acid analysis. The polypeptide chain is generally hydrolyzed to the constituent amino acids by 6 M HCl at 110° C for 24 to 72 hours. With this procedure, most of the amino acids are recovered quantitatively, but tryptophan is generally destroyed, so that it must be detected by other techniques. Alkaline or enzymatic hydrolysis can give good yields of tryptophan, or it may be measured by spectral or chemical techniques.

Serine and threonine are slowly destroyed, with half-times of about 50 to 100 hours; accurate determinations of their residues require different periods of hydrolysis and extrapolation to zero-time of hydrolysis. On the other hand, some peptide bonds involving valine and isoleucine are slowly hydrolyzed, owing to steric hindrance by the branched side chain, so that quantitative measurement requires long periods of hydrolysis. Cysteine is partially destroyed and oxidized; therefore, it is best determined after oxidation to the stable cysteic acid. The amino acids asparagine and glutamine are deaminated and converted to aspartic and glutamic acids, respectively. The amount of ammonia released during acid hydrolysis may be used to determine the sum of asparagine and glutamine. Alternatively, they may be determined directly after hydrolysis of the peptide bonds in ways that do not hydrolyze the amides; enzymatic digestion with proteases is the usual method. However, whether a small peptide contains the acidic or amide residues may often be determined readily by the net charge of the peptide as determined by its electrophoretic mobility at neutral pH.

If whether a residue is Asn or Asp is not known, it is designated Asx; similarly, Glx refers to either Gln or Glu.

In spite of these shortcomings of acid hydrolysis, it is usually adequate for determining the amino acid composition. In other cases, peptides and proteins may be broken down into the constituent amino acids without racemization, hydrolysis of side chain amides, and destruction of some amino acids, by the use of appropriate mixtures of proteolytic enzymes.

The nature of the amino acids present and their quantities are now readily determined by automatic amino acid analyzers. The amino acids are separated by ion-exchange chromatography and detected quantitatively by reaction with ninhydrin. A complete and quantitative amino acid analysis may be routinely performed in an hour with less than 1 nanomole of polypeptide chain.

Amino acid analysis does not give directly the number of residues of each amino acid per polypeptide chain; it merely supplies the ratio of the different amino acids. If the molecular weight of the polypeptide chain and the amount hydrolyzed are known exactly, the molar ratio of each amino acid may be calculated. However, there are generally substantial uncertainties in each of these values, so that the molar ratios calculated are seldom close to integral values. With small peptides in favorable cases, the relative amounts of each amino acid may be close to integral values, giving a minimal value for the amino acid composition of the peptide. For example, the minimum composition of a peptide having the amino acids Ala, Gly, Val, and Phe in the molar ratios of $2:3:4:2$ would be $(Ala_2, Gly_3, Val_4, Phe_2)$, with a molecular weight of 1022. However, the peptide could also consist of integral multiples of that composition, e.g., $(Ala_4, Gly_6, Val_8, Phe_4)$, $(Ala_6, Gly_9, Val_{12}, Phe_6)$, and so on. Unfortunately, quantitative amino acid analysis is generally subject to errors of at least ± 3 per cent, so uncertainty about the correct integral value arises when 16 or more residues of any amino acid are present. These procedures are limited to use with relatively small proteins.

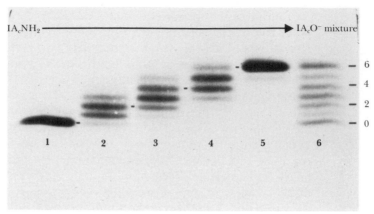

Figure 1–6
Counting integral numbers of Cys residues. Reduced bovine pancreatic trypsin inhibitor, with six Cys residues, was reacted completely with iodoacetamide (IAcNH$_2$), in lane 1, iodoacetate (IAcO$^-$), in lane 5, and varying ratios of the two in lanes 2 through 4. The unfolded protein was subjected to electrophoresis under conditions where its mobility is determined by its net charge, i.e., by the number of acidic carboxymethyl groups introduced by reaction of Cys residues with iodoacetate. In lane 6 is a mixture of the other samples.

Varying extents of reaction with iodoacetate generated six new electrophoretic bands, the number of Cys residues present on the protein. (Adapted from T. E. Creighton, Nature 284:487–489, 1980.)

There is a different approach, which can give integral values for the number of residues of one or more of the 20 amino acids, independent of any other properties of the polypeptide chain, including its molecular weight. To determine the value of n, the number of residues of the amino acid in question, the $n + 1$ species having $0, 1, 2, \cdots, n$ residues modified chemically are generated by varying extents of reaction with a reagent specific for that amino acid. The $n + 1$ species are separated by a procedure sensitive to the number of groups modified; electrophoresis or isoelectric focusing is an obvious candidate for modifications where the modification alters the charge on the amino acid side chain. The number of bands are counted, which is one more than the number of residues of the amino acid.

The procedure for counting the number of Cys residues is illustrated in Figure 1–6. Here the $n + 1$ electrophoretic bands were generated by using competition between the neutral and acidic thiol reagents, iodoacetamide and iodoacetate, respectively. Amino groups have also been counted using progressive modification with succinic anhydride, converting them to acidic residues. Other amino acids for which there are specific modification reactions should also be measurable in this way.

After the number of residues of one or more amino acids has been determined in this way, the relative amounts of all the amino acids determined by standard amino acid analysis may be converted directly to values of number of residues per polypeptide chain. Values for the other amino acids will not necessarily be integers, owing to experimental errors, but they may be sufficiently close.

The molecular weight of the polypeptide chain may then be calculated directly from the amino acid composition.

Hydrolysis of proteins. R. L. Hill. Adv. Protein Chem. 20:37–107, 1965.
Spectroscopic determination of tryptophan and tyrosine in proteins. H. Edelhoch. Biochemistry 6:1948–1954, 1967.
Determination of tryptophan. T.-Y. Liu. Methods Enzymol. 25:44–55, 1972.
Counting integral numbers of amino acid residues per polypeptide chain. T. E. Creighton. Nature 284:487–489, 1980.
Counting integral numbers of amino groups per polypeptide chain. M. Hollecker and T. E. Creighton. FEBS Letters 119:187–189, 1980.
Amino acid analysis at the picomole level. J.-Y. Chang, et al. Biochem. J. 199:547–555, 1981.

The Amino Acid Sequence

The amino acid sequence identifies a given protein, distinguishing it from all other proteins, even those with the same amino acid composition. It also determines all the chemical and biological properties of the protein. Appropriately the sequence also is known as the **primary structure,** since it is the most basic level of structure and also defines the higher levels of structure (secondary, tertiary, and quaternary) that will be encountered in further discussions of proteins. The most basic step in characterizing a protein is thus to determine its amino acid sequence.

The arrangement of amino acids in proteins. F. Sanger. Adv. Protein Chem. 7:1–67, 1952.
The Primary Structure of Proteins. W. A. Schroeder. New York, Harper & Row, 1968.
Recent developments in chemical modification and sequential degradation of proteins. G. R. Stark. Adv. Protein Chem. 24:261–308, 1970.
Strategy and tactics in protein chemistry. B. S. Hartley. Biochem. J. 119:805–822, 1970.
Recent advances in the chemical modification and covalent structural analysis of proteins. R. L. Heinrikson and K. J. Kramer. Prog. Bioorg. Chem. 3:141–250, 1974.
Strategy and methods of sequence analysis. W. H. Konigsberg and H. H. Steinmann. In The Proteins, 3rd ed. H. Neurath and R. L. Hill (eds.). Vol. 3, pp. 1–178, New York, Academic Press, 1977.
Advances in protein sequencing. K. A. Walsh, et al. Ann. Rev. Biochem. 50:261–284, 1981.

Amino-terminal and Carboxyl-terminal Residues

Several chemical methods are available for identifying the amino-terminal residue. They involve labeling chemically the α-amino group, hydrolyzing the peptide to the constituent amino acids, and then determining which amino acid is labeled. The two most widely used procedures employ fluoro-2,4-dinitrobenzene and dansyl chloride to label the α-amino group:

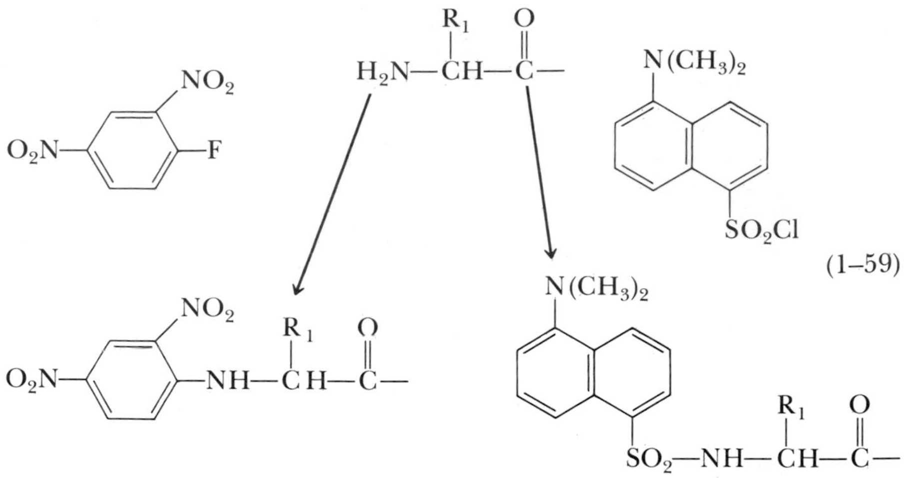

$$(1–59)$$

The linkages to the α-amino group withstand hydrolysis of the peptide. The labeled amino acids are identified by their yellow color in the fluoro-2,4-dinitrobenzene procedure, and by fluorescence when the dansyl derivative is used.

A related procedure involves reaction of the α-amino group with cyanate. Upon acid hydrolysis, the amino-terminal residue is converted to the hydantoin:

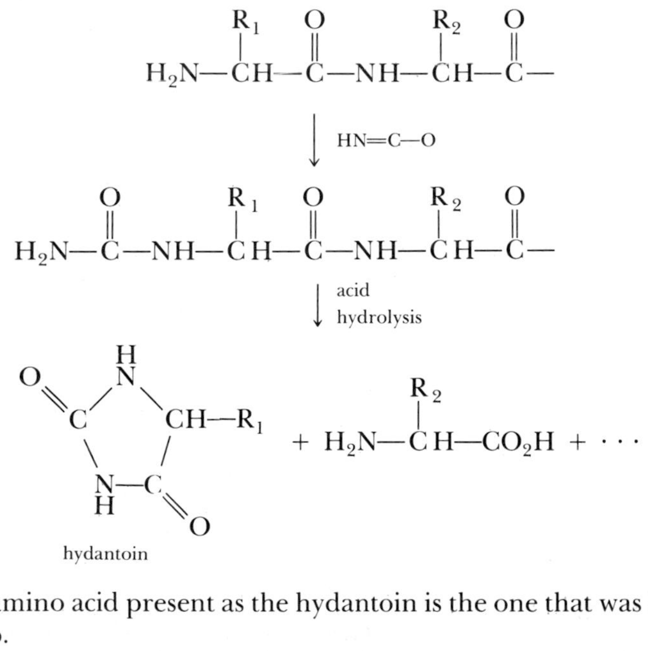

$$(1–60)$$

The only amino acid present as the hydantoin is the one that was the amino end group.

Comparable procedures for identifying the carboxyl-terminal residues are not as satisfactory. The most widely used of the available methods is hydrazinolysis. All the α-carboxyl groups involved in peptide bonds are converted into hydrazides:

$$H_2N-\underset{\underset{R_1}{|}}{C}H-\underset{\underset{O}{\|}}{C}-\cdots-NH-\underset{\underset{R_{n-1}}{|}}{C}H-\underset{\underset{O}{\|}}{C}-NH-\underset{\underset{R_n}{|}}{C}H-CO_2H$$

$$\downarrow \quad H_2N-NH_2$$

$$H_2N-\underset{\underset{R_1}{|}}{C}H-\underset{\underset{O}{\|}}{C}-NHNH_2 + \cdots + H_2N-\underset{\underset{R_{n-1}}{|}}{C}H-\underset{\underset{O}{\|}}{C}-NHNH_2$$

$$+ H_2N-\underset{\underset{R_n}{|}}{C}H-CO_2H \qquad (1-61)$$

Only the carboxyl-terminal residue is released as the unmodified amino acid, permitting its identification. However, the procedure has a number of shortcomings. Carboxyl-terminal Asn and Gln residues are not recovered as the free amino acids, because the side-chain amides are converted to hydrazides, and Arg residues are converted to ornithine (Eq. 1–22). The recovery rates for other amino acids range between 20 and 80 per cent, owing to an array of complicating side reactions. Nevertheless, it is the most satisfactory procedure currently available.

Short sequences of amino acids at both the amino and the carboxyl ends of peptides may often be determined by using exopeptidases that sequentially liberate single amino acids from either end, such as leucine aminopeptidase or carboxypeptidase. The amino acids liberated are detected by amino acid analysis. The sequence of the residues can be determined only by following the kinetics of amino acid release, since the second residue will be released only after the first has been removed, the third residue then only after the second, and so on. Generally, some luck is required for there to be a sequential release of each amino acid, as the nature of the terminal residue often determines the rate at which it is cleaved. For example, if the second residue is cleaved significantly more rapidly than the first, the apparent cleavage rate for each will be virtually the same, as determined by the first cleavage; it will be impossible to determine in this way the relative order of the two amino acids.

End group analysis using dansyl chloride. W. R. Gray. Methods Enzymol. 25:121–138, 1972.

Use of cyanate for determining NH_2-terminal residues in proteins. G. R. Stark. Methods Enzymol. 25:103–120, 1972.

Hydrazinolysis. W. A. Schroeder. Methods Enzymol. 25:138–143, 1972.

Enzymatic hydrolysis with carboxypeptidases. R. P. Ambler. Methods Enzymol. 25:143–154, 1972.

Stepwise sequence determination from the carboxyl terminus of peptides. J. L. Meuth, et al. Biochemistry 21:3750–3757, 1982.

The Edman Degradation

The most successful and widely used method of determining the sequence of amino acids in a protein is the **Edman degradation procedure,** which

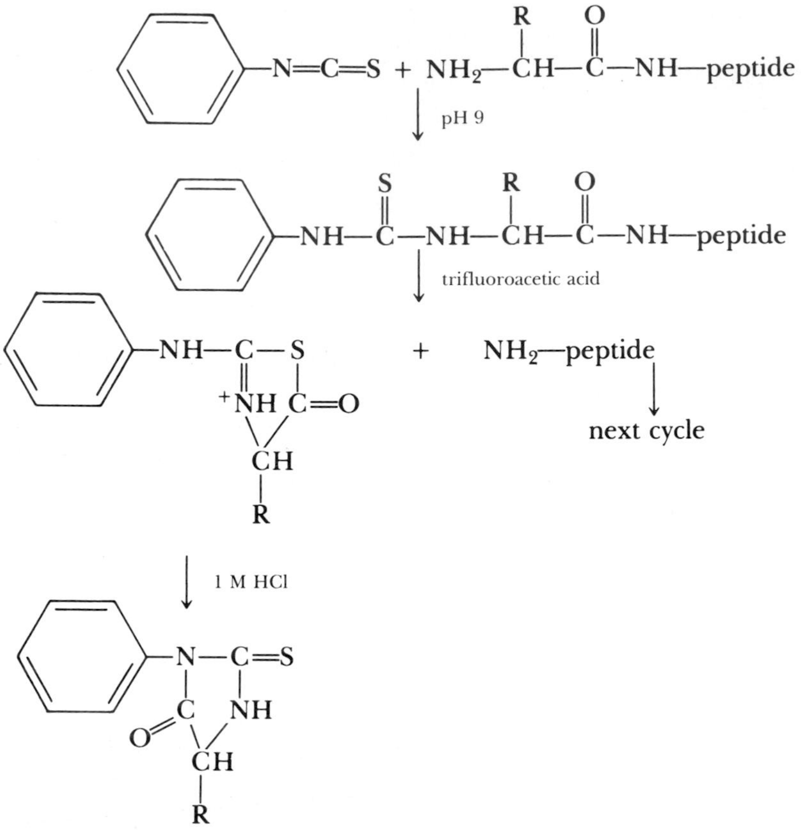

PTH-amino acid

Figure 1–7
One cycle of the Edman degradation for sequencing peptides from the amino end.

removes one residue at a time sequentially from the amino end. The cycle may be repeated many times to identify the sequence of amino acids (Figure 1–7).

The terminal α-amino group reacts with phenylisothiocyanate in alkaline medium to give the phenylthiocarbamyl (PTC) derivative of the peptide:

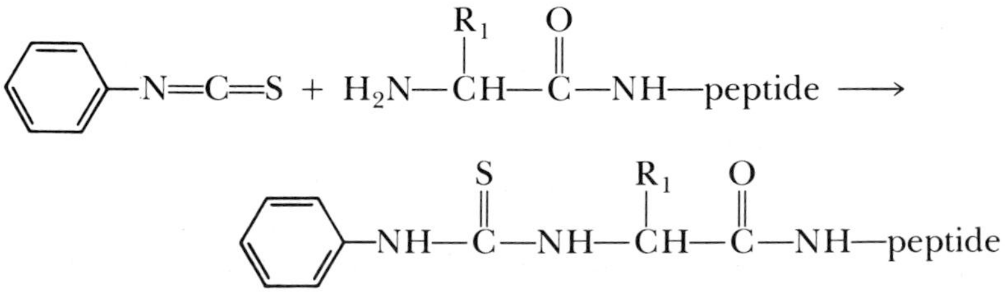

(1–62)

The PTC-peptide is then degraded by treatment with a strong acid, e.g., trifluoroacetic acid:

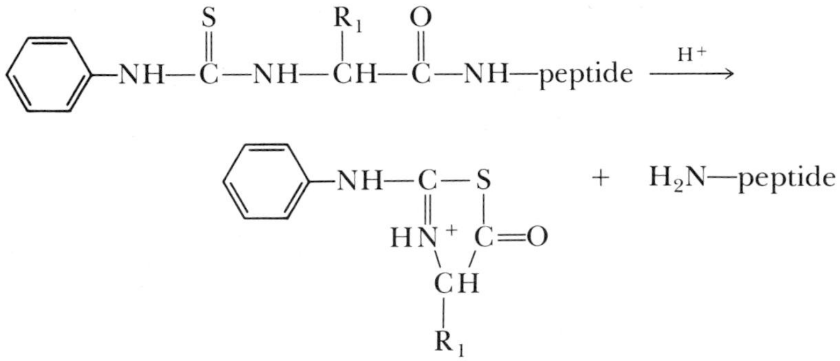

$$(1-63)$$

The PTC group introduces instability at the first peptide bond, causing it to be hydrolyzed under conditions where the other peptide bonds are stable. The liberated thiazolinine derivative of the amino-terminal residues then rearranges in aqueous solution to the phenylthiohydantoin (PTH) derivative:

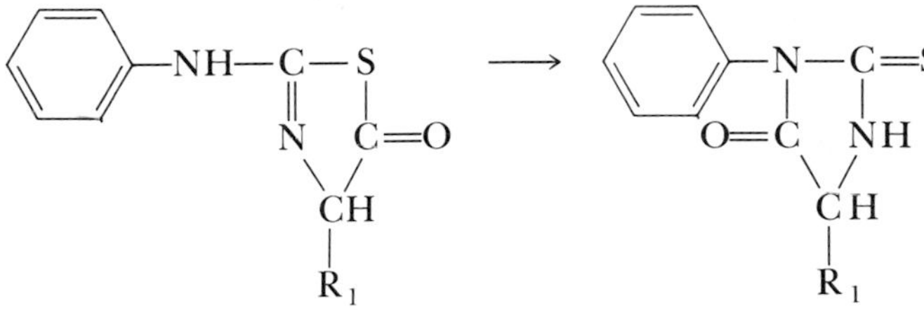

$$(1-64)$$

Identification of the amino acid PTH derivative, i.e., of the original amino-terminal residue, is possible with a variety of chromatographic techniques.

The beauty and great virtue of the method lie in the regeneration of the peptide, altered only by the loss of the amino-terminal residue. This shortened peptide is then ready for a second cycle of the same procedure, which identifies the second residue in the original sequence and generates the peptide shortened further, in which the next residue may be identified by a third cycle, and so on. A considerable number of carefully controlled manipulations are necessary for each cycle, but the method has now been automated so that sequences may be determined relatively rapidly. By this procedure, under favorable circumstances, the sequence of 30 to 60 residues from the amino end may be determined with a peptide or protein.

The primary difficulty with the Edman degradation approach to amino acid sequence determination is that, as with all repetitive procedures, whether degradative or synthetic, errors are cumulative. If at any stage a significant amount of unreacted amino-terminal residue remains, it will appear at the next cycle. As the same error is repeated, the amount and diversity of the background material rises until the residue at the cycle being processed can no longer be identified. Such admixture also occurs from random breaks of the peptide chain, produced by its subjection to not entirely innocuous reagents. Possibly the most serious cause of the decrease in yield at each step is oxidative desulfuration of the PTC-peptide to isocyanate derivatives, which can then no longer undergo cyclization and cleavage;

such chains are now merely alkylated and no longer take part in further cycles. Some peptides tend to have sufficient solubility in the organic solvents used for extraction to be lost from the reaction container. To overcome this problem, the peptide can be attached to an insoluble matrix. Use of such a matrix also has the benefit of simplicity of the degradation procedure, as the peptide–resin can be packed in a column through which the reagents and solvents are passed. However, there is no perfect way of attaching the peptides to a solid support; ideally, the linkage should be only through the terminal carboxyl group.

A protein sequenator. P. Edman and G. Begg. Eur. J. Biochem. 1:80–91, 1967.
Automated Edman degradation: the protein sequenator. H. D. Niall. Methods Enzymol. 27:942–1010, 1973.

Mass Spectrometry

Structure determination of large polar molecules by mass spectrometry is not an obvious approach, as the method involves evaporation of the sample into a vacuum, nonspecific fragmentation with an electron beam into an assortment of species, and identification of the cationic fragments solely by their mass-to-charge ratio. Yet the technique has very great advantages of sensitivity and accuracy; only micrograms of sample are required, and masses are measured to an accuracy of five decimal places. Consequently, it has been developed into a very useful technique for sequencing peptides of from 2 to 15 residues.

The polar groups of both the backbone and the side chains must be modified chemically to abolish their polarity and to make the peptide sufficiently volatile. A number of different procedures are available, and which is used will depend upon the individual case. The preferred technique appears to be that in which amino groups are first acetylated with acetic anhydride, and then carboxyl, hydroxyl, and NH groups are permethylated (Figure 1–8). If any Arg residues are present, they must first be treated with hydrazine to cleave the guanido group (Eq. 1–22). The peptide is then usually sufficiently volatile to be used in the mass spectrometer.

The fragmentation patterns of a peptide are potentially very complex, but the modifications can induce fragmentation of the backbone primarily at the peptide bond (Figure 1–9). Other fragmentations are possible, and the side chains also may be broken. Fortunately, however, the usual fragmentation patterns are now known, so that all the ion fragments produced with a peptide may be used to identify its sequence.

The fragments are identified only by their mass; therefore, it is impossible to distinguish between the isomeric Leu and Ile in this way. But the accuracy of the technique does make possible the recognition, and often the identification, of unusual amino acids in a peptide: In this way, for example, the novel amino acid γ-carboxyglutamic acid (Chapter 2) was found in thrombin; it had been identified as Glu during normal sequencing, as it is readily decarboxylated during these procedures.

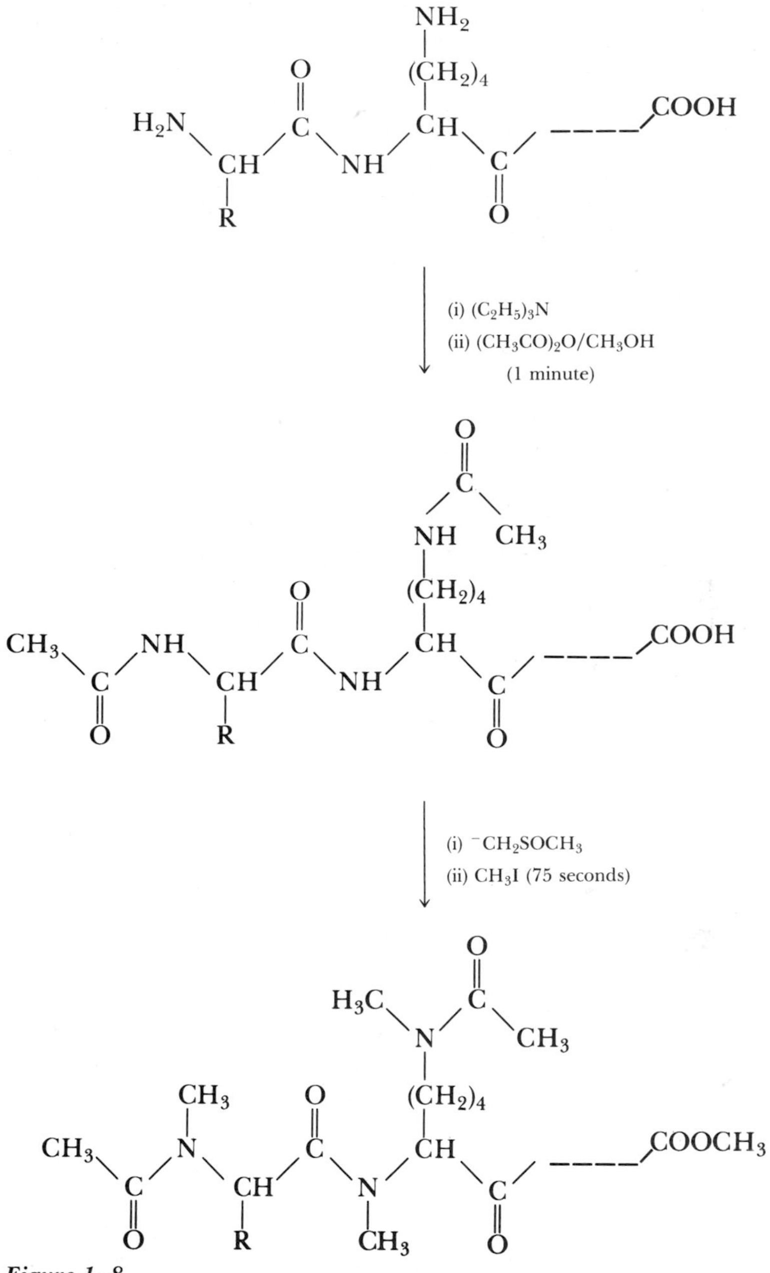

Figure 1–8
Modification reactions used to make peptides volatile for sequencing by mass spectrometry. In the first step, the peptide is treated with triethylamine to ensure alkaline conditions and with acetic anhydride in methanol to acetylate all amino groups. In the second step, the peptide is permethylated by treating it with the strong base methyl sulfinyl carbanion to abstract the hydrogen atoms and with methyl iodide to replace them with methyl groups. Any peptide containing Arg residues must first be treated by hydrazinolysis to remove the guanido group before being acetylated. (Adapted from H. R. Morris, Nature 286:447–452, 1980.)

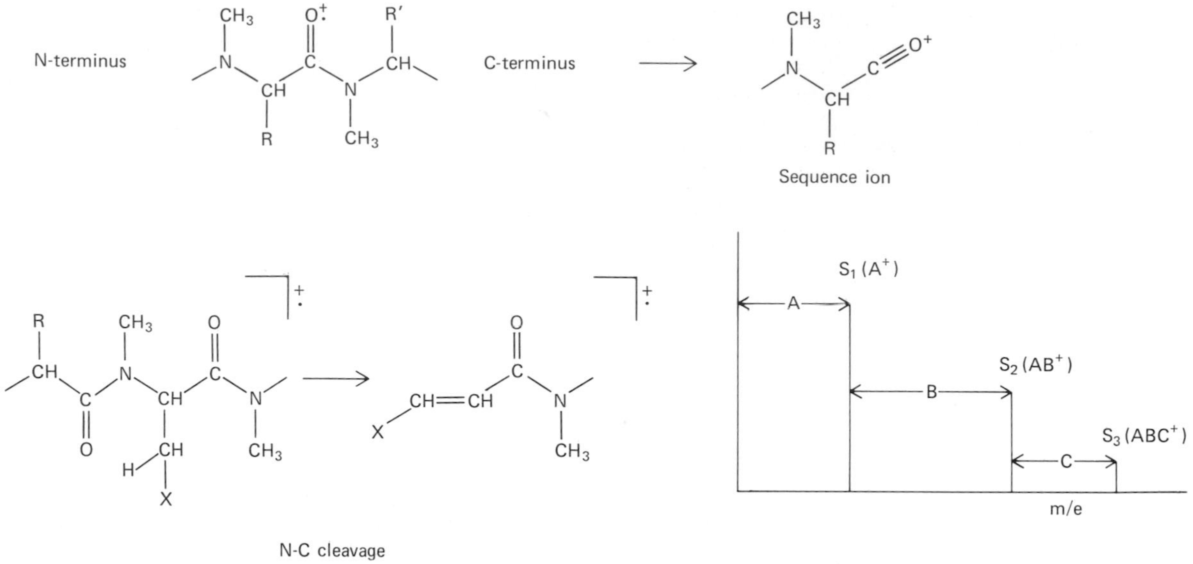

Figure 1–9

Fragmentation of the volatile modified peptide in mass spectrometry. At the top is shown the usual fragmentation of the peptide bond to generate the positively charged "sequence ion" that is detected. However, when the side chain R is that of Phe, Tyr, His, Trp, Asn, or Asp, it occurs by N—C cleavage, as illustrated on the lower left. On the lower right is shown the three sequence ions with increasing ratio of mass to charge, m/e, that would be expected in the mass spectrum with a peptide A—B—C. The three peaks originate from cleavage at each of the three peptide bonds. (Adapted from H. R. Morris, Nature 286:447–453, 1980.)

The greatest advantage of mass spectrometry is that peptides need not be pure. Mixtures of up to 5 peptides may be effectively subjected to fractional distillation within the spectrometer by gradually increasing the temperature. The signals in the mass spectrum due to any one peptide will rise and fall together, so that the complex spectra may be resolved into those of the individual components. As peptide purification is usually the most difficult step in protein sequencing, this property of mass spectrometry makes it extremely useful, in spite of the necessary chemical modification.

Technical advances are currently being made in this field. In particular, **Fast Atom Bombardment mass spectrometry** does not require chemical modification of the peptide; here the sample need not be volatile, because it is loaded in a glycerol matrix onto a metallic target and bombarded with accelerated argon atoms. However, this technique has the disadvantage that fragmentation of the peptide does not always occur.

Determination of the sequences of protein-derived peptides and peptide mixtures by mass spectrometry. H. R. Morris, et al. Biochem. J. 125:189–201, 1971.

Mass-spectrometric identification and sequence location of the ten residues of the new amino acid (γ-carboxyglutamic acid) in the N-terminal region of prothrombin. H. R. Morris, et al. Biochem. J. 153:663–679, 1976.

Mass spectrometric sequencing of peptides and proteins. K. Biemann. Pure and Appl. Chem. 50:149–158, 1978.

Biomolecular structure determination by mass spectrometry. H. R. Morris. Nature 286:447–452, 1980.

Fast Atom Bombardment—high field magnet mass spectrometry of 6000 dalton polypeptides. A. Dell and H. R. Morris. Biochem. Biophys. Res. Commun. 106:1456–1462, 1982.

Fragmentation of a Protein into Peptides

For most proteins, with more than about 50 amino acid residues, the amino acid sequence cannot be determined solely by Edman degradation. The protein must be cleaved into smaller peptides, the amino acid sequences of which can be determined individually. The basic approach used today is virtually the same as that originally developed by Sanger for insulin:

1. fragmentation of the polypeptide chain into peptides
2. isolation and purification of these smaller peptides
3. determination of the amino acid sequence of each peptide by chemical and enzymic means
4. repetition of steps 2 and 3 following a second type of fragmentation of the intact protein in order to obtain peptides that overlap peptides from the first fragmentation, thus making it possible to establish unambiguously the order of peptides in the entire protein chain

The approach first used by Sanger in determining the sequence of the B chain of insulin is useful to illustrate the basic strategy of sequence determination, even though present-day techniques would permit unambiguous sequence determination of larger peptides, thereby requiring the use of fewer peptides. Sanger used partial acid hydrolysis to generate a very large number of small peptides. However, peptides overlapping certain segments were not present; these were identified by generating larger, overlapping peptides by enzymic hydrolysis (Figure 1–10). Current methodology is significantly more advanced than that developed by Sanger for determining the first amino acid sequence of a protein, but it is a sobering thought that the original sequence of insulin appears to be the only one put forward that has been substantiated in all respects and has not required revision.

Partial acid hydrolysis was the prime method of fragmenting the polypeptide chain used by Sanger, because it produced a large number of small overlapping peptides that could be sequenced. However, the Edman degradation has made it much easier to sequence longer peptides; therefore, it is no longer desirable to fragment the chain nonspecifically into so many

	1	2	3	4	5	6	7	8	9	10	11	12	13	14
Peptides from acid hydrolyzates	Phe ·	Val		Glu ·	His		CysO₃H ·	Gly		His ·	Leu		Glu ·	Ala
		Val ·	Asp		His ·	Leu					Leu ·	Val		Ala ·
			Asp ·	Glu		Leu ·	CysO₃H		Ser ·	His		Val ·	Glu	
	Phe ·	Val ·	Asp			Leu ·	CysO₃H ·	Gly				Val ·	Glu ·	Ala
				Glu ·	His ·	Leu			Ser ·	His ·	Leu			
		Val ·	Asp ·	Glu							Leu ·	Val ·	Glu	
					His ·	Leu ·	CysO₃H							Ala ·
	Phe ·	Val ·	Asp ·	Glu					Ser ·	His ·	Leu ·	Val		
					His ·	Leu ·	CysO₃H ·	Gly			Leu ·	Val ·	Glu ·	Ala
	Phe ·	Val ·	Asp ·	Glu ·	His				Ser ·	His ·	Leu ·	Val ·	Glu	
				Glu ·	His ·	Leu ·	CysO₃H			His ·	Leu ·	Val ·	Glu	
									Ser ·	His ·	Leu ·	Val ·	Glu ·	Ala
Sequences deduced from the above peptides	Phe ·	Val ·	Asp ·	Glu ·	His ·	Leu ·	CysO₃H ·	Gly						
									Ser ·	His ·	Leu ·	Val ·	Glu ·	Ala
Peptides from pepsin hydrolyzate	Phe ·	Val ·	Asn ·	Gln ·	His ·	Leu ·	CysO₃H ·	Gly ·	Ser ·	His ·	Leu			
												Val ·	Glu ·	Ala ·
					His ·	Leu ·	CysO₃H ·	Gly ·	Ser ·	His ·	Leu			
Peptides from chymotrypsin hydrolyzate	Phe ·	Val ·	Asn ·	Gln ·	His ·	Leu ·	CysO₃H ·	Gly ·	Ser ·	His ·	Leu ·	Val ·	Glu ·	Ala ·
Peptides from trypsin hydrolyzate														
Structure of B chain of oxidized insulin	Phe ·	Val ·	Asn ·	Gln ·	His ·	Leu ·	CysO₃H ·	Gly ·	Ser ·	His ·	Leu ·	Val ·	Glu ·	Ala ·

Figure 1–10

Elucidation of the amino acid sequence of the B chain of insulin from the sequences of the peptides generated by partial acid, pepsin, chymotrypsin, and trypsin hydrolysis. At the bottom is the inferred sequence. (Adapted from E. O. P. Thompson, Sci. Amer., p. 36, May 1955.)

peptides, as the purification of the peptides is usually the most difficult part of sequence determination. It is now preferable to fragment the polypeptide chain at a few specific positions to generate a small number of longer peptides that may be sequenced directly. The positions of these peptides are determined by the overlapping peptides produced by a second method of fragmentation.

One of the most widely used methods of fragmenting a polypeptide chain remains the use of specific proteolytic enzymes. Trypsin is the most

1	2	3	4	5	6	7	8	9	10	11	12	13	14	15	16
				CysO₃H	Gly		Arg	Gly						Lys	Ala
Leu		Leu	Val		Gly	Glu		Gly	Phe			Thr	Pro		
			Val	CysO₃H		Glu	Arg								
	Tyr	Leu	Val		Gly	Glu	Arg						Pro	Lys	Ala
			Val	CysO₃H	Gly										
		Leu	Val	CysO₃H											
Leu	Tyr														
	Tyr	Leu	Val	CysO₃H								Thr	Pro	Lys	Ala
		Leu	Val	CysO₃H	Gly										
	Tyr	Leu	Val	CysO₃H	Gly							Thr	Pro	Lys	Ala
					Gly	Glu	Arg	Gly							
		Leu	Val	CysO₃H	Gly	Glu	Arg	Gly	Phe		Tyr	Thr	Pro	Lys	Ala
Leu															
Leu	Tyr										Tyr	Thr	Pro	Lys	Ala
		Leu	Val	CysO₃H	Gly	Glu	Arg	Gly	Phe	Phe					
								Gly	Phe	Phe	Tyr	Thr	Pro	Lys	
															Ala
Leu	Tyr	Leu	Val	CysO₃H	Gly	Glu	Arg	Gly	Phe	Phe	Tyr	Thr	Pro	Lys	Ala

specific protease found to date, cleaving the peptide bonds on the carboxyl-terminal side of Lys and Arg residues. Trypsin cleavage may also be restricted to Arg residues by reversibly blocking the ε-amino group of Lys residues by maleylation or citraconylation. Additional sites of cleavage may be introduced at Cys residues by reacting the thiol side chain with ethyleneimine to give the basic side chain S-aminoethyl Cys. An extracellular protease from *Staphylococcus aureus* is complementary to trypsin in that it cleaves only after the acidic residues Asp and Glu. Chymotrypsin is relatively specific for the peptide bonds following the aromatic residues, Tyr, Phe, and Trp, but it has significant activity with other hydrophobic residues, especially Leu.

Considerable progress has been made in the chemical cleavage of the polypeptide chain at specific residues. The most successful is the cleavage at Met residues by cyanogen bromide (CNBr). The nucleophilicity of the sulfur atom permits reaction with CNBr to the presumed sulfonium salt:

$$—CH_2—CH_2—S—CH_3 + CNBr \longrightarrow$$

$$\overset{\displaystyle CH_3}{\underset{\displaystyle +}{—CH_2—CH_2—\overset{|}{S}}}—C\equiv N + Br^- \qquad (1–65)$$

The particular stereochemistry of the methionine side chain favors the rapid intramolecular rearrangement of the sulfonium salt:

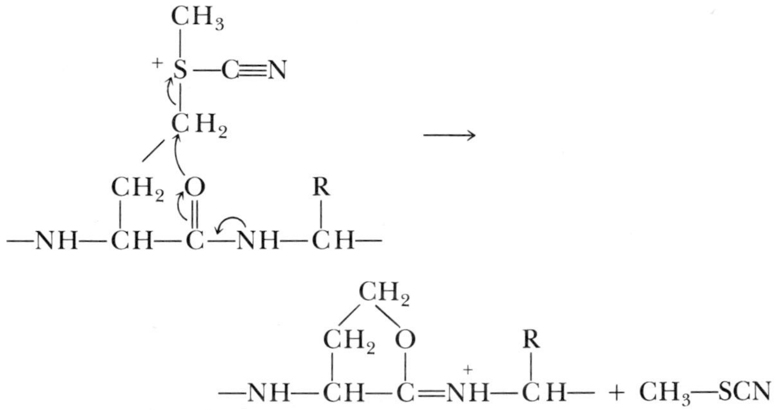

$$\longrightarrow$$

$$—NH—CH—\overset{R}{\underset{|}{C}}H— + CH_3—SCN \qquad (1–66)$$

The iminolactone generated is readily hydrolyzed by water to cleave the polypeptide chain:

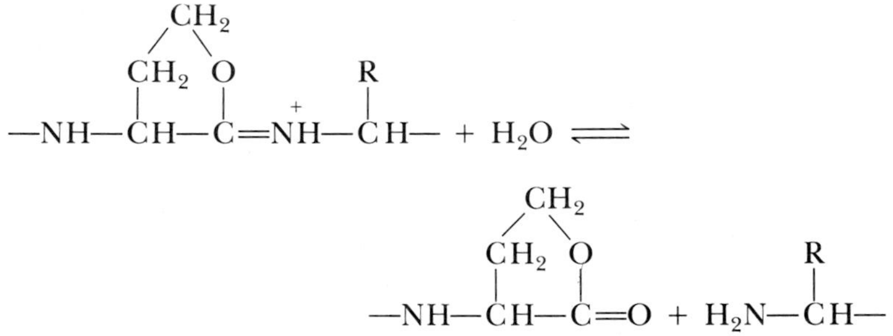

$$—NH—CH—C=\overset{+}{N}H—\overset{R}{\underset{|}{C}}H— + H_2O \rightleftharpoons$$

$$—NH—CH—C=O + H_2N—\overset{R}{\underset{|}{C}}H— \qquad (1–67)$$

to yield the amino terminal fragment with a terminal homoserine lactone residue, plus the carboxyl terminal peptide. The homoserine lactone is hydrolyzed reversibly to the free homoserine:

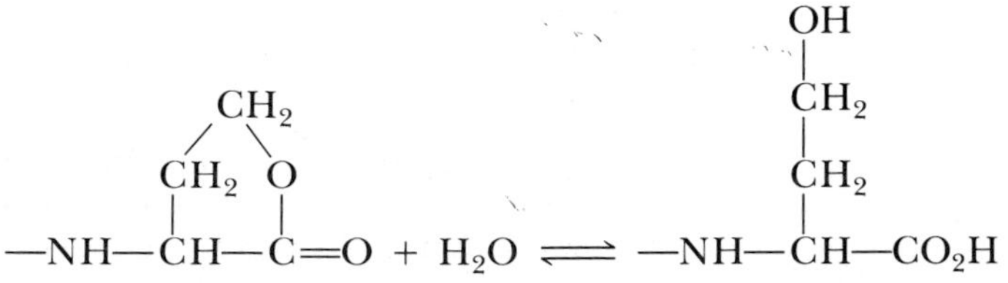

$$—NH—CH—C=O + H_2O \rightleftharpoons —NH—CH—CO_2H \qquad (1–68)$$

This equilibrium can complicate separation of CNBr-generated peptides, as the lactone moiety is uncharged whereas the free carboxyl group is ionized at neutral or alkaline pH. But the facility and completion of the CNBr cleavage more than compensate for this complication, and this procedure has become extremely useful in protein sequence determination.

A second chemical method that is specific and has had some success is the cleavage next to Cys residues produced after cyanylation. The thiol groups of Cys residues are cyanylated by reaction with 2-nitro-5-thiocyanobenzoic acid:

$$—CH_2SH + NC—S—\langle\rangle—NO_2 \longrightarrow$$
$$\qquad\qquad\qquad CO_2H$$

$$—CH_2—S—C{\equiv}N + HS—\langle\rangle—NO_2$$
$$\qquad\qquad\qquad\qquad\qquad CO_2H \qquad (1{-}69)$$

Peptide bond cleavage occurs upon incubation at alkaline pH:

$$\overset{\displaystyle N}{\underset{\displaystyle S}{\overset{\displaystyle \|}{C}}}$$

$$\begin{array}{cccc} R & O & CH_2 & O \\ | & \| & | & \| \\ —CH— & C—NH— & CH— & C—NH— \end{array} \xrightarrow{\ OH^-\ }$$

$$\begin{array}{cc} R & \qquad HN{=}C \diagup S \\ | & \\ —CH—CO_2^- & + \ HN—CH—\underset{\displaystyle \|}{C}—NH— \end{array} \qquad (1{-}70)$$

This reaction is very specific for Cys residues, but the disadvantage is that the new amino terminal group generated is blocked and not amenable to sequence analysis by the Edman procedure.

Nonenzymatic methods for the preferential and selective cleavage and modification of proteins. B. Witkop. Adv. Protein Chem. 16:221–321, 1961.

The cyanogen bromide reaction. E. Gross. Methods Enzymol. 11:238–255, 1967.

Selective cleavage and modification of peptides and proteins. T. F. Spande, et al. Adv. Protein Chem. 24:97–260, 1970.

Cleavage at cysteine after cyanylation. G. R. Stark. Methods Enzymol. 47:129–132, 1977.

Cleavage of tryptophanyl peptide bonds in cytochrome b_5 by cyanogen bromide. J. Ozols and C. Gerard. J. Biol. Chem. 252:5986–5989, 1977.

Oxidation of methionine to methionine sulfoxide as a side reaction of cyanogen bromide cleavage. R. Joppich-Kuhn, et al. Anal. Biochem. 119:73–77, 1982.

Purification of Peptides

After a protein has been cleaved into smaller peptides to be sequenced, the peptides must usually be purified. A wide variety of electrophoretic and chromatographic (e.g., thin-layer, ion-exchange, or high-pressure liquid chromatography) techniques are available, too numerous to describe here. Electrophoretic techniques are especially useful, as the electrophoretic mobility at certain pH values of a peptide may be predicted quite accurately from its amino acid composition. The mobility is proportional directly to the net charge of the peptide (ϵ) and inversely to the $\frac{2}{3}$ power of its molecular weight (M):

$$\text{mobility} \approx \epsilon\, M^{-2/3} \tag{1–71}$$

The observed mobility may also be used to determine the net charge, often making it possible to distinguish between Asn and Asp or Gln and Glu residues.

Often it is desirable to purify all the peptides containing one particular amino acid. This is most readily accomplished using "diagonal" techniques. If a mixture of peptides is separated in one dimension by some procedure and then subjected to the same procedure a second time, but at right angles to the first, all the peptides will lie on a diagonal, as they will have the same mobilities in both directions. However, peptides may be altered specifically after the first separation, so that their mobilities in the second dimension will be different from that in the first. The modified peptides will then lie off the diagonal of the unmodified peptides. If only a few peptides are altered, they may be purified by this procedure alone (Figure. 1–11). A modification reaction specific for one particular amino acid side chain may then be used to purify selectively all the peptides containing that amino acid.

For example, peptides containing Cys residues may be purified by blocking the thiol groups initially with iodoacetic acid. A diagonal map is then prepared by electrophoresis at pH 3.5, exposing the peptides to performic acid between the two separations. This oxidizes the carboxymethyl-Cys residues to the sulfones:

$$-CH_2-S-CH_2-CO_2H \xrightarrow{\text{HCO}_3\text{H}} -CH_2-\overset{\displaystyle O}{\underset{\displaystyle O}{\overset{\|}{\underset{\|}{S}}}}-CH_2-CO_2H \tag{1–72}$$

This lowers the pK_a value of the carboxyl groups, so that the peptides containing this residue are more acidic in the second dimension; consequently, they lie somewhat off the diagonal (Figure 1–11A). In the case illustrated, the six peptides off the diagonal contain the six Cys residues of the protein. The procedure may also be used with Cys residues blocked by mixed disulfides with neutral or basic groups; the performic acid cleaves the disulfide and converts the Cys side chain to that of the very acidic cysteic acid, $CysO_3H$:

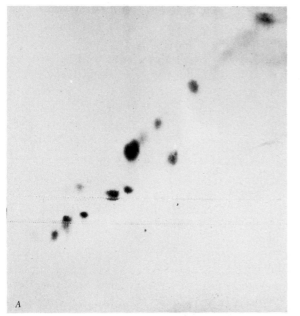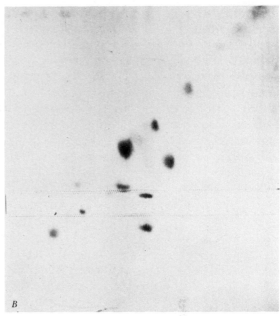

Figure 1–11

A, *Diagonal map of reduced carboxymethylated bovine pancreatic trypsin inhibitor. B, Diagonal map of the major one-disulfide intermediate trapped during refolding of the reduced protein. See Figure 1–6 for the demonstration that this protein has six Cys residues, Figure 6–9 for its structure, and Figure 7–18 for its pathway of refolding.*

The proteins were digested with trypsin, then chymotrypsin. The resulting peptides were separated by electrophoresis at pH 3.5 in the horizontal direction (anode at the right). After exposure to performic acid vapor, the electrophoresis was repeated in the vertical direction (anode at the top), then stained with ninhydrin.

The six numbered peptides in A contain the six carboxymethyl-Cys (CM-Cys) residues; they had slightly lower mobilities is the second dimension owing to the somewhat greater acidity of the CM-Cys residues upon oxidation to the sulfones. The normal diagonal of unmodified peptides is above the CM-Cys peptides; many normal peptides have migrated off the paper. The peptides contain the following residues of the primary structure: 1, 36 to 39 (Cys 38); 2, 1 to 15 (Cys 5 and Cys 14); 3, 47 to 53 (Cys 51); 4, 5 to 15 (Cys 5 and Cys 14); 5, 27 to 33 (Cys 30); 6, 54 to 58 (Cys 55).

In B, peptides 3 and 5 had altered mobilities, with the same in the first dimension, indicating that they were originally linked by a disulfide bond. Consequently, this protein had a single intramolecular disulfide bond linking Cys 30 and Cys 51. (Adapted from T. E. Creighton, J. Mol. Biol. 87:603–624, 1974.)

$$-CH_2-S-S-CH_2-CH_2-NH_3^+ \xrightarrow{\text{HCO}_3\text{H}}$$

Cys mixed disulfide

$$-CH_2-SO_3^- + {}^-O_3S-CH_2-CH_2-NH_3^+ \qquad (1\text{–}73)$$

Cys O₃H

Similar techniques have been developed for peptides containing Lys, Met, His, and Arg residues. For Lys-containing peptides, the amino groups are initially blocked with trifluoroacetyl or maleyl groups, which are then easily removed after the first electrophoresis. Met residues are first alkylated with iodoacetamide, to produce the charged sulfonium derivative:

$$-CH_2-CH_2-S-CH_3 \xrightarrow{\;ICH_2CONH_2\;}$$

$$-CH_2-CH_2-\overset{\overset{\displaystyle CH_3}{|}}{\underset{+}{S}}-CH_2CONH_2 \qquad (1-74)$$

After the first electrophoresis, heating causes this derivative to cleave the chain, by a reaction analogous to that which occurs with CNBr (see Eqs. 1–66 and 1–67):

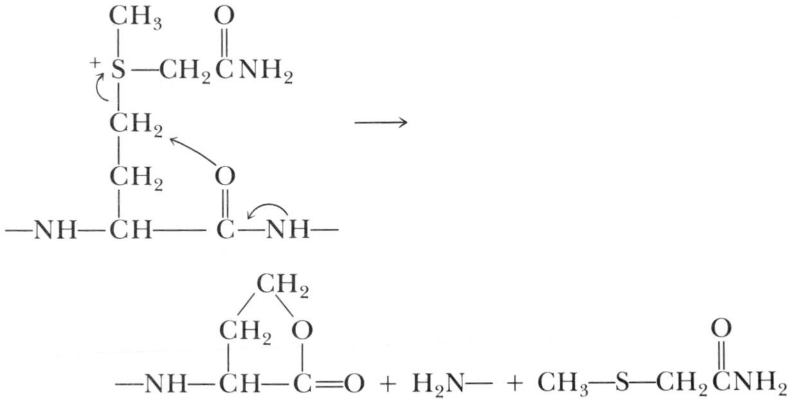

$$(1-75)$$

His residues are first blocked by dinitrophenylation and then restored by exposure to thiol after the first electrophoresis. Arg residues are first blocked by reaction with cyclohexanedione and then regenerated at alkaline pH.

Peptides containing methionine and carboxymethyl cysteine. J. I. Harris. Methods Enzymol. 11:390–398, 1967.

The use of logarithmic plots of electrophoretic mobilities of peptides. R. E. Offord. Methods Enzymol. 47:51–69, 1977.

Disulfide Bonds

Disulfide bonds between Cys residues are not strictly part of the amino acid sequence, or primary structure, but which Cys residues are linked in this manner may be determined with the same techniques. Disulfide bonds are covalent and are stable at most pH values in the absence of reductants, oxidants, or agents that catalyze disulfide interchange. However, disulfide interchange has been encountered unexpectedly by many workers, so care must be exercised to avoid any such side reactions. Nevertheless, proteins may be cleaved into peptides, and the peptides separated, under conditions where the disulfides remain intact. The identities of the peptides linked by disulfides can then be determined directly, or after breaking the disulfide bond and separating them.

A very elegant procedure for determining which Cys residues are paired was designed by Brown and Hartley as the original use of the diagonal method just described. A protein with disulfides is cleaved into peptides, and the peptides are separated by electrophoresis under conditions where any disulfides connecting them are stable. The peptides are then exposed to performic acid, which cleaves the disulfide bonds and converts the Cys residues to $CysO_3H$. The peptides originally linked by the disulfides are now independent and also more acidic. Upon electrophoresis in the second dimension, they usually migrate differently than in the first; thus they lie off the diagonal. Peptides originally linked by each disulfide are identified by their common mobility in the first dimension. A simple example is illustrated in Figure 1–11*B*.

Location of disulphide bridges by diagonal paper electrophoresis. J. R. Brown and B. S. Hartley. Biochem. J. 101:214–228, 1966.

NATURE OF AMINO ACID SEQUENCES

As of 1978, the *Atlas of Protein Sequence and Structure* listed the amino acid sequences of 1081 proteins, comprising 119,006 amino acid residues. This wealth of data has been extensively analyzed, using sophisticated statistical techniques.

Most proteins have very similar amino acid compositions. The frequencies with which the amino acids appear are tabulated in Table 1–1. Substantial deviation from this average composition is very rare for globular proteins, i.e., those that appear to function dissolved in an aqueous milieu. Proteins that are normally embedded in membranes have somewhat higher levels of the hydrophobic amino acids and lower levels of those normally ionized. The most atypical compositions are of those proteins with special structures, especially fibrous proteins such as collagen. This structural protein has Gly as one-third of its residues, and Pro as about one-fourth. The reason for this will be apparent from its structure, described in Chapter 5.

The lengths of the polypeptide chains of proteins vary greatly, in an essentially continuous distribution of lengths up to more than 2000 residues. The longest chain sequenced to date is that of the β-galactosidase from *Escherichia coli*, with 1021 residues. Of course, those proteins for which the sequence has been determined are not representative, as there is a very high bias in favor of smaller proteins. The average molecular weights of polypeptide chains in prokaryotic and eukaryotic cells have been measured to be about 24,000 and 31,700, respectively. Based on an average molecular weight for a residue of 115, these values correspond to respective lengths of 209 and 276 residues. The measurements were made with the total soluble protein extracted from these cells, so that each protein's contribution is weighted by its abundance.

Proteins generally have unique amino acid sequences, even though there are an astronomical number of sequences possible. For a chain of n residues, there are 20^n possibilities; this number is 10^{130} for a short chain of 100 residues and 10^{390} for $n = 300$.

In view of the astronomical number of sequences possible, the possibility that two will be similar just by chance is negligible. Consequently, similarities invariably reflect common ancestry of the structural genes, and comparisons of such sequences may be used to trace the process of evolution. Likewise, the presence of repeated sequences within a polypeptide chain indicates that its gene arose during evolution by a tandem duplication event. The evolutionary origins of protein sequences are discussed in Chapter 3.

The distribution of 27,506 amino acid residues within the known sequences of 207 unrelated proteins has been shown to be very nearly random, in that the frequency of occurrence of each pair of amino acids, separated by zero, one, or two residues, is close to the product of their frequencies of occurrence in proteins (Table 1–1). Exceptions occurred primarily at the ends of the chains. Met occurs at the amino terminal seven times more frequently than would be expected; this will be shown in Chapter 2 to be due to the way in which proteins are synthesized. Tyr at the carboxyl-terminal is extremely rare, occurring only once in 235 chains, whereas eight instances would be expected. This may be due to the existence of an enzyme that removes this residue.

These results do not imply that proteins are random and that there are no constraints on their amino acid sequences. The remainder of this volume will present ample evidence for the importance of nonrandom structure. The apparently random distribution of amino acids throughout the sequences of very many proteins implies simply that constraints imposed generally on all protein structures are negligible. Therefore, there are no simple, general rules of protein structure.

A limiting law relating the size and shape of protein molecules to their composition. H. F. Fisher. Proc. Natl. Acad. Sci. U.S.A. 51:1285–1291, 1964.

On the average hydrophobicity of proteins and the relation between it and protein structure. C. C. Bigelow. J. Theor. Biol. 16:187–211, 1967.

Size distribution of polypeptide chains. E. D. Kiehn and J. J. Holland. Nature 226:544–545, 1970.

Atlas of Protein Sequence and Structure. M. O. Dayhoff (ed.). Washington, D.C., National Biomedical Research Foundation, 1972 (and subsequent supplements).

The low polarity of many membrane proteins. R. A. Capaldi and G. Vanderkooi. Proc. Natl. Acad. Sci. U.S.A. 69:930–932, 1972.

The nicotinic cholinergic receptor: different compositions evidenced by statistical analysis. F. J. Barrantes. Biochem. Biophys. Res. Commun. 62:407–414, 1975.

The independent distribution of amino acid near neighbor pairs into polypeptides. M. H. Klapper. Biochem. Biophys. Res. Commun. 78:1018–1024, 1977.

PEPTIDE SYNTHESIS

The chemical synthesis of peptides and proteins by the methods of organic chemistry is important for the verification of structures of naturally occurring small peptides as determined by the techniques described earlier; for the preparation of analogues of biological peptides and proteins for experimental study of the relationship between structure and biological activity; and for the synthesis of peptides in large quantities for medical purposes. The chemical linkage of individual amino acids in peptide bonds is not in itself a complex chemical problem, but its execution with the 20 normal amino acids taxes the ingenuity of synthetic chemists, owing to the diversity of functional groups on the amino acid side chains. To prevent their participation in undesirable reactions during a peptide synthesis, all potentially reactive groups must be completely blocked during the synthesis by chemical groups that can then be removed completely at the end of the synthesis. A large number of protecting groups have been developed, each with its own advantages and disadvantages. No perfect, or even universally accepted, set has been developed, and each synthesis requires careful consideration of the tactics to be used.

Peptide bond formation between a carboxyl and an amino group does not occur spontaneously under normal conditions. One of these two groups, therefore, must be converted into a more reactive form; almost invariably, it is the carboxyl group. The common driving force in such carboxyl activation is the augmentation of the electrophilic character of the carbon atom; its intrinsically low electron density is further decreased in the activated derivatives by a negative inductive effect of the activating substituent, X:

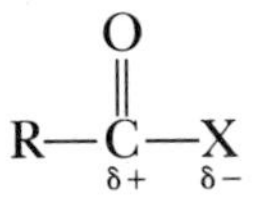

$$\underset{\delta+ \quad \delta-}{R-\overset{\overset{\textstyle O}{\|}}{C}-X} \tag{1-76}$$

The resulting electrophilic center permits attack by a nucleophilic nonionized amino group:

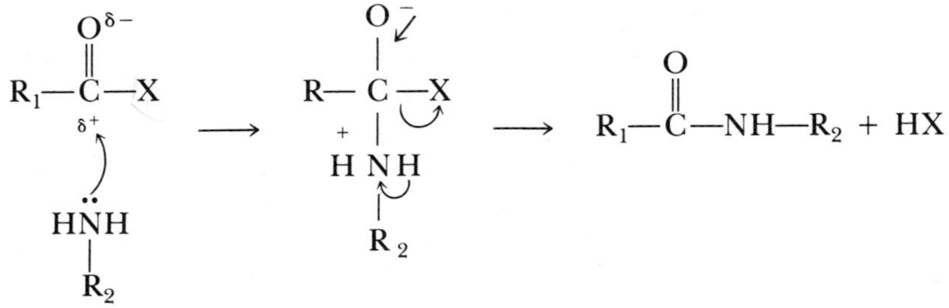

$$\tag{1-77}$$

Examples of activated carboxyl groups are

1. azides: $-\overset{\overset{\textstyle O}{\|}}{C}N_3;$

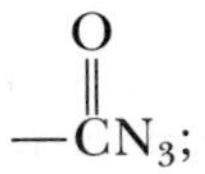

2. acid chlorides: $-\overset{\overset{\displaystyle O}{\|}}{C}Cl$;

3. mixed anhydrides: $-\overset{\overset{\displaystyle O}{\|}}{C}-O-\overset{\overset{\displaystyle O}{\|}}{C}-R$; and

4. activated esters: $-\overset{\overset{\displaystyle O}{\|}}{C}OR'$, where R′ is often *p*-nitrophenyl.

Peptide bond formation between free amino and carboxyl groups may also take place upon addition of appropriate coupling reagents. The reactive intermediates formed in these coupling reactions are invariably those just described, e.g., acid chlorides, active esters, and so on. The most successful reagents have been the carbodiimides, especially dicyclohexylcarbodiimide, which react with a free carboxyl group to make it reactive with amines:

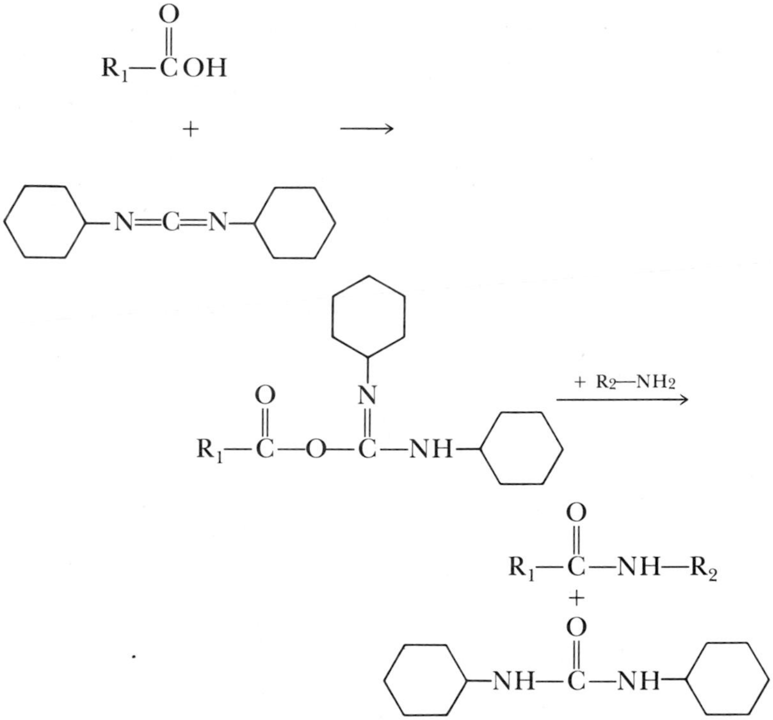

$$R_1-\overset{\overset{\displaystyle O}{\|}}{C}OH$$

$$+ \qquad \longrightarrow$$

$$+ R_2-NH_2 \longrightarrow$$

$$R_1-\overset{\overset{\displaystyle O}{\|}}{C}-NH-R_2$$
$$+$$

(1–78)

One of the most serious side reactions of peptide synthesis is racemization at the C^α atom of the amino acid. This racemization is a result primarily of the activation of the carboxyl group, which assists transient loss of the C^α hydrogen atom. Consequently, initially pure L isomer of the amino acid will gradually revert to the racemic DL mixture. Because racemization has a dramatic effect on protein conformation and biological activity, much effort has gone into minimizing it. However, every known method used for joining peptides causes racemization to some extent.

In order to couple only two amino acids, the amino group of the carboxyl-activated one or the carboxyl group of the amino component, or both, must be blocked so that polymerization does not occur. There is, then, only one way in which peptide bond formation may take place:

$$Y-NH-\underset{\underset{R_1}{|}}{CH}-\underset{\underset{}{\overset{O}{||}}}{C}-X \; + \; H_2N-\underset{\underset{R_2}{|}}{CH}-\underset{\underset{}{\overset{O}{||}}}{C}-Z \longrightarrow$$

$$Y-NH-\underset{\underset{R_1}{|}}{CH}-\underset{\underset{}{\overset{O}{||}}}{C}-NH-\underset{\underset{R_2}{|}}{CH}-\underset{\underset{}{\overset{O}{||}}}{C}-Z \qquad (1-79)$$

Before the peptide may be extended further, the protecting group on either the amino or the carboxyl end must be removed. Considerable ingenuity is required for the choice of such a protecting group, as it must be removed quantitatively under conditions where the other protecting groups on the side chains and on the other end are not affected.

Although the biochemical synthesis of proteins on ribosomes (Chapter 2) proceeds by assembly of the peptide chain from the amino to the carboxyl end, this strategy is not favorable for the chemical synthesis of peptides. Most stepwise syntheses have started with the carboxyl terminal amino acid and have added residues to its amino group.

Success in a peptide synthesis is often limited by the low solubility of intermediates during coupling reactions or by difficulties in the separation of the peptide products from the other reagents. The longer the peptide chain becomes, the more severe these problems tend to be. One solution to these problems is peptide synthesis in the solid phase, devised by Merrifield. The basic idea in this procedure is to attach the amino acid corresponding to the carboxyl terminal group of the desired peptide chain to an insoluble support and to extend the chain toward the amino end by stepwise coupling of activated amino acid derivatives. Filtration and thorough washing of the solid phase removes soluble by-products and excess reagents but retains the extended peptide chain. The functional groups of the main chain and of the side chain not to be involved in peptide bond formation must still be protected in the usual manner. One cycle of the synthesis consists of deprotecting the terminal α-amino group of the main chain and coupling it with the carboxyl group of the next protected amino acid. Successive residues are added to the chain by repeating this cycle until the polymer with the desired sequence is assembled. Finally, the assembled polymer is released into the solution by cleavage of the link to the solid support, and the protecting groups on the side chains are removed. A solid-phase synthesis of the peptide Gly-Asp-Ser-Gly is illustrated in Figure 1–12.

In exchange for the increased simplicity, speed, efficiency, and convenience offered by the solid-phase method, certain limitations and disadvantages appear. The intermediates in the synthesis cannot be purified, since they remain bound to the resin. Any impurities attached to the resin—that is, incomplete and wrong sequences—must be removed at the end of

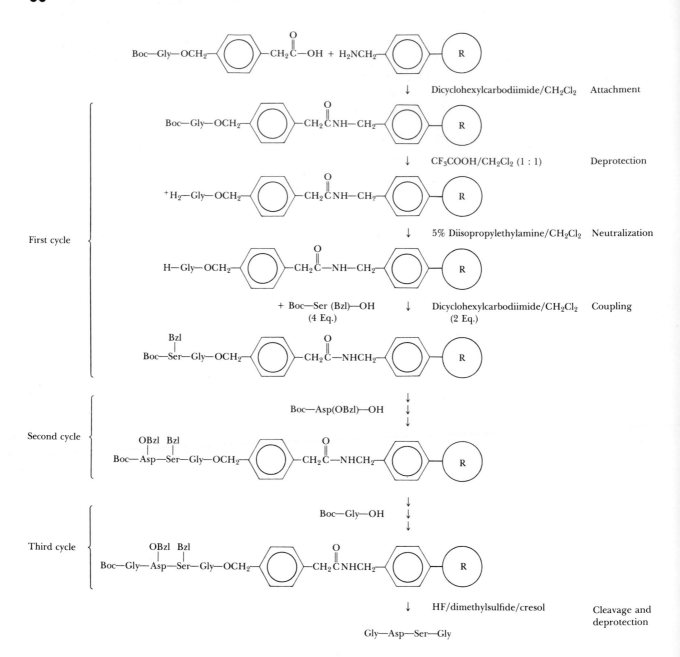

Figure 1–12
Solid-phase synthesis of the tetrapeptide Gly-Asp-Ser-Gly. The initial resin is shown at the top as (R), with one functional group indicated. The three steps of deprotection, neutralization, and coupling are shown only for the first cycle, even though they are required for each subsequent cycle. The protecting groups are Boc, t-butyloxycarbonyl-; Bzl, benzyl. (Courtesy of R. B. Merrifield.)

the synthesis. The efficiency of each coupling cycle is then crucial for the success of such a synthesis. Incomplete deprotection or coupling in each cycle causes a fraction of each of the nascent polypeptides to be irretrievably lost to the correct sequence. The cumulative effects of such incomplete coupling are best illustrated by simple calculations. If the overall efficiency of addition of each successive amino acid is 90 per cent, which would be considered a very successful chemical synthesis, the proportion of polypeptide chains with the correct sequence of amino acid residues would be 35 per cent after 10 residues, 12 per cent after 20, and only 0.003 per cent after 100. Even with an extremely proficient synthesis with 99 per cent efficiency at each step, the yield of a correct polypeptide chain of 100 residues would be only 37 per cent.

The general problem of the inefficiency of stepwise synthesis is circumvented most readily by preparing several moderate-sized peptides by the stepwise procedure, either in solution or on solid supports. These peptides are purified and then joined. Purification of small peptides is considerably easier than that of long peptides, as a single amino acid difference in a polypeptide is much more significant with 10 than with 100 residues. The products of the condensation of the fragments are also readily purified; for example, the 20-residue condensation product of two decapeptides is considerably different chemically from decapeptides that were either unreacted or involved in other side reactions. Separation of sizable fragments represents a considerably easier task than separation of a family of closely similar large peptides. On the other hand, technical difficulties are often encountered in the coupling step.

Virtually all peptides of known biological significance up to 20 or 30 residues in length have been synthesized, as have scores of analogues of many of them. The most appealing frontier in this field has naturally been the synthesis of larger and larger polypeptides, especially those of known proteins with interesting biological activities. However, the criteria of purity of the final product have necessarily been an inverse function of the length of the polypeptide chain, which has tended to limit the significance of the synthesis of larger peptides. For example, the synthesis of the polypeptide corresponding to the sequence of the protein ribonuclease was reported in 1969 simultaneously by two groups, one using stepwise synthesis of fragments in solution followed by their condensation, the other stepwise assembly of the entire polypeptide chain by the solid-phase method. As ribonuclease consists of 124 amino acid residues, the final products were not homogeneous, even after extensive purification, and it is not possible to determine chemically the proportion of correct sequence. Not surprisingly, the final products had lower biological activity than that found with the natural protein, but the use of biological activity as a criterion for purity, or for use in purification, is of questionable validity, since a synthetic polypeptide corresponding to ribonuclease but lacking 61 residues—nearly half the sequence—is also reported to have significant levels of ribonuclease activity. Furthermore, a synthetic human growth hormone was reported to possess 10 per cent of the growth-promoting potency of the 188-residue natural hormone, even though the amino acid sequence of the hormone according to which the synthetic material had been synthesized was subsequently found to be grossly incorrect; a 15-residue sequence occurring at positions 77–91 had been erroneously assigned to positions 17–31. It is inconceivable that the synthetic material had any molecules with the correct sequence.

The synthesis of polypeptides and proteins with unique amino acid sequences will undoubtedly be important in the future in studies of protein structure and function and for preparation of biologically active peptides of pharmaceutical use. At the present time, however, such contributions of large synthetic polypeptides remain limited by the technology of peptide synthesis.

Peptide Synthesis. M. Bodanszky and M. A. Ondetti. New York, John Wiley & Sons, 1966.

Solid Phase Peptide Synthesis. J. M. Stewart and J. D. Young. San Francisco, W. H. Freeman, 1969.

The total synthesis of an enzyme with ribonuclease A activity. B. Gutte and R. B. Merrifield. J. Am. Chem. Soc. 91:501–502, 1969.

Studies on the total synthesis of an enzyme. V. The preparation of enzymatically active material. R. Hirschmann, et al. J. Am. Chem. Soc. 91:507–508, 1969.

The synthesis of a protein possessing growth-promoting and lactogenic activities. C. H. Li and D. Yamashiro. J. Am. Chem. Soc. 92:7608–7609, 1970.

Revised primary structure for human growth hormone. H. D. Niall. Nature New Biol. 230:90–91, 1971.

2

PROTEIN BIOSYNTHESIS

The unique primary structure of a protein arises from its synthesis using the genetic information encoded in the nucleotide sequence of messenger RNA, which was transcribed from the segment of DNA comprising the gene for that protein. This process is so well-known, being thoroughly explained in all introductory textbooks of biology, that a description here would be inappropriate. Instead, let us merely review some of the aspects most relevant to the protein product.

SELECTED ASPECTS OF PROTEIN SYNTHESIS

The nucleotide sequence of a gene encodes a single protein amino acid sequence, which is transcribed and translated with remarkable accuracy, so that each gene has a unique protein product. There is usually a single gene for each protein, which most individuals of a species will have, for reasons to be explained in the next chapter. There are exceptions, where two or more allelic forms of a gene occur with significant frequencies in a population and where two or more genes within the same genome code for similar protein products, such as the H and the M polypeptide chains of lactate dehydrogenase, which account for the isozymes of this enzyme. Otherwise, it is generally possible to speak of a human hemoglobin, a horse cytochrome c, or a bovine chymotrypsin; these are just three examples of proteins to be described here, each with a well-defined structure.

Immunoglobulin Biosynthesis

A remarkable exception to gene-specific protein products is the case of the immunoglobulins, an important class of proteins to be discussed in Chapter 8. Every human is believed to have the potential for generating 10^6 to 10^8 different antibody molecules, even though the corresponding number of

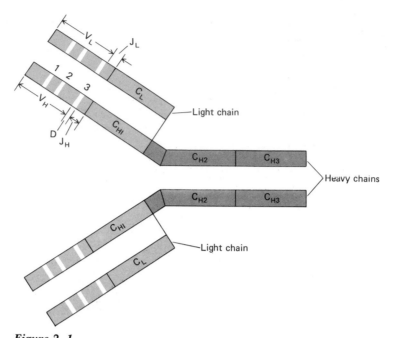

Figure 2–1
Schematic illustration of an immunoglobulin. There are two identical heavy chains, linked to each other by disulfide bonds, and each linked similarly to one of two identical light chains. The constant domains are labeled. The variable domain of each of the chains is at the amino end, at the far left. The three hypervariable regions are white; the polypeptide segments coded by the V, D, and J regions of the genome are indicated. See also Figure 8–14. (Adapted from H. V. Molgaard, Nature 286:657–659, 1980.)

structural genes is apparently lacking. The heterogeneity of normal antibody preparations greatly hindered the structural characterization of specific molecules, but studies making use of their basic similarity in gross structure have determined that each is composed of equal numbers of light (L) and heavy (H) polypeptide chains (Figure 2–1). More detailed studies became possible when it was realized that myelomas, tumors of immunoglobulin-producing cells, secrete large quantities of homogeneous immunoglobulins—sometimes only L or H chains, sometimes both. In particular, amino acid sequence determinations demonstrated that all H and L chains are significantly homologous, both with each other and internally; the H chain contains four regions of internal homology, and the L chain two, each consisting of about 110 amino acid residues. The amino terminal regions of both H and L were the most variable in different proteins, suggesting that they were responsible for antibody diversity, and were designated V_H and V_L, respectively. The others were much less variable, each being almost constant within a single class, and were designated C_L, C_{H1}, C_{H2}, and C_{H3}.

Within the variable regions, three segments were found to be hypervariable, namely, residues 29 to 34, 50 to 56, and 89 to 94 in V_L, and the

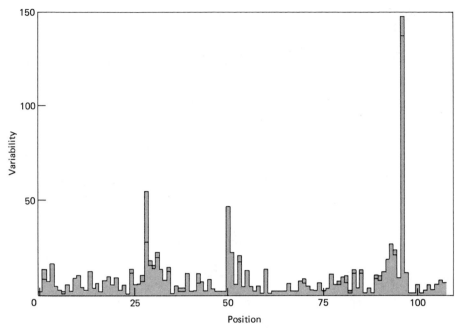

Figure 2–2

Variability of the primary structures of immunoglobulin light chains. The variability of each position is defined as the number of different amino acids occurring at that position in the different immunoglobulins divided by the frequency of the most common amino acid. This parameter may vary in value, ranging between 1, for an absolutely invariant residue, and 400, where all 20 amino acids occur with equal frequency. (Adapted from T. T. Wu and E. A. Kabat, J. Exp. Med. 132:211–250, 1970.)

homologous regions in V_H (Figure 2–2). These residues are primarily involved in antigen recognition; their variation is responsible for the different antigen specificities of the different immunoglobulins (see Chapter 8). There is further variability in that there are different classes of immunoglobulins. Each unique pair of V_H and V_L regions, which defines the antigen specificity, is produced successively within the different classes, first as immunoglobulin M, then immunoglobulin D, and later immunoglobulin G, or sometimes A or E. The primary structure of the constant regions determines the type and class of the immunoglobulin. The L chain of humans is generally of the κ and λ class; H chains can be α, δ, ε, γ, or μ, giving rise to the IgA, IgD, IgE, IgG, or IgM classes, respectively, which differ in their physiological functions. Furthermore, each of the different classes is expressed first as a membrane-bound receptor molecule and later, upon terminal differentiation of the antibody-producing lymphocyte, as a secreted protein that is identical to the receptor molecule in all but a few residues at its carboxyl end.

The basis for the remarkable variability is becoming apparent from current studies of genome structure. The variability of the V_L and V_H

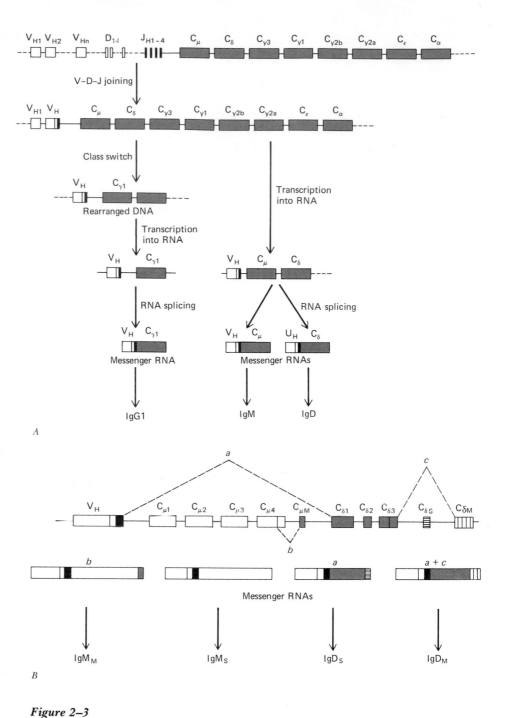

Figure 2–3

Probable mechanism of synthesis of immunoglobulin heavy chains. At the top of A is shown the arrangement of the gene segments coding for polypeptide chains within the germ-line DNA. There are n different V_{Hi} regions and $1 D_j$ region, four J_{Hk} regions, and eight constant regions, which specify the different classes of immunoglobulins (Ig). During differentiation of an antibody-producing lymphocyte, one V_{Hi} segment, one D_j segment, and one J_{Hk} segment are joined together by DNA rearrangements to produce a mosaic V_H region. At an early stage of development, a cell may make both IgM and IgD heavy chains by transcribing the V_H, regions arises from their being formed by different combinations of different gene segments (V, J, and D) during differentiation of a lymphocyte, including some variation in the way that they are joined (Figure 2–3). As there are a substantial number (perhaps up to 500 in some classes) of different such gene segments in the germ-line genome, which may be joined together in a large number of combinations, a large number of different V_L and V_H regions may be produced. Additional variability may arise by somatic mutations of these different gene segments during lymphocyte development. Moreover, different V_L and V_H regions may be combined within a single immunoglobulin molecule; if there are 10^3 to 10^4 of each, the postulated 10^6 to 10^8 different immunoglobulin molecules could be produced.

The different classes of immunoglobulins arise because the V_L and V_H genes are attached to different gene segments coding for the different classes of constant regions of the light and heavy chains. The joining of different regions occurs at the level of the DNA in the lymphocyte and by splicing of messenger RNAs (Figure 2–3A). An added complication is that each of the constant regions is coded by a distinct segment of DNA—an **exon**—separated from those of the other regions by intervening sequences—**introns**—which are removed by splicing of the messenger RNA (Figure 2–3B). Consequently, the final messenger RNA and amino acid sequence of each immunoglobulin polypeptide chain is a mosaic patched together from a number of different genome segments.

An analysis of the sequences of the variable regions of Bence-Jones proteins and myeloma light chains and their implications for antibody complementarity. T. T. Wu and E. A. Kabat. J. Exp. Med. 132:211–249, 1970.

Structural studies of immunoglobulins. R. R. Porter. Science 180:713–716, 1973.

Antibody structure and molecular immunology. G. M. Edelman. Science 180: 830–840, 1973.

C_μ, and C_δ regions into RNA and then splicing it to produce the messenger RNAs for either IgM or IgD heavy chains, as depicted on the right. At a later stage of development, there is a deletion of some of the constant regions, as shown on the left, where C_μ, C_δ, and $C_{\gamma3}$ were deleted. The V_H region is then transcribed attached to a $C_{\gamma1}$ region to produce an IgG1 heavy chain. (Modified from J. L. Marx, Science 212:1015–1017, 1981.)

The constant regions of A are each composed of multiple coding regions (exons) and intervening sequences (introns). These are shown in B for the C_μ and C_δ regions, illustrating the RNA-splicing events that take place during simultaneous production of both IgM and IgD, either secreted or membrane-bound. Splicing that removes the segment a will produce an IgD heavy chain; otherwise an IgM heavy chain will be produced. If segment b is removed by splicing, the IgM_M will be membrane-bound, owing to a hydrophobic portion of chain coded for by region $C_{\mu M}$; otherwise, the polypeptide chain will terminate at the end of $C_{\mu4}$ and the IgM_S will be secreted. A similar mechanism is believed to operate for switching between membrane-bound and secreted IgD, depending upon whether the secretory ($C_{\delta S}$) or membrane-binding ($C_{\delta M}$) regions are translated. (Modified from M. Robertson and M. Hobart, Nature 290:543–544, 1981, according to the results of H.-L. Cheng, et al., Nature 296:410–415, 1982.)

Unusual distribution of amino acids in complementarity-determining (hypervariable) segments of heavy and light chains of immunoglobulins and their possible roles in specificity of antibody-combining sites. E. A. Kabat, et al. J. Biol. Chem. 252:6609–6616, 1977.

Antibody diversity. J. G. Seidman, et al. Science 202:11–17, 1978.

Antibodies, introns and biosynthetic versatility. M. Robertson and M. Hobart. Nature 290:543–544, 1981.

Antibodies: getting their genes together. J. L. Marx. Science 212:1015–1017, 1981.

Genes of lymphocytes: diverse means to antibody diversity. M. Robertson. Nature 290:625–627, 1981.

Gene rearrangement can extinguish as well as activate and diversify immunoglobulin genes. N. Gough. Trends Biochem. Sci. 6:300–302, 1981.

Intervening Sequences and Protein Structural and Functional Units

The nucleotide sequences of the gene and messenger RNA are simply colinear with the amino acid sequence of the protein in prokaryotes, but most eukaryotic genes have unexpectedly been found to be interrupted by intervening sequences, or introns, as in the case of the immunoglobulins (Figure 2–3*B*). They are removed by splicing of the transcribed RNA to link the often much smaller amino acid–coding segments (exons) into the colinear messenger RNA. A function for these introns has still not been found, but an attractive possibility was that individual exons correspond to functional or structural units of the protein. Genes for proteins might have then evolved by the genetic shuffling of different exons to generate different proteins composed of different combinations of protein structural units (see Chapter 3).

This hypothesis was prompted by studies on immunoglobulins, where the spliced genetic segments code for individual domains of the intact protein. That exons do correspond to individual parts of proteins that have arisen by gene duplication of each part is the case with collagen and ovomucoid. However, within a nonrepeated segment, there are instances where introns clearly occur within regions coding for protein structural units, as in alcohol dehydrogenase and ovomucoid. Results of attempts to divide other proteins into structural units that correspond to their exons seem sufficiently dubious to conclude at the present time that there is no well-defined, universal relationship between exons and protein structure. However, this conclusion may be premature, as there are only a few examples known, and may be shown to be incorrect by the time this volume is published. One difficulty is in defining protein structural units, as will become apparent in Chapter 6.

Why genes in pieces? W. Gilbert. Nature 271:501, 1978.

Exons encode functional and structural units of chicken lysozyme. A. Jung, et al. Proc. Natl. Acad. Sci. U.S.A. 77:5759–5763, 1980.

Ovomucoid intervening sequences specify functional domains and generate protein polymorphism. J. P. Stein, et al. Cell 21:681–687, 1980.

Organization and expression of eucaryotic split genes coding for proteins. R. Breathnach and P. Chambon. Ann. Rev. Biochem. 50:349–383, 1981.

Alcohol dehydrogenase gene of *Drosophila melanogaster:* relationship of intervening sequences to functional domains in the protein. C. Benyajati, et al. Proc. Natl. Acad. Sci. U.S.A. 78:2717–2721, 1981.

Correlation of DNA exonic regions with protein structural units in haemoglobin. M. Gō. Nature 291:90–92, 1981.

O₂ binding properties of the product of the central exon of β-globin gene. C. S. Craik, et al. Nature 291:87–90, 1981.

Genes pieced together—exons delineate homologous structures of diverged lysozymes. P. J. Artymiuk, et al. Nature 290:287–288, 1981.

Structure and regulation of a collagen gene. B. deCrombrugghe and I. Pastan. Trends Biochem. Sci. 7:11–13, 1982.

Intron—exon splice junctions map at protein surfaces. C. S. Craik, et al. Nature 299:180–182, 1982.

Protein Sequences from Gene Sequences

The four-nucleotide language of the nucleic acids is converted to the 20-amino-acid language of proteins using the well-known genetic code (Figure 2–4). The code was long thought to be universal, but some minor variations have appeared in mitochondria from the fungus *Neurospora crassa*, from yeast, and from mammals. In each, UGA codes for Trp rather than for

First position	Second position				Third position
	U	C	A	G	
U	Phe	Ser	Tyr	Cys	U
	Phe	Ser	Tyr	Cys	C
	Leu	Ser	Terminate	Terminate	A
	Leu	Ser	Terminate	Trp	G
C	Leu	Pro	His	Arg	U
	Leu	Pro	His	Arg	C
	Leu	Pro	Gln	Arg	A
	Leu	Pro	Gln	Arg	G
A	Ile	Thr	Asn	Ser	U
	Ile	Thr	Asn	Ser	C
	Ile	Thr	Lys	Arg	A
	Met	Thr	Lys	Arg	G
G	Val	Ala	Asp	Gly	U
	Val	Ala	Asp	Gly	C
	Val	Ala	Glu	Gly	A
	Val	Ala	Glu	Gly	G

Figure 2–4
The normal genetic code used in amino acid biosynthesis. The nucleotides of the three positions of the codons are those of the messenger RNA.

5'---AAGAGAACGAAGGGAAGAAAGUGACGGAGAGGCGGUGAACAUCUCCCGCCCGGCGGGGAGCGGCCGCGGCCUGCGGCUUCCCGUGACAGAGCCUCAGCCUGCCUGGAAG
    -500   -450   -400

```
Met Pro Arg Leu Cys Ser Ser Arg Ser Gly Ala Ser Met Val Arg Gly Trp Cys Leu Glu
AUG CCG AGA CUG UGC AGC AGU CGU UCG GGC GCC UCC AUG GUG CGU GGU UGG UGC CUG GAG
        -100        -120                              -80
```

```
Ser Gln Cys Gln Asp Leu Thr Thr Glu Ser Ala Cys Ile Arg Ala Cys Lys Asp Leu Thr Pro Val
AGC CAG UGU CAG GAC CUC ACC ACG GAA AGU GCU UGC AUC CGG GCC UGC AAG GAC CUC ACG CCG GUG
-300                                                -250                    -80
```

```
Phe Pro Gly Asp Asn Leu Arg Thr Gln Pro Leu Phe Gly His Phe Arg Trp Asp Arg Phe Gly Arg Arg
UUC CCC GGC GAU AAC CUG ACU CAG CCG CUC CGU UUC GGC CAU UUC CGC UGG GAC CGC UUC GGC CGU CGG
-60                          -200                            -150
```

```
Asn Ser Ser Gly Val Gly Ala Ala Val Ala Ala Gln Lys Arg Glu Glu Val Met Gly Val Ala Asp
AAU AGC AGC GGU GUU GGA GCG GCG GUG GCG CAG AAG CGC GAG GAG GUG AUG GGC GUG GCG GCG GAU
-40                              -100                        -20
```

```
Asp Lys Arg Pro Asn Gly Ala Glu Asp Glu Lys Arg Tyr Ser Met Glu Lys Pro Val Gly Glu Arg Pro
GAC AAG CGC CCG AAC GGC GCC GAG GAC GAG AAG CGG UAC UCU AUG GAA AAG AAG AGG GUG GGC CGC CCG
-1  1                                          -50
```

```
Val Lys Val Tyr Pro Asn Gly Glu Ala Gln Ala Phe Ala Ala Arg Ala Glu Thr Gly Glu Leu Arg Leu Glu
GUG AAG GUG UAC CCC AAC GGC GAG GCC CAG GCC UUU GCC CGG GCG GAG AGG ACC GGG GAG AGG CUC GAG
20                              60              100                          40
```

```
Gln Ala Arg Gly Glu Ala Pro Glu Ala Leu Glu Leu Val Leu Tyr Gly Glu Ala Glu Ala
CAG GCG CGC GGC GAG GCC CGG CCG GCU GAG GCU CUG GUG UAU GGG GAG GAG GCU GAG GCG
150                                          200
```

```
Ala Glu Lys Lys Asp Pro Ser Gly Trp Arg Pro Pro Lys Asp Tyr Arg Gly Phe Met Thr
GCC GAG AAG AAG GAC CCG AGC UGG CGC UUC CCC AAG GAC UAC CGC GGG UUC AUG ACG ACC
80                      250          300              100
```

```
Ser Lys Gln Thr Gln Ser Leu Val Thr Leu Pro Phe Lys Asn Ala Ile Asn His Lys Lys Lys Gly Gln
UCC AAG CAA ACG CAG AGC CUG GUC ACG CUG CCC UUC AAG AAC GCC AUC AAC GCC CAC AAG AAG AAG CAG UGA
120                  350                              500                              400
```

GGGCCUCUCGGCGGGAAAGUUGACCGGAAGGCCCUCUUCUGCCCUCACGCGCCUCUGCUGGCGGGCAGCCCAGGCAGGAGGAUUCGCCCAGCAGUGCGGCCGCCAGGCAGUGAUGGCGCCAGGUGUGCCCAGGUAUCCGCCGACUCUUAAAGC
                   450   500

UGUCUGUAGUUAAGAAAUAAAACCUUUCAAGUUUCACGAAAAAAAA---3'
               550

chain termination. This makes their code more symmetric, in that UGG also codes for Trp. Similarly, AUA codes for Met, rather than Ile, in mammalian mitochondria; Neurospora and yeast retain the usual code. A further change, specific to yeast, is that the four codons specified by the first two bases CU all code for Thr rather than Leu. The significance of these changes and how they occurred during evolution is currently a mystery.

The genetic code is of very practical importance for the study of proteins in that the elucidation of the nucleotide sequence of a gene, of its messenger RNA, or of a cDNA (complementary DNA) copy of the latter, provides the amino acid sequence of the polypeptide chain synthesized from that gene, if the genetic code is known. The recent revolution in the technology of nucleic acid manipulations and sequencing has produced a large number of protein sequences that were previously unknown and would have been impractical to determine by protein sequencing directly, owing either to scarcity of the protein, to the length of the polypeptide chain, or to its chemical nature, as with water-insoluble membrane proteins. It is usually much easier to sequence a nucleic acid copy of a gene rather than the protein, for simple technical reasons. For example, recent studies revealed the sequence of the DNA complementary to the messenger RNA for the hormone precursor pro-opiomelanocortin (Figure 2–5); this sequence provided the previously unknown amino acid sequence of that protein, which is a precursor to a number of important peptide hormones (see p. 71).

However, the nucleic acid sequence provides only the amino acid sequence of the polypeptide chain during its synthesis; many alterations of the polypeptide chain may be produced during and after biosynthesis. Consequently, the structure of a protein must also be examined for such posttranslational modifications. The nature of the most common modifications will be described in the following section.

Biosynthesis of a polypeptide chain also is not sufficient for production of a biologically active protein, for it must be folded into its active conformation and must be directed to its appropriate site of action. Many proteins must be incorporated into cell membranes and into organelles such as mitochondria, lysosomes, and chloroplasts; others must be secreted from the cell in which they were synthesized. Some aspects of this problem of topogenesis will be described later in this chapter. Virtually nothing is known about how proteins fold in vivo after biosynthesis; this topic will be

---

Figure 2–5
Nucleotide sequence of the messenger RNA for bovine pro-opiomelanocortin, as determined from the sequence of complementary DNA, plus the inferred amino acid sequence of the protein product. The numbering systems for both amino acid residues and nucleotides begin with the amino-terminal residue of corticotropin, with preceding residues having negative numbers. The processing of this protein is described in Figure 2–6. (Adapted from S. Nakanishi, et al., Nature 278:423–427, 1979.)

discussed in Chapter 7 with regard to refolding in vitro of the unfolded, completed polypeptide chain.

Finally, some small peptides with unusual structures are synthesized by entirely different mechanisms, independent of ribosomes and nucleic acids, which will be described briefly in the final section of this chapter.

The amino acid code. T. H. Jukes. Adv. Enzymol. 47:375–432, 1978.
Novel features in the genetic code and codon reading patterns in *Neurospora crassa* mitochondria based on sequences of six mitochondrial rRNAs. J. E. Heckman, et al. Proc. Natl. Acad. Sci. U.S.A. 77:3159–3163, 1980.
Different pattern of codon recognition by mammalian mitochondrial tRNAs. B. Barrell, et al. Proc. Natl. Acad. Sci. U.S.A. 77:3164–3166, 1980.
Codon recognition rules in yeast mitochondria. S. G. Bonitz, et al. Proc. Natl. Acad. Sci. U.S.A. 77:3167–3170, 1980.

COVALENT MODIFICATION OF POLYPEPTIDES

Release of the completed polypeptide chain is not necessarily the last step in the formation of a biologically active protein; many proteins are modified covalently during or after biosynthesis. Consequently, the primary structure may be altered, and novel amino acid side chains may be introduced. Over 200 variant amino acid residues have been detected in proteins thus far. Many are easy to overlook, and only recently has their characterization become commonplace; consequently, the list of modifications is constantly growing. Some modifications are common, others extremely rare. Some occur in all of the molecules of a protein, others in just a small fraction. Some are introduced while the polypeptide chain is being assembled; others are manifestations of old age. Some are produced by specific enzymes, whereas others probably result from nonenzymatic, chemical modification. Some are reversible, others not. Some are of obvious physiological importance, whereas others probably arise simply by mistreatment during isolation of the protein. Unequivocal determination of which attributes apply to a modification is not yet possible, and it is difficult to assess the importance of many that have been reported. We will consider here only those that are produced enzymatically and seem most important to our present understanding of protein structure and function.

Even with those posttranslational modifications of obvious importance, there are very many unanswered questions about how and when the modification is introduced. Whether or not a specific residue of a protein is modified probably depends upon both the structural properties of the protein and its accessibility to the modifying enzymes, which appear often to be strictly compartmentalized within cells. There is probably a complex interplay between the two parameters, in that both may change with time and with sequential modification of the protein. These intriguing aspects are the current subject of intensive study.

Posttranslational covalent modification of proteins. R. Uy and F. Wold. Science 198:890–896, 1977.
In vivo chemical modification of proteins (post-translational modification). F. Wold. Ann. Rev. Biochem. 50:783–814, 1981.

**Proteolytic
Cleavage of the
Polypeptide Chain**

The formyl groups present on the initiating Met residue of polypeptides in prokaryotes are generally removed rapidly during biosynthesis by a "deformylase" enzyme. In many cases, the Met residue is also removed, by a specific aminopeptidase, especially if the second residue is small and uncharged, e.g., Ala, Ser, Thr, or Gly. Consequently, not all mature proteins have Met as the amino-terminal residue, even though all are synthesized this way. The amino terminal leader sequence, which causes a polypeptide to be secreted (see below), is invariably removed by a specific protease.

Many proteins are processed further by proteolytic cleavage of the polypeptide chain at one or more peptide bonds. The polypeptide chain before such further processing is generally designated the **"pro"** protein. Such processing is generally necessary for the biological function of the protein, as it is usually inactive before cleavage. Often, only the processed form of a protein is isolated, so that the occurrence of processing may be easily overlooked; discrepancies between the amino acid sequence of the protein and that indicated by its gene sequence are often the first indication of the existence of intermediate forms or precursors.

Many peptide hormones are produced by proteolytic processing of much larger protein precursors. This mode of biosynthesis appears to help to coordinate the separate hormonal actions that combine to produce complex behaviors and responses of the individual. An excellent example is provided by the single polypeptide chain of pro-opiomelanocortin (Figure 2–5), which is the inactive precursor for the opioid peptides corticotropin (ACTH), lipotropin, melanocyte-stimulating hormones, and endorphin, as illustrated in Figure 2–6. As in many such instances, the sites of cleavage at both ends are flanked by pairs of Lys and Arg residues. Because the precursor is split at different sites in different tissues, different combinations of peptides are generated from a single gene product.

A simpler and more familiar sequence of events occurs in the synthesis of insulin. It is synthesized as the 110-residue pre-proinsulin, which has the initiating Met residue removed during biosynthesis. Upon secretion,

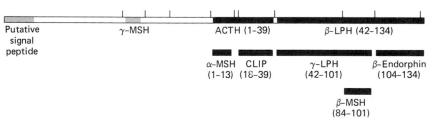

Figure 2–6
Proteolytic products of pro-opiomelanocortin. The amino acid sequence of the precursor is given in Figure 2–5 and is represented here by the horizontal bar; positions of pairs of sequential basic residues are indicated by vertical lines. Products of known structure are indicated by the dark portions, with their names and residue numbers given below. They are corticotropin (ACTH), β and γ lipotropins (LPH), α and β melanocyte-stimulating hormones (MSH), corticotropin-like intermediate-lobe peptide (CLIP), and β-endorphin. The putative signal peptide and presumed γ-MSH are indicated by gray, but their terminal groups are not known for certain. (Adapted from S. Nakanishi, et al., Nature 278:423–427, 1979.)

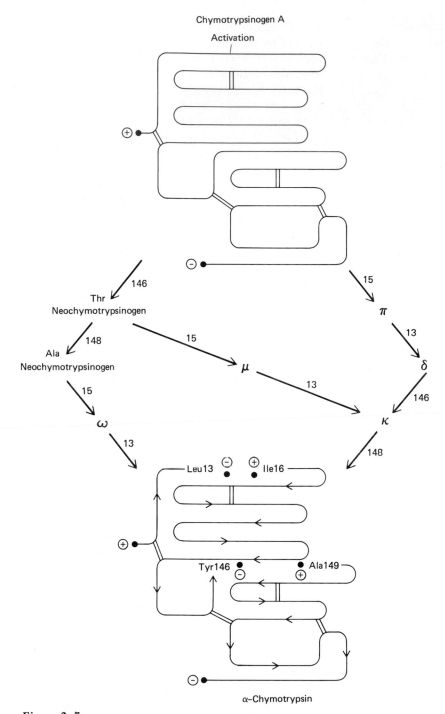

Figure 2–7
Known steps in the activation of bovine chymotrypsinogen A by proteolytic cleavage. The single polypeptide chain cross-linked by five disulfide bonds (double lines), shown at the top, undergoes four sequential proteolytic cleavages at the peptide bonds following residues 13, 15, 146, and 148 to generate the three chain α-chymotrypsin shown at the bottom. The

the 23-residue "pre" peptide is removed from the amino end. The resulting proinsulin is later cleaved further, but in this instance the 35-residue peptide removed proteolytically (the C peptide) is in the middle of the chain. The final two-chain insulin molecule has only 51 of the original 110 residues. The proinsulin precursor appears to be necessary for folding of the polypeptide chain, because after removal of the C peptide, the A and B chains of insulin are not able to recombine efficiently.

Many enzymes, especially proteases, are synthesized as inactive "pro" forms, which have come to be known as **zymogens.** Examples are the well-studied trypsinogen, chymotrypsinogen, and pepsinogen, which, as described more fully in Chapter 9, are converted to the respective active proteases trypsin, chymotrypsin, and pepsin by proteolytic cleavage of the chain at one or more peptide bonds (Figure 2–7). This type of activation process is now known to be a basic part of blood coagulation and fibrinolysis, of complement reaction of the immune response, of hormone action, and of supramolecular assembly, as with collagen and viruses. In all of these instances, the "pro" protein folds to a specific conformation, which is very important for the specificity of the activation cleavage; the limited proteolysis is invariably directed toward very specific loops of the polypeptide chain that are on the surface of the folded protein.

The specificity of the activation process made possible by both the sequence and the folded conformation of the protein is probably why this type of activation process has been used so widely in biological systems. For example, trypsinogen is secreted from the pancreas (via the precursor pre-trypsinogen) but is cleaved by the very specific enzyme enterokinase, which is secreted from the brush border of the small intestine. Active trypsin is therefore generated only where the two secretory streams converge. This is extremely important to prevent premature activation of this proteolytic enzyme. In turn, trypsin catalyzes the conversion of other pancreatic zymogens to their active forms.

Several stages of proteolytic processing appear to occur in multiplication of poliovirus and several related viruses. Here the entire viral RNA molecule functions as the messenger RNA, which appears to be translated as a single unit to yield a large precursor protein of 2207 residues (Figure 2–8). This precursor protein is not released from the ribosome but is cleaved during biosynthesis to yield three major polypeptides. These are then cleaved further at various stages in reproduction and assembly of the virus. All the viral polypeptides have Gly as the amino-terminal residue, because cleavage occurs predominantly at Gln-Gly sequences, presumably

---

number of the residue preceding each bond cleaved is indicated for each step. The two dipeptides of residues 14–15 and 147–148 are released, but the three large fragments are linked by disulfides and by the folded structure of the protein (see Figure 9–10). Only cleavage of the peptide bond after residue 15 is required for enzyme activity. The process may occur by any of the paths depicted, through the various intermediate forms. (Adapted from S. K. Sharma and T. R. Hopkins, Bioorg. Chem. 10:357–374, 1981.)

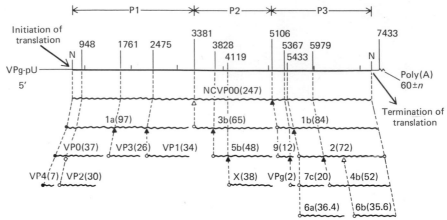

Figure 2–8

Organization of poliovirus RNA and its numerous gene products. The RNA sequence of 7433 nucleotides, plus the poly(A) tail, is shown as a straight line, with the single amino acid–coding segment between nucleotides 743 and 7367 drawn thicker. The RNA has a small polypeptide product of 22 residues, VPg, attached covalently to it by a phosphoryl linkage through a Tyr side chain. The protein products of this RNA are shown as wavy lines, with the molecular mass in kilodaltons given in parentheses. The initial product is the full polypeptide chain of 2207 residues, NCVPOO, which is initially cleaved into the three fragments 1a, 3b, and 1b. The smaller fragments are generated by further cleavages of uncertain order. Many of the proteolytic cleavages (those sites marked with closed triangles) are between residues -Gln-Gly- and are catalyzed by the viral protein P3-7c. Other common cleavage sites (those marked with open triangles) are between -Tyr-Gly- residues. The open diamond indicates cleavage between -Asn-Ser- residues. Not all such pairs of residues are cleaved. (Redrawn from N. Kitamura, et al., Nature 291:547–553, 1981; kindly provided by E. Wimmer.)

by the same proteinase, which is produced from the viral precursor protein; the initial cleavage may then be autocatalytic. Use of this process of protein biosynthesis may be related to the fact that eukaryotic mRNAs are monocistronic, coding for a single polypeptide chain.

One essential characteristic of this type of biological control of protein structure and function is that it is usually irreversible. Only in the special case of protease inhibitors (Chapter 9) are proteolytic cleavages generally observed to be reversible.

Polypeptide cleavages in the formation of poliovirus proteins. M. F. Jacobson and D. Baltimore. Proc. Natl. Acad. Sci. U.S.A. 61:77–84, 1968.

Proinsulin, a biosynthetic precursor of insulin. P. T. Grant and T. L. Coombs. Essays Biochem. 6:69–92, 1970.

Post-translational cleavage of polypeptide chains: role in assembly. A. Hershko and M. Fry. Ann. Rev. Biochem. 44:775–797, 1975.

Role of proteolytic enzymes in biological regulation. H. Neurath and K. A. Walsh. Proc. Natl. Acad. Sci. U.S.A. 73:3825–3832, 1976.

Insulin today. D. F. Steiner. Diabetes 26:322–340, 1977.

Nucleotide sequence of cloned cDNA for bovine corticotropin-β-lipotropin precursor. S. Nakanishi, et al. Nature 278:423–427, 1979.

Primary structure, gene organization and polypeptide expression of poliovirus RNA. N. Kitamura, et al. Nature 291:547–553, 1981.

Discovery of pro-opiomelanocortin—a cellular polyprotein. E. Herbert. Trends Biochem. Sci. 6:184–188, 1981.

Nucleotide sequence and the encoded amino acids of human serum albumin mRNA. A. Dugaiczyk, et al. Proc. Natl. Acad. Sci. U.S.A. 79:71–75, 1982.

Nucleotide sequence of cloned cDNA encoding bovine arginine vasopressin—neurophysin II precursor. H. Land, et al. Nature 295:299–305, 1982.

Synthesizing the opioid peptides. J. L. Marx. Science 220:395–397, 1983.

Biochemistry of the enkephalins and enkephalin-containing peptides. S. Udenfriend and D. L. Kilpatrick. Arch. Biochem. Biophys. 221:309–323, 1983.

Vasopressin and oxytocin are expressed as polyproteins. D. Richter. Trends Biochem. Sci. 8:278–281, 1983.

Modification of the Amino and Carboxyl Terminal Groups

Once released from the ribosome, amino acid residues may be removed from the polypeptide, as discussed above, but generally are not added. Nevertheless, an exception was believed to be the α-chain of tubulin, a protein that is part of microtubules; one residue of tyrosine appeared to be added via a peptide bond to the carboxyl end of the polypeptide by a specific enzyme. Incorporation of the Tyr residue by this enzyme does not require tRNA, ribosomes, or other components of the protein biosynthetic machinery but does require ATP. The reaction is reversible, and the Tyr residue may also be removed by the enzyme in the presence of ADP and inorganic phosphate, P_i:

$$\alpha + \text{tyrosine} + \text{ATP} \rightleftharpoons \alpha\text{-Tyr} + \text{ADP} + P_i \qquad (2\text{--}1)$$

However, sequencing of the messenger RNA for this protein demonstrated that Tyr is the carboxyl-terminal residue of the synthesized chain, so the posttranslational modification is its removal, not its addition. Why the removal is coupled to ATP formation, which makes it energetically reversible, is not known.

The only other known instances of addition of residues to complete chains involve their transfer from activated transfer RNAs to the α-amino groups of some peptides and proteins. For example, a single enzyme from *E. coli* transfers Leu and Phe residues from Leu-tRNA[Leu] and Phe-tRNA[Phe], respectively, to the α-amino groups of proteins with Arg or Lys aminoterminal residues. A similar mammalian enzyme catalyzes the transfer of Arg to peptides with Glu or Asp amino-terminal residues. No components of the protein biosynthetic apparatus are required, other than the charged tRNA. There is no known function for these reactions.

More common modifications of the terminal amino and carboxyl groups are acetylation and amidation, respectively, but little is known about their significance. Acetylation is usually produced early in translation, when the nascent chain is still attached to the ribosome, but it clearly cannot occur before any other processing of the amino terminal group. Amidation of the carboxyl terminal group occurs in peptides cleaved out of larger precursors, in which the next residue is Gly; this Gly residue is degraded enzymatically to leave only the amino group. To the protein chemist, the major effect of these modifications is to make the chain end groups resistant to sequence determination.

Peptide acceptors in the leucine, phenylalanine transfer reactions. R. L. Soffer. J. Biol. Chem. 248:8424–8428, 1973.

Amino-terminal arginylation of chromosomal proteins by arginyl tRNA. H. Kaji. Biochemistry 15:5121–5125, 1976.

Enzyme which specifically adds tyrosine to the α chain of tubulin. D. Raybin and M. Flavin. Biochemistry 16:2189–2194, 1977.

Preferential action of a brain detyrosinolating carboxypeptidase on polymerized tubulin. N. Kumar and M. Flavin. J. Biol. Chem. 256:7678–7686, 1981.

Molecular cloning and nucleotide sequence of full-length cDNA coding for porcine gastrin. O. J. Yoo, et al. Proc. Natl. Acad. Sci. U.S.A. 79:1049–1053, 1981.

Nucleotide and corresponding amino acid sequences encoded by α and β tubulin mRNA's. P. Valenzuela, et al. Nature 289:650–656, 1981.

NH$_2$-terminal processing of *Dictyostelium discoideum* actin *in vitro*. K. Redman and P. A. Rubenstein. J. Biol. Chem. 256:13226–13229, 1981.

Mechanism of C-terminal amide formation by pituitary enzymes. A. F. Bradbury, et al. Nature 298:686–688, 1982.

Glycosylation

Glycoproteins, in which one or more carbohydrate units have been attached covalently to the protein, are the most widely distributed and varied products of posttranslational modification. Most secretory proteins, including the immunoglobulins, are glycoproteins, as are important components of membranes such as cell receptors, where the carbohydrates are involved in cell-to-cell adhesion. Glycosylation is intimately involved in protein secretion, as described in the next section.

The monomer units that are commonly attached to proteins include galactose, mannose, glucose, *N*-acctylglucosamine, *N*-acetylgalactosamine, sialic acids, fucose, and xylose. The carbohydrate chains occur with a bewildering variety of lengths and structures, but some typical structures encountered are illustrated in Figure 2–9.

The carbohydrate chains are generally attached to proteins via the hydroxyl groups of Ser or Thr residues, the amide N atom of Asn side chains or through hydroxy-Lys residues in the special case of collagen. The particular Ser and Thr residues *O*-glycosylated do not appear to occur in unique amino acid sequences, but glycosylation of hydroxy-Lys residues (Hyl) in collagen occurs in those residues in a characteristic sequence -Gly-X-Hyl-Y-Arg-, where X and Y are any amino acids. The Asn residues N-glycosylated occur in the sequences -Asn-X-Ser- or -Asn-X-Thr-, where X may be any of the normal amino acids, other than Pro.

The initial *N*-glycosylation of Asn side chains appears to occur on the nascent chain during translation. However, it cannot be only the amino acid sequence that determines whether glycosylation occurs at appropriate residues. Some proteins, such as elastase and carboxypeptidase, contain one or more potential glycosylation sites but are not glycosylated. Others, such as bovine pancreatic ribonuclease, are secreted in both glycosylated and nonglycosylated forms. Still others, such as ovalbumin and deoxyribonuclease, are essentially fully glycosylated at a single site but contain

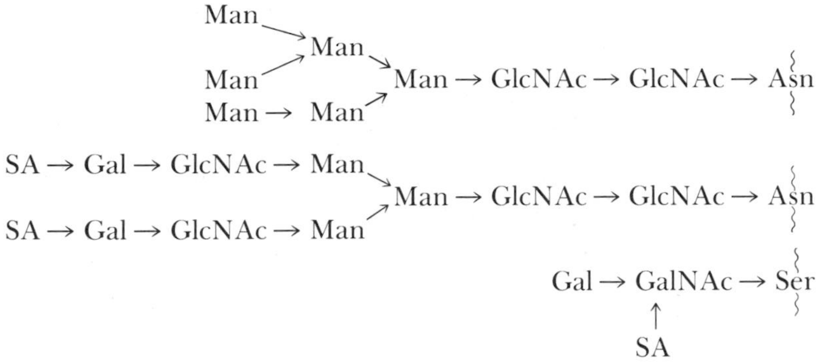

$$\text{Man} \searrow$$
$$\text{Man} \rightarrow \text{Man} \searrow$$

Man ↘
　　　Man ↘
Man ↗　　　Man → GlcNAc → GlcNAc → Asn　　　I
Man → Man ↗

SA → Gal → GlcNAc → Man ↘
　　　　　　　　　　　　　Man → GlcNAc → GlcNAc → Asn　　II
SA → Gal → GlcNAc → Man ↗

Gal → GalNAc → Ser　　III
　　　　↑
　　　　SA

Figure 2–9

Some typical structures of oligosaccharide chains found attached to proteins. I is the poly-mannose "core" oligosaccharide transferred to the protein from the lipid pyrophosphoryl-dolichol; the lipid-linked oligosaccharide originally contained an additional three glucose residues that are removed after transfer to the protein. II is a typical product of processing of core structure I. III is a typical O-glycosidically linked structure. The abbreviations of the sugar residues are Man, mannose; Gal, galactose; GlcNAc, N-acetylglucosamine; GalNac, N-acetylgalactosamine; SA, sialic acid. The positions and anomeric configurations of the carbohydrate linkages are omitted for simplicity. (From J. A. Hanover and W. J. Lennarz, Arch. Biochem. Biophys. 211:1–19, 1981; © Academic Press.)

at least one additional nonglycosylated sequence of -Asn-X-Ser- or -Asn-X-Thr-. Some other property of the nascent polypeptide chain, such as its conformation or site of synthesis, must also determine whether glycosylation takes place.

The initial *N*-glycosylation during biosynthesis inserts the "core" carbohydrate (I in Figure 2–9). This is a complex oligosaccharide that is assembled as a lipid-linked intermediate and then transferred as a unit to the nascent polypeptide chain. This preassembly mechanism is presumably used because the nascent chain to be glycosylated is only transient. The core oligosaccharide often is subsequently enlarged by addition of further carbohydrate units or is otherwise modified (as II in Figure 2–9). The enzymes that catalyze these further additions appear to be localized in the Golgi apparatus and to act farther along the processing and secretion pathways. *O*-glycosylation of Ser and Thr residues also occurs in the Golgi apparatus by sequential addition of sugar residues to amino acid side chains thought to be exposed on a completed and folded protein. Assembly presumably can occur sequentially because the protein is stable.

Oligosaccharide groups may be important for routing a protein to its final destination, serving as a "ticket" recognized by the sorting apparatus. This has been demonstrated in the case of mannose-6-phosphate groups on proteins destined for lysosomes. If these groups are missing, as in I-cell disease, where the enzyme that catalyzes their incorporation is missing, the proteins are secreted rather than transferred to lysosomes. This is just one example of the probable importance of glycosylation of proteins. Many other biochemical functions are suspected and likely to be documented in the near future.

Glycoproteins. R. D. Marshall. Ann. Rev. Biochem. 41:673–702, 1972.

Glycoproteins as components of cellular membranes. R. C. Hughes. Prog. Biophys. Mol. Biol. 26:189–267, 1973.

Glycoproteins. R. G. Spiro. Adv. Protein Chem. 27:349–467, 1973.

The role of surface carbohydrates in the hepatic recognition and transport of circulating glycoproteins. G. Ashwell and A. Morell. Adv. Enzymol. 41:99–128, 1974.

Inherited disorders of lysosomal metabolism. E. F. Weufeld, et al. Ann. Rev. Biochem. 44:357–376, 1975.

The role of polyprenol-linked sugars in glycoprotein synthesis. C. J. Waechter and W. J. Lennarz. Ann. Rev. Biochem. 45:95–112, 1976.

Comparative aspects of glycoprotein structure. R. Kornfeld and S. Kornfeld. Ann. Rev. Biochem. 45:217–237, 1976.

Synchronised transmembrane insertion and glycosylation of a nascent membrane protein. R. E. Rothman and H. F. Lodish. Nature 269:775–780, 1977.

Enzymatic conversion of proteins to glycoproteins. D. D. Pless and W. J. Lennarz. Proc. Natl. Acad. Sci. U.S.A. 74:134–138, 1977.

Localization of the enzyme system for glycosylation of proteins via the lipid-linked pathway in rough endoplasmic reticulum. U. Czichi and W. J. Lennarz. J. Biol. Chem. 252:7901–7904, 1977.

Glycosylation of ovalbumin nascent chains. The spatial relationship between translation and glycosylation. C. G. Glabe, et al. J. Biol. Chem. 255:9236–9242, 1980.

Sugar residues on proteins. P. V. Wagh and O. P. Bahl. Crit. Rev. Biochem. 10:307–377, 1981.

Synthesis and processing of asparagine-linked oligosaccharides. S. C. Hubbard and R. J. Ivatt. Ann. Rev. Biochem. 50:555–583, 1981.

Transmembrane assembly of membrane and secretory glycoproteins. J. A. Hannover and W. J. Lennarz. Arch. Biochem. Biophys. 211:1–19, 1981.

Passage of viral membrane proteins through the Golgi complex. J. Green, et al. J. Mol. Biol. 152:663–698, 1981.

The biosynthetic pathway of the asparagine-linked oligosaccharides of glycoproteins. R. J. Staneloni and L. F. Leloir. Crit. Rev. Biochem. 12:289–326, 1982.

γ-Carboxyglutamic Acid

Proteins involved in blood clotting and calcium metabolism have been found recently to contain the unusual amino acid γ-carboxyglutamic acid, abbreviated **Gla:**

$$
\begin{array}{c}
\text{HO}_2\text{C} \qquad\qquad \text{CO}_2\text{H} \\
\diagdown \text{CH} \diagup \\
| \\
\text{CH}_2 \quad \text{O} \\
| \qquad\quad \| \\
-\text{NH}-\text{CH}-\text{C}- \\
\text{Gla}
\end{array}
\qquad (2\text{--}2)
$$

The two adjacent carboxyl groups are able to bind ions such as calcium very tightly, and the presence of this amino acid residue is invariably associated with calcium binding. The formation of this novel amino acid has been studied most thoroughly in the blood-clotting enzymes, including prothrombin, where it is essential for the Ca^{2+}-mediated activation to thrombin. It is now known that this amino acid residue results from post-translational modification of Glu residues, probably of the nascent polypeptide chain during biosynthesis, because the enzyme responsible is associated with the microsomal fraction of cells. Ten specific Glu residues in prothrombin are converted to Gla; these are the first ten Glu residues synthesized in the first 33 residues of prothrombin (not including any "pre" sequence in pre-prothrombin). The other blood-clotting enzymes that are modified have amino-terminal sequences homologous to that of thrombin, and the Glu residues converted to Gla also are homologous to those in prothrombin. Gla residues have also been observed in other proteins, all apparently involved in calcium metabolism. It is probable that they are formed in a manner similar to that for the blood-clotting enzymes, but little is known about them.

The enzyme that converts the Glu residues to Gla requires O_2, CO_2, an energy source, a reducing agent, and vitamin K. This is the first biochemical function found for this vitamin, which had been known to be necessary for thrombin-catalyzed blood coagulation. Its role in the conversion of Glu to Gla residues is probably involved in labilization of the γ-hydrogen atom to be replaced by a carboxyl group. The CO_2 required by the enzyme is inserted as a carboxyl group into the Glu side chain. This modification is vital for the function of these proteins, as they are functionally inactive if not modified. This can be caused by the absence of the vitamin or by the presence of antagonists such as warfarin and dicumarol.

Vitamin K–dependent carboxylation of blood coagulation proteins. J. Stenflo. Trends Biochem. Sci. 1:256–258, 1976.

Vitamin K–dependent formation of γ-carboxyglutamic acid. J. Stenflo and J. W. Suttie. Ann. Rev. Biochem. 46:157–172, 1977.

Vitamin K, prothrombin, and γ-carboxyglutamic acid. J. Stenflo. Adv. Enzymol. 46:1–31, 1978.

Mechanism of action of vitamin K: synthesis of γ-carboxyglutamic acid. J. W. Suttie. Crit. Rev. Biochem. 8:191–223, 1980.

Vitamin K–dependent carboxylation. J. W. Suttie. Trends Biochem. Sci. 5:302–304, 1980.

Vitamin K–dependent carboxylase: evidence for cofractionation of carboxylase and epoxidase activities, and for carboxylation of a high-molecular-weight microsomal protein. R. Wallin and J. W. Suttie. Arch. Biochem. Biophys. 214:155–163, 1982.

Hydroxylation of Pro and Lys

An important step in the maturation and secretion of collagen is the hydroxylation by specific enzymes of certain Pro and Lys residues, to place hydroxyl groups on the γ-carbon and δ-carbon atoms of the respective side chains:

γ-OH-Pro (4-hydroxy-Pro) δ-OH-Lys (5-hydroxy-Lys)

(2–3)

These residues are hydroxylated only when they appear in the sequences -Pro-Gly- and -Lys-Gly-. A few Pro residues of certain types of collagen are hydroxylated at the β-carbon:

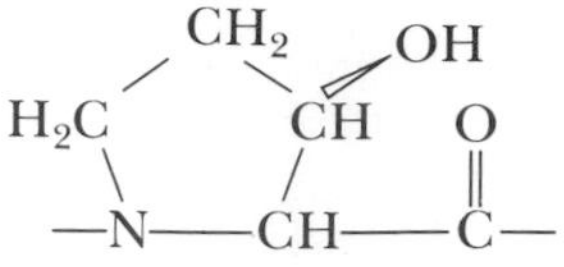

β-OH-Pro

(3-hydroxy-Pro)

These modifications require ascorbic acid and apparently occur on the nascent polypeptide chains during biosynthesis. They are vital for the folding and secretion of the collagen: γ-OH-Pro stabilizes the collagen triple helix (see pp. 194–197), and δ-OH-Lys is necessary for formation of intermolecular cross-links and for attachment of glycosyl groups. These amino acid residues are generally not found in other proteins.

Biochemistry of the hydroxyprolines. R. Kutton and A. N. Radhakrishnan. Adv. Enzymol. 37:273–347, 1973.

Prolyl hydroxylase. G. J. Cardinale and S. Udenfriend. Adv. Enzymol. 41:245–300, 1974.

The biosynthesis of collagen. P. Bornstein. Ann. Rev. Biochem. 43:567–603, 1974.

Metabolism of proline and hydroxyproline. E. Adams and L. Frank. Ann. Rev. Biochem. 49:1005–1061, 1980.

Regulation of collagen synthesis by ascorbic acid. S. Murad, et al. Proc. Natl. Acad. Sci. U.S.A. 78:2879–2882, 1981.

Iodination

The iodination of Tyr residues is a very specific posttranslational modification that occurs essentially only in the protein thyroglobulin in the thyroid gland of vertebrates. Iodine is accumulated by the thyroid and used by the membrane-associated enzyme thyroid peroxidase to iodinate fewer than one fourth of the more than 100 Tyr residues of thyroglobulin to ε-iodo-Tyr and then to ε1,ε2-diiodo-Tyr:

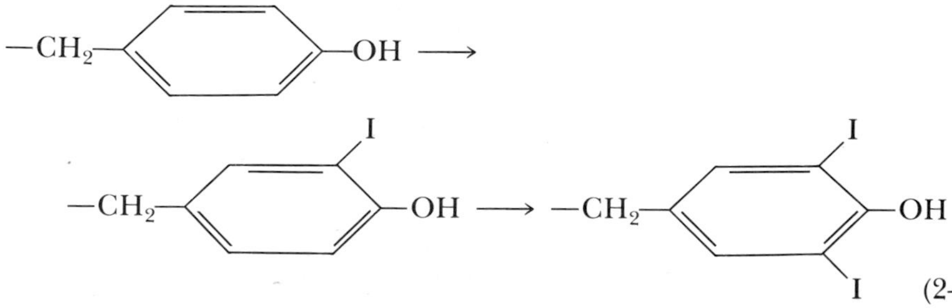

The side chain of one such iodinated Tyr residue is then transferred to another iodinated Tyr side chain in several instances to give residues of thyroxine. This new amino acid, the active hormone, is liberated by proteolytic cleavage by thyroid proteases:

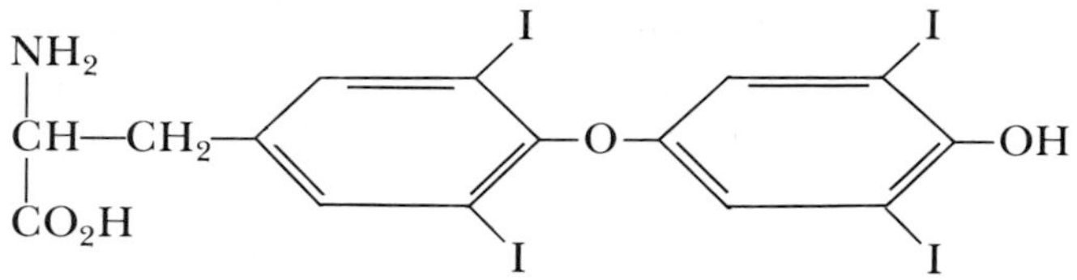

$$\begin{array}{c}
NH_2 \\
| \\
CH-CH_2- \\
| \\
CO_2H
\end{array}
\quad \text{(ring)}-I,\ O,\ (ring)-OH \quad (2\text{--}6)$$

thyroxine

Only about four to eight specific Tyr residues of the original protein are so converted to thyroxine residues. The liberated hormone is then secreted into the bloodstream. Some of the thyroxine is subsequently deiodinated to triiodothyronine.

Tyr residues of other proteins may be iodinated enzymatically. All that is required is iodide, hydrogen peroxide, and a peroxidase enzyme:

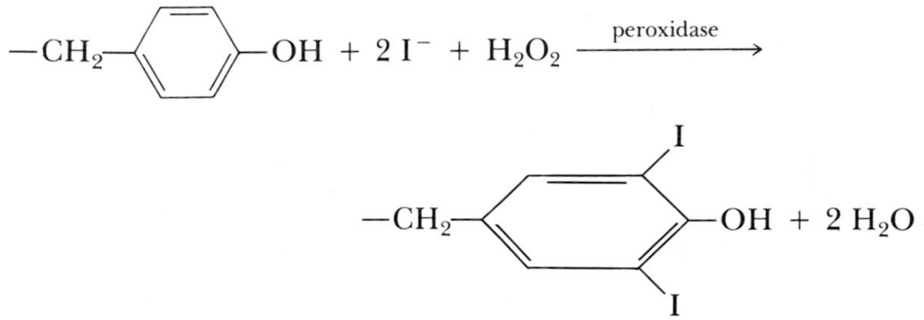

$$-CH_2-\text{(ring)}-OH\ +\ 2\,I^-\ +\ H_2O_2\ \xrightarrow{\text{peroxidase}}$$

$$-CH_2-\text{(ring, }I\text{, }I\text{)}-OH\ +\ 2\,H_2O \quad (2\text{--}7)$$

No physiological significance of such reactions is known, but this one provides a useful means of radiolabeling proteins with ^{125}I or ^{131}I, especially for radioimmunoassay.

Natural and artificial iodoproteins. J. Roche and R. Michel. Adv. Protein Chem. 6:253–297, 1951.

An approach to the structure of thyroglobulin. Hormone-forming sequences in porcine thyroglobulin. C. Marriq, et al. Eur. J. Biochem. 111:33–47, 1980.

Thyroglobulin structure–function. S. B. Chernoff and A B. Rawitch. J. Biol. Chem. 256:9425–9430, 1981.

Iodine and the thyroid hormones. E. Frieden. Trends Biochem. Sci. 6:50–53, 1981.

Covalent Attachment of Prosthetic Groups

The functional properties of many proteins involve non–amino acid moieties that are bound to the protein. In most cases they are bound noncovalently by the folded protein (Chapter 8), but in a few they are attached irreversibly. The most abundant examples of these covalently bonded moieties are the heme groups of cytochromes. The irreversible bonding presumably reflects the need for modulation of their electron affinity without the possibility that an energetically unfavorable redox would escape from the constraints of the protein through dissociation.

The most thoroughly studied heme protein, cytochrome c, is encoded by a gene in the nucleus and synthesized on ribosomes in the cytoplasm. The unmodified polypeptide chain is then taken up by a mitochondrion and transported across its outer membrane by a specific transport system. A necessary step in this translocation process is the covalent attachment of heme to the protein by thioether bonds between two Cys residues:

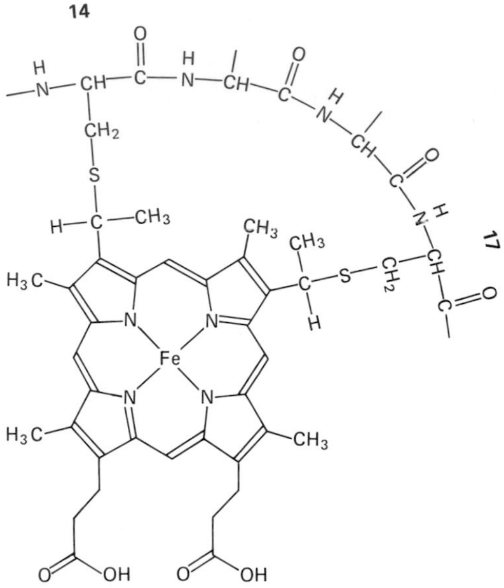

(2–8)

Assembly of cytochrome *c*. Apocytochrome *c* is bound to specific sites on mitochondria before its conversion to holocytochrome *c*. B. Hennig and W. Neupert. Eur. J. Biochem. 121:203–212, 1981.

Formation of a cytochrome *c*–like species from horse apoprotein and hemin catalyzed by yeast mitochondrial cytochrome *c* synthetase. D. Veloso, et al. J. Biol. Chem. 256:8646–8651, 1981.

Phosphorylation

A large number of proteins are now known to be phosphorylated at specific sites by protein kinase enzymes using ATP as phosphoryl donor, with subsequent dramatic effects on their biological properties. This notable method of regulation of protein activity will be described in Chapter 9.

The sites of phosphorylation are usually the hydroxyl groups of specific Ser or Thr residues, but Tyr, His, and Lys residues may also be used:

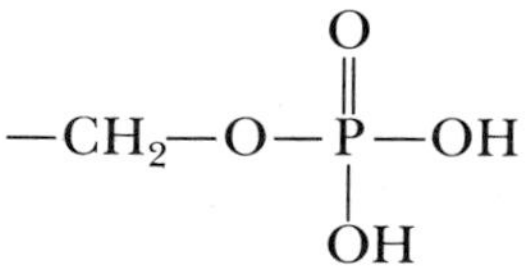

Ser P

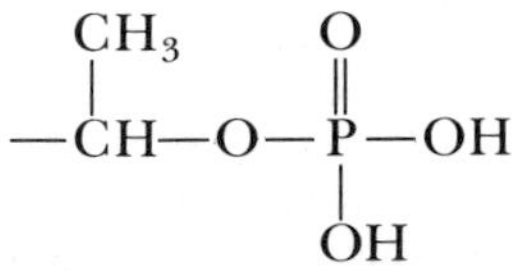

Thr P

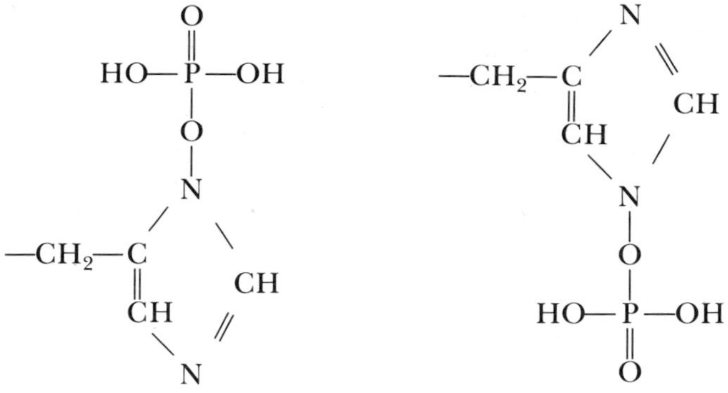

His P

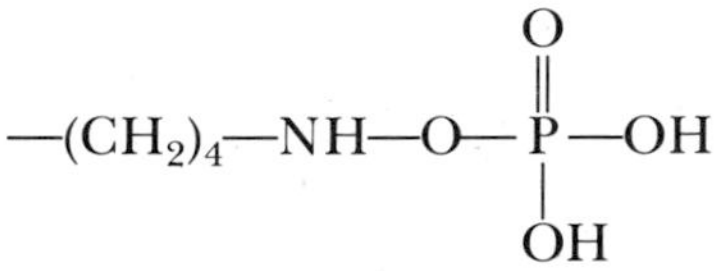

$-(CH_2)_4-NH-O-\overset{\displaystyle O}{\underset{\displaystyle OH}{\overset{\|}{P}}}-OH$

Lys P

$-CH_2-\langle\text{benzene ring}\rangle-O-\overset{\displaystyle O}{\underset{\displaystyle OH}{\overset{\|}{P}}}-OH$

Tyr P

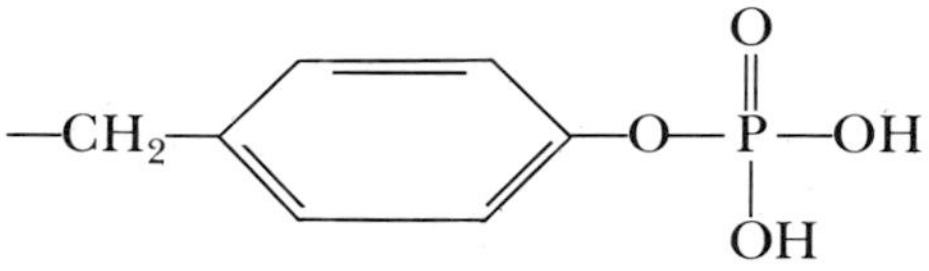

(2–9)

Which residues are phosphorylated depends upon the primary structure around the residue and upon its accessibility to the kinase, as these modifications are produced in the folded forms of the proteins. The influential cyclic AMP-dependent kinase has a strong preference for Ser residues located 2 to 5 residues to the carboxyl side of one or two basic amino acids, usually Arg. The nutritionally significant phosphorylation of the milk protein casein usually occurs 2 residues to the amino side of an acidic residue. In none of these cases are the rules of recognition fully understood.

Phosphoproteins. G. Taborsky. Adv. Protein Chem. 28:1–20, 1974.

Protein phosphorylation. C. S. Rubin and O. M. Rosen. Ann. Rev. Biochem. 44:831–887, 1975.

Phosphorylated proteins as physiological effectors. P. Greengard. Science 199: 146–152, 1978.

Phosphorylation–dephosphorylation of enzymes. E. G. Krebs and J. A. Beavo. Ann. Rev. Biochem. 48:923–959, 1979.

Protein Phosphorylation in Regulation. P. Cohen (ed.). Amsterdam, Elsevier/North Holland, 1980.

Analysis of the sequence of amino acids surrounding sites of tyrosine phosphorylation. T. Patschinsky, et al. Proc. Natl. Acad. Sci. U.S.A. 79:973–977, 1982.

Methylation

Proteins such as histones, myosin, actin, rhodopsin, and cytochrome c and those of ribosomes have been found in a variety of instances to be methylated at the α-amino group or on the side chains of Lys, Arg, and His. The enzymes catalyzing these methylation reactions generally use S-adenosylmethionine as methyl donor to add one, two, or three methyl groups to the N^{ζ} atom of Lys, one or two methyl groups on either of the $N^{\eta 1}$ or $N^{\eta 2}$ atoms of Arg, or a methyl group on either the $N^{\delta 1}$ or $N^{\epsilon 2}$ atoms of His. These modifications have been shown to occur as posttranslational modifications of very specific residues in the intact proteins, but neither the time nor the cellular site of methylation is generally known. Once incorporated, the methyl groups do not appear to be removed, but the extent of methylation is generally not complete in that specific residues are generally found to be methylated in only a fraction of the molecules of the protein.

Reversible methylation of Glu residues

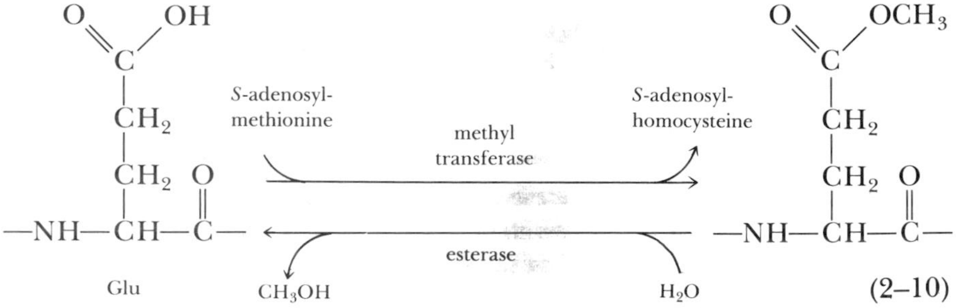

$$(2\text{--}10)$$

has recently been found to be involved in the chemotactic response of bacteria and of leukocytes. Specific proteins are methylated by a methyl transferase enzyme in response to chemotactic attractants. The methyl groups are removed by an esterase in response to repellants. It is anticipated that further such examples of reversible methylation will be found.

Biological methylation: selected aspects. G. L. Cantoni. Ann. Rev. Biochem. 44:435–451, 1975.

Protein methylation: chemical, enzymological and biological significance. W. K. Paik and S. Kim. Adv. Enzymol. 42:227–286, 1975.

Occurrence of methylated amino acids as N-termini of proteins from *E. coli* ribosomes. R. Chen, et al. J. Mol. Biol. 111:173–181, 1977.

Bacterial motility and chemotaxis: the molecular biology of a behavioral system. R. M. Macnab. Crit. Rev. Biochem. 5:291–341, 1978.

Protein methylation in behavioural control mechanisms and in signal transduction. M. S. Springer, M. F. Goy, and J. Adler. Nature 280:279–284, 1979.

Biochemistry of sensing and adaptation in a simple bacterial system. D. E. Koshland, Jr. Ann. Rev. Biochem. 50:765–782, 1981.

Adenylylation and ADP-ribosylation

A modification related to phosphorylation is one in which the adenylyl moiety of ATP is transferred to the hydroxyl group of a tyrosine residue in a phosphodiester linkage:

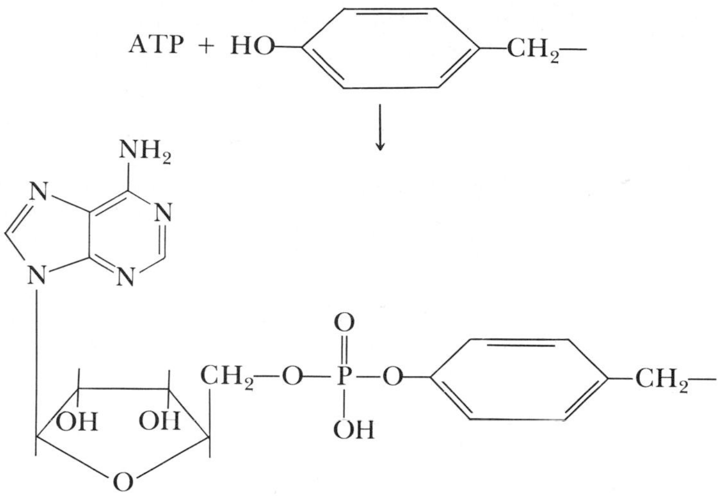

(2–11)

This modification of one Tyr residue of the polypeptide chain is used to modulate the activity of the enzyme glutamine synthetase from *E. coli;* the adenylylated enzyme is inactive. A specific enzyme catalyzes the modification, and another is present to remove the adenylyl group. With other enzymes, UTP replaces ATP, to give the uridylylated derivatives.

A similar modification to inactivate enzymes occurs in some instances using NAD. The nicotinamide portion is split off and the remainder ADP-ribosyl moiety transferred to an Arg residue in some instances, but to other unknown residues in other proteins:

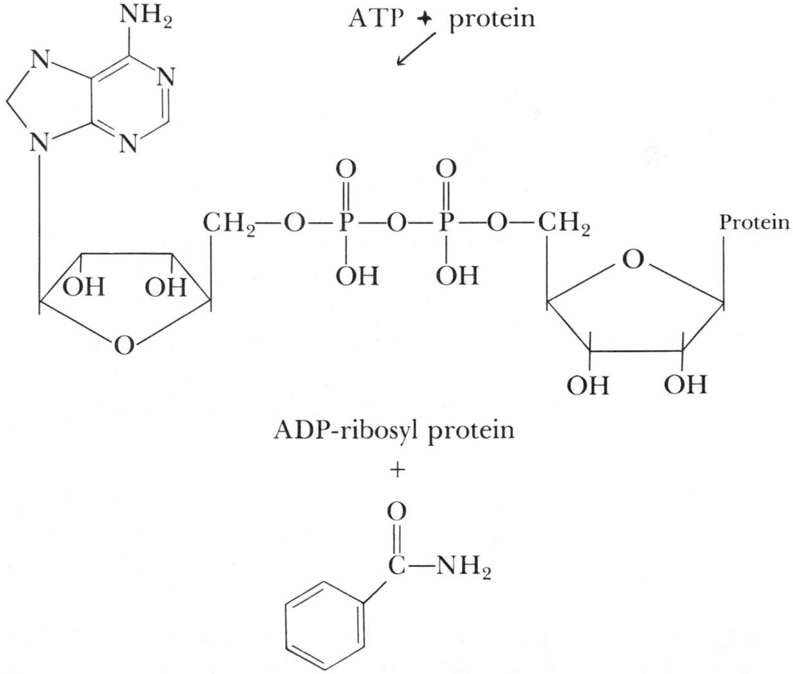

ADP-ribosyl protein

+

nicotinamide

(2–12)

In some instances, poly(ADP-ribose) chains of up to 65 residues may be added.

Such modifications are known to be caused by enzymes produced by toxins and bacteriophage in order to inactivate proteins of their hosts.

Adenylyl transfer reactions. E. R. Stadtman. In The Enzymes, 3rd ed. P. D. Boyer (ed.). Vol. 8, pp. 2–49. New York, Academic Press, 1973.

Poly(adenosine diphosphate ribose). T. Sugimura. Prog. Nucleic Acid Res. Mol. Biol. 13:127–151, 1973.

Poly(ADP-ribose) and ADP-ribosylation of proteins. O. Hayaishi and K. Ueda. Ann. Rev. Biochem. 46:95–116, 1977.

ADP-ribosylation of nuclear proteins. M. R. Purnell, et al. Biochem. Soc. Trans. 8:215–227, 1980.

ADP-ribosylation of proteins—a multifunctional process. H. Hilz. Hoppe Seylers Z. Physiol. Chem. 1415–1425, 1981.

Covalent Cross-links within or between Polypeptide Chains

The only type of covalent cross-link normally found in globular proteins is the disulfide bond between two Cys residues:

$$
\begin{array}{ccc}
\mid & & \mid \\
NH & & C{=}O \\
\mid & & \mid \\
CH{-}CH_2{-}S{-}S{-}CH_2{-}CH & & \\
\mid & & \mid \\
C{=}O & & NH \\
\mid & & \mid \\
\end{array}
\qquad (2\text{--}13)
$$

Which Cys residues are so linked depends upon which are brought into appropriate proximity by the conformational properties of the polypeptide chain, as will be described in Chapter 7. The many ways in which disulfide bonds may be made and broken have been discussed in Chapter 1 in the description of the Cys residue.

Other types of cross-links are found in proteins of the connective tissue, especially collagen and elastin. In these proteins, covalent cross-links are formed primarily between Lys residues after oxidative deamination of one of the amino groups. There may also be ester linkages between Asp or Glu residues and the hydroxyl groups of other side chains. A great variety of cross-links are formed, too many to enumerate here; the interested reader is referred to the article by Gallop, Blumenfeld, and Seifter.

Amide cross-links between the side chains of Gln and Lys residues in fibrin are produced in the course of blood clotting catalyzed by a specific transglutaminase, Factor XIII of blood coagulation:

$$
\begin{array}{cccc}
\mid & & & \mid \\
NH & O & & C{=}O \\
\mid & \parallel & & \mid \quad \overset{NH_3}{\nearrow} \\
CH{-}CH_2{-}CH_2{-}CNH_2 & + & H_2N{-}(CH_2)_4{-}CH & \\
\mid & & & \mid \\
C{=}O & & & NH \\
\mid & & & \mid \\
\text{Gln} & & \text{Lys} & \\
\end{array}
$$

$$
\begin{array}{ccc}
\mid & & \mid \\
\mathrm{NH} & \mathrm{O} & \mathrm{C}{=}\mathrm{O} \\
\mid & \parallel & \mid \\
\mathrm{CH}{-}(\mathrm{CH_2})_2{-}\mathrm{C}{-}\mathrm{NH}{-}(\mathrm{CH_2})_4{-}\mathrm{CH} \\
\mid & & \mid \\
\mathrm{C}{=}\mathrm{O} & & \mathrm{NH} \qquad (2{-}14)\\
\mid & & \mid
\end{array}
$$

The physiological role of such cross-links is to stabilize the blood clot. Their medical importance has been demonstrated directly in patients lacking the ability to make them. Although such cross-links are not generally observed in other proteins, they may be present in structural proteins of hair, wool, and the epidermis and are suspected in some other specific systems.

Structure and metabolism of connective tissue proteins. P. M. Gallop, et al. Ann. Rev. Biochem. 41:617–672, 1972.

The biosynthesis of collagen. P. Bornstein. Ann. Rev. Biochem. 43:567–603, 1974.

Isopeptide crosslinks—their occurrence and importance in protein structure. R. S. Asquith, et al. Angew. Chem. [Engl.] 13:514–520, 1974.

The ε-(γ-glutamyl)lysine crosslink and the catalytic role of transglutaminase. J. E. Folk and J. S. Finlayson. Adv. Protein Chem. 31:1–133, 1977.

Isopeptide linkage between *N*-α-monomethyl alanine and lysine in ribosomal protein S11 from *Escherichia coli*. R. Chen and U. Chen-Schmeisser. Proc. Natl. Acad. Sci. U.S.A. 74:4905–4908, 1977.

Transglutaminases. J. E. Folk. Ann. Rev. Biochem. 49:517–531, 1980.

TOPOGENESIS

During or after biosynthesis, polypeptide chains must be directed to their functional positions within the organism. Very many proteins probably function in solution within the cytoplasm in which they were synthesized; there may then be no necessity to direct them to specific sites, as diffusion may suffice. However, other proteins must be incorporated within specific organelles of the cell or within membranes. Others must be secreted from the cells in which they are synthesized, to function in various extracellular fluids. How the various polypeptides are directed to their specific sites is known as the "sorting problem."

The sorting problem is well illustrated by the components of the eukaryotic ribosome. Ribosomal proteins are synthesized in the cytoplasm and transported to the nucleolus. There they are assembled into the large and small subunits, probably by attachment to a large nascent precursor ribosomal RNA, while it is still being transcribed from the DNA. The precursor must be processed, and the small 5 S and 5.8 S ribosomal RNAs, which are synthesized at other sites in the nucleus, have to be incorporated into the large subunit. The large and small subunits must then be transported from the nucleus to the cytoplasm. But the situation is even more complicated for the ribosomes of mitochondria and chloroplasts, which are different from those of the cytoplasm. Yet most, if not all, of these ribosomal proteins are encoded in nuclear DNA and synthesized on cytoplasmic ribosomes, as are many other mitochondrial proteins. These proteins must then be transported to the appropriate organelle, pass through the membrane, and assemble into functional ribosomes within the organelle, where the ribosomal RNA molecules have been synthesized from organelle-specific DNA. Exactly how this is accomplished is unknown.

Many proteins are secreted from cells. How this occurs specifically now seems clear, at least in outline. Such secretory proteins have long been known to be synthesized in the rough endoplasmic reticulum by membrane-bound ribosomes. They are secreted through this membrane into the lumen and then pass to the Golgi area, where they are condensed into coated vesicles. These secretory granules eventually fuse with the plasma membrane of the cell, releasing their contents outside.

It now appears that all such secretory proteins contain specific **signal sequences** of 15 to 30 amino acid residues. These cause the nascent polypeptide chain, plus the attached ribosome–mRNA complex, to bind at specific recognition sites on the microsomal membrane, to form the rough endoplasmic reticulum. Translation of all mRNA molecules is believed to occur initially on soluble ribosomes; only after synthesis of the signal sequence do ribosomes translating secretory proteins become attached to the membrane. Binding at specific sites on the membrane then causes the nascent chain to be transferred vectorially through the membrane, into the lumen (Figure 2–10).

Most signal sequences are at the amino end of the polypeptide chain, the one synthesized first. Upon emergence from the ribosome, this signal sequence binds to a **signal recognition particle** composed of six distinct polypeptide chains plus a small 7 S RNA molecule. Further translation is then inhibited until the ribosome binds to a "docking protein" in the membrane of the endoplasmic reticulum. The signal recognition particle is released, the inhibition of further translation is removed, and the nascent polypeptide chain is translocated through the membrane. The signal sequence is usually cleaved rapidly from the nascent chain and thus does not appear in the mature protein. The initial chain with the signal sequence is generally designated the **"pre"** protein. It is generally detected in cell-free biosynthetic systems lacking the peptidase to cleave off the signal sequence.

The known signal sequences of different proteins are not identical, but they do have some common features. In particular, each is composed of relatively hydrophobic amino acids, as might be expected for a polypeptide to be inserted into the membrane. However, the remainder of the residues of such secretory proteins are quite polar, and how they traverse the membrane is a mystery. Mutational alteration of the signal sequence, particularly replacement of a hydrophobic residue by one ionized, can abolish secretion. Yet a functional signal sequence is not sufficient for secretion; a hybrid protein composed of a normally cytoplasmic protein plus the amino-terminal portion of a secretory protein, including the entire signal sequence, was found not to be secreted.

Membrane proteins also appear to be synthesized with signal se-

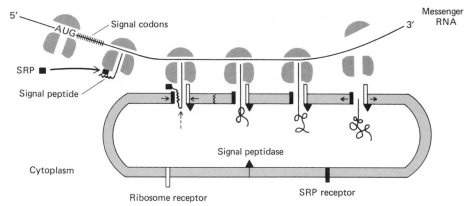

Figure 2–10

Mechanism of protein secretion across membranes according to the signal hypothesis. Ribosomes in the cytoplasm are shown traversing the messenger RNA, starting at the AUG initiation codon. The following hatched portion contains the codons for the signal peptide, which is shown as a wavy line. Upon translation of the signal sequence, the signal recognition particle (SRP) binds to the complex and inhibits further synthesis of the polypeptide chain. This prevents completion of any secretory protein in the cytoplasm on free ribosomes. Only when the ribosomal complex binds to a suitable membrane, containing ribosome and SRP receptors, does translation proceed. The nascent chain is then extruded through the membrane, presumably through a channel. At some time, the signal peptide is usually removed by a membrane-bound signal peptidase and the SRP released. The polypeptide chain may then be further processed by disulfide bond formation, attachment of oligosaccharides, and so on. (Adapted from G. Blobel, et al., Proc. Natl. Acad. Sci. U.S.A. 77:1496–1500, 1980; and P. Walter and G. Blobel, J. Cell Biol. 91:557–561, 1981.)

quences. This may then cause them to be inserted into the membrane, but "stop" sequences would also seem to be required to cause the completed polypeptide chain to remain within the membrane rather than to be secreted.

There are many important questions yet to be answered about how proteins are directed to their targets, and this area is currently one of intensive study.

A possible precursor of immunoglobulin light chains. C. Milstein, et al. Nature New Biol. 239:117–120, 1972.

Transfer of proteins across membranes. I. Presence of proteolytically processed and unprocessed nascent immunoglobulin light chains on membrane-bound ribosomes of murine myeloma. G. Blobel and B. Dobberstein, J. Cell. Biol. 67:835–851, 1975.

Intracellular aspects of the process of protein synthesis. G. E. Palade. Science 189:347–358, 1975.

Membrane assembly *in vitro:* synthesis, glycosylation, and asymmetric insertion of a transmembrane protein. F. N. Katz, et al. Proc. Natl. Acad. Sci. U.S.A. 74:3278–3282, 1977.

The assembly of proteins into biological membranes: the membrane trigger hypothesis. W. Wickner. Ann. Rev. Biochem. 48:23–45, 1979.

Transmembrane translocation of protein. G. Von Heijne and C. Blomberg. Eur. J. Biochem. 97:175–181, 1979.

Mutations which affect the function of the signal sequence of the maltose binding protein of *Escherichia coli*. H. Bedouelle, et al. Nature 285:78–81, 1980.

A signal sequence is not sufficient to lead β-galactosidase out of the cytoplasm. F. Moreno, et al. Nature 286:356–359, 1980.

Secretion and membrane localization of proteins in *Escherichia coli*. M. Inouye and S. Halegoua. Crit. Rev. Biochem. 7:339–371, 1980.

The mechanism of protein secretion across membranes. B. D. Davis and P.-C. Tai. Nature 283:433–438, 1980.

Intracellular protein topogenesis. G. Blobel. Proc. Natl. Acad. Sci. U.S.A. 77: 1496–1500, 1980.

Translocation of proteins through biological membranes. A. Waksman, et al. Biochim. Biophys. Acta 604:249–296, 1980.

Assembly of proteins into membranes. W. Wickner. Science 210:861–868, 1980.

Transfer of proteins across membranes. G. Kreil. Ann. Rev. Biochem. 50:317–348, 1981.

Imported mitochondrial proteins cytochrome b_2 and cytochrome c_1 are processed in two steps. S. M. Gasser, et al. Proc. Natl. Acad. Sci. U.S.A. 79:267–271, 1982.

Compartmentation of newly synthesized proteins. A. W. Strauss and I. Boime. Crit. Rev. Biochem. 12:205–235, 1982.

Secretory protein translocation across membranes—the role of the "docking protein." D. I. Meyer, et al. Nature 297:647–650, 1982.

Signal recognition particle contains a 7 S RNA essential for protein translocation across the endoplasmic reticulum. P. Walter and G. Blobel. Nature 299:691–698, 1982.

Nucleocytoplasmic segregation of proteins and RNAs. E. M. DeRobertis. Cell 32: 1021–1025, 1983.

BIOSYNTHESIS OF UNUSUAL PEPTIDES

Relatively small, naturally occurring peptides are known in which there are D-amino acids; amino acids other than the normal 20, even if covalently modified; covalent linkages other than through peptide bonds; and cyclic rings of peptide linkages. An example is the ubiquitous peptide glutathione, γ-Glu-Cys-Gly, in which the amino-terminal Glu residue is linked through the side-chain carboxyl rather than the normal α-carboxyl group:

$$\underset{\substack{| \\ H_2N-CH-CH_2-CH_2-}}{CO_2H} \overset{O}{\underset{||}{C}} -NH- \underset{\substack{| \\ CH}}{\overset{\substack{SH \\ | \\ CH_2}}{}} \overset{O}{\underset{||}{C}} -NH-CH_2-CO_2H \quad (2-15)$$

The tyrocidines are cyclic peptides with normal peptide bonds, but containing the D as well as the L isomers of phenylalanine, plus L-ornithine (Orn), an amino acid that occurs naturally as an intermediate in biosynthesis of arginine but is normally not used in protein biosynthesis:

$$
\begin{array}{ccc}
\text{LPro} & \longrightarrow & \text{LPhe} \\
\uparrow & & \downarrow \\
\text{DPhe} & & \text{DPhe} \\
\uparrow & & \downarrow \\
\text{LLeu} & & \text{LAsn} \\
\uparrow & & \downarrow \\
\text{LOrn} & & \text{LGln} \\
\uparrow & & \downarrow \\
\text{LVal} & \longleftarrow & \text{LTyr}
\end{array}
\qquad (2\text{--}16)
$$

The arrows denote the polarity of the polypeptide chain, being the direction of the
$$-\overset{\overset{\displaystyle O}{\|}}{C}-NH-$$
peptide linkage. The linear gramicidins have residues of DLeu and DVal; the amino-terminal residue is formylated, and the carboxyl-terminal residue is linked in a peptide bond to ethanolamine:

$$H\overset{\overset{\displaystyle O}{\|}}{C}-\text{LVal-Gly-LAla-DLeu-LAla-DVal-LVal-DVal-LTrp-}$$
$$\text{DLeu-LTrp-DLeu-LTrp-DLeu-LTrp}-NH-CH_2-CH_2OH \qquad (2-17)$$

All such peptides have been observed to be synthesized in a manner basically different from that for normal proteins described above, but having certain similarities: The amino acids are generally activated by ATP to form the aminoacyl adenylates, but they are then bound to the enzyme complex carrying out the synthesis, through the —SH group of the cofactor 4′-phosphopantetheine, rather than to tRNA. Covalent linkage of the amino acids occurs on the enzyme complex, and the sequence of residues linked together is determined by the enzyme. Such a mechanism of synthesis has been designated the **"thiotemplate" mechanism.**

This type of mechanism of peptide synthesis would be impractical for most proteins, as the protein template to direct the synthesis requires a protein even larger than that to be synthesized. If the template were likewise to be synthesized in such a manner, an even larger template would be required; an ever-increasing, noncoverging series of protein-synthesizing enzymes would be necessary. Nevertheless, such a mechanism was apparently used for biosynthesis of a few small peptides of unnatural structure. Such peptides are usually antibiotics and probably serve to inhibit competing or dangerous species. In this regard, peptides somewhat different from normal proteins—and thus resistant to proteases and other defense mechanisms—probably were quite useful during evolution in biological warfare between species.

Polypeptide synthesis on protein templates: the enzymatic synthesis of gramicidin S and tyrocidine. F. Lipmann, et al. Adv. Enzymol. 35:1–34, 1971.

Nonribosomal polypeptide synthesis on polyenzyme templates. F. Lipmann. Acc. Chem. Res. 6:361–367, 1973.

The protein thiotemplate mechanism of synthesis for the peptide antibiotics produced by *Bacillus brevis*. S. G. Laland and T. L. Zimmer. Essays Biochem. 9:31–57, 1973.

Biosynthesis of small peptides. K. Kurahashi. Ann. Rev. Biochem. 43:445–459, 1974.

3

EVOLUTIONARY AND GENETIC ORIGINS OF PROTEIN SEQUENCES

There is a remarkable biochemical unity in the living world that belies the incredible diversity that we see at the macroscopic level. All living organisms, from the smallest viruses, to unicellular organisms like bacteria and algae, to plants, and up to the primates, use the same 20 amino acids in proteins and the same bases of DNA and RNA. Moreover, the amino acids are always the L isomer, and the ribose moieties of DNA and RNA are exclusively the D isomer. The basic properties of proteins described in Chapter 1 pertain equally well to *E. coli* and to elephants. The ways in which these basic building blocks are used are also very similar, as was shown in the previous chapter on protein biosynthesis. The biochemical similarities extend to higher levels, such as biochemical pathways and metabolism, but here differences between different groups become more apparent. In general, the higher the organizational level, the greater the differences among species.

This general pattern can be satisfactorily rationalized with our present knowledge only by postulating that all present living organisms arose from a single common ancestor, which had already acquired the basic biochemical features now common to all of them. It is not at all clear how this common ancestor arose, but at some early stage of evolution, some 3×10^9 years ago, there must have existed a population of interbreeding organisms that possessed DNA genetic material and a biosynthetic apparatus for expressing the genetic information basically like that present today.

The diversity of living organisms present today must then have arisen by the process of evolutionary divergence. Functional differences would have evolved by Darwinian evolution, in response to the selective pressures on each species. However, the most basic biochemical processes would have been conserved, as any change in them would have been lethal.

The macroscopic evidence for the Darwinian scheme of divergent evolution comes both from the fossil record and from the present morphological similarities between related species. But the recent information from the molecular level of the sequences of proteins and nucleic acids is even more compelling and gives new insights into the process of evolution. The genetic information contained at the present time in the nucleic acids and proteins of each organism is so vast that the probable process of evolutionary divergence may be traced back in time. This is due to the high information content of such sequences, which may be illustrated by calculating the number of possible sequences of a single gene of average size: A string of 1000 nucleotides, each being any one of four possible bases, may occur in 4^{1000}, or 10^{600}, different combinations. As pointed out previously, a polypeptide chain of 100 amino acid residues could exist in 10^{130} combinations. Just one molecule of each of these different proteins would fill the entire universe 10^{27} times over, even if packed together in the most efficient manner. Clearly, not all these possible proteins exist today, nor are they likely to have ever existed on Earth, as it is only about 10^{17} seconds old.

As will become apparent, a particular protein present today is merely one of the evolutionary products of many millenia. Its particular amino acid sequence is primarily determined by its ancestry. It probably represents only one of many related sequences that could serve equally as well. Within the constraint that it be functional, it is the product of the stochastic process of mutation, and of the haphazard accidents of history, rather than a finely tuned, unique entity. It has changed in the evolutionary past and may be expected to change further in the future. Its presence today must be considered a somewhat fortuitous and transient occurrence in a constantly changing world. This flexibility of form varies in degree with different proteins, so that collectively they provide a wealth of information about the evolutionary past.

Equally, the evolutionary variation of a specific protein provides us with much information about the protein itself. Evolutionary divergence has resulted in very many forms of essentially the same protein, all with the same biological function. The nature of the evolutionary divergence permitted in the amino acid sequence reflects the constraints of structure and function in each instance.

The Molecular Basis of Evolution. C. B. Anfinsen. New York, John Wiley & Sons, 1959.

The Haemoglobins in Genetics and Evolution. V. M. Ingram. New York, Academic Press, 1963.

Molecules as documents of evolutionary history. E. Zuckerkandl and L. Pauling. J. Theor. Biol. 8:357–366, 1965.

Evolving Genes and Proteins. V. Bryson and H. J. Vogel (eds.). New York, Academic Press, 1965.

Molecules and Evolution. T. H. Jukes. New York, Columbia University Press, 1966.

Mechanisms of protein evolution. G. H. Dixon. Essays Biochem. 2:147–204, 1966.

Evolution of enzymes. E. L. Smith. In The Enzymes, 3rd ed. P. D. Boyer (ed.). Vol. 1, pp. 267–339. New York, Academic Press, 1970.

On some principles governing molecular evolution. M. Kimura and T. Ohta. Proc. Natl. Acad. Sci. U.S.A. 71:2848–2852, 1974.

The information content of protein amino acid sequences. T. T. Wu, et al. Ann. Rev. Biochem. 43:539–566, 1974.

Molecular Evolution. F. J. Ayala (ed.). Sunderland, Mass., Sinauer Associates, 1976.

Biochemical evolution. A. C. Wilson, et al. Ann. Rev. Biochem. 46:573–639, 1977.

Protein evolution. R. F Doolittle. In The Proteins, 3rd ed. H. Neurath and R. L. Hill (eds.). Vol. 4, pp. 1–118. New York, Academic Press, 1979.

EVOLUTION AND THE GENETICS OF POPULATIONS

The information present in protein amino acid sequences, and in the nucleic acid sequences of the corresponding gene and messenger RNA, is relevant to both evolution and protein function. The latter topic is pertinent to our concerns here, but interpretation of this information requires some understanding of evolution and of how it is reflected in protein structures. This section, therefore, is devoted to some basic principles of genetics, population biology, and evolution.

The reader should be forewarned that it is impossible to prove any account of evolution, because the necessary information about past events is largely unavailable. There is only the very sparse fossil record. The alternative approach is to speculate about the past on the basis of present-day biological species. Consequently, the field of evolution is one of much debate and controversy. There is no consensus of opinion on many of the most important questions. The account presented here must also be very abbreviated, so it is not possible to present all points of view. Instead, an objective explanation of what seems most plausible is given.

Populations, Species and Evolution. E. Mayr. Cambridge, Mass., Harvard University Press, 1970.

Genetics of the Evolutionary Process. T. G. Dobzhansky. New York, Columbia University Press, 1970.

The Genetic Basis of Evolutionary Change. R. C. Lewontin. New York, Columbia University Press, 1974.

Molecular Population Genetics and Evolution. M. Nei. New York, Elsevier/North Holland, 1975.

The Principles of Human Biochemical Genetics, 3rd ed. H. Harris, Amsterdam, Elsevier/North Holland, 1980.

Prebiological Evolution

How life first arose on this planet will always be something of a mystery, but there are some plausible explanations of how it might have occurred. The most frequently evoked scenario is that proposed by Oparin and by Haldane some 50 years ago. The environment on this planet at the probable time of the origin of life, some 3 to 4 billion years ago, was probably hot and rich in the simple compounds CO_2, CO, N_2, and H_2O. By strictly chemical processes, catalyzed by the action of ultraviolet (UV) light and electrical discharges of lightning, these compounds may have combined to produce larger molecules, including the amino acids, nucleotides, and other basic building blocks of biochemistry. The surface of the planet would then have become a moderately concentrated organic "soup." By further chemical processes, these small molecules may have become polymerized into polypeptides and polynucleotides.

The fact that many of these reactions may be simulated in the laboratory proves that the scheme is plausible. However, polymerization reactions are invariably dehydrations and therefore require the removal of water in order to be favorable energetically. Such condensation reactions are unlikely to have occurred with an aqueous solution but are favored only in rather specialized environments.

Somehow, an "organism" arose that consisted of nucleic acids and proteins, was capable of replicating the nucleic acids, which coded for at least some of the proteins, and was able to translate this genetic information. There is no single satisfactory explanation of how this could have occurred; some even suggest that it arrived on a space ship, but this theory merely transfers the location of where life first arose. The present-day replication and translation of nucleic acids requires proteins, and the production of proteins requires nucleic acids; the question is which came first, just as in the case of the chicken and the egg. There is no contemporary mechanism for translating information from the protein primary structure into nucleic acids—in fact, the reverse flow of information is the natural one—so it is most plausible that the nucleic acids were incorporated first into the living organism. However, the initial replication and transcription of such nucleic acids must have been quite different from those present today.

In any case, there must have arisen an organism that used the 20 L-amino acids to produce proteins like those present today. The choice of the particular amino acids and nucleotides used, especially the particular isomers, is likely to have been largely accidental, as no other plausible explanation has been found. Since the organism could duplicate itself and express its genetic information, it must have been a relatively complex entity, although probably less so than present-day bacteria.

Most likely, the original organism used the amino acids, nucleotides, sugars, and other biochemical building blocks that were present in the primeval soup and produced by solely chemical processes. This natural supply of organic materials would have become limiting at some stage, so there would have been selective pressure for the evolution of biochemical pathways. Organisms that could synthesize an essential amino acid, or other constituent, from some other, nonutilized constituent of the organic soup would have had a great selective advantage. But then this constituent may also have become depleted from the soup, so that the ability to use a different constituent would have been selected. By this backward process, one step at a time, the biochemical pathways are believed to have evolved. Ultimately, photosynthesis would have appeared, as the natural supply of all organic materials became negligible. The atmosphere would then have

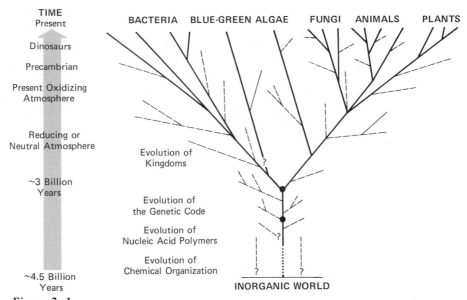

Figure 3–1

A general phylogenetic tree of how life on Earth may have evolved. The dashed lines represent hypothetical lineages that have not survived to the present day. (Adapted from M. O. Dayhoff (ed.), Atlas of Protein Sequence and Structure. Washington, D.C., National Biomedical Research Foundation, 1972.)

become less reducing, as O_2 was produced, and aerobic metabolic pathways would have evolved.

The original organisms may have been rather similar to the methanogenic bacteria present today in anaerobic environments. As evolution progressed and the Earth changed, to present a large number and variety of ecological habitats, the wide variety of species existing today would have arisen by the process of divergent evolution (Figure 3–1).

In spite of this wide divergence of biological life, the basic biochemical unity of life would have been established by its presence in the common ancestor. To have changed dramatically the 20 L-amino acids, the bases of the nucleic acids, or the genetic code would have proved lethal. Yet there are exceptions, such as the slightly different genetic code used in mitochondria (Chapter 2). Mitochondria have separate genetic systems with a DNA genome of 16,569 base pairs in humans. This genome codes for 12 S and 16 S ribosomal RNAs, for at least 19 transfer RNAs, and for polypeptide chains of a number of proteins. There has been considerable speculation that mitochondria evolved from ancient bacteria that had entered into a symbiotic relationship with a host eukaryotic cell. It is then possible that the mitochondrial ancestor was a more primitive cell that had diverged before the occurrence of the last common ancestor of other prokaryotic and eukaryotic cells. Other descendants of this primitive organism, which did not become mitochondria, would have disappeared by the present time. Likewise, any other primitive organisms predating the putative common ancestor would not have survived to the present day, although it is possible that they have yet to be discovered. If this were to happen, the common ancestor would be assigned to an even more primitive organism.

There is some suggestion that one of the 20 amino acids may have changed since the common ancestor—that originally ornithine was used in place of arginine. The frequencies with which the amino acids occur in proteins are not the same, but this set of frequencies is similar in almost all proteins (Table 1–1). The distribution is not random; the frequency of each amino acid corresponds rather closely to the number of codons possessed by each amino acid in the genetic code (Figure 2–4). For example, Trp has only one codon and the lowest frequency in proteins, while amino acids like Ser that have six codons occur in high frequencies (Figure 3–2). A conspicuous exception is Arg, which occurs at less than half the expected frequency. It has been postulated that the amino acid ornithine

$$\begin{array}{l} NH_2 \\ | \\ CH-CH_2-CH_2-CH_2-NH_2 \\ | \\ CO_2H \end{array} \tag{3-1}$$

was used initially and that arginine was not available in the primeval soup: After the development of the urea cycle in metabolism

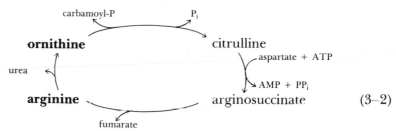

$$\tag{3-2}$$

arginine was produced and perhaps competed with ornithine for incorporation into proteins. It won this evolutionary competition and became the accepted amino acid. However, its presence in proteins was somewhat deleterious, perhaps owing to the highly bulky and basic guanido group. Its occurrence was then decreased through selection; perhaps it was replaced mutationally by Lys, which is more like the original Orn and also occurs in somewhat higher-than-expected frequencies. This scheme is highly speculative but illustrates the types of arguments that can, and must, be used when considering evolution.

Life: Its Origin, Nature, and Development. A. I. Oparin. New York, Academic Press, 1964.

The origin of life and the origin of enzymes. A. I. Oparin. Adv. Enzymol. 27:347–380, 1965.

The origin of the genetic code. F. H. C. Crick. J. Mol. Biol. 38:367–379, 1968.

Evolution of the genetic apparatus. L. E. Orgel. J. Mol. Biol. 38:381–393, 1968.

Self-organization of matter and the evolution of biological macromolecules. M. Eigen. Naturwiss. 58:465–523, 1971.

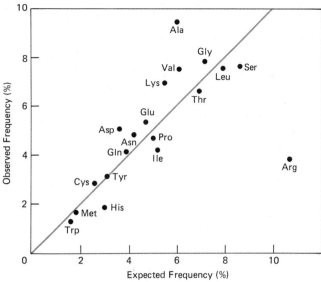

Figure 3–2
Correlation between the observed frequencies of amino acid residues in 20 unrelated proteins and those expected on the basis of random permutations of DNA nucleotides and the genetic code. (From J. R. Jungck, Currents Modern Biol. 3:307–318, 1971; © Elsevier Biomedical Press B.V.)

Origin of the cell: experiments and premises. S. W. Fox. Naturwiss. 60:359–369, 1973.
The Origins of Life: Molecules and Natural Selection. L. E. Orgel. New York, John Wiley & Sons, 1973.
The Origins of Life on Earth. S. L. Miller and L. E. Orgel. Englewood Cliffs, N.J., Prentice-Hall, 1974.
A proposal concerning the origin of life on the planet earth. C. R. Woese. J. Mol. Evol. 13:95–101, 1979.
Reasons for the occurrence of the twenty coded protein amino acids. A. L. Weber and S. L. Miller. J. Mol. Evol. 17:273–284, 1981.
How life began. P. Cloud. Nature 296:198–199, 1982.

Population Genetics

It is now pertinent to inquire how genes and their protein products might have changed during evolution. We shall consider first what factors influence the genetic constitution of species, then how new species might arise, and finally how evolutionary divergence might have occurred.

The basic biological unit of evolution is the **species,** which comprises one or more populations of individuals interbreeding to at least some extent with each other, but not with individuals of other species. Usually there are a large number of **populations,** separated to some extent geographically, but with sufficient migration between them so that there is a common gene pool characteristic of the species as a whole. Where there are considerable barriers to interbreeding, distinct races occur, but they have the potential for interbreeding.

All species and populations consist of a finite number of individuals, which in turn limits the genetic variability that may occur. Each gene within an individual has only a moderate probability of being passed on to the next generation; where the population is stable in size, this probability is 0.75 on average. It is then very improbable that any copy of a gene will be passed on for very many generations, even in the absence of selective pressures. As a consequence, all copies of a particular gene present at any instant of time in the gene pool of a population are likely to have been replicated from a single common copy of that gene present a finite number of generations previously. In an ideal population, this number of generations is equal to four times the number of breeding individuals in that population. Therefore, the genetic variation between the copies of a gene in the individuals of a population will be limited to that arising by mutation over the time since they shared a common ancestor.

This limitation, then, is the major reason that species are on the whole genetically homogeneous; it explains why we can speak of human insulin, bovine ribonuclease, or *E. coli* RNA polymerase. The chances are that most individuals will have the gene specifying that protein which is representative of its species.

Nevertheless, variants do arise by the process of mutation; they are found for every gene in every species. The various forms of a gene in a species are known as **alleles.** The most extensively studied species is *Homo sapiens,* for obvious reasons. For example, over 300 hemoglobin variants are now known. Most are exceedingly rare, producing little or no clinical effect, and are known only because of extensive public health screening programs. Most allelic forms of hemoglobin are identified by an altered electrophoretic mobility. Approximately 1 in 800 humans is estimated to have a variant hemoglobin detectable in this way; there is probably an equal number thus far undetectable. Similar variants are found in other proteins, but the data are not as extensive.

Most of these variant proteins differ from the normal at only one position in the polypeptide chain, one amino acid being substituted for another. The positions and identities of these changes have usually been identified by two-dimensional peptide maps, or "fingerprints." The protein is cleaved into small peptides using a relatively specific protease, such as trypsin. The peptides are then separated on paper or thin-layer supports by means of two different separation procedures sequentially and at right angles to each other, usually chromatography and electrophoresis. A variant protein generally has most of the peptides of its fingerprint in the same positions as the normal protein, but the peptide with one residue different should have some difference in mobility in either or both dimensions. The variant peptide may then be identified by its amino acid composition or sequence. Fingerprints of normal and sickle cell hemoglobin are shown in Figure 3–3, as this particular variant was the first to be characterized.

The most common type of amino acid replacement observed is one that results from alteration of one of the nucleotides of the corresponding codon of the structural gene. There are exceptions, such as substitutions

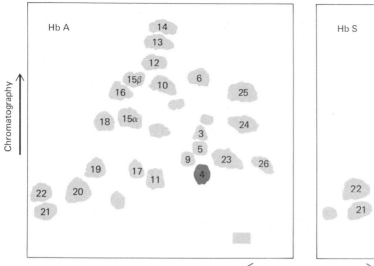
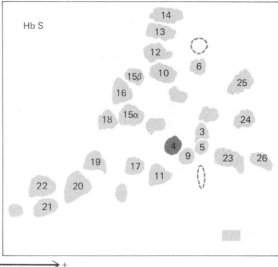

Electrophoresis

Figure 3–3

Two-dimensional maps, or fingerprints, of normal (Hb A, left) and sickle cell (Hb S, right) hemoglobins. The two proteins were digested with trypsin, and the tryptic digests applied at the rectangular origin of the chromatography paper. The peptides were separated in the horizontal dimension by electrophoresis in pyridine acetate buffer, pH 6.4. After drying, the paper was subjected to chromatography in the second dimension, using a 35:35:30 mixture of isoamyl alcohol, pyridine, and water. The peptides were then detected by staining with ninhydrin.

The two fingerprints differ significantly only in the position of peptide 4. It consists of residues 1 to 8 of the β chain, with the sequence Val-His-Leu-Thr-Pro-Glu-Glu-Lys in Hb A, but Val-His-Leu-Thr-Pro-Val-Glu-Lys in Hb S. The Val residue, rather than Glu, in Hb S makes this peptide more basic at pH 6.4, so it migrates electrophoretically faster toward the cathode. It also is less polar so it migrates more rapidly during chromatography. Consequently, peptide 4 from Hb S lies further to the left and above that from Hb A.

The other peptides are identical in the two proteins. (Adapted from C. Baglioni, Biochim. Biophys. Acta 48:392–396, 1961.)

of two nucleotides within a codon or within the gene, as well as deletions and insertions of nucleotides. For example, hemoglobin Grady has the sequence of residues 116 to 118 (-Glu-Phe-Thr-) duplicated in tandem. Insertions or deletions of one or two nucleotides alter the reading frame of the gene, so that subsequent amino acids are incorrect. Three examples are hemoglobins Wayne, Cranston, and Tak, in which the insertions occur near the carboxyl end of the β chain (Figure 3–4). Because the terminal residues are altered, the normal termination codon UAA is missed, and translation continues until another termination codon is encountered. Those β chains then consist of 157 residues, rather than the normal 147. A similar situation occurs when the termination codon is mutated; four such variants are known for the hemoglobin α chain, in which the normal α chain of 141 residues is elongated to 172 residues. The added sequences are iden-

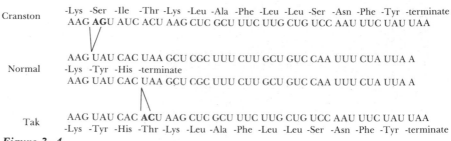

Figure 3–4

*Insertion mutations that alter the reading frame of the gene for the β chain of hemoglobin.
The sequences of the mRNA and of the protein product are shown for the normal and two
mutants. The inserted nucleotides in the mutants are in bold type. Note that they appear to
arise by duplication of the preceding nucleotides. (Sequences from N. Proudfoot and G.
Brownlee, Brit. Med. Bull. 32:251–256, 1976.)*

tical, except for residue 142, which is variously Gln, Glu, Ser, or Lys; each
of these amino acids may be derived by a single nucleotide substitution in
the normal UAA termination codon.

Many of the variant proteins do not function properly, and affected
individuals demonstrate some clinical deficit. Such mutations are clearly
detrimental and are no doubt selected against by the classical Darwinian
process. Where the individual does not survive long enough to reproduce,
none of the genes of that individual will be passed on to the next generation. Smaller selective pressures occur when there is simply a lower probability of reproducing.

Many mutant proteins are functionally indistinguishable from the normal; if they have no effect on the probability of reproduction in affected
individuals and their progeny, such a mutation is said to be **neutral.** Many
evolutionists dispute that any mutation may be truly neutral, but the molecular evidence is that most are, at least to a first approximation. A neutral
mutation will not be subjected to selective pressure, but the likelihood that
it will disappear after a few generations is strong.

Mutations may conceivably produce a more suitable protein that would
be beneficial to the individual. Such mutations are readily observed in
artificial procedures with microorganisms, where the environment may be
drastically manipulated, but they are not obvious in most natural populations, and their role in the evolution of species is not clear. Even with
favorable variants, it is unlikely that they will survive the random processes
of gene loss that occur in populations of finite size and that maintain
genetic homogeneity. For example, a mutant allele that increases the genetic fitness of an individual by 5 per cent has only a 5 per cent probability
of replacing the former, less fit allele.

A single normal allele generally predominates in a species, but many
instances are known in which two or more alleles occur in high frequencies.
This is possible with neutral alleles by random processes, known as **genetic
drift.** There are also genetic means by which natural selection can maintain

two or more particular alleles in a population. The best established is that of **heterosis,** when diploid individuals having one copy of each of the two alleles, or **heterozygotes,** are more fit than individuals having the same two copies of either of them, or **homozygotes.** The best example appears to be sickle cell hemoglobin, a human variant present in high frequency in African populations (Figure 3–3). The change of Glu to Val at residue 6 of the β chain causes the deoxy form of this hemoglobin to polymerize, thereby producing the characteristic "sickling" of the erythrocytes. This causes blockage of the capillaries carrying blood to vital regions, producing a clinical crisis for the affected person, who is unlikely to survive beyond the age of 30 years. Yet the sickle cell gene is present in African populations with a frequency of about 0.05.

This rather high frequency of an apparently detrimental variant is not without genetic advantage: Only the homozygotes for this allele suffer from sickle cell anemia. Heterozygotes do not, because about half of their hemoglobin is the normal type, which inhibits polymerization of the sickle cell molecules; moreover, the presence of the sickle cell hemoglobin gives them resistance to malaria. The responsible parasite, *Plasmodium falciparum,* spends part of its reproductive cycle in human erythrocytes, and the sickle cell variety has been shown to resist infection for reasons that are not clear. Therefore, persons heterozygous for sickle cell hemoglobin have moderately normal blood and are resistant to malaria, whereas sickle cell homozygotes suffer from the anemia and normal homozygotes suffer from malaria. The heterozygotes are consequently fitter than homozygotes in areas where malaria is prevalent, so selective pressures maintain both alleles in such populations.

It is tempting to invoke similar selective arguments to account for the other instances of multiple alleles, but it seems unlikely that there could be selection for many such genes in any one species. For example, if many other human genes had polymorphisms such as sickle cell hemoglobin, all humans would tend to be homozygous for an average number of these genes and would be very ill indeed. It seems more plausible that most multiple alleles are neutral, but argument about this question continues.

Mutational change of some species can occur very rapidly. The most dramatic example, and one of great practical importance, is the human influenza virus. The amino acid sequences of its coat proteins change very rapidly by unknown mechanisms, so that it is no longer recognized by antibodies produced in victims of an earlier strain. Influenza epidemics occur in this way every few years.

Protein variation in natural populations of animals. J. R. Powell. Evol. Biol. 8:79–119, 1975.

Molecular genetics of human hemoglobin. D. J. Weatherall and J. B. Clegg. Ann. Rev. Genetics 10:157–178, 1976.

Polymorphism and the subunit structure of enzymes: a contribution to the neutralist–selectionist controversy. H. Harris, et al. Proc. Natl. Acad. Sci. U.S.A. 74:698–701, 1977.

Genetically controlled variation in the shapes of enzymes. G. Johnson. Prog. Nucleic Acid Res. Mol. Biol. 22:293–326, 1978.

Heterosis as an explanation for large amounts of genic polymorphism. R. Lewontin, et al. Genetics 88:149–170, 1978.

Cellular mechanism for the protective effects of haemoglobin S against *P. falciparum* malaria. G. Pasrol, et al. Nature 274:701–703, 1978.

Erythrocytic mechanism of sickle cell resistance to malaria. M. J. Friedman. Proc. Natl. Acad. Sci. U.S.A. 75:1994–1997, 1978.

Reevaluation of level of genic heterozygosity in natural populations of *Drosophila melanogaster* by two-dimensional electrophoresis. A. J. LeighBrown and C. H. Langley. Proc. Natl. Acad. Sci. U.S.A. 76:2381–2384, 1979.

Antigenic drift between the haemagglutinin of the Hong Kong influenza strains A/Aichi/2/68 and A/Victoria/3/75. M. Verhoeyen, et al. Nature 286:771–776, 1980.

Cloning and DNA sequence of double-stranded copies of haemagglutinin genes from Hs and H3 strains elucidates antigenic shift and drift in human influenza virus. M. J. Gething, et al. Nature 287:301–306, 1980.

The complete amino acid sequence of the three alcohol dehydrogenase alleloenzymes (Adh^{N-11}, Adh^S, and Adh^{UF}) from the fruitfly *Drosophila melanogaster*. D. R. Thatcher. Biochem. J. 187:875–886, 1980.

Hemin lyses malaria parasites. A. U. Orjih, et al. Science 214:667–669, 1981.

Formation of New Species

Evolution has proceeded primarily by divergence, to produce the wide variety of species present today. It is important to examine how species diverged from each other in order to consider the significance of the differences between them present today.

The answer is not known with certainty, but such divergence is believed to occur by geographic separation of populations of a species for a sufficiently long period of time. During that time, they may evolve independently in such a way that interbreeding does not occur when the two populations subsequently come into contact. **Speciation** is then considered to have occurred. Alternatively, the isolation may be merely genetic, preventing productive interbreeding.

The mechanisms that prevent interbreeding are not at all clear, but closely related species are known to be very similar at the molecular level; extensive genetic mutation, therefore, is not required. Genetic divergence may occur at a higher level, such as in the case of gross chromosomal rearrangements.

A new species might arise by interbreeding of only a few individuals—for example, following geographic isolation of a lone pregnant female. Whatever variant alleles her offspring happened to possess in their chromosomes now have a very high probability of being fixed in the new species, even if such alleles were of no physiological significance, or even if slightly deleterious. This extreme example of a random evolutionary process is known as the **"founder effect."**

Once a new species has arisen, its genes are free to evolve independently of the species from which it arose. Speciation marks the time at which a new evolutionary branch arises (Figure 3–1), but whether that branch will ever appear on a phylogenetic tree requires that descendants from both the original species and the new species either survive to the present time or be represented in the fossil record.

Evolutionary Divergence

Each gene present today in any individual of any species has arisen by the process of evolutionary divergence, as it has been duplicated at each generation of the past, following one particular pathway of an evolutionary tree like that shown in Figure 3–1. During that time, it has inevitably accumulated mutations, unless natural selection acted to eliminate every mutation. This seems unlikely, and if the possibility of neutral mutations is accepted, a gene present today should have changed during evolution at the rate at which neutral mutations occurred during that gene lineage. The overall mutation rate is probably similar for each gene, approximately 10^{-8} per nucleotide per year, but the neutral mutation rate will be some fraction of this and will vary with each nucleotide and each gene. It will be proportional to the fraction of mutations that are acceptable to the function of the gene. Mutations that adversely affect the gene are assumed to have been selected against and not to have survived. Many genes appear to tolerate a large fraction of mutational changes, whereas others tolerate very few without a deleterious effect. Therefore, even in the absence of any natural selection for favorable mutations, a gene should have changed during evolution at its particular neutral mutation rate.

Which neutral mutations would accumulate in this way would be a random process. They might become fixed in a species by their presence in one of its founders or, after speciation, as a result of the finite number of individuals in the species. The exact process is immaterial for neutral mutations. The important point is that each gene must evolve at the neutral mutation rate per individual gene, even in the absence of positive selection.

SPECIES VARIATION OF PROTEIN PRIMARY STRUCTURE

The proteins of species known to be closely related are invariably very similar in their amino acid sequences. For example, the c-type cytochromes, the hemoglobin α, β, and γ chains, and the fibrinopeptides of chimpanzees are identical to those of humans, while their myoglobin and hemoglobin δ chains each have one amino acid difference. As less closely related species are compared, the differences between their proteins increase. A difference matrix giving the number of amino acid differences in cytochrome c among a few selected eukaryotic species is shown in Figure 3–5.

Any proteins or genes having a significant number of similarities are said to be **homologous,** implying that they have descended from a common ancestor. Closely related proteins generally differ only by replacements of one amino acid by another at one or more positions within the polypeptide chain, but with more widely separated species, small differences also occur in the total number of residues in the chain, owing to the deletion or addition of residues within the polypeptide chain or at either end. A composite of the amino acid sequences for cytochrome c from widely divergent

	Chimpanzee	Sheep	Rattlesnake	Carp	Snail	Moth	Yeast	Cauliflower	Parsnip
Human	0	10	14	18	29	31	44	44	43
Chimpanzee		10	14	18	29	31	44	44	43
Sheep			20	11	24	27	44	46	46
Rattlesnake				26	28	33	47	45	43
Carp					26	26	44	47	46
Garden snail						28	48	51	50
Tobacco hornworm moth							44	44	41
Baker's yeast (iso-1)								47	47
Cauliflower									13

Figure 3–5
Number of amino acid differences for cytochrome c *of a few representative species. These proteins all consist of at least 104 amino acid residues, so each pair have at least half their residues identical. (Sequences from M. O. Dayhoff (ed.), Atlas of Protein Sequences and Structure. Washington, D.C., National Biomedical Research Foundation, 1972.)*

eukaryotic species is shown in Figure 3–6, where every amino acid that has been observed to occur at each position is listed. Wide variation has been found at some positions, as at 60, 89, and 92. At others only amino acids with chemically similar side chains are found, as with the large non-polar residues at positions 35 and 36. No variation is found at other positions, such as the Cys residues at positions 14 and 17, to which the hemc is attached (Eq. 2–8), and the invariable stretch of residues 76 through 80. The surplus of basic over acidic residues is also conserved within rather narrow limits in this protein. There do appear to have been constraints on the evolutionary changes made in cytochrome c, but the variation that *has* occurred is perhaps the more amazing.

Indeed, as more species that are less closely related are included, the apparent constraints on the variation seem to decrease. For example, cytochromes from a wide variety of bacteria are now recognized as being descended from a common ancestor of c-type cytochromes in Figure 3–5. These bacterial cytochromes had been given the designations c_2, c_5, c_6, c_{550}, c_{551}, and c_{555}, as they were detectably different in their spectral properties from cytochrome c. These physical differences arise from the much wider variation they have undergone, but they are clearly all homologous. When 29 of these sequences are compared with those of cytochrome c, only 3 residues are found to be invariant: Cys 17, His 18, and Met 80, using the numbering system of Figure 3–6. The invariance of these residues is due to their functional interactions with the heme group (see Figure 6–28). Some pairs of the c-type cytochromes differ at 88 per cent of the residues.

```
                                        Glu
                                        Lys       Arg                     Glu
                        Gly             Asn Ala Lys     Lys               Met
                        Asp     Tyr Lys Ser Thr Asn     Gly               Val Leu
                Ala Arg Glu Pro Phe Lys Asp Ile Ser Ser Ala Asn Thr       Ile Gln        Asn              Gly
                Lys Gly Ala Ser Leu Glu Gln     Ser Val Asp Thr Gln Thr Val   Thr Asn        Ser Glu          Ala
                Ser Pro Thr Gly Val Ser Ala     Asn Ala Glu Val Asp Arg Leu   Glu Ser Arg Ala Leu Gln      Ser Ile
Acetyl-Ala-Thr-Phe-Ser-Glu-Ala-Pro-Pro-Gly-Asp-Pro-Lys-Ala-Gly-Glu-Lys-Ile- Phe-Lys-Thr-Lys-Cys-Ala-Glx-Cys-His-Thr-Val-
     -8                    -1 +1                        10                              20
```

```
Gly Ala                                   Trp                         Asp                Lys
Ala Leu         Lys                        Tyr                         Glu                Glu
Asp Asn         Pro         Ile            Phe     Phe Ile     Lys     Gln                Asp
Glu Glu     Glu Val     Gly Val            Ser     Val Val Asn Thr Gln Gln Ala Val   Ala      Asn
Gln Arg Ala Gly Ala Gln Ser Thr   Ser     Asn     Ile Tyr Ser His Thr Ser Val Pro Phe Thr   Thr Ala
Glx-Lys-Gly-Ala-Gly-His-Lys-Gln-Gly-Pro-Asn-Leu-His-Gly-Leu-Phe-Gly-Arg-Gln-Ser-Gly-Thr-Thr-Ala-Gly-Tyr-Ser-Tyr-Ser-Thr-
                            30                              40                                  50
```

```
                        Lys
            Gly         Glu
            Ala     Val Gln Tyr Lys
            Ser     Ile Asp Pro Ala     Arg
            Arg     Leu Asn Asn Pro     Leu Val
            Lys Ala Thr Thr Asp Asn Val Met Ile
        Ala Gln Ser Gly Gln Glu Ser Gln Asp Asp His Lys     Glu                Met
Ala Ile Asp Met Asn Ile Gln Ala Glu Glu Asn Met Phe Asp Phe Thr                Phe Val
Gly-Asn-Lys-Asn-Lys-Ala-Val-Asn-Trp-Gly-Glx-Glx-Thr-Leu-Tyr-Glu-Tyr-Leu-Leu-Asn-Pro-Lys-Lys-Tyr-Ile-Pro-Gly-Thr-Lys-Met-
                        60                              70                                  80
```

```
                        Gly
                        Ala     Lys
                        Ser     Glu                 Ser
                    Lys Thr     Gln             Thr Lys     Lys Ala
        Gly         Glu Lys     Asn             His Glu Leu Glu Ser
Ala     Ala         Asp Asn     Thr             Lys Asn Ala Gln Lys
Ser     Thr         Asn Gln Asp     Gly His Val Ala Leu Asp Thr Cys Asn Asp
Ile     Val     Ile Ser Ala Ala Glu Glu Val Asn Ile Leu Thr Phe Met Arg Glu Ser Ser Ala Glu
Val-Phe-Pro-Gly-Leu-Lys-Lys-Pro-Glx-Glx-Arg-Ala-Asp-Leu-Ile- Ser-Tyr-Leu-Lys-Gln-Ala-Thr-Ser-Gln-Glu
                90                              100
```

Figure 3–6

Composite amino acid sequence of cytochrome c *from 41 species. The continuous sequence of 113 residues is that of* Ginkgo biloba, *that with the longest polypeptide chain. The numbering of the residues is that of the mammalian proteins, which start at position 1 and end at 104; the proteins included here start at positions −8 to +1 and end at 103 to 105. Only two instances of internal deletions occur, of single residues at positions 99 and 100.*

The structural basis of the conservation of some residues is given in Table 6–7. (Sequences from M. O. Dayhoff (ed.), Atlas of Protein Sequences and Structure. Washington, D.C., National Biomedical Research Foundation, 1972.)

Additions and deletions also become more common, so that 162 residues are required to account for all the sequences, whereas only 113 are required in Figure 3–6. The greater divergence observed among more widely different species probably reflects primarily the greater time during which change has been permitted, but there may also be some differences in the functional constraints along different lineages, varying somewhat with what particular sequence is present and with the environment in which it must function.

With very large differences in amino acid sequences, the few remaining identities conceivably could have arisen by chance. In cases where there are only two slightly similar sequences, statistical procedures must be used to assess the significance of the similarity. However, in the case of the *c*-type cytochromes, representatives of each group are known to have very similar folded conformations (see Figure 6–28). It is becoming increasingly apparent that the folded conformations of proteins are much more highly conserved than are the amino acid sequences.

The implied homology of a number of proteins may also be examined by reconstructing the implied evolutionary divergence, the phylogenetic tree of the species. The current sequences of homologous proteins should be compatible with a reasonable scheme of divergence from a common ancestral gene.

Comparative aspects of primary structures of proteins. C. Nolan and E. Margoliash. Ann. Rev. Biochem. 37:727–790, 1968.

Atlas of Protein Sequence and Structure. M. O. Dayhoff (ed.). Washington, D.C., National Biomedical Research Foundation, 1972. (plus subsequent supplements)

Evolution at two levels in humans and chimpanzees. M. C. King and A. C. Wilson. Science 188:107–116, 1975.

The molecular evolution of cytochrome *c* in eukaryotes. W. M. Fitch. J. Mol. Evol. 8:13–40, 1976.

Decoding the pattern of protein evolution. M. Goodman. Prog. Biophys. Mol. Biol. 37:105–164, 1981.

Reconstruction of the Phylogenetic Tree

The evolutionary process of divergence may be traced by reconstructing the phylogenetic tree. This may often be accomplished with the amino acid sequences of a single protein such as cytochrome *c* present in contemporary species, as is illustrated in Figure 3–7. However, such evolutionary divergence is the product of a stochastic process, so variations are expected when there is a small number of differences; moreover, it is not surprising that such a tree may show discrepancies with those calculated using other proteins or with other evolutionary evidence. Nevertheless, the phylogenetic tree is usually remarkably similar to that expected from other data and may be made most reliable by using a number of different proteins.

Phylogenetic trees are generally derived by generating that tree requiring the minimum number of mutations. Sophisticated computing techniques are required, as there are more than 2 million trees possible with only 10 species. The most widely used method employs a difference matrix like that in Figure 3–5; it relies solely on the sequence differences among pairs of present-day species. Nevertheless, this method can generate the correct tree if the rates of divergence are roughly equal in all branches of the tree, although statistical variations in the number of differences will lead to mistakes.

The other major method infers the ancestral sequences of the proteins that occurred at the branch points. Each ancestral sequence is chosen to minimize the number of mutations that occurred throughout the entire

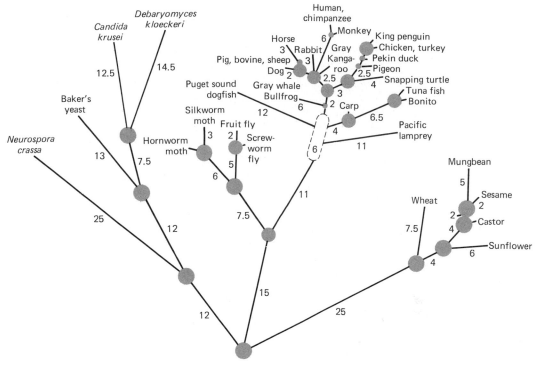

Figure 3–7
Phylogenetic tree constructed from the sequences of cytochrome c. *The numbers of amino acid changes along each segment are given. (Adapted from M. O. Dayhoff (ed.), Atlas of Protein Sequence and Structure. Washington, D.C., National Biomedical Research Foundation, 1972.)*

phylogenetic tree. One such inferred sequence for the common ancestor of cytochrome *c* in the plants, animals, and fungi (i.e., the lowest node in Figure 3–7) is compared in Figure 3–8 with that for the current human cytochrome *c*. Only about a dozen amino acid replacements, plus small changes at the ends of the chain, are required to have occurred during the descent of humans from the common eukaryotic ancestor.

More accurate phylogenetic trees should be generated if the folded conformation of the proteins (Chapter 6) are taken into account, since homologous residues are believed to occupy homologous conformational positions. Minimizing the number of mutational events may not always be a valid method of aligning protein sequences.

Construction of phylogenetic trees. W. M. Fitch and E. Margoliash. Science 155:279–284, 1967.

The origin and evolution of protein superfamilies. M. O. Dayhoff. Fed. Proc. 35:2132–2138, 1976.

Congruency of phylogenies derived from different proteins. E. M. Prager and A. C. Wilson. J. Mol. Evol. 9:45–57, 1976.

	1	5	10	15	20	25
Ancestral	Pro-Ala-Gly-Asp- ? -Lys-Lys-Gly-Ala-Lys-Ile-Phe-Lys-Thr- ? -Cys-Ala-Gln-Cys-His-Thr-Val-Glu- ? -Gly-Gly- ? -
Human	Gly-Asp-Val-Glu-Lys-Gly-Lys-Lys-Ile-Phe-Ile -Met-Lys-Cys-Ser-Gln-Cys-His-Thr-Val-Glu-Lys-Gly-Gly-Lys-

	30	35	40	45	50
His-Lys-Val -Gly-Pro-Asn-Leu-His-Gly-Leu-Phe-Gly-Arg-Lys- ? -Gly-Gln-Ala- ? -Gly-Tyr-Ser-Tyr-Thr-Asp-
His-Lys-Thr-Gly-Pro-Asn-Leu-His-Gly-Leu-Phe-Gly-Arg-Lys-Thr-Gly-Gln-Ala-Pro-Gly-Tyr-Ser-Tyr-Thr-Ala -

	50	60	65	70	75
Ala-Asn-Lys-Asn-Lys-Gly- ?- ?-Trp- ? -Glu-Asn-Thr-Leu-Phe-Glu-Tyr-Leu-Glu-Asn-Pro-Lys-Lys-Tyr-Ile -
Ala-Asn-Lys-Asn-Lys-Gly-Ile-Ile-Trp-Gly-Glu-Asp-Thr-Leu-Met-Glu-Tyr-Leu-Glu-Asn-Pro-Lys-Lys-Tyr-Pro-

	80	85	90	95	100
Pro-Gly-Thr-Lys-Met- ?-Phe- ? -Gly-Leu-Lys-Lys- ? - ? -Asp-Arg-Ala-Asp-Leu-Ile-Ala-Tyr-Leu-Lys- ? -
Pro-Gly-Thr-Lys-Met-Ile-Phe-Val-Gly-Ile -Lys-Lys-Lys-Glu-Glu-Arg-Ala-Asp-Leu-Ile-Ala-Tyr-Leu-Lys-Lys-

Ala-Thr-Ala
-Ala-Thr-Asn-Glu

Figure 3–8

Comparison of the sequences of contemporary human cytochrome c *and that of the common ancestor of the species of Figure 3–7, the lowest node of the phylogenetic tree, inferred from the present-day sequences. Positions for which it is impossible to decide between two or more alternatives are indicated by "?". (From M. O. Dayhoff (ed.), Atlas of Protein Sequence and Structure. Washington, D.C., National Biomedical Research Foundation, 1972.)*

Origins of prokaryotes, eukaryotes, mitochondria, and chloroplasts. R. M. Schwartz and M. O. Dayhoff. Science 199:395–403, 1978.

The phylogeny of prokaryotes. G. E. Fox, et al. Science 209:457–463, 1980.

Testing the theory of evolution by comparing phylogenetic trees constructed from five different protein sequences. D. Penny, et al. Nature 297:197–200, 1982.

Mutations Implied by Divergence

The mutational events responsible for the genetic divergence of proteins may be inferred from the phylogenetic trees and the amino acid replacements, using the genetic code. The mutations that have accumulated in a number of different proteins are very nonrandom, both in the positions in a protein that are altered and in the amino acids that replace each other. The range of variability for the different positions of eukaryotic cytochrome *c* is illustrated in Figure 3–6. The nonrandom nature of the pairs of amino acids that replace each other in a number of different proteins is shown in the matrix of Figure 3–9. The most prevalent replacements occur in amino acids with chemically similar amino acid side chains: Gly/Ala, Ala/Ser, Ser/Thr, Ile/Val/Leu, Asp/Glu, Lys/Arg, Tyr/Phe, and so forth.

This bias in amino acid replacements does not arise from the nature of the genetic code, even though it is not random and even though amino acids differing by single nucleotides in their codons are most likely to replace each other. The second nucleotide of the codons is the most important in specifying the chemical nature of the amino acid; changes in the first interconvert rather similar amino acids, whereas changes in the third usually produce no change in amino acid (see Figure 2–4). The expected distribution with random base-pair changes and random use of the genetic code is shown in the upper half of Figure 3–9; this distribution is remarkably different from that observed.

Expected Number of Random Mutational Events

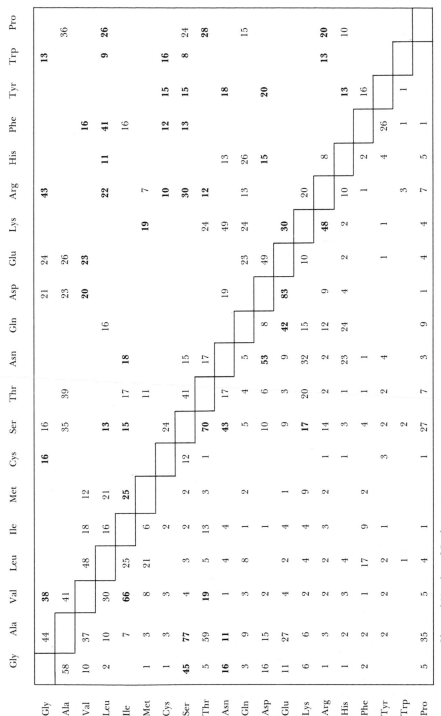

Observed Number of Replacements

Figure 3-9
Relative frequencies of amino acid replacements observed in a total of 1572 examples between closely related proteins (bottom, left) and those expected for random single-site mutations (top, right), calculated from the genetic code and the frequencies of occurrence of the different amino acid residues. The greatest discrepancies are shown in bold type. The observed replacements between chemically similar amino acids are generally much more frequent than expected for random mutations. (Observed replacements compiled by M. O. Dayhoff (ed.), Atlas of Protein Sequence and Structure. Vol. 5, Suppl. 3. Washington, D.C., National Biomedical Research Foundation, 1978.)

The bias for chemically conservative amino acid replacements is not apparent in the rare variants in contemporary populations, like those of human hemoglobin described earlier. These variants reflect primarily the process of mutation, so their occurrence cannot account for the evolutionary divergence.

It is most likely that the bias has been produced by the process of natural selection, which has permitted the survival of only those mutations that did not disrupt the function of the protein. Selection is still in progress with the new mutations present in current populations and will undoubtedly ensure that adverse mutations do not survive. This conclusion is supported by the finding from gene sequences that mutations that do not alter the amino acid, owing to degeneracy of the genetic code, have accumulated during evolution much more frequently than even the chemically conservative mutations. Genetic changes that do not alter the function of the protein have been preserved to a greater extent than those that do.

The numbers of the amino acid replacements shown in Figure 3–9 also reflect the frequency with which each amino acid occurs in proteins. This varies over a wide range (Table 1–1), but the relative evolutionary variation of each amino acid may be calculated from the number of times it has been changed, divided by the number of times it occurs. Normalized values for the 20 amino acids are shown in Table 3–1. The acidic and hydrophilic amino acids are those most frequently changed, whereas the large residues are replaced least. The other conserved residues are Gly, Pro, and Cys, which will be shown to have special roles in protein structure; this factor can largely explain the evolutionary changes that have occurred (Chapter 6).

Preponderance of synonymous changes as evidence for the neutral theory of molecular evolution. M. Kimura. Nature 267:275–276, 1977.
Evolution of the three overlapping gene systems in G4 and ØX174. J. C. Fiddes and G. N. Godson. J. Mol. Biol. 133:19–43, 1979.

Table 3–1   *Relative Mutabilities of the Amino Acid Residues During Evolution*

The number of times a given amino acid residue has been changed during evolution was divided by the number of times it occurred in the proteins considered. The values were then normalized, the highest being set at 100.

Asn	100	His	49
Ser	90	Arg	49
Asp	79	Lys	42
Glu	76	Pro	42
Ala	75	Gly	37
Thr	72	Tyr	31
Ile	72	Phe	31
Met	70	Leu	30
Gln	69	Cys	15
Val	55	Trp	13

Adapted from M. O. Dayhoff (ed.), Atlas of Protein Sequence and Structure. Vol. 5, Suppl. 3. Washington, D.C., National Biomedical Research Foundation, 1978.

Evolutionary nucleotide replacements in DNA. T. H. Jukes and J. L. King. Nature 281:605–606, 1979.

Silent nucleotide substitutions and the molecular evolutionary clock. T. H. Jukes. Science 210:973–978, 1980.

Estimation of evolutionary distances between homologous nucleotide sequences. M. Kimura. Proc. Natl. Acad. Sci. U.S.A. 78:454–458, 1981.

Rates of Evolutionary Divergence

If a number of different proteins from several species are compared, a wide range of interspecies differences is found (Table 3–2). For example, humans and rhesus monkeys differ in 1 per cent of residues for cytochrome c, but in 3 to 5 per cent for hemoglobin α and β polypeptide chains, and in 30 per cent for fibrinopeptides. This same ranking of proteins applies with other species comparisons. Apparently, therefore, proteins have evolved at different rates. Rapidly changing proteins are useful for comparing closely related proteins, whereas slowly changing proteins can reveal the earliest events in evolution.

The specific rate at which a protein has evolved appears to have been approximately constant throughout the evolutionary past (Figure 3–10). Evolutionary change at the molecular level has therefore been proposed to serve as an evolutionary clock, and the times of divergence of other groups may be calculated from such data.

The approximate rates at which different proteins have evolved are given in Table 3–3. The values are expressed as number of single base-pair mutations per position accepted as a function of time. It is analogous to a mutation rate.

How can we explain both this wide range of evolutionary rates in different proteins and the nearly constant rate with time for each protein? It seems most plausible that the observed variation is primarily due to neutral mutations that have not affected significantly the function of the protein and have not been subject to selection. The constant rate of evolutionary divergence would then be the neutral mutation rate per gene copy for that gene—i.e., the rate at which the present-day gene accumulated neutral mutations during its descent from the earliest common ancestor.

This rate would differ for each gene or protein to the extent that neutral mutations would be possible. The possibility of neutral mutations would depend upon how much variation in amino acid sequence was permitted. If the precise amino acid sequence is not critical for the function of the protein, a large fraction of the total mutations would be neutral,

Table 3–2   Relative Extent of Divergence for Several Proteins

| PROTEIN | PER CENT OF RESIDUES DIFFERING FROM HUMAN PROTEIN | | | | | | | | |
	Rhesus monkey	Cow	Pig	Rabbit	Chicken	Frog	Fish	Fruit fly	Yeast
Cytochrome c	1	10	10	9	13	17	20	27	41
Hemoglobin α chain	3	12	13	18	25		50		
Hemoglobin β chain	5	17	16	10	26	46			
Fibrinopeptides A and B	30	70	67	70					

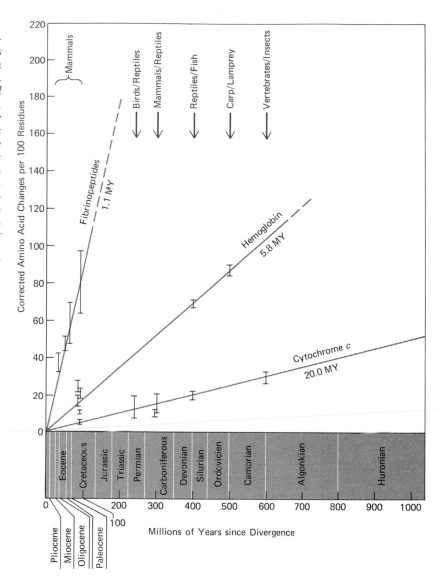

and the protein would have evolved relatively rapidly. Fibrinopeptides appear to be examples, as they seemingly serve primarily to block the aggregation of the precursor fibrinogen. They are cleaved proteolytically by thrombin from the amino ends of two of the three fibrinogen polypeptides in the first step of blood clotting and play no further known role. As a consequence of their removal, the fibrinogen is converted to fibrin, which aggregates and forms the basis of the blood clot. The only known functional constraints on the amino acid sequences of the fibrinopeptides are a carboxyl-terminal Arg residue, required for proteolytic splitting by thrombin, and a somewhat acidic nature, possibly to inhibit aggregation of fibrinogen. Within these minor limitations, many different amino acid

Table 3–3 Rate of Evolution for Some Proteins

PROTEIN	ACCEPTED POINT MUTATIONS PER 100 RESIDUES PER 10^8 YEARS
Histones	
H4	0.25
H3	0.30
H2A	1.7
H2B	1.7
H1	12
Fibrous proteins	
Collagen (α-1)	2.8
Crystallin (αA)	4.5
Intracellular enzymes	
Glutamate dehydrogenase	1.8
Triosephosphate dehydrogenase	5.0
Triosephosphate isomerase	5.3
Lactate dehydrogenase H	5.3
Lactate dehydrogenase M	7.7
Carbonic anhydrase B	25
Carbonic anhydrase C	48
Electron carriers	
Cytochrome c	6.7
Cytochrome b_5	9.1
Plastocyanin	14
Ferredoxin	17
Hormones	
Glucagon	2.3
Corticotropin	4.2
Insulin	7.1
Thyrotropin β chain	11
Lipotropin β chain	13
Lutropin α chain	14
Proparathyrin	14
Prolactin	20
Growth hormone	25
Lutropin β chain	33
Insulin C peptide	53
Oxygen-binding proteins	
Myoglobin	17
Hemoglobin	
α chain	27
β chain	30
Secreted enzymes	
Trypsinogen	17
Lysozyme	40
Ribonuclease	43
Immunoglobulins	
κ chains (V region)	100
κ chains (C region)	111
λ chains (V region)	125
γ chains (V region)	143
Snake venom toxins	
Long neurotoxins	111
Cytotoxins	111
Short neurotoxins	125
Other proteins	
Parvalbumin	20
Albumin	33
Lactalbumin	43
Fibrinopeptide A	59
Casein κ chain	71
Fibrinopeptide B	91

From A. C. Wilson, et al., Ann. Rev. Biochem. 46:573–639, 1977.

sequences are functional, and changes during evolution have occurred rather uniformly and extensively at all positions.

At the other extreme, a protein for which very few amino acid replacements are acceptable would have evolved at a very slow rate. An example of a slowly varying protein is cytochrome c, which must interact with a number of other proteins in its biological function of transferring electrons (Chapter 8). Evolutionary variation has occurred at only some sites (Figure 3–6), and there have been considerable constraints upon which changes were possible.

This is not to say that natural selection has not been important, as it has undoubtedly selected against all adverse mutations. The total mutation rate appears to be relatively constant for all genes in all species. The neutral mutation rate would be the fraction of this total mutation rate that generated neutral, rather then deleterious, mutations.

The neutral mutation rate would be different for each position within the gene or protein. The third nucleotide base pair in many codons is not specific for the amino acid coded and is found to have evolved most rapidly. Some amino acids within a protein are likely to have vital roles for the function of the protein, so that they cannot be changed; the neutral mutation rate here would be zero. Examples are the invariant Cys 17, His 18 and Met 80 of c-type cytochromes (Figure 3–6). Each position within a polypeptide chain is likely to have a unique probability of undergoing a neutral mutation.

Proinsulin illustrates this principle of different rates within a polypeptide chain. The C peptide is removed proteolytically from the middle of the proinsulin chain after it has folded to the correct conformation. The role of the C peptide is presumably to ensure the correct assembly of the molecule; it plays no further known role and is discarded. Other crosslinks are able to serve in the refolding of insulin in vitro. Accordingly, the C-peptide portion of proinsulin has been found to have evolved an order of magnitude more rapidly than the remainder of the molecule, the A and B chains, which make up the functional hormone.

This evolutionary scheme has included no role for positive selection for advantageous changes in proteins during evolution. Their occurrence would be expected to have increased the rate of evolutionary change. Such exceptions are remarkably few, but cytochrome c in the rattlesnake and insulin in the guinea pig and the coypu have diverged significantly faster than expected. Functional differences in guinea pig insulin are known, so there may have been positive selection for favorable mutations. These may then be the exceptions that prove the rule. There are some other known examples of functionally significant evolutionary changes in proteins, as will be noted later; these, however, appear to be a minor fraction of the total evolutionary change.

Evolutionary rate at the molecular level. M. Kimura. Nature 217:624–626, 1968.
The rate of molecular evolution considered from the standpoint of population genetics. M. Kimura. Proc. Natl. Acad. Sci. U.S.A. 63:1181–1188, 1969.
Non-Darwinian evolution. J. L. King and T. H. Jukes. Science 164:788–798, 1969.

Non-Darwinian evolution and biological progress. J. M. Thoday. Nature 255:675–677, 1975.

Is the evolution of insulin Darwinian or due to selectively neutral mutation? T. L. Blundell and S. P. Wood. Nature 257:197–203, 1975.

Molecular evolution: the neutralist–selectionist controversy. H. Harris. Fed. Proc. 35:2079–2082, 1976.

Progress in the neutral mutation–random drift controversy. J. L. King. Fed. Proc. 35:2087–2091, 1976.

Protein evolution and the molecular clock. W. M. Fitch and C. H. Langley. Fed. Proc. 35:2092–2097, 1976.

Are evolutionary rates really variable? J. H. Gillespie and C. H. Langley. J. Mol. Evol. 13:27–34, 1979.

The untranslated regions of β-globin mRNA evolve at a functional rate in higher primates. S. L. Martin, et al. Cell 25:737–741, 1981.

Biochemical peculiarities of the guinea pig and some examples of convergent evolution. J. C. Wriston. J. Mol. Evol. 17:1–9, 1981.

GENE REARRANGEMENTS AND THE EVOLUTION OF COMPLEXITY

Species differ widely in their biological complexity and in their content of DNA; present-day species are also considerably more complex than their ancestors are likely to have been. Most likely, therefore, both genetic and biochemical complexity have increased during evolution. It is difficult to imagine that useful genes could arise de novo; this probably happened once to give rise to the original living organism, but there are no indications that it has happened a second time.

Instead, additional genetic material probably arose by duplication of preexisting genes. With an additional copy of a gene available, the first may continue to provide the original function while the second accumulates mutations that might alter this function. If the second gene evolves to serve yet another useful function, it would tend to be retained in the genome.

The origin by duplication of genes will be apparent from their present sequences only if divergence has been limited, either by a short time since duplication or by constraints imposed by the retention of similar function. In the absence of such contraints, distantly related genes and their protein products are unlikely to be recognizable today as descendants from a common ancestral gene.

Products of Gene Duplication

A wide variety of related proteins are known that obviously are the products of gene duplication. The most striking of these are the globin polypeptide chains of myoglobin and of hemoglobin, which have very similar functions in binding oxygen (Chapter 8). Humans and most other higher organisms have individual genes coding for these very similar polypeptides (Figure 3–11). The single polypeptide chain of myoglobin functions in storing oxygen in muscle, whereas the hemoglobins are tetrameric molecules that carry oxygen within erythrocytes. The polypeptide chains of hemoglobin function at different times of life, and they aggregate to form comparable tetrameric molecules. The ϵ and ζ chains are used in the very young embryo, and the α and γ chains by the fetus at a later stage; the β and δ chains replace the γ chain in the adult. The similarities between the amino acid sequences of the polypeptide chains (Figure 3–11) are suffi-

```
             1                              10                         20
Myoglobin  Gly- - -Leu-Ser -Asp-Gly-Glu-Trp-Gln -Leu-Val-Leu-Asn-Val -Trp-Gly-Lys-Val -Glu-Ala-Asp-Ile -Pro -Gly -His -Gly-Gln-Glu-Val-
       α   Val- - -Leu-Ser -Pro-Ala-Asp-Lys -Thr-Asn-Val-Lys -Ala -Trp-Gly-Lys-Val -Gly -Ala-His -Ala-Gly -Gln-Tyr-Gly-Ala-Glu-Ala-
       β   Val-His-Leu-Thr-Pro -Glu-Glu-Lys -Ser -Ala -Val-Thr-Ala -Leu-Trp-Gly-Lys-Val -Asn- - - -   -Val-Asp-Glu-Val -Gly-Gly-Glu-Ala-
      Gγ   Gly-His-Phe-Thr-Glu-Glu-Asp-Lys -Ala -Thr-Ile -Thr-Ser -Leu-Trp-Gly-Lys-Val -Asn- - - -   -Val-Glu-Asp-Ala -Gly-Gly-Glu-Thr-
       δ   Val-His-Leu-Thr-Pro -Glu-Glu-Lys -Thr-Ala -Val-Asn-Ala -Leu-Trp-Gly-Lys-Val -Asn- - - -   -Val-Asp-Ala -Val -Gly-Gly-Glu-Ala-
       ε   Val-His-Phe-Thr-Ala -Glu-Glu-Lys -Ala -Ala -Val-Thr-Ser -Leu-Trp-Ser-Lys-Met-Asn- - - -   -Val-Glu-Glu-Ala -Gly-Gly-Glu-Ala-
```

```
  30                        40                          50                             60
-Leu-Ile  -Arg-Leu-Phe-Lys -Gly-His -Pro-Glu -Thr-Leu-Glu -Lys -Phe-Asp-Lys-Phe-Lys-His -Leu-Lys-Ser -Glu-Asp-Glu-Met-Lys -Ala-Ser -Glu-
-Leu-Glu -Arg-Met-Phe-Leu-Ser-Phe-Pro-Thr-Thr-Lys -Thr-Tyr-Phe-Pro-His-Phe- -  -Asp-Leu-Ser-His - - - - - - - - -Gly-Ser -Ala-
-Leu-Gly -Arg-Leu-Leu-Val -Val-Tyr-Pro-Trp-Thr-Gln -Arg-Phe-Phe-Glu-Ser-Phe-Gly-Asp-Leu-Ser-Thr-Pro-Asp-Ala-Val -Met-Gly-Asn-Pro-
-Leu-Gly -Arg-Leu-Leu-Val -Val-Tyr-Pro-Trp-Thr-Gln -Arg-Phe-Phe-Asp-Ser-Phe-Gly-Asn-Leu-Ser-Ser -Ala-Ser -Ala-Ile -Met-Gly-Asn-Pro-
-Leu-Gly -Arg-Leu-Leu-Val -Val-Tyr-Pro-Trp-Thr-Gln -Arg-Phe-Phe-Glu-Ser-Phe-Gly-Asp-Leu-Ser-Ser -Pro -Asp-Ala-Val -Met-Gly-Asn-Pro-
-Leu-Gly -Arg-Leu-Leu-Val -Val-Tyr-Pro-Trp-Thr-Gln -Arg-Phe-Phe-Asp-Ser-Phe-Gly-Asn-Leu-Ser-Ser -Pro -Ser -Ala-Ile -Leu-Gly-Asn-Pro-
```

```
            70                            80                          90
-Asp-Leu-Lys-Lys-His-Gly-Ala-Thr-Val-Leu-Thr-Ala-Leu-Gly -Gly -Ile -Leu-Lys-Lys -Lys -Gly -His -His -Glu-Ala -Glu -Ile -Lys-Pro -Leu-Ala-
-Gln -Val-Lys -Gly-His-Gly-Lys-Lys -Val-Ala -Asp-Ala-Leu-Thr-Asn-Ala-Val -Ala-His -Val -Asp-Asp-Met-Pro-Asn-Ala -Leu-Ser-Ala -Leu-Ser-
-Lys -Val-Lys -Ala-His-Gly-Lys-Lys -Val-Leu-Gly -Ala-Phe-Ser -Asp-Gly-Leu-Ala-His -Leu-Asp-Asn-Leu-Lys-Gly -Thr-Phe-Ala-Thr-Leu-Ser-
-Lys -Val-Lys -Ala-His-Gly-Lys-Lys -Val-Leu-Thr-Ser-Leu-Gly -Asp-Ala-Ile -Lys-His -Leu-Asp-Asp-Leu-Lys-Gly -Thr-Phe-Ala-Gln -Leu-Ser-
-Lys -Val-Lys -Ala-His-Gly-Lys-Lys -Val-Leu-Gly -Ala-Phe-Ser -Asp-Gly-Leu-Ala-His -Leu-Asp-Asn-Leu-Lys-Gly -Thr-Phe-Ser-Gln -Leu-Ser-
-Lys -Val-Lys -Ala-His-Gly-Lys-Lys -Val-Leu-Thr-Ser-Phe-Gly -Asp-Ala-Ile -Lys-Asn-Met-Asp-Asn-Leu-Lys-Pro -Ala -Phe-Ala-Lys -Leu-Ser-
```

```
              100                      110                           120
-Gln -Ser -His-Ala-Thr-Lys-His -Lys -Ile -Pro -Val-Lys-Tyr-Leu-Glu-Phe-Ile -Ser-Glu-Cys-Ile -Ile -Gln -Val -Leu-Gln-Ser -Lys -His -Pro-Gly-
-Asp-Leu-His-Ala-His -Lys-Leu-Arg-Val-Asp-Pro-Val-Asn-Phe-Lys -Leu-Leu-Ser-His -Cys-Leu-Leu-Val -Thr-Leu-Ala-Ala -His -Leu-Pro-Ala-
-Glu-Leu-His-Cys-Asp-Lys-Leu-His -Val-Asp-Pro-Glu-Asn-Phe-Arg-Leu-Leu-Gly-Asn-Val-Leu-Val -Cys -Val -Leu-Ala-His -His -Phe-Gly-Lys-
-Glu-Leu-His-Cys-Asp-Lys-Leu-His -Val-Asp-Pro-Glu-Asn-Phe-Lys -Leu-Leu-Gly-Asn-Val-Leu-Val -Thr-Val -Leu-Ala-Ile -His -Phe-Gly-Lys-
-Glu-Leu-His-Cys-Asp-Lys-Leu-His -Val-Asp-Pro-Glu-Asn-Phe-Arg-Leu-Leu-Gly-Asn-Val-Leu-Val -Cys -Val -Leu-Ala-Arg-Asn-Phe-Gly-Lys-
-Glu-Leu-His-Cys-Asp-Lys-Leu-His -Val-Asp-Pro-Glu-Asn-Phe-Lys -Leu-Leu-Gly-Asn-Val-Met-Val -Ile -Ile -Leu-Ala-Thr-His -Phe-Gly-Lys-
```

```
               130                      140                          150
-Asp-Phe-Gly -Ala-Asp-Ala -Gln-Gly-Ala-Met-Asn-Lys-Ala -Leu-Glu -Leu-Phe-Arg-Lys -Asp-Met-Ala -Ser-Asn-Tyr-Lys -Glu-Leu Gly-Phe-Gln-Gly
-Glu-Phe-Thr-Pro-Ala -Val -His -Ala-Ser-Leu-Asp-Lys-Phe-Leu-Ala -Ser -Val -Ser  Thr-Val -Leu-Thr-Ser-Lys -Tyr-Arg
-Glu-Phe-Thr-Pro-Pro -Val -Gln-Ala  Ala-Tyr-Gln -Lys-Val -Val -Ala -Gly -Val -Ala-Asn-Ala -Leu-Ala -His-Lys -Tyr-His
-Glu-Phe-Thr-Pro-Gln -Val -Gln-Ala-Ser-Trp-Gln  Lys-Met-Val -Thr-Gly-Val -Ala -Ser -Ala -Leu-Ser  Ser-Arg-Tyr-His
-Glu Phe-Thr-Pro-Glu-Met-Gln-Ala-Tyr-Gln  Lys-Val -Val -Ala -Gly -Val -Ala-Asn-Ala -Leu-Ala -His-Lys -Tyr-His
-Glu-Phe-Thr-Pro-Glu-Val -Gln-Ala-Ala-Trp-Gln -Lys-Leu Val -Ser -Ala -Val-Ala-Ile -Ala -Leu-Ala -His-Lys -Tyr-His
```

Figure 3–11
Amino acid sequences of human myoglobin and the α, β, γ, δ, and ε chains of hemoglobin. The ^{A}γ chain differs from ^{G}γ only in having Ala rather than Gly at position 138. The sequences have been aligned to maximize the homology between them, introducing deletions (-) where necessary. (Sequences of myoglobin and α and β chains from M. O. Dayhoff (ed.), Atlas of Protein Sequence and Structure. Washington, D.C., National Biomedical Research Foundation, 1972. Sequences of γ, δ, and ε chains obtained from the gene sequences given in J. L. Slightom, et al., Cell 21:627–638, 1980; in R. A. Spritz, et al., Cell 21:639–646, 1980; and in F. E. Baralle, et al., Cell 21:621–626, 1980.)

ciently striking to make it obvious that they must have arisen by gene duplication from a common ancestor.

The probable sequence of events may be inferred from the relative similarities of the sequences (Figure 3–12). The myoglobin sequence is least like the others, so it probably diverged first by duplication of the ancestral globin gene, to give the ancestral myoglobin and ancestral hemoglobin genes (Figure 3–13). The latter probably duplicated again at a later

	α	β	γ	δ	ε
Myoglobin	38	36	36	37	34
α		64	59	63	55
β			107	136	110
γ				105	116
δ					106

Figure 3–12
Numbers of amino acid identities between the amino acid sequences of the human myoglobin and hemoglobin α, β, ^{G}γ, δ, and ε polypeptide chains.

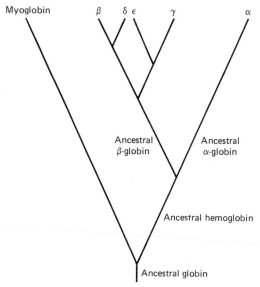

Figure 3–13
Probable evolutionary scheme for the divergence of the globin chains, inferred from the relative similarities of their gene and protein sequences. The separation of the ^{A}γ and ^{G}γ genes, which differ at seven nucleotides and one amino acid residue, would be too recent to be visible on this scheme. (Adapted from A. Efstratiadis, et al., Cell 21:653–668, 1980.)

time and gave rise to the ancestral α-globin gene and that of the β-type genes (β, γ, δ, and ε). The next duplication of the latter would have produced the line to the γ and ε genes, which diverged by a further duplication later on, and to the β and δ genes, which also diverged later.

The evolutionary scheme of Figure 3–13 is for the evolutionary divergence of duplicated genes *within* a species, whereas that of Figure 3–7 is of divergence *between* species. The two phenomena of gene duplication and species divergence must be kept separate, although there are many confusing ancestral trees in the literature with both types of divergence included. It seems more appropriate to use different dimensions for the two processes, as is illustrated in Figure 3–14.

When species diverge, they appear to retain initially most, if not all,

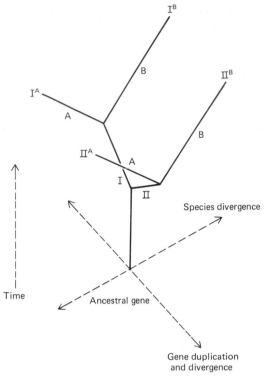

Figure 3–14

Simple example of a three-dimensional phylogenetic tree showing both gene duplication and species divergence. The ancestral gene in the ancestral species was first duplicated to give two homologous genes, I and II. The ancestral species then diverged to give two species A and B. Both genes I and II would also separate at this stage, although either could be lost at any time. Further gene duplication also could occur along any branch. Each gene will independently accumulate mutational alterations along its branch; the differences between any two contemporary genes would be proportional to the length of the branches joining them. This example, with just one duplication followed by one species divergence, and no loss of genes, would yield at the present time four genes: I^A and II^A (species A) and I^B and II^B (species B).

of the genome. Therefore, each of the globin genes present in Figure 3–14 at the time of species divergence must be separated in that dimension, although it is possible that one or more of the genes present initially might be lost in either species during its subsequent evolution. Comparisons of species and gene divergence are good checks on the consistency of both schemes.

Such comparisons also serve to confirm that evolutionary divergence of genes depends primarily upon the time since they diverged, irrespective of whether they have been present within the same or different species. For example, the human and dog α chains are nearly equally similar to the β, γ, and myoglobin chains of both species (Table 3–4). Other such comparisons give the same result—i.e., that expected if all these genes

Table 3–4   Paralogous Comparisons of the Globin Chains

COMPARISON PROTEIN	NUMBER OF DIFFERENCES IN AMINO ACID SEQUENCE	
	Human α-globin	Dog α-globin
Human β-globin	84	84
Human γ-globin	89	83
Human myoglobin	115	119
Dog β-globin	88	87
Dog myoglobin	115	118

The globin genes duplicated and diverged long before the human and dog species did, so comparisons between the different globins are relevant to the time since the genes duplicated. Consequently, very similar results are obtained irrespective of which species is used.

accumulated mutations at the same constant rate since the time of gene duplication.

The presence of several related genes and proteins within an individual presents potential pitfalls in evolutionary studies. For example, the α and β chains of hemoglobin are the products of a gene duplication that was fixed long before mammals diverged; since the time of the gene duplication, both sequences have been evolving independently in each species. Comparison of the α chain of one mammal with the β chain of another would involve comparison of two sequences whose time of divergence was much greater than that of the species being compared. Genes, as well as their protein products, that occupy the same genetic locus in different species are said to be **orthologous,** whereas those at different genetic sites are **paralogous.** Phylogenies can be reconstructed only by comparing orthologous genes. Comparisons of paralogous genes or proteins from different species are related not to the time since divergence of the species (Table 3–4) but to the time since the gene duplication.

The preceding discussion was somewhat simplified, in that it ignored the existence of two distinct human γ chains, one with Gly at position 136 (^{G}γ) and the other with Ala (^{A}γ); there are two corresponding genes on each chromosome. Recent analysis of the structures of all the hemoglobin genes has found that there are also two α genes, but they are probably identical and certainly produce identical polypeptide chains. Duplication of the α genes is common in primates and appears to have occurred before they diverged. Yet the duplicate genes in individual species are indistinguishable, and the sequences of the intervening regions, or **introns,** and of the flanking noncoding regions are also very similar. They have diverged much less than would be expected if they had accumulated mutations since duplication, so mechanisms for comparing and correcting duplicate genes have been suggested. Consequently, duplicated genes may begin to diverge only at a time considerably after duplication, when they escape from such correction mechanisms.

The DNA sequence studies have also uncovered **pseudogenes,** extra genes that are not expressed. They are distinctly homologous to the various globin genes but have a variety of nucleotide changes that must prevent formation of an identifiable globin-like polypeptide chain. In particular, there are insertions and deletions that would produce the wrong reading frame and termination codons. These pseudogenes have apparently arisen in many cases by integration of complementary DNA copies of messenger RNAs. Such duplicated genes, even if inactive, may provide a wealthy genetic store of genes that might evolve to produce useful products.

The globins are examples of gene duplications in which the functional properties of the proteins have been closely conserved (Chapter 8); such preservation of function is not necessary, however, and varying degrees of functional divergence are being uncovered in other examples of duplicated genes. For example, the serine proteases trypsin, chymotrypsin, elastase, and thrombin have undoubtedly arisen by gene duplication; yet they have markedly different substrate preferences (Chapter 8). Human haptoglobin is also homologous to the serine proteases, even though it has no proteolytic activity; its function is to bind to $\alpha\beta$ complexes of the hemoglobin chains.

Chicken lysozyme from eggs is quite homologous to α-lactalbumin from bovine milk: Of 123 residues, 47 are identical, and many others are similar in chemical properties. But lysozyme functions to degrade the polysaccharides of bacterial cell walls, whereas α-lactalbumin regulates lactose synthesis by binding to a membrane-bound galactosyl transferase. The latter enzyme normally glycosylates proteins in the Golgi apparatus but, when complexed with α-lactalbumin, synthesizes lactose. The only apparent similarity in the functions of these two proteins is that they both act in reactions involving saccharides, although it is not clear that α-lactalbumin actually binds them.

Many other proteins may be related in this way, but both time and selective pressures may have eroded their original amino acid sequence similarities so that their common ancestry is not obvious today.

Evolution of structure and function of proteases. H. Neurath, et al. Science 158:1638–1644, 1967.

Homologies in serine proteases. B. S. Hartley. Philos. Trans. R. Soc. Lond. [Biol.] 257:77–87, 1970.

Evolution by Gene Duplication. S. Ohno. London, Allen & Unwin, 1970.

Lactose synthetase. R. L. Hill and K. Brew. Adv. Enzymol. 43:411–490, 1975.

The appearance of new structures and functions in proteins during evolution. E. Zuckerkandl. J. Mol. Evol. 7:1–57, 1975.

The structure and evolution of the human β-globin gene family. A. Efstratiadis, et al. Cell 21:653–668, 1980.

The molecular genetics of human hemoglobins. T. Maniatis, et al. Ann. Rev. Genetics 14:145–178, 1980.

Rapid duplication and loss of genes coding for the α chains of hemoglobin. E. A. Zimmer, et al. Proc. Natl. Acad. Sci. U.S.A. 77:2158–2162, 1980.

Covalent structure of human haptoglobin: a serine protease homolog. A. Kurosky, et al. Proc. Natl. Acad. Sci. U.S.A. 77:3388–3392, 1980.

```
      1                              10
  Ala-Tyr-Lys-Ile-    -Ala-Asp-Ser-Cys-Val-Ser-Cys-Gly-Ala-
  -Ile-Phe-Val-Ile-Asp-Ala-Asp-Thr-Cys-Ile-Asp-Cys-Gly-Asn-
   30                              40
```

```
               20
  Cys-Ala-Ser-Glu-Cys-Pro-Val-Asn-Ala-Ile-Ser-Gln-Gly-Asp-Ser-
  Cys-Ala-Asp-Val-Cys-Pro-Val-Gly-Ala-Pro-Val-Gln-Glu
               50
```

Figure 3–15
Homology between the two halves of the amino acid sequence of Clostridium pasteurianum
ferredoxin. Identical residues are enclosed in solid lines; similar residues, in dashed lines.

Table 3–5   Proteins with Internal Duplications

PROTEIN	TOTAL NUMBER OF RESIDUES IN POLYPEPTIDE CHAIN	NUMBER OF RESIDUES IN REPEATED UNIT	NUMBER OF REPETITIVE UNITS IN POLYPEPTIDE
Mammalian collagen	1052	3	337
Human lipid-binding proteins			
A-1	245	11	18
C-1	57	25	2
Sheep keratin, high-sulfur	171	10	13
fractions	131	10	12
Emu β-keratin	102	10	3
Rabbit troponin C	159	38	4
Rabbit myosin A2 light chain	149	38	4
Immunoglobulin heavy chain (μ)	452	108	4
Bovine sperm histone	47	8	3
Rat testis–specific basic protein	54	17	2
Human serum albumin	584	195	3
Prothrombin	582	105	2
Chlorobium ferredoxin–1	60	28	2

A history of the human fetal globin gene duplication. S.-H. Shen, et al. Cell
26:191–203, 1981.

Evolution of the β-globin gene cluster in man and the primates. P. A. Barrie, et
al. J. Mol. Biol. 149:319–336, 1981.

Similar amino acid sequences: chance or common ancestry? R. F. Doolittle. Science
214:149–159, 1981.

Gene Elongation by Duplication

If duplication occurs within a gene, it will be elongated, with part or all of
the amino acid sequence duplicated so long as the reading frame of the
gene sequence is maintained. The two parts will then diverge, accumulating mutational changes. The most striking instance of this is seen in the
ferredoxins, where the homology between the two halves of the amino acid
sequence is quite apparent (Figure 3–15). Other instances of probable gene
duplication are given in Table 3–5.

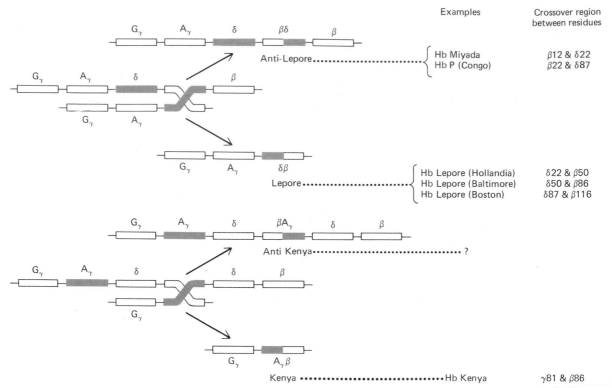

Figure 3–16
Genetic mechanism for production of hybrid proteins by recombination between homologous genes. The chromosomal arrangements of the β-like globin genes are depicted, with recombination between the δ and β genes at the top and between the ^Aγ and β genes at the bottom. These events are inferred from the amino acid sequences of the hybrid proteins produced. (From D. J. Weatherall and J. B. Clegg, Cell 16:467–479, 1979; © The MIT Press.)

Partial gene duplications in which the reading frame is altered will not be apparent from the protein primary structure. They could be evident from the gene sequence if divergence had not proceeded too far, but there are no known examples.

Repeating sequences and gene duplications in proteins. A. D. McLachlan. J. Mol. Biol. 64:417–437, 1972.
Origin of periodic proteins. M. Ycas. Fed. Proc. 35:2139–2140, 1976.
Gene duplications in the structural evolution of chymotrypsin. A. D. McLachlan. J. Mol. Biol. 128:49–79, 1979.
Original domain for the serum albumin family arose from repeated sequences. S. Ohno. Proc. Natl. Acad. Sci. U.S.A. 78:7657–7661, 1981.

Genetic Consequences of Duplicated Genes

With identical or very similar nucleotide sequences along the chromosome, mispairing between them on two such chromosomes often occurs, leading to unequal crossing-over. The number of genes will be increased on one of the chromosomes but decreased on the other. Duplicated chromosomal segments are considered to be relatively unstable genetically for this reason.

Hybrid proteins may also result from such nonhomologous crossing-over. Two well-known examples occur in hemoglobins Lepore and Kenya. In the first, crossing-over between β and δ genes produced a gene and a polypeptide chain that is β-like at one end and δ-like at the other. In the case of hemoglobin Kenya, the gene and protein are hybrids of β and $^A\gamma$ (Figure 3–16).

Gene Fusion and Division

Gross rearrangements of the order of genes on chromosomes also occur during evolution, but this is primarily of genetic interest. More subtle changes also have occurred, such as the fusion of genes to produce a single polypeptide chain with the functions of both of the original gene products. The reverse process of gene splitting may have occurred instead, and it is impossible to determine which sequence of events took place; however, gene fusion is more compatible with the gradual evolution of genetic complexity.

For example, the final enzyme in the tryptophan biosynthetic pathway, tryptophan synthetase, is composed of two polypeptide chains, α and β, in prokaryotes. The individual proteins, α and the β_2 dimer, catalyze two sequential reactions:

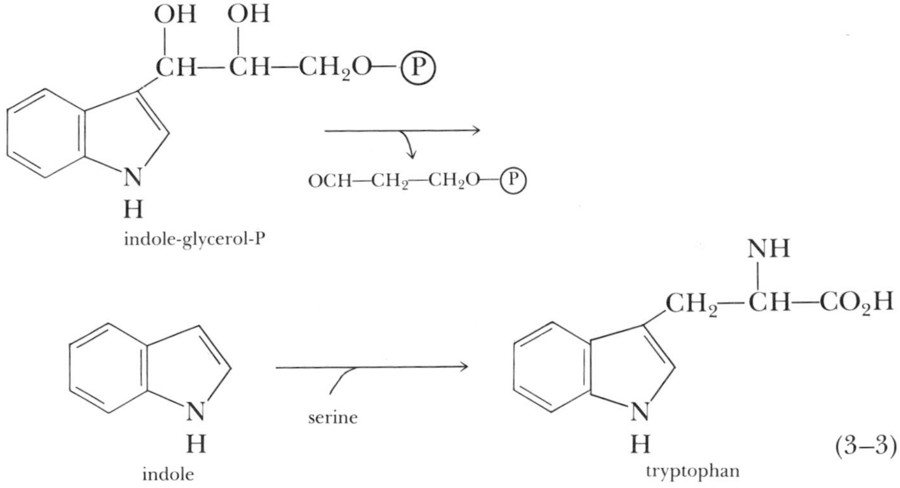

$$\tag{3-3}$$

However, the normal functional tryptophan synthetase is an $\alpha_2\beta_2$ complex that catalyzes both steps in concert; indole is not a detectable intermediate, probably because this very nonpolar substance would tend to leak from the cells through the membrane. Fungi, such as yeast, have refined this reaction even further, apparently having fused the structural genes for the α and β polypeptides to produce a single polypeptide chain. This chain dimerizes to form an enzyme with functions like those of the bacterial enzyme. One end of the fused polypeptide chain appears to be like the bacterial α chain; the other end, like the β chain.

Various bacteria differ similarly with respect to the enzymes for the earlier steps in this pathway. There are five further enzymatic functions required for these steps in tryptophan biosynthesis, which are catalyzed by five different polypeptide chains in some bacteria, encoded by five different genes. However, in bacteria like *E. coli,* two pairs of these genes are fused to produce two different bifunctional polypeptides. In both instances, the two functions are carried out by the corresponding two regions of the single polypeptide chains. In at least one instance, the two proteins of the chain may be cleaved by treatment of the native bifunctional protein with proteases; the two fragments each carry out one of the enzymatic functions. It seems a general rule that the individual polypeptide segments maintain their functional and structural integrity; the fused proteins consist of two independent units linked merely by a bit of polypeptide chain. The functional reason for such gene fusions is usually not obvious; they may tend to keep two sequential enzymes in close proximity in the cell, in order to utilize efficiently the intermediate. However, other instances are known in which the two enzymes fused do not act sequentially.

Gene rearrangements in the evolution of the tryptophan biosynthetic pathway. I. P. Crawford. Bacteriol. Rev. 39:87–120, 1975.
Gene fusion during the evolution of the tryptophan operon in Enterobacteriaceae. G. F. Miozzari and C. Yanofsky. Nature 277:486–489, 1979.
Tryptophan synthetase: structure, function, and subunit interaction. E. W. Miles. Adv. Enzymol. 49:127–186, 1979.
Limited proteolysis of *N*-(5'-phosphoribosyl)anthranilate isomerase: indoleglycerol phosphate synthase from *Escherichia coli* yields two different enzymically active, functional domains. K. Kirschner, et al. J. Mol. Biol. 143:395–409, 1980.
Favin versus concanavalin A: circularly permuted amino acid sequences. B. A. Cunningham, et al. Proc. Natl. Acad. Sci. U.S.A. 76:3218–3222, 1979.

GENETIC MANIPULATION OF PROTEIN STRUCTURE

Genetic mutation provides an extremely specific method of modifying protein structures. A combination of genetic analysis and structural and functional characterization of the protein product may provide a wealth of data about the roles of the individual amino acid residues in the function of the protein.

Genetic probes of enzyme structure. M. J. Schlesinger. In The Enzymes, 3rd ed. P. D. Boyer (ed.). Vol. 1, pp. 241–266. New York, Academic Press, 1970.

Mutant Proteins

The mutant forms of a protein that inevitably occur in a large population can provide a wide variety of modifications in the protein molecule, as in the case of the more than 300 human hemoglobin variants discovered thus far. These mutations have a wide range of effects on the hemoglobin molecule, including instability and precipitation, polymerization as in the case of sickle cell hemoglobin, dissociation of the normal $\alpha_2\beta_2$ tetramer into $\alpha\beta$ dimers, decreased affinity for the heme group, increased susceptibility to oxidation of the heme iron atom, and altered oxygen affinities. On the other hand, many mutant forms have no apparent change in their functional properties. This notable diversity in effect indicates that the various residues have greatly varying roles in the function of hemoglobin, as will be described in Chapter 8.

The practical difficulty with any study using such natural mutants is that they occur very rarely, the normal mutation rate being quite low. The large number of hemoglobin variants detected is due to their medical importance and to widespread screening programs that include persons with no clinical evidence of hemoglobin dysfunction. Such a selection is unlikely to become available for many other proteins.

Genetic studies are much more profitable with cell cultures and with microorganisms. The mutation rate may be increased dramatically, and in specific directions, by using physical and chemical mutagens. Mutants with the desired properties may also be selected. Such extensive genetic screening has been carried out with the tryptophan synthetase α subunit of *E. coli* by Yanofsky and colleagues. A substantial number of mutant α subunits in which one of the 268 amino acid residues has been changed and the enzymatic activity consequently altered have been isolated; the amino acid changes that have occurred are summarized in Figure 3–17. Only eight residues have been found to be altered individually, but these changes have been observed to recur in independent experiments, using different mutagens; the limited number of mutants, therefore, does not reflect a limited number of mutations. Instead, it is only when one of these eight residues has been changed to certain other amino acids that the α subunit is inactivated. "Mis-sense" mutations are likely to have occurred at all other positions, but the mutant proteins must be sufficiently active to escape detection by the selection procedure. Mutations that produce chain termination, insertions, and deletions that shift the reading frame are all found throughout the gene. Such analysis implies that these eight amino acid residues normally are vital for the function of the protein, whereas the other 260 residues may be altered with little effect.

The bacteria that produce these inactive mutant proteins may be further mutated, to determine which amino acid replacements will restore the activity of the α subunit. Some of these revertant mutations restore the original amino acid, whereas many insert others, usually with chemical properties similar to those of the original (Figure 3–17). Somewhat surprisingly, some of the second mutations occur at sites different from that of the first, to produce proteins with two amino acid changes; the second mutation must compensate in some way for the first (Table 3–6).

Other properties may be selected in mutant proteins, such as enhanced susceptibility to high temperatures. Many genetic manipulations may also be performed combining or separating individual mutations. Hybrid proteins may also be constructed genetically, using recombination between genes for different proteins or for the same protein from different species. Hybrid tryptophan synthetase molecules of *E. coli* and *Salmonella typhimurium* parentage have been found to be fully functional in vivo, suggesting that the considerable evolutionary divergence between these two species was of little functional significance.

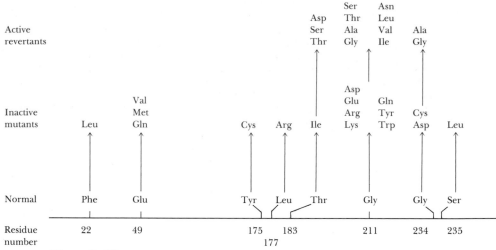

Figure 3–17
Genetic amino acid replacements that affect substantially the activity of the tryptophan syn-
thetase α *subunit of* Escherichia coli. *The positions in the primary structure of the eight*
positions which may be altered to abolish the enzyme activity are shown at the bottom, with
the normal amino acid residues indicated. The replacements that individually abolish enzyme
activity are shown above. Further individual amino acid replacements at these positions that
restore activity are shown at the top. (Data kindly provided by C. Yanofsky; see J. Mol. Biol.
86:775–784, 1974.)

Table 3–6   Mutant and Revertant Tryptophan Synthetase α Subunits

STRAIN	AMINO ACID SEQUENCE	ENZYME ACTIVITY
	175 177 211 213	
Wild type	-Tyr-Leu-Leu-.................-Gly-Phe-Gly-	+
↓		
Mutant	-Tyr-Leu-Leu-.................**-Glu**-Phe-Gly-	−
↓		
Revertant	-Tyr-Leu-Leu-.................-**Val**-Phe-Gly-	+
↓		
Mutant	-Tyr-Leu-Leu-.................-Val-Phe-**Val**-	−
↓		
Revertant	-Tyr-Leu-Leu ....-Val-Phe-**Gly**-	+
Revertant	-Tyr-Leu-Leu-.................-Val-Phe-**Ala**-	+
Revertant	-Tyr-Leu-Leu-.................-**Gly**-Phe-Val-	+
Revertant	-Tyr-Leu-Leu-.................-**Ala**-Phe-Val-	+
Revertant	-Tyr-Leu-**Arg**-.................-Val-Phe-Val-	+
Recombinant	-Tyr-Leu-**Arg**-.................-Gly-Phe-Gly-	−
Wild type	-Tyr-Leu-Leu-.................-Gly-Phe-Gly-	+
↓		
Mutant	-Tyr-Leu-Leu-.................**-Glu**-Phe-Gly-	−
↓		
Revertant	**-Cys**-Leu-Leu-.................-Glu-Phe-Gly-	+
Recombinant	**-Cys**-Leu-Leu-.............-Gly-Phe-Gly-	−

Data from C. Yanofsky, JAMA 218:1026–1038, 1971.

Molecular pathology of human haemoglobin. M. F. Perutz and H. Lehmann. Nature 219:902–909, 1968.

Tryptophan synthetase. C. Yanofsky and I. P. Crawford. In The Enzymes, 3rd ed. P. D. Boyer (ed.). Vol. 7, pp. 1–31, New York, Academic Press, 1972.

Tryptophan synthetase α chain positions affected by mutations near the ends of the genetic map of trpA of *Escherichia coli.* C. Yanofsky and V. Horn. J. Biol. Chem. 247:4494–4498, 1972.

Functional significance of the evolutionary divergence between the tryptophan operons of *Escherichia coli* and *Salmonella typhimurium.* T. E. Creighton. J. Mol. Evol. 4:121–137, 1974.

Structure and properties of a hybrid tryptophan synthetase α chain produced by genetic exchange between *Escherichia coli* and *Salmonella typhimurium.* C. Yanofsky, S. S. L. Li, et al. Proc. Natl. Acad. Sci. U.S.A. 74:286–290, 1977.

Procedure for production of hybrid genes and proteins and its use in assessing significance of amino acid differences in homologous tryptophan synthetase α polypeptides. W. P. Schneider, et al. Proc. Natl. Acad. Sci. U.S.A. 78: 2169–2173, 1981.

Genetic assignment of resonances in the NMR spectrum of a protein: *lac* repressor. M. A. C. Jarema, et al. Proc. Natl. Acad. Sci. U.S.A. 78:2707–2711, 1981.

Simulation of Evolution

The use of microorganisms, with their rapid growth rates and short generation times, permits evolutionary-type processes to be simulated in the laboratory. Mutagens can be used to increase the mutation rate, and intense selective conditions can be applied. Population studies are generally made in a chemostat, where a growing culture is constantly diluted with fresh medium, so that a steady state is attained. When the culture is not growing under optimal conditions, better-adapted mutant strains with shorter generation times will take over.

One such study by Hartley and colleagues has attempted to alter the substrate specificity of the enzyme ribitol dehydrogenase of *Klebsiella aerogenes* by growing the bacteria on xylitol as sole carbon source. This enzyme normally catalyzes the conversion of ribitol to D-ribulose:

$$
\begin{array}{ccc}
\begin{array}{c}
CH_2OH \\
| \\
HCOH \\
| \\
HCOH \\
| \\
HCOH \\
| \\
CH_2OH \\
\text{ribitol}
\end{array}
&
\xrightarrow[]{NAD \qquad NADH}
&
\begin{array}{c}
CH_2OH \\
| \\
C{=}O \\
| \\
HCOH \\
| \\
HCOH \\
| \\
CH_2OH \\
\text{D-ribulose}
\end{array}
\end{array}
\qquad (3\text{--}4)
$$

Ribulose can be metabolized further, so that ribitol can serve as sole carbon source for growth. No specific enzyme is present in *K. aerogenes* for the utilization of the similar sugar xylitol, but ribitol dehydrogenase can use it as a substrate, although very poorly:

$$\begin{array}{ccc}
\text{CH}_2\text{OH} & & \text{CH}_2\text{OH} \\
| & & | \\
\text{HCOH} & \xrightarrow{\text{NAD} \quad \text{NADH}} & \text{C}=\text{O} \\
| & & | \\
\text{HOCH} & \longleftrightarrow & \text{HOCH} \\
| & & | \\
\text{HCOH} & & \text{HCOH} \\
| & & | \\
\text{CH}_2\text{OH} & & \text{CH}_2\text{OH} \qquad (3\text{--}5)
\end{array}$$

xylitol   D-xylulose

Because the product xylulose is normally metabolized further, xylitol can be used as sole carbon source, but under these conditions the bacteria grow very slowly, because xylitol is such a poor substrate for ribitol dehydrogenase.

Since xylitol differs only slightly from ribitol, it was hoped to alter by mutation the substrate specificity of ribitol dehydrogenase. When the bacteria were grown slowly on xylitol in a chemostat, mutants with increased growth rates were found to take over the cultures. However, these mutants simply produced more of the normal ribitol dehydrogenase by duplicating its structural gene. After several steps, the bacteria produced 20 per cent of their total protein as this enzyme.

Mutant forms of ribitol dehydrogenase that utilized xylitol more readily were found when the mutation rate was increased so that multiple mutations within the gene became significant. It appears that more than one amino acid residue must be changed before the substrate specificity of this enzyme can be altered.

This study confirmed the importance of gene duplication in evolution. A similar finding of extensive gene duplication was made by Schimke and colleagues when eukaryotic cells were cultured in the presence of an anticancer drug, methotrexate, which is an inhibitor of the enzyme dihydrofolate reductase.

Another evolutionary study by Clarke and colleagues using a different enzyme found changing its substrate specificity somewhat easier. An amidase of normal *Pseudomonas aeruginosa* permits utilization of either acetamide or propionamide as sole nitrogen or carbon source, or both:

$$\text{R}-\overset{\displaystyle \text{O}}{\overset{\|}{\text{C}}}-\text{NH}_2 + \text{H}_2\text{O} \longrightarrow \text{R}-\overset{\displaystyle \text{O}}{\overset{\|}{\text{C}}}\text{OH} + \text{NH}_3 \qquad (3\text{--}6)$$

where R is either CH_3- or CH_3CH_2-. Altered amidases with different substrate specificities (i.e., larger R groups) were isolated by selecting for the ability to utilize the appropriate amides. One mutant that could utilize *N*-phenyl acetamide as carbon source had an altered amidase that catalyzed the reaction

$$\text{CH}_3-\overset{\displaystyle \text{O}}{\overset{\|}{\text{C}}}-\text{NH}-\hspace{-2pt}\bigcirc + \text{H}_2\text{O} \longrightarrow \text{CH}_3\overset{\displaystyle \text{O}}{\overset{\|}{\text{C}}}\text{OH} + \text{H}_2\text{N}-\hspace{-2pt}\bigcirc \quad (3\text{--}7)$$

This amidase differed from the normal by a single residue, Thr being replaced by Ile.

The examples of mutationally altered proteins given here have been described structurally only in terms of their amino acid sequences, as little else is known about them. For the present, we can only look forward to characterization of their three-dimensional structures and correlation of these with their functional properties.

Gene duplication in experimental enzyme evolution. P. W. J. Rigby, et al. Nature 251:200–204, 1974.

Evolution in action. J. L. Betz, et al. Nature 247:261–264, 1974.

Gene amplification and drug resistance in cultured murine cells. R. T. Schimke, et al. Science 202:1051–1055, 1978.

Evolution of a new enzymatic function by recombination within a gene. B. G. Hall and T. Zuzel. Proc. Natl. Acad. Sci. U.S.A. 77:3529–3533, 1980.

Changes in the substrate specificities of an enzyme during directed evolution of new functions. B. G. Hall. Biochemistry 20:4042–4049, 1981.

Properties of an altered dihydrofolate reductase encoded by amplified genes in cultured mouse fibroblasts. D. A. Haber, et al. J. Biol. Chem. 256:9501–9510, 1981.

4

PHYSICAL FORCES THAT DETERMINE THE PROPERTIES OF PROTEINS

The first three chapters of this volume have discussed proteins only in terms of their covalent structures. This is adequate for small molecules but not for macromolecules, especially proteins, because their large sizes permit them to fold back on themselves so that there may be numerous simultaneous interactions between different parts of the molecule. A complex, marginally stable three-dimensional structure results with proteins, providing the unique environments and orientations of functional groups that give them their special functional properties. It is important to have some feeling for the weak forces upon which the structures and functions of proteins are dependent.

The physical natures of noncovalent interactions between atoms are fairly well understood with individual molecules in a gas or a regular solid, but not in liquids, owing to the complexity of the constantly changing interactions between ensembles of many molecules in the liquid state. This is especially important with proteins, because their folded conformations are usually studied only in an environment of liquid water, and the most important characteristic of all intermolecular forces in water is that they are generally more dependent upon the properties of this extraordinary solvent than on the nature and strength of the intermolecular forces themselves. The interactions of water with ions, dipoles, and hydrogen bond donors or acceptors are so strong as to cause a leveling or disappearance of most of the forces that would produce a strong intermolecular interaction between such groups in a vacuum or in a nonpolar solvent. The most important noncovalent force in water, the hydrophobic interaction, is almost certainly a consequence of the strong interaction of the water molecules with each other, rather than of a direct interaction between the solute molecules.

This chapter will describe briefly the types of noncovalent interactions that exist between atoms, review the structure and properties of liquid water, and then examine how the various physical interactions may act in an aqueous environment and within a polypeptide chain.

THE PHYSICAL NATURE OF NONCOVALENT INTERACTIONS

Electrostatic Forces

The most familiar physical interaction is the **electrostatic force** that occurs between any two charged particles. The change in energies of two point charges as they are brought together as a function of their distance of separation is given by Coulomb's law:

$$\Delta E = \frac{Z_A Z_B \epsilon^2}{r_{AB}} \qquad (4-1)$$

where r_{AB} is the distance between the two charges, A and B; Z_A and Z_B are their respective number of unit charges; and ϵ is one unit of electronic charge. If the charges are of the same sign, the interaction is repulsive, with an increase in energy, but it is attractive, with a lowering of energy, if they are of opposite signs. However, this is applicable only to a vacuum and to point charges; also, the electrostatic interaction between two ions does not become infinitely great at small values of r_{AB}, as implied by Equation 4–1, because they become subject to strong repulsions when in close contact. Coulomb's law as expressed in Equation 4–1 is not appropriate when the ions are in solution, but it can give a reasonable approximation of the actual energetic situation in dilute solutions if account is taken of the solvent molecules, by including a factor D, known as the **dielectric constant** of the solvent:

$$\Delta E = \frac{Z_A Z_B \epsilon^2}{D\, r_{AB}} \qquad (4-2)$$

The dielectric constant is an independently measurable property of the bulk solvent that reflects the polarizability of the solvent molecules, which diminishes the electrostatic interaction between ions. Equation 4–2 should be valid when the average number of solvent molecules between the two interacting charges is large enough that the properties of this "micro" region of solvent can be expected to be similar to those of the bulk solvent. It should be noted that nonpolar liquids have very low dielectric constants of about 2, while those of polar liquids, such as water (80) and formamide (109), are very high. Consequently, electrostatic interactions are considerably reduced in the latter solvents. Nevertheless, such interactions are unique in that they are significant over relatively large distances, varying only inversely with the linear distance.

Electrostatic forces also occur with uncharged, but polar, molecules: There is no net charge, but the electrons are distributed within the molecule so that there is an excess of negative charge in one part of the molecule, with a corresponding excess of positive charge in another part. This is expressed as the **dipole moment,** μ, which is defined as

$$\mu = z\,d \qquad\qquad (4\text{--}3)$$

where z is the amount of the excess charge separated in the molecule and d is the distance between the centers of the excess charge. The measured value of μ may be compared with that expected from Equation 4–3 for a completely ionic molecule. For example, a completely ionic molecule of HCl would consist of a proton at one end and a chloride ion at the other, 1.27 Å apart, each with a charge of that of the electron, 4.8×10^{-10} esu (electrostatic units). The dipole moment would then be 6.1×10^{-18} esu-cm, or 6.1 Debye units. The experimental value is 1.03 Debye units, or 17 per cent of the fully ionic molecule.

The energy of the electrostatic interaction involving such dipoles is more complex than that between two simple ions, because the extent of the dipolar nature of such molecules is often changed, i.e., induced, by the other molecules with which it is interacting. This is the **polarizability** of a molecule, which is defined as the increase in its dipole moment with an applied electric field. Consequently, the energy of interaction generally varies with the inverse of the distance between such molecules raised to some power, generally greater than 1 but less than 6.

Most of the charged groups of proteins have pK_a values (Table 1–1) within the pH range likely to be encountered in studies with proteins, so their ionization is pH-dependent. Ionization is also dependent upon the presence of other charged groups within proximity and upon the environment, in particular, the accessibility to water and the polarity of the solvent. These effects on simple compounds are illustrated in Table 4–1 and Figure 4–1. Other polar groups have significant dipole moments, such as the peptide bond with $\mu = 3.5$ Debye units approximately (see Eq. 1–7, p. 5).

Table 4–1   Effect of Nonaqueous Environment on Ionization

ACID OR BASE	pK_a VALUE (weight per cent dioxane in solvent)				
	0	20	45	50	70
Acetic acid	4.76	5.29	6.31		8.34
$(HOCH_2)_3C\text{---}NH_2$	8.0	8.0	8.0	8.0	8.0
Benzoylarginine	3.34			4.59	4.60
Glycine $\{$ ---CO_2H	2.35	2.63	3.11		3.96
---NH_2	9.78	9.29	8.49		7.42

From A. Fersht, Enzyme Structure and Mechanism. Reading, England, W. H. Freeman, 1977.

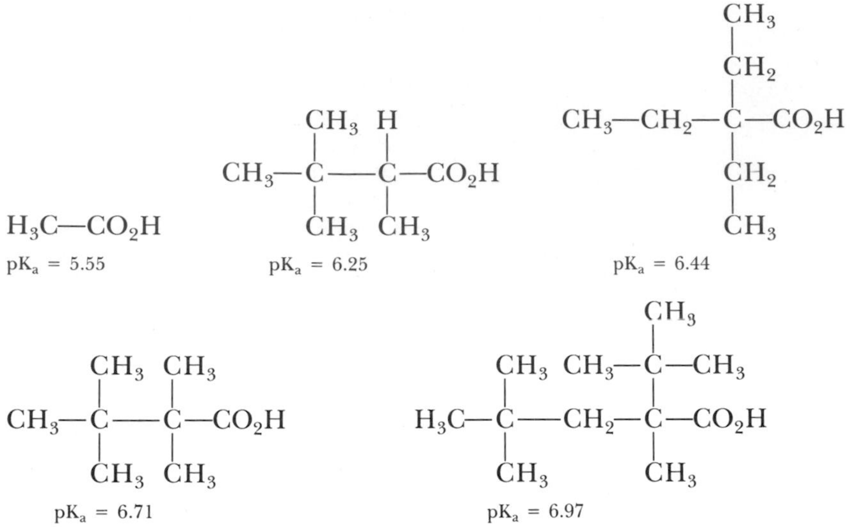

Figure 4–1
Steric effects on the ionization of carboxyl groups. The pK$_a$ values increase with bulkiness of the aliphatic groups surrounding the carboxyl group, indicating a decreased tendency to ionize, presumably due to decreased solvation of the ion. The pK$_a$ values were measured in equal volumes of methanol and water at 40° C. (Values from G. S. Hammond and D. H. Hogle, J. Am. Chem. Soc. 77:338–340, 1955.)

Hydrogen Bonds

One particular type of primarily electrostatic interaction between polar molecules has come to be known as the **hydrogen bond.** This interaction is intermediate between covalent bonding and ionic interactions, because a hydrogen atom is shared between a potential acid (proton donor) and a base (proton acceptor). The exact nature of such bonds is not clear, but it is in many ways useful to consider the hydrogen bond as an intermediate stage in the transfer of a proton from an acid, AH, to a base, B:

$$A-H + B \longleftrightarrow {}^{\delta-}A\text{---}{}^{\delta+}H \cdots B \longleftrightarrow A^- \cdots H-B^+ \qquad (4\text{--}4)$$

The acid and the base interact in such a way that a partial bond is formed between the proton and the base, with a resultant stretching and weakening of the A—H bond. Because of this, the geometric arrangement of A, H, and B is very important; optimal hydrogen bonds have linear geometry, and bent hydrogen bonds are weaker, but the exact relationship is still not certain. The extent of the proton transfer—i.e., the position of equilibrium in Equation 4–4—depends upon the relative acidities of the acid and base: If the base B is weak, little interaction will occur; a strong base will abstract the proton completely. The relative affinities of A$^-$ and B determine both the relative position of the hydrogen atom and the strength of the hydrogen bond. Values for the energy of a typical hydrogen bond in a vacuum are believed to lie in the range of -3 to -6 kcal/mole.

Table 4–2 *Typical Hydrogen Bonds*

TYPE OF BOND	TYPICAL COMPOUND	OPTIMAL DISTANCE BETWEEN NONHYDROGEN ATOMS (Å)
O—H $\cdots$ O	Water	2.8
	Primary alcohols	2.7
	Carboxylic acids	2.6
N—H $\cdots$ N	Ammonia, amines	3.1
N—H $\cdots$ O	Urea, amides, peptides	2.9

A list of the predominant types of hydrogen bonds involving atoms found in proteins is given in Table 4–2. Essentially only the oxygen and nitrogen atoms of proteins participate in forming significant hydrogen bonds; residues of the type —OH and —NH can act as good hydrogen bond donors, while virtually any oxygen atom and any nonprotonated nitrogen atom can serve as hydrogen bond acceptors. It should be noted that the polypeptide backbone has an excellent hydrogen bond donor,

$$—\overset{\displaystyle H}{N}—,$$ and an excellent acceptor, $$—\overset{\displaystyle O}{\underset{\displaystyle \|}{C}}—,$$ in each residue (except Pro).

Van der Waals Forces

Weak interactions occur between all atoms, whether polar or nonpolar—even between two noble gas atoms in the gas phase. This is an essentially electrostatic interaction that arises because the distribution of electronic charge will, at any instant of time, be localized in some specific asymmetric geometry dependent upon the distribution of electrons in the various electronic orbitals. This transient dipole induces a complementary dipole in a neighboring molecule, which will in turn help to stabilize the original dipole. An instant later, the dipoles in the two molecules are likely to be reversed, but these oscillating dipoles give rise to a net attractive force between any two atoms, often designated the **van der Waals** or **London dispersion force.** For small molecules, it is small, generally of the order of -0.3 kcal/mole.

However, when the two atoms come very close together, the repulsion between their electron clouds becomes very strong and eventually counterbalances the attractive forces. The attractive force is very sensitive to the distance between the two atoms, r, varying as r^{-6}, but the repulsive force is even more sensitive, varying approximately as r^{-12}. The energy of these van der Waals or London dispersion forces may generally be approximated by the Lennard-Jones potential energy function

$$\Delta E = -\frac{A}{r^6} + \frac{B}{r^{12}} \qquad (4\text{–}5)$$

to give an energy of interaction as a function of distance, as illustrated in Figure 4–2. The distance of separation of the two atoms at which the

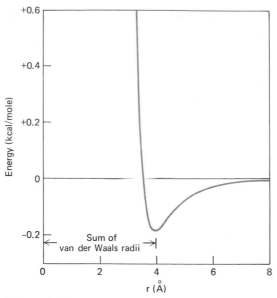

Figure 4–2
Representative profile of the energy of Van der Waals interaction as a function of the distance,
r, *between the centers of two atoms. The energy was calculated using the empirical equation*

$$E = \frac{B}{r^{12}} - \frac{A}{r^6}$$

*(Values for the parameters B = 2.75 × 10⁶ kcal Å¹²/mole and A = 1425 kcal Å⁶/mole,
for the interaction between two carbon atoms, from M. Levitt, J. Mol. Biol. 82:393–420,
1974.)*

interaction is most favorable is known as the **van der Waals contact distance.** Each atom has a characteristic van der Waals radius, the contact
distance being the sum of the radii for a particular pair of of atoms. The
van der Waals radii for atoms that occur in proteins are listed in Table
4–3.

Biophysical Chemistry. J. T. Edsall and J. Wyman. New York, Academic Press,
1958.

The Nature of the Chemical Bond. L. Pauling. Ithaca, N.Y., Cornell University
Press, 1960.

The Hydrogen Bond. G. C. Pimentel and A. L. McLellan. San Francisco, W. H.
Freeman, 1960.

Kinetic and thermodynamic studies of hydrogen bonding. G. G. Hammes and A.
C. Park. J. Am. Chem. Soc. 91:956–961, 1969.

Hydrogen bonding. G. C. Pimentel and A. L. McLellan. Ann. Rev. Phys. Chem.
22:347–385, 1971.

Energy functions for peptides and proteins. II. The amide hydrogen bond and
calculation of amide crystal properties. A. T. Hagler and S. Lifson. J. Am.
Chem. Soc. 96:5327–5335, 1974.

A model for the hydrogen bond. L. C. Allen. Proc. Natl. Acad. Sci. U.S.A.
72:4701–4705, 1975.

Table 4–3   Representative Values for van der Waals Radii of Atoms Found in Proteins

ATOM	COVALENT BONDING	VAN DER WAALS RADIUS (Å)
Hydrogen		1.00
Carbon	double bonds	1.60
	aromatic	1.70
	amide	1.50
Oxygen		1.35
Nitrogen	aromatic	1.70
	amide	1.45
Sulfur		1.70

From Handbook of Biochemistry. Cleveland, Chemical Rubber Publishing Co., 1968.

Analysis of O—H $\cdots$ O hydrogen bonds. J. Mitra and C. Ramakrishnan. Int. J. Pept. Protein Res. 9:27–48, 1977.
Noncovalent interactions. P. A. Kollman. Acc. Chem. Res. 10:365–371, 1977.
Studies on hydrogen bonds. IV. Proposed working criteria for assessing qualitative strength of hydrogen bonds. J. Mitra and C. Ramakrishnan. Int. J. Pept. Protein Res. 17:401–411, 1981.

THE PROPERTIES OF LIQUID WATER AND OF NONCOVALENT INTERACTIONS IN THIS SOLVENT

The properties of liquid water derive ultimately from the physical properties of the water molecule, which are fairly well known from both experimental and theoretical studies. The structure of H_2O in its lowest energetic ground state is known accurately:

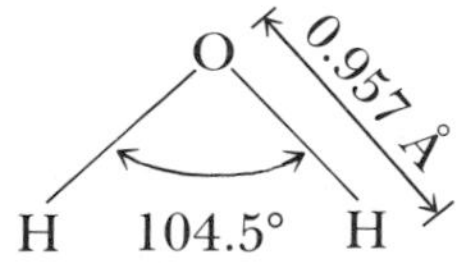

$$\text{H} \quad 104.5° \quad \text{H} \qquad \text{O} \quad 0.957\,Å \tag{4–6}$$

The most prominent electronic feature of the water molecule is the two lobes of negative charge formed by the lone-pair electrons on the oxygen atom. These lobes project above and below the plane of the three atoms in Equation 4–6 and probably project away from the hydrogens to some extent. They undoubtedly contribute to the very significant dipole moment (1.84 Debye units) and the polarizability of the water molecule; this dipole moment would correspond to O—H bonds with 33 per cent ionic character. The two lone-pair electrons on the oxygen atom permit it to serve simultaneously as hydrogen acceptor in two hydrogen bonds; the two hydrogen atoms can be donated in two further hydrogen bonds.

The structural importance of the ability of water to be both an acceptor and a donor in hydrogen bonds is illustrated by the normal crystalline structure of ice (ice I), where this potential is utilized fully, each molecule of water being involved in four hydrogen bonds (Figure 4–3). Every oxygen atom is at the center of a tetrahedron formed by four oxygen atoms, each about 2.76 Å away. Every water molecule is hydrogen-bonded to its

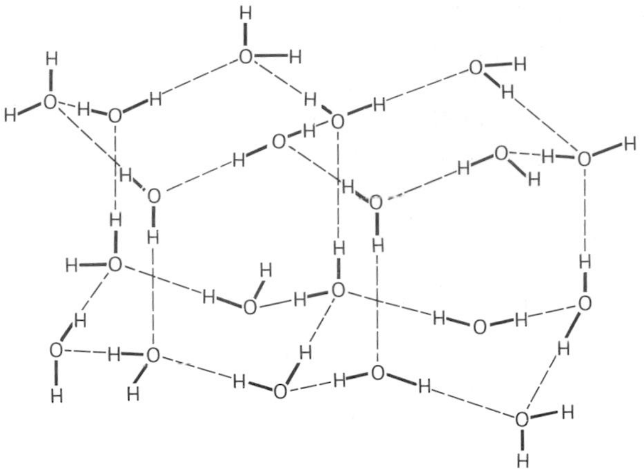

Figure 4–3
The structure of normal ice. Each H_2O molecule is involved in four hydrogen bonds, each 2.76 Å long, two as hydrogen donor and two as acceptor. As a consequence, substantial empty channels run between the molecules. (Adapted from F. H. Stillinger, Science 209:451–457, 1981.)

four nearest neighbors; its hydrogen atoms are directed toward the lone-pairs of electrons on two of these neighbors, to form two O—H $\cdots$ O hydrogen bonds. In turn, each of its lone-pair electrons is directed toward a hydrogen atom of one of the other two neighbors, forming two O $\cdots$ H—O hydrogen bonds. This is the most stable arrangement at normal pressures, even though it results in shafts of unoccupied space running between molecules in the crystal lattice. This open structure is the reason that ice has a density lower than that of liquid water at 0° C and hence floats. More compact crystalline forms are possible and are obtained at high pressures, but they have bent hydrogen bonds; the favorable energy of the linear hydrogen bond then obviously is sufficient to determine the structure of the normal ice lattice, in spite of the resulting open structure. The energy of the hydrogen bond in ice is not known with certainty but probably lies within the range of -4 to -7 kcal/mole per bond.

The structural properties of liquid water are much less certain, owing to the basically disordered nature of liquids in general. For example, the dielectric relaxation time of water indicates that a molecule experiences a displacement on average about once every 10^{-11} second at 0° C. However, it does appear that the time-averaged distribution of nearby molecules in the liquid is close to that expected with a tetrahedral arrangement of molecules like that in ice. X-ray scattering of water shows that the positions of nearest neighbors are not random: The average number of nearest neighbor molecules is about 4.4, being on average 2.9 Å away. The relative positions of molecules up to about 8 Å away are still close to those expected from the ice lattice but less fixed. There is little or no evidence for structural ordering of molecules over a range greater than 8 Å. Computer simulations of liquid water suggest that it consists of a random network of hydrogen bonds, many very strained, that is continually undergoing re-arrangement, with bonds being strained, broken, and then re-formed. Roughly half the potential hydrogen bonds are present at any instant.

In spite of the uncertainty about the precise nature of liquid water, the strong tendency of water molecules to participate as both hydrogen bond donors and acceptors is clearly the basic explanation of the many peculiar properties of this solvent. It also appears to cause thermodynamic studies of phenomena in this solvent to be particularly complex, because mutually compensating changes in entropy and enthalpy tend to occur, with little or no change in free energy. For example, the favorable formation of hydrogen bonds in water should produce a favorable decrease in the enthalpy, H. However, hydrogen bonds require that the participating molecules become relatively fixed; therefore, the entropy, S, also decreases. These two contributions to the Gibbs free energy, G, thus tend to cancel out, since

$$\Delta G = \Delta H - T\,\Delta S \tag{4-7}$$

where T is the temperature. Consequently, even with large changes in enthalpy and entropy, relatively little or no change in free energy may result. Other relatively small effects may thus be predominant in determining the free energy of any such transition. These properties make rationalization of thermodynamic phenomena in water a very tricky business.

The strength of noncovalent interactions between different molecules or between different portions of a polypeptide chain dissolved in water depend upon the net difference between their interactions with each other and with water. The interactions of neutral small molecules with water may be measured by their relative tendencies to equilibrate between the vapor and aqueous phases. The peptide backbone is very hydrophilic, being found in the aqueous environment at a concentration 10^7 times greater than in the vapor phase. This partition coefficient was measured using N-methylacetamide

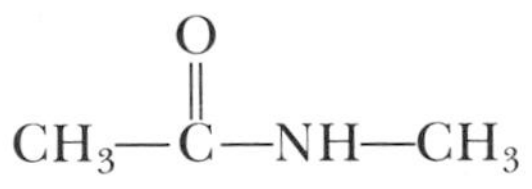

$$CH_3-\overset{\overset{\textstyle O}{\|}}{C}-NH-CH_3 \tag{4-8}$$

as a model of the peptide backbone. Analogous small molecules were used to measure the same parameter for 19 of the 20 amino acid side chains. This measure of the hydrophilicity of the amino acids can be expressed energetically as $-RT \ln$ (partition coefficient, H_2O: vapor), and the values are given in Table 4–4.

The net interaction between two groups in solution is usually measured by the equilibrium constant, K_{AB}, for the association of two appropriate individual molecules, A and B:

$$A + B \underset{}{\overset{K_{AB}}{\rightleftharpoons}} A \cdot B \tag{4-9}$$

*Table 4–4   Preferences of Amino Acid Side Chains
for Aqueous or Nonpolar Environments*

AMINO ACID	RELATIVE HYDROPHILICITY[a] (kcal/mole)	RELATIVE HYDROPHOBICITY[b] (kcal/mole)
Gly	0	0
Leu	0.11	1.8
Ile	0.24	(2.5)
Val	0.40	1.5
Ala	0.45	0.5
Phe	3.15	2.5
Cys	3.63	(−2.8)
Met	3.87	1.3
Thr	7.27	0.4
Ser	7.45	−0.3
Trp	8.28	3.4
Tyr	8.50	2.3
Gln	11.77	−0.3
Lys	11.91	(−4.2)
Asn	12.07	−0.2
Glu	12.58	(−9.9)
His	12.62	0.5
Asp	13.31	(−7.4)
Arg	22.31	(−11.2)
Pro		(−3.3)

[a]Measured from the distribution coefficient, K_{eq}, between H_2O and vapor for a small molecule representing the side chain of the amino acid:

$$K_{eq} = [X]_{H_2O}/[X]_{vapor}$$

Ionized groups have negligible tendency to vaporize, so the observed values for nonionized groups were corrected where necessary to consider the fraction, α, of nonionized form present at pH 7. The hydrophilicity at pH 7 is then defined as $RT \ln (K_{eq}/\alpha)$ and is expressed relative to that of Gly.

Adapted from R. V. Wolfenden, et al., Science 206:576–577, 1979.

[b]The hydrophobicities of the side chains were measured as for hydrophilicities, but by their distribution between a nonpolar solvent, either ethanol or dioxane, and water. Positive values indicate preference for the nonpolar solvent. Values in parentheses were calculated from their surface area (see Figure 4–5), subtracting 2.5 kcal/mole for each polar group, and correcting for the fraction of nonionized form present at pH 7.

Values in parentheses, from G. Von Heijne and C. Blomberg, Eur. J. Biochem. 97:175–181, 1979; other experimental data, from Y. Nozaki and C. Tanford, J. Biol. Chem. 246:2211–2217, 1971.

$$K_{AB} = \frac{[A \cdot B]}{[A][B]} \tag{4–10}$$

Some values of K_{AB} for interactions pertinent to proteins are given in Table 4–5.

Electrostatic interactions in water are considerably diminished by its large dielectric constant. The high polarizability of water molecules also results in strong interactions with ions. A small, highly charged ion will orient the water molecules immediately adjacent into a firm hydration

Table 4–5   Association in Water of Small Molecules
Typical of Noncovalent Interactions in Proteins

TYPE OF INTERACTION	EXAMPLE	ASSOCIATION CONSTANT, K_{AB} (M^{-1})	
Salt-bridge	NH$_2$ | NH$_2$—C=NH$_2^+$ · $^-$O$_2$C—CH$_3$	0.5	[a]
		0.37	[b]
	CH$_3$(CH$_2$)$_3$NH$_3^+$ · $^-$O$_2$C—CH$_3$	0.31	[b]
Hydrogen bond	Formic acid dimers	0.04	[c]
	Urea dimers	0.04	[d]
	N-methylacetamide dimers	0.005	[e]
	δ-Valerolactam dimers	0.013	[f]

[a]From C. Tanford, J. Am. Chem. Soc. 76:945–946, 1954.

[b]From B. Spriggs and P. Haake, Bioorg. Chem. 6:181–190, 1977.

[c]From A. Katchalsky, et al., J. Am. Chem. Soc. 73:5889–5890, 1951.

[d]From J. A. Schellman, Compt. Rendu Trav. Lab. Carlsberg Ser. Chim. 29:223–229, 1955.

[e]From I. M. Klotz and J. S. Franzen, J. Am. Chem. Soc. 84:3461–3466, 1962.

[f]From H. Susi, et al., J. Biol. Chem. 239:3051–3054, 1964.

sphere, which will remain with the ion for an appreciable time and will have properties that are quite different from those of the bulk water molecules. These water molecules constitute the immediate hydration shell of the ion and are more "structured" than the bulk molecules of liquid water. Outside this firm hydration shell, the electrical field will not be strong enough to bind additional water molecules, but it will be able to compete effectively with the dipole–dipole forces that normally provide nonrandom orientation among the molecules of liquid water. However, in larger ions the charge density may be so low that there is no significant influence on the solvent. This strong hydration makes the removal of ions from aqueous solution to a nonpolar environment very unfavorable energetically.

The exact magnitude of the energy of solvation or hydration of individual ions is not known, but it is certainly of the order of -50 to -200 kcal/mole. This interaction with water is so strong that the tendency of ions to interact with each other in water is negligible in comparison. The net energy of an interaction between two ions in water depends upon the small difference between their energies when they are interacting, when they may need to lose some of their surrounding water of hydration, and their energies when they are interacting separately with water. Consequently, small ionized molecules associate only weakly in water (Table 4–5).

Interactions between ions in water are entropically driven, the enthalpy change being unfavorable; therefore they increase in magnitude with higher temperatures. This entropy change is due to the liberation of some of the hydration shell upon ion-pair formation.

Individual molecules with polar groups capable of forming hydrogen bonds also do not interact substantially in water, with $K_{AB} = 10^{-2}\,M^{-1}$

(Table 4–5), since the polar groups form apparently equivalent hydrogen bonds with water. They do interact in nonpolar solvents, where 10^3 times greater values of K_{AB} are found.

Van der Waals attractive forces occur between all atoms; consequently, assessment of their possible importance for noncovalent interactions between molecules dissolved in water involves estimation of the difference between the energies of such interactions plus water–water interactions, on the one hand, and that of the interactions of the separated molecules with the aqueous solvent on the other. The sensitivity of the van der Waals interaction with at least the sixth power of the contact distance and the complex structure of water make this a formidable theoretical problem that has not been solved. Nevertheless, the relatively open structure of water imposed by the tendency to form hydrogen bonds produces a relatively low density, so there may be correspondingly less opportunity for van der Waals interactions with the aqueous solvent. Water molecules depicted by their van der Waals radii occupy only 58 per cent of the volume of liquid water; the remainder is empty space.

The properties of pure water are sufficiently complex in themselves to be only partially understood, but the situation becomes even more complex when significant concentrations of other molecules are dissolved in it. Yet the normal environment of proteins is water with a relatively high ionic strength produced by a variety of salts. Some proteins are not stable in pure water but require a moderate ionic strength for stability. One of the most obvious properties of all salts is to compete in electrostatic interactions, but they also increase the solubility of the peptide bond in water. The basis for such effects is not always clear; they can be due to direct interactions between the two added components of the aqueous solution or to indirect effects of the added component on the properties of water. It is often difficult to rationalize the effects of added third components on aqueous solutions, but they are often used in studies of proteins, either to stabilize or to denature them.

Some factors in the interpretation of protein denaturation. W. Kauzmann. Adv. Protein Chem. 14:1–63, 1959.

Stability of an amide–hydrogen bond in an apolar environment. I. M. Klotz and S. B. Farnham. Biochemistry 7:3879–3882, 1968.

The Structure and Properties of Water. D. Eisenberg and W. Kauzmann. Oxford, Oxford University Press, 1969.

Enthalpy–entropy compensation phenomena in water solutions of proteins and small molecules: a ubiquitous property of water. R. Lumry and S. Rajendar. Biopolymers 9:1125–1227, 1970.

The effects of salts on the free energy of the peptide group. P. K. Nandi and D. R. Robinson. J. Am. Chem. Soc. 94:1299–1308, 1972.

Water—A Comprehensive Treatise. F. Franks (ed.). New York, Plenum Press, 1972.

Interaction of the peptide bond with solvent water: a vapor phase analysis. R. Wolfenden. Biochemistry 17:201–204, 1978.

Water revisited. F. H. Stillinger. Science 209:451–457, 1980.

Affinities of amino acid side chains for solvent water. R. Wolfenden, et al. Biochemistry 20:849–855, 1981.

THE HYDROPHOBIC INTERACTION

Within the polar aqueous environment, noncovalent interactions between different parts of a polypeptide chain, by electrostatic interactions, hydrogen bonds, or van der Waals forces, are not expected to be especially favorable energetically, because similar interactions will occur between the polypeptide and the water surrounding it. One of the most significant interactions encountered in water is the relatively *unfavorable* one between water and nonpolar molecules, as the latter cannot participate in the hydrogen bonding that appears to be so important in liquid water. It now appears that it is the *absence* of hydrogen bonding between nonpolar molecules and water, rather than a favorable interaction between the nonpolar groups themselves, that is a major factor in the stability of proteins, nucleic acids, and membranes, and which has come to be known as the **hydrophobic interaction.**

It is well known that water, in comparison with organic liquids, is a very poor solvent for apolar molecules and that aqueous solutions of such molecules have many anomalous physical properties. This may be illustrated by considering the solvation of small molecules in water. The first step may be regarded as the formation of a cavity in the water into which the solute will fit. The free energy required for the formation of such a cavity in liquid water will be large because it requires the separation of strongly interacting water molecules; this property is reflected in the high surface tension of water. Once a cavity is formed and the solute placed in it, the solvent will undergo any further changes that reduce the free energy of the system. With a polar solute, hydrogen bonds and other electrostatic interactions will occur between the solute and the water; these favorable interactions will compensate for the initial energy required to form the cavity. However, a nonpolar solute will gain only the minor van der Waals forces with the solvent. To compensate for the absence of favorable electrostatic interactions with the solute, the solvent surrounding it will rearrange to form the most extensive number of hydrogen bonds between the water molecules. Extreme examples of ordered water cages, or **clathrates,** are observed around apolar gases at low temperatures and high gas pressures; the water molecules are fully hydrogen-bonded, with each oxygen surrounded by four nearest neighbor hydrogen atoms. However, the degree of ordering of water molecules around a nonpolar solute under normal conditions is much less than in the case of the clathrates.

The energetics of dissolving a nonpolar molecule in water can be illustrated by the transfer of methane from an inert solvent, such as carbon tetrachloride, to water, either directly or through the gas phase (Figure 4–4). It must be kept in mind that negative changes in free energy, ΔG, and in enthalpy, ΔH, but positive changes in entropy, ΔS, give favorable contributions to a transition. That transfer of methane from the gas or nonpolar solvent to water is energetically unfavorable is demonstrated by the positive values of ΔG. Even though the enthalpy changes are favorable, probably resulting from favorable van der Waals contacts with the water, they are more than offset by the unfavorable changes in entropy. The basis of these values may be rationalized by the changes that occur during trans-

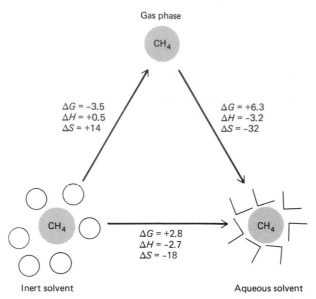

Gas phase

$\Delta G = -3.5$
$\Delta H = +0.5$
$\Delta S = +14$

$\Delta G = +6.3$
$\Delta H = -3.2$
$\Delta S = -32$

$\Delta G = +2.8$
$\Delta H = -2.7$
$\Delta S = -18$

Inert solvent   Aqueous solvent

Figure 4–4

Thermodynamics of the hydrophobic effect. The transfer of one molecule of methane to water from an inert, nonpolar solvent, such as carbon tetrachloride, is illustrated both directly and through the gas phase. The changes in Gibbs free energy, ΔG, and in enthalpy, ΔH, are expressed in kcal/mole; those of entropy, ΔS, in entropy units, kcal/mole/°K. (Adapted from W. P. Jencks, Catalysis in Chemistry and Enzymology. New York, McGraw-Hill, 1969.)

fer to the gas phase. The unfavorable change in enthalpy of $+0.5$ kcal/mole upon transfer of methane to the gas phase reflects primarily the loss of van der Waals interactions with the inert solvent, but this is more than offset by the increase in entropy due to the increase in freedom upon going to the gas phase. The transfer of methane to the gas phase is therefore favorable, as would be expected, since it normally is a gas. Upon transferring methane from the gas to aqueous solution, the favorable enthalpy change of -3.2 kcal/mole is significantly greater than that going to the inert solvent, -0.5 kcal/mole. Methane would hardly be expected to have correspondingly better van der Waals interactions with water; this favorable ΔH, therefore, must be due to a change in the enthalpy of the solvent, caused by an increase in the hydrogen bonding between water molecules, presumably those around the methane molecule. This results in a loss of entropy of the water molecules; the total entropy loss is thus over twice that occurring upon dissolution of methane in the inert solvent.

With nonpolar solutes, the interactions between the water and the solute appear to be negligible relative to the initial unfavorable energy required to form the initial cavity in the water. Consequently, an inverse linear relationship is found between the solubility of hydrocarbons in water and the surface area of the cavity required to accommodate the solute (Figure 4–5). The same relationship holds for apolar molecules as diverse

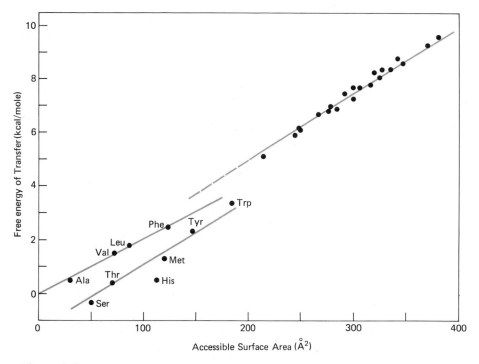

Figure 4–5

Correlation between hydrophobicity and accessible surface area. The unlabeled dots are for various hydrocarbons; the line extrapolates back to the origin and has a slope of 25 cal/Å². The labeled dots refer to the side chains of the amino acids; the line passing through the nonpolar Ala, Val, Leu, and Phe has a slope of 22 cal/Å². The other amino acids have polar groups and consequently lower hydrophobicities than those expected from their surface areas. (From F. M. Richards, Ann. Rev. Biophys. Bioeng. 6:151–176, 1977; © Annual Reviews Inc.)

as the noble gases, halocarbons such as CCl_4 and CF_4, linear and cyclic hydrocarbons, and aromatic rings.

Since the interactions of nonpolar molecules with water are so unfavorable, there is a strong force causing them to aggregate in water, as the surface area of the cavity then decreases. This means of minimizing unfavorable interactions with water is the origin of the apparent favorable hydrophobic interaction between nonpolar molecules in water. Direct measurement of this interaction between nonpolar molecules in aqueous solution is difficult, owing to their low solubility, but it may be pictured as being the transfer of a nonpolar molecule from an aqueous environment to one nonaqueous, the reverse of that depicted in Figure 4–4. The hydrophobic interaction thus has a negative ΔG, owing to a favorable, positive ΔS, but has a positive, unfavorable ΔH.

Because of the positive value of ΔH for the hydrophobic interaction, the free energy of this interaction actually becomes *more* favorable with increasing temperature within a certain range. However, the thermodynamics of the hydrophobic interaction are much more complex than indicated above in that the values of both the enthalpy and entropy changes are markedly dependent upon the temperature. At low temperatures, both the enthalpy and entropy changes upon removal of apolar molecules from contact with water are positive, but both decrease with increasing temperature; the hydrophobic interaction becomes maximal about when the entropy change reaches zero. Both the enthalpy and entropy changes become negative at higher temperatures, and the hydrophobic interaction consequently becomes weaker.

The change in ΔH as a function of temperature is characteristic of hydrophobic interactions; it is indicative of a change in the heat capacity of the system, C_p, since

$$\Delta C_p^\circ = \frac{\partial(\Delta H^\circ)}{\partial T} \qquad (4-11)$$

The change in C_p of the system has been attributed to the excess heat needed to "melt" the ice-like water around the nonpolar molecules in solution. These observations illustrate the very important point that thermodynamic studies in aqueous solution may be dominated by contributions from the solvent. The significance of measured values of ΔH and ΔS for a macromolecular transition in water is very seldom clear.

These very anomalous properties of liquid water, the poor solubilities of nonpolar molecules, and the favorable opposite effect—the hydrophobic interaction between such nonpolar molecules—have been well known and well studied for over 30 years; yet their physical basis is still the subject of much uncertainty. In the absence of full understanding of the hydrophobic interaction, recourse has been made to empirical measurement. A **hydrophobicity** scale for the nonionized amino acid side chains was derived by Nozaki and Tanford by measuring the apparent free energy of transfer of the side chains to water from a nonpolar solvent, generally ethanol or dioxane. This was measured by the relative solubilities of the free amino acids in aqueous and nonpolar solvents, designated S_{H_2O} and S_{np}, respectively:

$$\Delta G_{tr} = RT \ln \frac{S_{np}}{S_{H_2O}} \qquad (4-12)$$

Amino acids are very insoluble in nonpolar solvents, owing to the polar amino and carboxyl groups, so the value of S_{np} was generally obtained by extrapolation of the solubilities measured in various mixtures of water and nonpolar solvent to the pure nonpolar solvent. The measured values were then made relative to that of glycine, which was assigned $\Delta G_{tr} = 0$, because it has no side chain but otherwise is the same as the other amino acids. The relative values, ΔG_{tr}, are assumed to measure the hydrophobicities of the side chains and are given in Table 4–4.

As was found for other nonpolar molecules, there is a linear relationship between the accessible surface area of the nonpolar side chains and their hydrophobicities (Figure 4–5). The free energy of transfer from nonaqueous to aqueous solution increases by about 24 calories for each additional $Å^2$ of accessible surface area. For the side chains with single polar atoms, such as those of the hydroxyl groups of serine, threonine, and tyrosine, there is a similar relationship, but the hydrophobicities are lower by 1 to 2 kcal/mole. In aqueous solution these polar atoms undoubtedly form hydrogen bonds with water, but hydrogen bonds to ethanol or dioxane are probably less favorable, by 1 to 2 kcal/mole. For side chains with two polar atoms, such as those of asparagine, glutamine, and histidine, the hydrophobicities are even less than those of the side chains with single polar atoms. The differences are on average of about the right magnitude for forming two or more hydrogen bonds, but there is a greater degree of variation, which suggests other, more specific interactions with the solvents.

It has been attempted to measure in a similar way the hydrophobicity of the polypeptide backbone, but the various possible model compounds studied have given disturbingly varied values for a peptide unit, suggesting again more specific interactions with water.

The hydrophobicity of an acidic or basic side chain is difficult to assess. It is unlikely that an ionized group will partition significantly into a nonpolar solvent; such a group probably will do so only as the nonionized form. The hydrophobicities of these side chains at pH 7 may then be estimated from the fraction of nonionized form present at that pH and from their surface areas and numbers of polar groups. Such values are given in parentheses in Table 4–4.

Not surprising, the hydrophobicities of the amino acid side chains are related to their hydrophilicities, since the former is a measure of their lack of affinity for water and the latter just the opposite. However, there is no quantitative relationship. The hydrophilicity is primarily a function of the number of polar groups, which are necessary for interaction with water. The hydrophobicity is only slightly affected by polar groups, probably owing to the ability of the relatively nonpolar solvents used to form hydrogen bonds. It seems likely that the hydrophobicity of a side chain with polar groups has been overestimated in these measurements. All the measured values depend somewhat on the nature of the nonpolar solvent.

One important aspect of hydrophobicity, which follows from its dependence on surface area, is that it is additive, in that the hydrophobicity of a large molecule is close to the sum of that of its constituent parts. This aspect has been utilized by measuring the hydrophobicity constants of individual structural moieties. The hydrophobicity of any molecule may then be estimated from the sum of the constants of its individual substituents. Some pertinent values are given in Table 4–6.

The hydrophobic interaction is also sensitive to other components added to the aqueous solution. For example, the protein denaturants urea and guanidinium chloride (also known as guanidine hydrochloride) decrease the hydrophobic interaction (Figure 4–6), and this is likely to be the basis of their denaturing action (see p. 293). Salts increase the hydrophobic

Table 4–6   *Hydrophobicity Constant (π) of Some Groups*[a]

GROUP	π	GROUP	π
—CH$_3$	0.56	—OH	−0.67
—SCH$_3$	0.61	—SH	0.39
—CH$_2$OH	−1.03	—NH$_2$	−1.23
—CH(CH$_3$)(CH$_3$)	1.53	—NH$_3^+$	−4.19
		—CO$_2$H	−0.32
		—CO$_2^-$	−4.36
—CH(CH$_3$)(CH$_2$CH$_3$)	2.04	—C(=O)—NH$_2$	−1.49
—CH$_2$—C$_6$H$_5$ (benzyl)	2.01		
indole	2.14	imidazole	−0.25

[a]The parameter π is defined as the logarithm of the ratio of the partition coefficient between *n*-octanol and water of a molecule containing the pertinent group, *P*, to that of an otherwise identical molecule lacking it and with an —H atom instead, P_0:

$$\pi = \log (P/P_0)$$

The hydrophobicity is 2.3 RT π, or 1.36 π kcal/mole at 25° C.

From C. Hansch and A. Leo, Substituent Constants for Correlation Analysis in Chemistry and Biology. New York, John Wiley & Sons, 1979.

interaction; the effect depends upon the nature of the ions, following the Hofmeister series:

Cations: $Mg^{++} > Li^+ > Na^+ > K^+ > NH_4^+$

Anions: $SO_4^{2-} > HPO_4^{2-} >$ acetate $>$ citrate $>$ tartrate $> Cl^- > NO_3^-$

$$> ClO_3^- > I^- > ClO_4^- > SCN^- \tag{4–13}$$

This order was observed by Hofmeister over 90 years ago to be the order of effectiveness as precipitants of serum globulins. The same series has been observed in effectiveness in decreasing the solubility in water (i.e., "salting-out") of nonpolar molecules, of affecting the stability of specific conformations of proteins and nucleic acids, and of stabilizing detergent micelles. The Hofmeister series also governs the effect of guanidinium salts as protein denaturants: Guanidinium thiocyanate is a more potent denaturant than the chloride salt, whereas guanidinium sulfate actually stabilizes proteins.

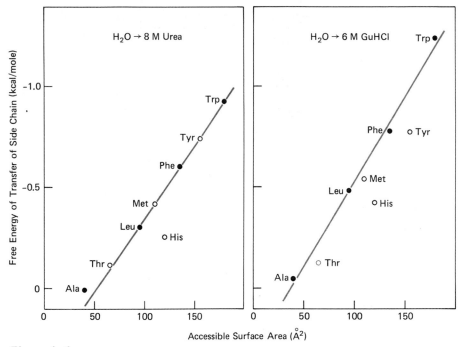

Figure 4–6
Correlation between free energy of transfer of the amino acid side chains from water to either 8 M urea or 6 M guanidinium chloride (GuHCl) and their accessible surface areas (Table 1–1). The solid lines have slopes of 7.1 and 8.3 cal/mole/Å², respectively. (Adapted from T. E. Creighton, J. Mol. Biol. 129:235–264, 1979; experimental transfer data from Y. Nozaki and C. Tanford, J. Biol. Chem. 238:4074–4081, 1963, and 245:1648–1652, 1970.)

It is most likely that all of these effects on the hydrophobic interaction arise indirectly from effects on the structure of water. For example, the Hofmeister series generally correlates best with the effect of the salts on the surface tension of water: Those salts that increase most the hydrophobic interaction also increase most the surface tension of water. This correlation is to be expected if hydrophobicity arises from the energy required to make a cavity in water, and the hydrophobic interaction is of about the expected magnitude. For example, the surface tension at a water–hexane interface is about 51 dynes/cm, which is 1.22×10^{-6} cal/cm² or 1.22×10^{-22} cal/Å². Multiplying this by Avogadro's number to obtain the value per mole indicates that forming 1 Å² of contact surface area between water and a nonpolar solute should cost energetically about 76 cal/mole. This is only somewhat larger than the values of hydrophobicity measured experimentally (Figure 4–5), 24 cal/mole per 1 Å² of accessible surface area (the accessible surface is larger than the contact surface, both of which are described in Figure 6–20).

Some consider this correlation between hydrophobicity and the surface area and surface tension to be fortuitous, as water is treated as a continuum and the surface tension is a macroscopic property of the liquid. With very small systems, such as the interaction between two small, spherical molecules, the buried surface is difficult to define and may be very dependent upon the molecular structures of both water and the interacting molecules. Nevertheless, the empirical observations with the model systems are sufficiently impressive to indicate that they provide a semiquantitative estimation of the hydrophobic interaction in large molecules, such as proteins. On the other hand, care must be exercised to ensure that microscopic aspects do not invalidate the approach.

Neutral salts: the generality of their effects on the stability of macromolecular conformations. P. H. Von Hippel and K.-Y. Wong. Science 145:577–580, 1964.

The solubility of amino acids and two glycine peptides in aqueous ethanol and dioxane solutions. Establishment of a hydrophobicity scale. Y. Nozaki and C. Tanford. J. Biol. Chem. 246:2211–2217, 1971.

Theory of hydrophobic bonding. II. The correlation of hydrocarbon solubility in water with solvent cavity surface area. R. B. Hermann. J. Phys. Chem. 76:2754–2759, 1972.

The apolar bond—a re-evaluation. M. H. Klapper. Prog. Bioorg. Chem. 2:55–132, 1973.

Hydrophobic bonding and accessible surface area in proteins. C. H. Chothia. Nature 248:338–339, 1974.

The thermodynamic basis of the stability of proteins, nucleic acids, and membranes. H. Edelhoch and J. C. Osborne, Jr. Adv. Protein Chem. 30:183–250, 1976.

Salt effects on hydrophobic interactions in precipitation and chromatography of proteins: an interpretation of the lyotropic series. W. Melander & C. Horvath. Arch. Biochem. Biophys. 183:200–215, 1977.

Theory of the hydrophobic effect. L. R. Pratt and D. Chandler. J. Chem. Phys. 67:3683–3704, 1977.

Substituent Constants for Correlation Analysis in Chemistry and Biology. C. Hansch and A. Leo. New York, John Wiley & Sons, 1979.

Hydrophobic Interactions. A. Ben-Naim. New York, Plenum Press, 1980.

The Hydrophobic Effect, 2nd ed. C. Tanford. New York, Wiley-Interscience, 1980.

INTRA-MOLECULAR INTERACTIONS

Prior discussion of noncovalent interactions considered primarily the individual units that make up proteins. Although the interactions are basically the same, the observed effects are somewhat different when the interacting groups are part of the same molecule: In the first place, the individual interactions have been described as between different molecules, either as bimolecular association constants (Table 4–5) or as partition coefficients between two liquids (Table 4–4), whereas they are unimolecular processes in proteins. Second, multiple interactions may take place simultaneously, so that their effect on each other must be taken into account.

For example, individual ions do not associate very strongly in aqueous solution (Table 4–5), because of their favorable entropy and strong interaction with water. On the other hand, a polyelectrolyte with a number of such charged groups on a single molecule will bind ions of the opposite charge very tightly, owing to interactions between the charged groups. Being part of the same molecule, they are constrained by the covalent bonds to be close to each other, which is very unfavorable energetically when they are of the same sign. To compensate for this electrostatic repulsion, they attract counterions from the solution very tightly; just how tightly depends upon the charge density of the polyelectrolyte and the valence of the counterions. The degree of binding is essentially independent of the concentration of counterions in the bulk solvent, because the diminishment of the electrostatic repulsions in the polyelectrolyte by the bound counterions is more favorable energetically than is their equilibration with the bulk solvent. The counterions need not be bound at specific sites but may retain their water of hydration and move in an unrestricted and random way along the polyelectrolyte chain. This phenomenon can be very important for binding of other ligands with the same charge, because their displacement of the original counterions into a very dilute bulk solvent can provide a strong driving force for association.

Counterions may also be bound tightly and specifically, with loss of hydration shell, if two or more ionized groups on a molecule are in a suitable position, so that they can act simultaneously. This phenomenon is well known chemically as the **chelate effect.**

Besides attracting counterions, another means of compensating for unfavorable electrostatic repulsions in a polyelectrolyte is to suppress the ionization of a fraction of the groups. Consequently, groups on a polyelectrolyte may have pK_a values very different from those found when they are isolated. Ionic interactions within small molecules have traditionally been described with the simple Debye-Hückel procedure, but it has not been found to be adequate with many model polyelectrolytes; consequently, the analysis of electrostatic effects in proteins is very uncertain.

The other problem with interactions within the same molecule is how to consider those measured in model compounds by bimolecular association constants, K_{AB} (Eqs. 4–9 and 4–10). The association constant K_{AB} has units of (concentration)$^{-1}$; its value therefore depends upon which units are used. For the same reason, the free energy of the interaction cannot be simply taken as $-RT \ln K_{AB}$, although this is often done, when the standard state is said to be unity of whichever concentration units are employed. It is easily forgotten that this energy of interaction pertains only to that particular concentration; the degree of association of any two molecules will be essentially zero at very low concentrations but may be complete at very high concentrations. A more reasonable procedure is to measure the interaction by the relative proportion of the limiting reactant, e.g., A, associated at equilibrium:

$$\Delta G° = -RT \ln \frac{[A \cdot B]}{[A]} \tag{4–14}$$

It follows from Equation 4–10 that

$$\Delta G° = RT \ln (K_{AB} [B]) \qquad (4-15)$$

which emphasizes the dependence of the free energy of the interaction on the concentration of the free second reactant, B.

Bimolecular association constants of the type K_{AB} (Table 4–5) may be applied to the interaction between two such groups when on the same molecule only if their effective concentrations relative to each other are known. This will be designated here as $[A/B]_i$, where i is the particular molecule, emphasizing that it is of one group relative to the other and that its value will pertain only to a particular macromolecule. The equilibrium constant for an intramolecular interaction between groups A and B on molecule i will then be given by

$$K_{intra} = K_{AB} [A/B]_i \qquad (4-16)$$

The value of $[A/B]_i$ would be expected to be zero when the molecule to which groups A and B are attached keeps them apart, but rather large when they are held in proximity and in orientation appropriate for interaction. It was thought for a long time that the maximum value in aqueous solution was about 55 M, the concentration of pure water, when one group could be considered to be immersed in an environment of the second component. However, many experimental comparisons of intramolecular and intermolecular reactions in small molecules have found that effective concentrations are usually in the region of 10^3 to 10^{10} M if there is no strain in the intramolecular case. Some examples are given in Table 4–7. The lower values are observed for flexible molecules, where the two groups have a great deal of freedom when not interacting. Larger values are found when the two groups are constrained by the molecule of which they are a part to be in suitable proximity and orientation even when not interacting.

In the case of the isolated molecule, this is a purely entropic effect, reflecting the decreased entropy gained by the groups upon breaking an interaction. Maximum effective concentrations of 10^8 to 10^{10} M are expected theoretically on this basis alone, similar to the experimental values (Table 4–7). The maximum value also depends upon the type of interaction: Those where the proximity and orientation of the interacting groups are very important and, when a covalent bond is formed, have very high values. Where these factors are not so important, the interacting groups already have significant degrees of freedom and have less entropy to gain upon dissociating, so that lower effective concentrations apply. The values are much greater than 55 M, because the molecules of a liquid have a high degree of translational and rotational freedom; consequently, they are far from being in optimal conditions for interacting with a solute molecule. The very high effective concentrations possible in intramolecular cases can result in these interactions being considerably more stable than those with water; i.e., K_{intra} of Equation 4–16 can be greater than unity.

Unfortunately, the magnitudes of the maximum effective concentrations expected for interactions of the type observed in proteins are not

Table 4–7   Selected Examples of Measured Values of Effective Concentrations of Two Reactive Groups in Small Molecules

EQUILIBRIUM REACTION	EFFECTIVE CONCENTRATION
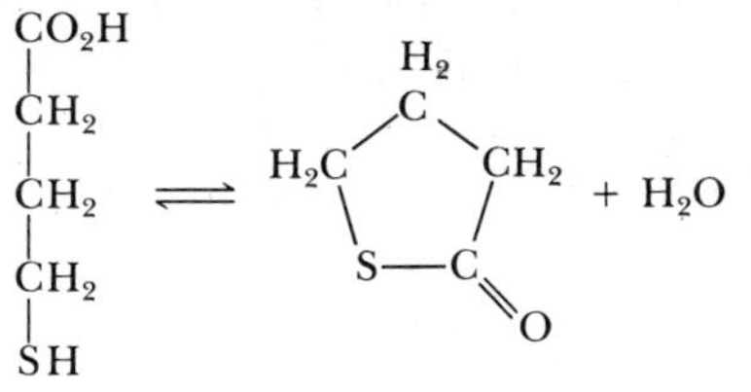 CO_2H–CH_2–CH_2–CH_2–CH_2–SH ⇌ (thiolactone ring) + H_2O	3.7×10^3 M[a]
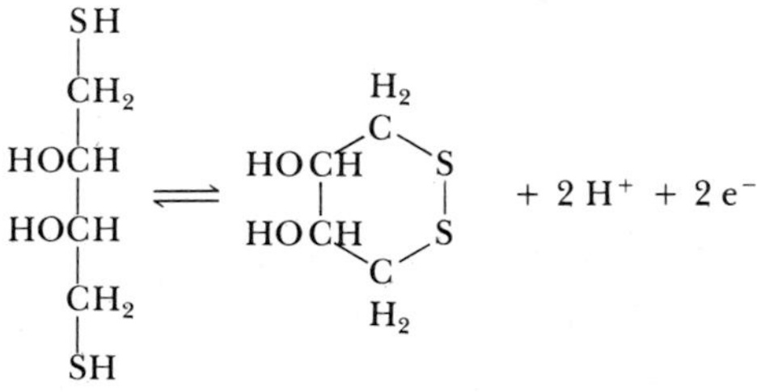 SH–CH_2–$HOCH$–$HOCH$–CH_2–SH ⇌ (dithiane ring) + $2\,H^+ + 2\,e^-$	1.3×10^4 M[b]
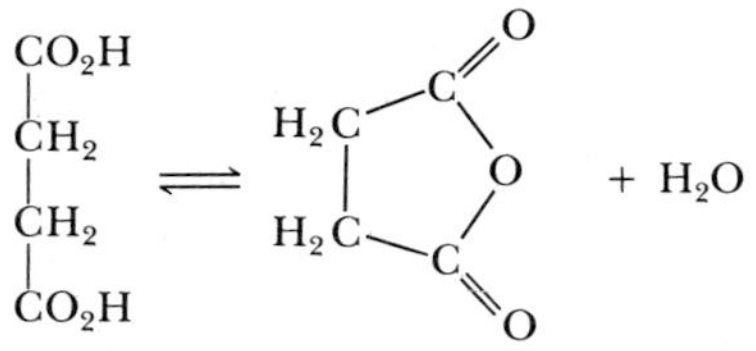 CO_2H–CH_2–CH_2–CO_2H ⇌ (succinic anhydride) + H_2O	1.9×10^5 M[a]
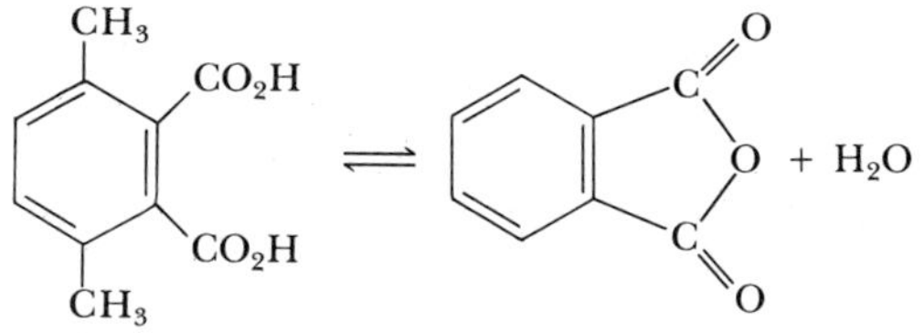 (dimethyl benzene di-CO_2H) ⇌ (phthalic anhydride) + H_2O	5.4×10^9 M[a]

[a]From A. J. Kirby, Adv. Phys. Org. Chem. 17:183–278, 1980.
[b]From W. W. Cleland, Biochemistry 3:480–482, 1964.

known. Hydrogen bonds are moderately sensitive to orientation, probably having partial covalent character, so substantial values would be expected, but probably less than those of 10^{10} M found with covalent interactions. Ionic and hydrophobic interactions are not very stringent stereochemically, so maximum values of 10^2 to 10^3 M apply in these instances.

The effective concentration that applies in the stability of an interaction is that of the two groups when they are not interacting. If there are no other interactions keeping the groups together, its value will be relatively low, but if other interactions are present simultaneously to keep the groups in proximity, their effective concentrations will be high. This is illustrated schematically in Figure 4–7, with three interactions, A · B,

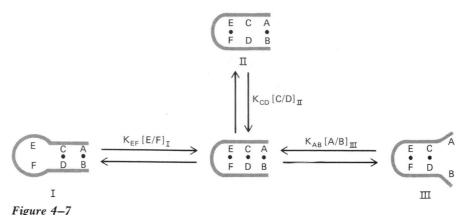

Figure 4–7
Simple schematic diagram of cooperativity among three simultaneous interactions occurring in groups A through F in a molecule. The strength of each interaction is determined by the effective concentration of the two groups when they are not interacting, as in conformations I, II, and III; this is inversely proportional to the degree of flexibility permitted. Therefore, the most stable interactions are between those groups that are held most rigidly by the other interactions, and the stability of each interaction depends upon that of all the others.

C · D, and E · F. It would be expected that $[C/D]_{II} > [E/F]_I > [A/B]_{III}$. Clearly, the effective concentration of each pair should be much greater than it would be were the other interactions not present; likewise, each depends upon the stability provided by the others, so that the overall stability of the three interactions is very cooperative. This will be shown to be a characteristic of folded proteins.

The other factor that enters into interactions between solute molecules in concentrated solutions is the **excluded volume effect,** where the molecules take up a significant fraction of the total volume. Presumably, such effects can be local and may be pertinent to intramolecular interactions. Mass-action equilibria such as Equation 4–10 should use the chemical activities of the reactants, not their concentrations. In dilute solutions the two are interchangable, but in any concentrated solution the activity can be very different from that expected with the particular concentration. Because the solvating water around a solute molecule becomes limited, favored equilibria are those in which the molecules have minimal interactions with the solvent. Associations between molecules that produce more spherical species are consequently more favorable energetically than would be expected from only the equilibrium constant calculated using concentrations. Such phenomena are believed to be important within cells, where the nonwater molecules may represent one third of the total volume.

Entropic contributions to rate accelerations in enzymic and intramolecular reactions and the chelate effect. M. I. Page and W. P. Jencks. Proc. Natl. Acad. Sci. U.S.A. 68:1678–1683, 1971.
The energetics of neighbouring group participation. M. I. Page. Chem. Soc. Rev. 2:295–323, 1973.

Electrolyte theory—improvements since Debye and Hückel. K. S. Pitzer. Acc. Chem. Res. 10:371–377, 1977.

The molecular theory of polyelectrolyte solutions with applications to the electrostatic properties of polynucleotides. G. S. Manning. Quart. Rev. Biophys. 11:199–246, 1978.

Thermodynamic analysis of ion effects on the binding and conformational equilibria of proteins and nucleic acids: the roles of ion association or release, screening, and ion effects on water activity. M. T. Record, Jr., et al. Quart. Rev. Biophys. 11:103–178, 1978.

Effective molarities for intramolecular reactions. A. J. Kirby. Adv. Phys. Org. Chem. 17:183–278, 1980.

Excluded volume as a determinant of macromolecular structure. A. P. Minton. Biopolymers 20:2093–2120, 1981.

5

CONFORMATIONAL PROPERTIES OF POLYPEPTIDE CHAINS

The covalent structure of a protein is also known as the **primary structure,** because it is only the first level of protein structure. The higher levels refer to the three-dimensional orientation in space of such large molecules. **Secondary structure** is the local arrangement of the polypeptide backbone; **tertiary structure** refers to the overall three-dimensional architecture of the polypeptide chain.

The complexity of three-dimensional structure arises from the intrinsic ability of single covalent bonds to be rotated. Rotation about several such bonds in a linear molecule will produce different nonsuperimposable three-dimensional arrangements of the atoms that are generally described as **conformations.** They are interconvertible solely by such bond rotations, without breaking covalent bonds or changing the chirality of atoms, which would give a different **configuration.** The two terms should not be interchanged or confused.

However, how to define a conformation is not obvious. Even a simple molecule might be considered to exist in an infinite number of conformations if the positions of the atoms were defined sufficiently accurately, since bond lengths tend to vary by ± 0.05 Å and bond angles by about 5 degrees at room temperature. Consequently, only the energetically most stable arrangements that are separated by distinct energy barriers are generally classified as individual conformations. For example, small molecules such as CH_4 are considered to have a single conformation, as does C_2H_6— that in which rotation about the C—C bond has the six equivalent H atoms staggered. This may be illustrated in a Newman projection, where the spatial arrangements of the groups attached to two atoms are viewed in projection down the bond joining them:

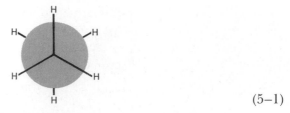

$$(5-1)$$

Although all rotations about the C—C bond are intrinsically permitted, repulsive interactions between the H atoms favor the staggered structure. A molecule such as 1,2-dichloroethane may be considered to have three conformations, in which the two Cl atoms are *trans* (*t*), *gauche⁺* (g^+), and *gauche⁻* (g^-), respectively.

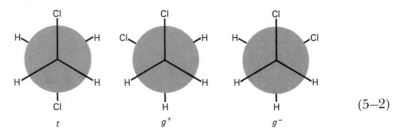

$$(5-2)$$

Each amino acid residue of a polypeptide chain contains three bonds of the polypeptide backbone, plus those of the side chain (Fig. 5–1). The peptide bond of the backbone has partial double-bond character and consequently is limited to planar *cis* or *trans* rotations. However, the other two

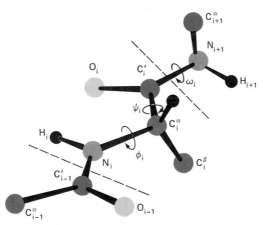

Figure 5–1
Perspective drawing of a section of polypeptide chain representing two peptide units. The limits of a residue (number i of the chain) are indicated by the dashed lines. The recommended notations for atoms and torsion angles are indicated. The polypeptide is shown in the fully extended conformation, where $\phi = \psi = \omega = 180°$; the amino acid residue is the L isomer.

have single-bond character, as do many of the bonds of the side chains. Each amino acid residue can exist in a number of conformations, perhaps 10 on average.

A relatively small polypeptide chain of 100 residues, then, may exist in up to 10^{100} conformations. The magnitude of this astronomic number may perhaps be appreciated by considering that it is 10^{31} times greater than one estimate of the number of atoms in the universe (10^{79}). A fraction of these theoretical conformations are not possible in practice, as some of the atoms of the polypeptide would be required to occupy the same space, but there is plenty of scope to allow for this excluded volume effect, or to be more conservative in the estimate of the number of conformations possible per residue and yet end up with an astronomic number of polypeptide conformations; for example, only 2 conformations per residue would still yield over 10^{30} theoretical conformations for a chain of 100 residues. Each conformation would have a very small probability of being adopted in a random polypeptide, and each molecule of a finite population is likely to have a unique conformation at any instant of time.

The potential for a large number of distinct conformations makes unlikely the existence of a single stable conformation, because a factor known as the **conformational entropy,** ΔS_{conf}, is lost on adopting a single conformation. It has the value $R \ln N$, where N is the number of conformations possible, and the contribution to the free energy of the disordered polymer is given by $-RT \ln N$. If $N = 10^{100}$ for a chain of 100 residues, this has the value of -136 kcal/mole at $25° C$. Any one conformation would have to have stabilizing interactions of at least this magnitude in order to be present in half the molecules of a population. This concept may be illustrated by simply writing down equilibria between N equally probable states, so that the equilibrium constant between any pair has the value of unity. If any one state has a more favorable stability, in that its equilibrium constant with any of the other equally probable states, K_{conf}, is greater than unity, this value would have to be at least N for it to have 0.5 probability of being present. Its more favorable free energy must then have the value

$$\Delta G^0 = -RT \ln K_{conf} = T\Delta S_{conf} \qquad (5-3)$$

As a consequence of the large conformational entropy of polypeptides, they will tend to exist in many different conformational states unless very many stabilizing interactions are possible simultaneously. Yet proteins and some regular polypeptides are able to do this, as will be described following a description of polypeptides that do not.

Because of the complexities of proteins, basic studies of polypeptide conformation have used synthetic polyamino acids, where all residues are identical, or simple repeating polymers of more than one amino acid, such as poly(Pro-Ala). The regularity of the covalent structure results in the acquisition of conformations either very regular or very random.

Principles of Polymer Chemistry. P. J. Flory. Ithaca, N.Y., Cornell University Press, 1953.

Physical Chemistry of Macromolecules. C. Tanford. New York, John Wiley & Sons, 1961.

Conformations of Macromolecules. T. M. Birshtein and O. B. Ptitsyn. New York, Wiley-Interscience, 1964.

Conformation of polypeptides and proteins. G. N. Ramachandran and V. Sasisekharan. Adv. Protein Chem. 23:283–437, 1968.

Biophysical Chemistry, Part III. C. Cantor and P. R. Schimmel. San Francisco, W. H. Freeman, 1980.

POLYPEPTIDES AS RANDOM POLYMERS

With the possibility of so many conformations of comparable probability, the conformational properties of disordered polypeptides are best calculated statistically by the mathematical procedures developed for synthetic polymers. Such calculation requires detailed knowledge of the conformational properties of the monomeric unit of the polymer, its relative energies in all possible conformational states.

Configurational Statistics of Polymeric Chains. M. V. Volkenstein. New York, John Wiley & Sons, 1963.

The configuration of random polypeptide chains. II. Theory. D. A. Brant and P. J. Flory. J. Am. Chem. Soc. 81:2791–2800, 1965.

Random coil configurations of polypeptide copolymers. W. G. Miller, et al. J. Mol. Biol. 23:67–80, 1967.

Conformational energy estimates for statistically coiling polypeptide chains. D. A. Brant, et al. J. Mol. Biol. 23:47–65, 1967.

Statistical Mechanics of Chain Molecules. P. J. Flory. New York, John Wiley & Sons, 1969.

Local Restrictions on Flexibility: The Ramachandran Plot

A portion of the backbone of a polypeptide chain is shown in Figure 5–1, illustrating the conventions used in describing protein conformation. The peptide bond is usually planar, owing to its partial double bond, and the group of atoms

$$-C_i^\alpha \diagdown \underset{\diagup}{N} \underset{H}{\diagup} \overset{\diagup}{-}C' \underset{\diagdown O}{\overset{\diagup C_{i+1}^\alpha -}{}} \qquad (5\text{--}4)$$

usually behaves as a rigid unit; this group is often designated a **peptide unit,** but this term will not be used here. The unit of **residue,** which refers to the atoms from the same amino acid, will be used instead.

Bond angles are denoted by the symbol τ, with a subscript i referring to the number of the residue in the chain and with the symbols of the atoms defining the bond angle placed in brackets following τ. For example, τ_i ($NC^\alpha C'$) denotes the angle formed by the $C^\alpha N$ and $C^\alpha C'$ bonds of the ith residue.

Rotation about the N—C^α bond of the peptide backbone is denoted by the torsion angle ϕ; rotation around the C^α—C' bond, by ψ; and that about the peptide (C'—N) bond, by ω. The choice of the values of these angles of rotation is made by giving the maximum value of $+180°$ (which is the same as $-180°$) to the maximally extended chain, as in Figure 5–1. In the fully extended polypeptide chain, the N, C^α, and C' atoms are all *trans* to each other. In the *cis* configurations, the angles ϕ, ψ, and ω are given the value of 0. Rotation from this position about the bonds, so that the atoms viewed behind the rotated bond move *counterclockwise*, are given *negative* values; those *clockwise* are given *positive* values. Thus, the values of the torsion angles lie in the range $-180°$ to $+180°$.

The torsion angles of the side chain are designated by χ_j where j is the number of the bond working outward from the C^α atom of the main chain. The accepted designations of the side-chain atoms are given in Figure 1–1.

The relative probability of the adoption of each particular three-dimensional arrangement of atoms by a residue depends upon its energy, which depends upon bond angles and length and upon the interactions between atoms not linked by covalent bonds.

The *trans* form of the peptide bond ($\omega = 180°$) is favored 10^3-fold over the *cis* form ($\omega = 0°$), since in the latter the C^α atoms, any side chains, and residue $i + 2$ are in too-close proximity:

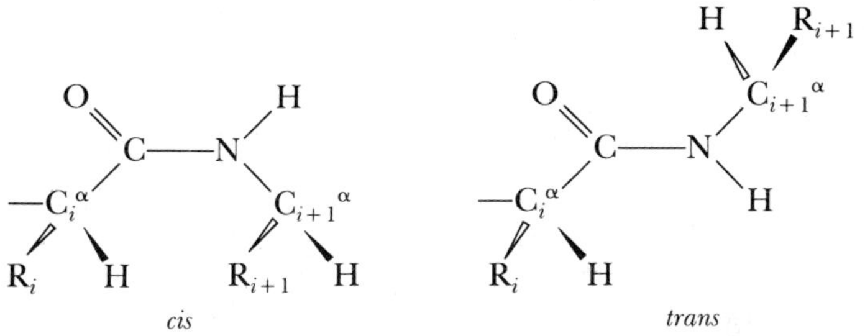

$$\tag{5-5}$$

However, when residue $i + 1$ is Pro, there is very little difference between the two:

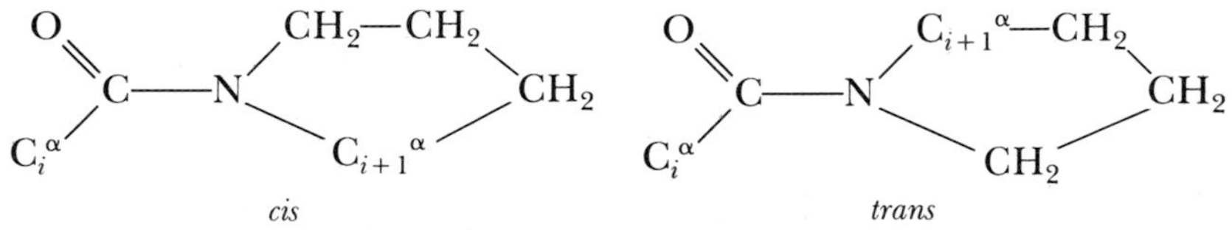

$$\tag{5-6}$$

and the *trans* form is only slightly favored by about 80 : 20. Small deviations from planarity ($\Delta\omega = -20°$ to $+10°$) are only slightly unfavorable energetically.

There are also geometrical restraints on possible values of ϕ and ψ, even though these are single covalent bonds. For example, the case with $\phi = \psi = 0°$

$$\text{(5-7)}$$

is clearly impossible, since the carbonyl oxygen atom O_{i-1} and the H atom of N_{i+1} would overlap. A detailed analysis of the possible values of ϕ and ψ was first made by Ramachandran and coworkers, using hard-sphere models of the atoms and fixed geometry of the bonds. The permitted values of ϕ and ψ are generally indicated on a two-dimensional map of the $\phi-\psi$ plane in what has become known as a **Ramachandran plot.** An example for an Ala residue is shown in Figure 5–2A. The normally allowed values are enclosed by the solid black lines; the extreme limits permitted, with some unfavorable contacts, by the solid gray lines. The connecting region indicated by the dashed gray lines becomes permitted if slight alterations of bond angles are allowed. The fraction of the total area that is fully allowed is only 7.5 per cent; the partially allowed region makes up 22.5 per cent. Consequently, the polypeptide backbone is relatively restricted in its flexibility.

When there is no C^β atom attached to the C^α atom, as in Gly residues, the restrictions on the allowed conformations are much less severe (Figure 5–2B), and 45 per cent of the total area is fully allowed; 61 per cent comes within the extreme limits. The Ramachandran plot is symmetric in this case, as Gly residues have two equivalent H atoms on C^α and no asymmetric center.

With longer, more complex side chains of the other amino acids, additional restrictions to the flexibility of the backbone arise, which are indicated by the step diagram of Figure 5–2C. Proline residues are a special case; here the side chain is covalently bonded to both C^α and N atoms. This fairly rigid five-membered ring drastically limits rotation about the N—C^α bond, and the value of ϕ is limited to approximately $-60°$.

More precise Ramachandran plots calculate the relative energies of each conformation, permitting flexibility of bond lengths and angles, and evaluating all favorable and unfavorable interactions. For realistic models, they should also include the solvent and all its interactions. The procedures and parameters used are primarily empirical and approximate, but they are constantly improving; they give results similar to those with the simple procedures (Figure 5–2D).

The mean geometry of the peptide unit from crystal structure data. G. N. Ramachandran, A. S. Kolaskev, et al. Biochim. Biophys. Acta 359:298–302, 1974.

An explanation for the rare occurrence of *cis* peptide units in proteins and polypeptides. G. N. Ramachandran and A. K. Mitra. J. Mol. Biol. 107:85–92, 1976.

Statistical Properties of Random Polypeptides

The calculated and observed properties of polymers are often compared to those calculated for the hypothetical **random flight chain,** or freely jointed chain. This is just a mathematical string of vectors of fixed length representing the bonds between atoms; the atoms are not included, and the chain has no volume. The angle between each pair of bond vectors is free to assume all values with equal probability, and all rotations about the bonds are equally likely.

A more realistic model of actual polymers is the unperturbed state, which takes into account the three-dimensional structure and the conformational preference of the monomer units (Figure 5–2), but not interactions with solvent or between distant parts of the chain. The conformation of each monomer is assumed to be independent of the other monomers. Statistical evaluation of the many conformations possible involves averaging over all of them, with weighting factors determined by the energies of the bond rotations. Most calculations with polypeptides simplify the detailed chemical architecture of the polypeptide backbone and use only the mathematically expedient **virtual bond,** a vector joining the adjacent C^{α} atoms; with a planar *trans* peptide bond it has a length of 3.8 Å.

Of greatest interest with random polymers are the average and the distribution of the dimensions between parts of the chain. The root-mean-square value of the distance, r, between two atoms of the hypothetical random flight chain separated by n bonds of length l is given by the relationship

$$\langle r^2 \rangle_0^{1/2} = n^{1/2} l \tag{5–8}$$

where the angle brackets are used to indicate the statistical mechanical average over all conformations; the subscript zero refers to the unperturbed state.

The other statistical measure often used is the average radius of gyration (R_G), which is defined as the root-mean-square (rms) distance of the collection of atoms from their common center of gravity. For the random flight chain,

$$\langle R_G \rangle_0^2 = \frac{nl^2}{6} \frac{n + 2}{n + 1} \tag{5–9}$$

For large values of n this becomes

$$\langle R_G \rangle_0^2 = \frac{nl^2}{6} = \frac{\langle r^2 \rangle_0}{6} \tag{5–10}$$

This simple relationship exists between the average radius of gyration and the average end-to-end distance in the unperturbed state of polymers.

All real polymers have relatively fixed bond angles and some restric-

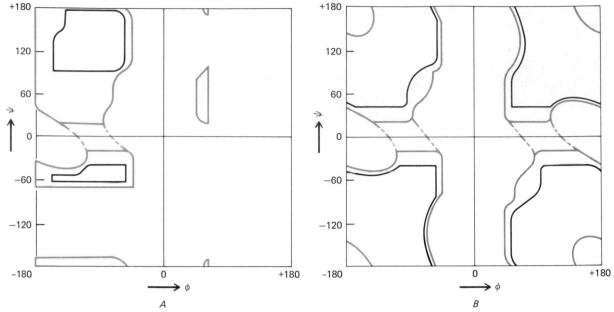

Figure 5–2

Ramachandran plots of the allowed values of φ and ψ for different residues. Those for Ala and Gly residues in a polypeptide chain are shown in A and B, respectively. The fully allowed regions are enclosed by the solid black lines; the partially allowed conformations, by a solid gray line. The connecting region enclosed by the dashed gray lines is permissible with slight flexibility of bond angles. This two-dimensional plot is continuous at the edges, since a rotation of − 180° is the same as +180°; hence, the partially allowed conformations at the bottom are continuous with those at the top. (Adapted from G. N. Ramachandran and V. Sasisekharan, Adv. Protein Chem. 23:283–437, 1968.)

C, Ramachandran plot for different amino acid side chains. Gly residues may occupy regions 1 to 4; Ala residues, 2 to 4. Amino acids with longer side chains are restricted to regions 3 and 4, but Val and Ile are permitted only in region 4, which includes only 5 per cent of the total area. (From G. Nemethy, et al., J. Phys. Chem. 70:998, 1969, © American Chemical Society.)

D, Conformational energy contour map calculated for N-acetyl-N′-methyl alanine amide, using empirical energy functions. The contour lines are labeled with the potential energy in kcal/mole above the minimum at φ = − 84°, ψ = 79°. (Adapted from S. S. Zimmerman, et al., Macromolecules 10:1–9, 1977.)

tions on rotation about bonds, which increase their dimensions relative to those of the random flight chain. This factor is usually expressed by the **characteristic ratio,** which is defined as

$$C_n = \frac{\langle r^2 \rangle_0}{nl^2} \qquad (5–11)$$

where $\langle r^2 \rangle_0$ is the observed average end-to-end distance squared, while nl^2 is this value for the random flight chain. The characteristic ratio increasingly deviates from unity as n increases, but it then approaches a limit for very long chains, which is designated as C_∞. This value is a measure of the inherent stiffness of the chain.

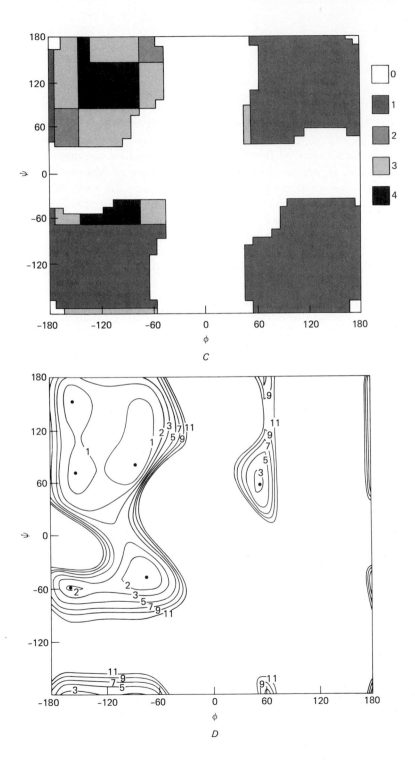

C

D

For long polypeptide chains of n residues of amino acids other than Gly and Pro, the rms end-to-end distance in angstroms is approximately given by

$$\langle r^2 \rangle_0^{1/2} = (130\,n)^{1/2} \tag{5-12}$$

which gives a value for the characteristic ratio, C_∞, of 9.0. This relatively large value for C_∞ is due to the steric restrictions on the rotations about ϕ and ψ and to the asymmetry of the backbone produced by the presence of only L-amino acid residues. For example, a polymer of alternating D and L Ala residues has a value for C_∞ of only 0.9, since the direction of the chain may be effectively reversed with each residue. The value of C_∞ for polyglycine is only 2.0, owing both to the increased flexibility possible with Gly residues and to the symmetric nature of the allowed torsion angles (Figure 5–2B). A small percentage of randomly distributed Gly residues in a polypeptide chain has a large effect on the chain dimensions of random polypeptides. The presence of proline residues also lowers the characteristic ratio of polypeptides, because the chain tends to change directions.

The distribution of end-to-end distances may be readily calculated for a freely jointed chain from either the Gaussian distribution function or the radial distribution function:

$$W(x,\,y,\,z)\,dx\,dy\,dz = (\beta/\sqrt{\pi})^3\,e^{-\beta^2 r^2}\,dx\,dy\,dz$$
$$W(r)\,dr = (\beta/\sqrt{\pi})^3\,e^{-\beta^2 r^2}\,4\pi r^2 dr \tag{5-13}$$

where

$$r^2 = x^2 + y^2 + z^2$$
$$\beta = \left(\frac{3}{2nl^2}\right)^{1/2}$$

These functions are illustrated in Figure 5–3 for a hypothetical random flight polypeptide chain. The Gaussian distribution function $W(x, y, z)\,dx\,dy\,dz$ gives the probability that the end of the chain is within the volume $dx\,dy\,dz$ at coordinates x, y, and z, taking the other end of the chain as the origin. Since the distribution is spherically symmetric, it is most often expressed as the radial distribution function $W(r)dr$, the probability that the two ends of the chain are within a distance r and $r + dr$ from each other.

It is clear from the Gaussian distribution function that the probability is at a maximum near the origin, but the radial distribution function is zero there. This is simply a mathematical consequence of the decreasing volume of the spherical shell between r and $r + dr$ as r decreases.

The probability that the ends of the chain will be near each other is most pertinent for the probability of interaction between two functional groups on the chain. This may be expressed as the effective concentrations of the two ends with respect to each other. Some calculated values for

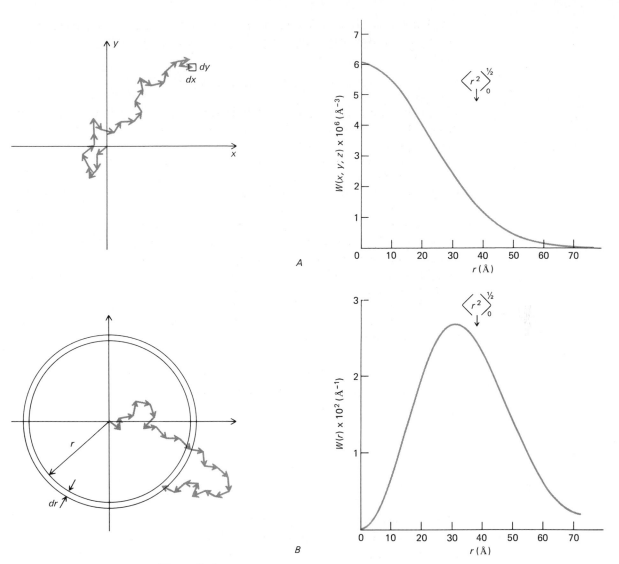

Figure 5–3

Illustration of Gaussian (A) and radial (B) distribution functions for the end-to-end distance of a freely jointed chain. On the left of each is a two-dimensional representation of how each is defined, showing the stippled area for which the probability is calculated that the end of the chain lies in it. On the right of each is the calculated distribution for unperturbed chains of 100 residues and a monomer length of 3.8 Å, corresponding to a freely jointed polypeptide chain. The root-mean-square distance, $\langle r_0^2 \rangle^{1/2}$ (= 38 Å), is indicated. The scale for r would be increased by the factor $C_n^{1/2}$ for unperturbed non–freely jointed chains.

unperturbed polypeptide chains are shown in Table 5–1, for the special case of cyclization of the backbone. Rather stringent stereochemical restraints apply here, giving rather low values for short polypeptides. For

Table 5–1   Theoretical Values for the Effective Molar Concentrations of the Ends of Random Polypeptides of n *Residues*

POLYPEPTIDE	$n = 6$	$n = 10$	$n = 20$
Poly(Gly)	3.6×10^{-2}	6.3×10^{-2}	3.3×10^{-2}
Poly(Ala)	8.5×10^{-4}	6.7×10^{-4}	4.3×10^{-4}
Poly(Pro-Ala)	1.2×10^{-2}	5.8×10^{-3}	2.8×10^{-3}

Adapted from M. Mutter, J. Am. Chem. Soc. 99:8307–8314, 1977.

longer chains, the effective concentration decreases by the factor $n^{-3/2}$. Polyglycine has the highest value, owing to the flexibility of this amino acid. In all cases, the effective concentrations of the chain ends are much less than 1 M.

With these low effective concentrations of groups attached to polypeptide chains, single interactions between hydrogen bond donors and acceptors or between ion pairs would not be expected to be very substantial. By Equation 4–16 and the association constants in Table 4–6 of 0.01 to 0.4 M^{-1}, effective concentrations of 10^{-1} to 10^{-4} M would produce intramolecular equilibrium constants for hydrogen bond or ion-pair formation of only 10^{-6} to 4×10^{-2}. That is, only 10^{-6} to 4×10^{-2} of the molecules would be expected to have these interactions at any instant of time. Consequently, individual weak interactions would not be expected to be sufficient to fix the polypeptide chain so that pairs of groups may interact. However, this is not to say that these weak interactions are insignificant, because the proportions of molecules in which these groups are in proximity might be much lower in their absence.

All of these calculations have included some conformations in which the hypothetical unperturbed chain has intersected itself. Such conformations are impossible; a real chain has a finite volume that excludes other parts of the chain. This excluded volume effect must require larger dimensions of the chain than those calculated, but it is very complex, and there is no satisfactory way of including it in the calculation.

Dimensions of protein random coils. W. E. Miller and C. V. Goebel. Biochemistry 7:3925–3935, 1968.

Macrocyclization equilibria of polypeptides. M. Mutter. J. Am. Chem. Soc. 99: 8307–8314, 1977.

Ring closure reactions of bifunctional chain molecules. G. Illuminati and L. Mandolini. Acc. Chem. Res. 14:95–102, 1981.

REGULAR CONFORMATIONS OF POLYPEPTIDES

The random coil might be considered the natural state of a polymer, favored by its conformational entropy and interactions with the solvent. However, other conformations will be adopted if sufficient interactions are possible, within or between molecules. Many synthetic polyamino acids, where one or a few amino acids are polymerized in a regular sequence,

Table 5–2 **Parameters for Regular Polypeptide Conformations**

	BOND ANGLE (degrees)			RESIDUES PER TURN	TRANSLATION PER RESIDUE (Å)
	ϕ	ψ	ω		
Antiparallel β-sheet	−139	+135	−178	2.0	3.4
Parallel β-sheet	−119	+113	180	2.0	3.2
Right-handed α-helix	−57	−47	180	3.6	1.50
3_{10} helix	−49	−26	180	3.0	2.00
π helix	−57	−70	180	4.4	1.15
Polyproline I	−83	+158	0	3.33	1.9
Polyproline II	−78	+149	180	3.00	3.12
Polyglycine II	−80	+150	180	3.0	3.1

Adapted from G. N. Ramachandran and V. Sasisekharan, Adv. Protein Chem. 23:283–437, 1968; and IUPAC–IUB Commission on Biochemical Nomenclature, Biochemistry 9:3471–3479, 1970.

have been found to adopt a few such regular conformations that are also found in natural proteins. The regularity of the conformation is a result of regularity of the primary structure. Each residue, or short sequence of residues that makes up the repetitive unit, will tend to adopt the same conformation; therefore, it may be specified by just a few dihedral angles. The polypeptide chain then will have some form of a helical conformation, which will be characterized by the number of residues per turn of helix and by the distance traversed along the helix axis per residue. These values for the regular conformations described here are given in Table 5–2.

X-ray analysis and protein structure. R. E. Dickerson. In The Proteins, 2nd ed. H. Neurath (ed.). Vol. 2, pp. 603–778. New York, Academic Press, 1964.

The α-Helix

The right-handed α-helix is the most well-known and prominent of the polypeptide regular structures (Figure 5–4). It has 3.6 residues per turn and a translation per residue of 1.50 Å, or 5.41 Å per turn. The torsion angles ϕ and ψ are favorable for most residues (see Figure 5–5), and the atoms of the backbone pack closely, making very favorable van der Waals contacts. Most conspicuously, the backbone carbonyl oxygen of each residue hydrogen bonds to the backbone NH of the fourth residue along the chain. These hydrogen bonds are 2.86 Å long from the O atom to the N atom and are very nearly straight, close to the optimal geometry for such an interaction (p. 137), and are nearly parallel to the helix axis. All the hydrogen bonds point in the same direction, so the dipoles of the peptide bonds probably interact favorably head to tail. Consequently, an α-helix has a considerable dipole moment, since the partial charges at the ends are well separated in space.

The side chains project out from the helix and do not interfere with it, except in the bulkiest examples. Only Pro residues are incompatible with this conformation, because the side chain is bonded to the backbone N atom, preventing its participation in hydrogen bonding and interfering

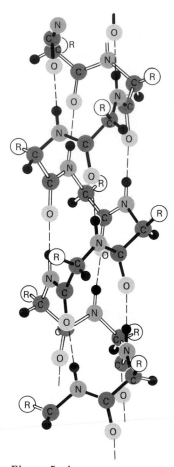

Figure 5–4
The right-handed α-helix. (Adapted from L. Pauling, The Nature of the Chemical Bond, 3rd ed. Ithaca, N.Y., Cornell University Press, 1960.)

in the packing. The stereochemical properties of the α-helix are so favorable that it is often considered the most natural conformation for a polypeptide. However, we will see that α-helices are only marginally stable in solution and that amino acids differ in their propensity to adopt this conformation.

A left-handed α-helix is also sterically possible, with the same values of ϕ and ψ but of opposite sign (Figure 5–5). However, such a conformation is not favorable energetically, as the side chains are in close contact with the backbone, and it is generally not observed.

The structure of proteins: two hydrogen-bonded helical configurations of the polypeptide chain. L. Pauling, et al. Proc. Natl. Acad. Sci. U.S.A. 37:205–211, 1951.

Refinement of bond angles of an α-helix. S. Arnott and S. D. Dover. J. Mol. Biol. 30:209–212, 1967.

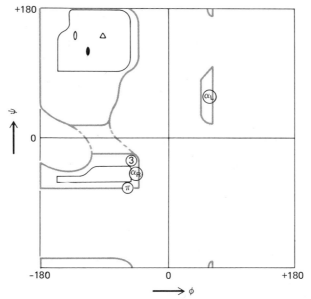

Figure 5–5
Regular conformations of polypeptides on a Ramachandran plot. The regular conformations are α_R, the right-handed α-helix; α_L, left-handed α-helix; $\lozenge$, antiparallel β-sheet; $\bullet$, parallel β-sheet; 3, right-handed 3_{10} helix; π, right-handed π helix; $\triangle$, polyproline I, polyproline II, and polyglycine II. (Adapted from G. N. Ramachandran and V. Sasisekharan, Adv. Protein Chem. 23:283–437, 1968.)

Conformational energy estimates for helical polypeptide molecules. D. A. Brant. Macromolecules 1:291–300, 1968.
The α-helix dipole and the properties of proteins. W. G. J. Hol, et al. Nature 273:443–446, 1978.

β-Sheets

In the β-sheet conformation, the polypeptide chain is nearly fully extended, and individual strands aggregate side by side, forming hydrogen bonds between the carbonyl and NH groups of the backbone (Figure 5–6). In addition to the hydrogen bonds, the dipoles of the peptide bonds alternate along the chain, providing favorable conditions for interaction. The adjacent strands may be either parallel or antiparallel; the two forms differ slightly in dihedral angles (Table 5–2 and Figure 5–5), but both are sterically favorable. Synthetic polymers generally adopt either one or the other orientation, but mixed sheets with both parallel and antiparallel strands are found in proteins (Chapter 6); the middle strand then takes on an intermediate conformation.

The extended strands of planar β-sheets do not appear to be helical, but they may be considered a rather special helix with 2.0 residues per turn and a translation of 3.2 or 3.4 Å per residue. Most sheets that have been observed in detail are not planar but have a twisted conformation. The values of ϕ and ψ are both somewhat more positive in value than those given in Table 5–2 and presumably are somewhat favored energetically, to give a right-handed twist to the backbone.

The extended conformation of the polypeptide chain results in the side chains protruding on alternating sides of the sheet. Most amino acids are stereochemically compatible with the β-sheet conformation, except for Pro, which has no NH group to participate in hydrogen bonding and cannot adopt the appropriate value of φ.

β-Sheets may involve aggregation of different molecules, or the polypeptide chain may loop back on itself to form an intramolecular sheet, known as the **cross-β conformation.**

Configurations of polypeptide chains with favoured orientations around single bonds: two new pleated sheets. L. Pauling and R. B. Corey. Proc. Natl. Acad. Sci. U.S.A. 37:729–740, 1951.

Structure of β-poly-L-alanine: refined atomic coordinates for an antiparallel beta-pleated sheet. S. Arnott, et al. J. Mol. Biol. 30:201–208, 1967.

Conformation of twisted β-pleated sheets in proteins. C. Chothia. J. Mol. Biol. 75:295–302, 1973.

β-pleated sheets in oligopeptide crystals. T. Ashida, et al. Int. J. Pept. Protein Res. 17:322–329, 1981.

Role of interchain interactions in the stabilization of the right-handed twist of β-sheets. K.-C. Chou, et al. J. Mol. Biol. 168:389–407, 1983.

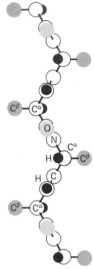

A

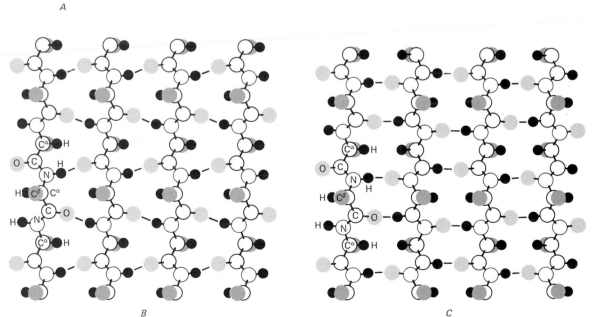

B

C

Figure 5–6
Drawings of a single, straight β*-strand* (A) *and its incorporation into flat parallel* (B) *and antiparallel* (C) *pleated sheets. (Adapted from L. Pauling, The Nature of the Chemical Bond, 3rd ed., Ithaca, NY, Cornell University Press, 1960.)*

Other Regular Conformations (Helices)

Other regular conformations have been proposed for polypeptide chains; many have not been observed in proteins and are seen only with certain polypeptides in special instances.

Variations on the α-helix in which the chain is either more tightly or more loosely coiled, so that hydrogen bonds between corresponding groups are closer or further apart in the primary structure by one residue, are designated the **3₁₀ helix** or the **π helix,** respectively. The packing of the backbone atoms is too tight in the 3_{10} helix; it has not been observed as a regular structure but occurs only at the ends of α-helices, where one turn may have this conformation locally. The π helix would have a hole down the middle, so the backbone atoms would not be in contact, and the values of φ and ψ are rather unfavorable. This helix has never been observed, nor have the more extreme helical variants that could be imagined.

Pro residues are incompatible with both α-helix and β-sheet conformations, so it is not surprising that poly(Pro) forms other regular conformations, known as poly(Pro) I and II. Form I contains all *cis* peptide bonds, whereas II has *trans;* Pro is the only amino acid where the *cis* form is generally significant (see Eq. 5–6). The former is a right-handed helix with 3.3 residues per turn, whereas the latter is a left-handed helix with 3.0 residues per turn (Figure 5–7). Which form is adopted depends primarily on the solvent: Form II predominates in water; it is encountered only locally in proteins when there are sequential Pro residues, except for collagen and some related structures (see p. 194).

Glycine residues also have unique conformational flexibility and poly(Gly) likewise forms two regular conformations, designated I and II. The former has a β-sheet conformation; the latter is a threefold helix quite like that of poly(Pro) II.

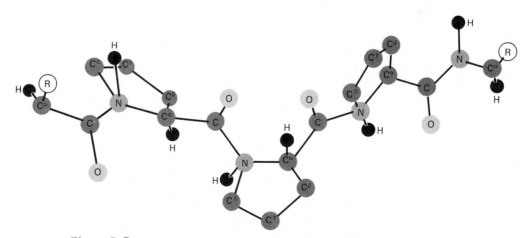

Figure 5–7
The polyproline II helix. (Adapted from A. G. Walton, Polypeptides and Protein Structure. New York, Elsevier-North Holland, 1981.)

In summary, the only regular conformations encountered with polypeptides under normal conditions are the α-helix, parallel and antiparallel β-sheets, and the poly(Pro) II helix. The last one occurs only with two specific polyamino acids, so only the α-helix and the β-sheets are regular conformations likely to be encountered with a typical protein sequence.

Structure of polyglycine II. F. H. C. Crick and A. Rich. Nature 176:780–781, 1955.
Structure of poly-L-proline II. V. Sasisekharan. Acta Crystallog. 12:897–903, 1959.
Molecular structure of polyglycine II. G. N. Ramachandran, et al. Biochim. Biophys. Acta 112:168–170, 1966.

EXPERIMENTAL CHARACTERIZATION OF POLYPEPTIDES IN SOLUTION

A large number of techniques has been developed for characterizing synthetic polymers. Many of these techniques are concerned with determining molecular weight averages and distributions, but they are not pertinent with proteins, where the polypeptide chain has a fixed length. Other techniques are concerned with chain dimensions and conformation, which are more pertinent and will be described very briefly.

The molecular state in real polymers differs from the hypothetical unperturbed state used in calculations in that such polymers have finite dimensions, as well as attractions and repulsions both within the molecule and with the solvent. The choice of solvent is of crucial importance for polymer conformation. For example, most synthetic polypeptides can be induced into the α-helical conformation by dissolving in dichloroacetic acid, whereas β-sheets are favored in solutions of formic acid. Where the side chain can ionize, the pH is of crucial importance, as the electrostatic repulsions drive the ionized form into the random coil state, where the distances between charges may be maximized. A good solvent interacts favorably with the polymer, tending to keep the polymer segments farther apart than expected for the unperturbed state. On the other hand, polymer–polymer contacts are preferred in a poor solvent, so the average dimensions are lower. A θ solvent is one in which the repulsive and attractive forces are exactly counterbalanced; the polymer dimensions are then those calculated for the unperturbed state.

However, proteins are much more complex than ordinary polymers, owing to the diversity of the amino acid side chains. A solvent that is good for polar side chains is likely to be poor for hydrophobic groups, so a θ solvent is virtually impossible. The best solvents for disordered proteins appear to be very concentrated solutions of urea or guanidinium salts, usually the chloride or thiocyanate (see Chapter 4).

Conformation and conformational transitions of poly-α-amino acids in solution. N. Lotan, et al. Ann. Rev. Biochem. 41:869–902, 1972.
Methods for the study of the conformation of small peptide hormones and antibodies in solution. L. C. Craig, et al. Ann. Rev. Biochem. 94:479–490, 1975.
Biophysical Chemistry, Part II. C. Cantor and P. R. Schimmel. San Francisco, W. H. Freeman, 1980.

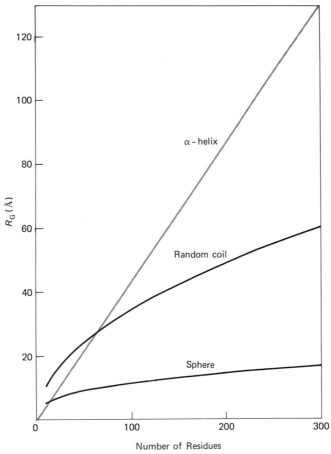

Figure 5–8
Radius of gyration, R_G, for polypeptide chains of various lengths in α-helical, random coil, and compact spherical conformations.

Hydrodynamic Properties

The dimensions of a protein depend upon its conformation (Figure 5–8) and determine its hydrodynamic properties. However, all hydrodynamic properties also depend upon the solvation of the molecule; solvent that is either bound to the surface or trapped in crevices or holes in the interior moves with the molecule and increases its effective mass. Only with a number of different measurements can the relative contributions of shape and viscosity be determined.

The **intrinsic viscosity,** [η], of a molecule is a measure of its effective hydrodynamic volume per unit of mass, determined by its resistance to flow. It is related to the radius of gyration and the mean end-to-end distance. For an ideal random polymer

$$[\eta] = \Phi\frac{\langle r^2\rangle^{3/2}}{M} = \Phi\frac{6^{3/2}R_G^{3}}{M} \qquad (5\text{–}14)$$

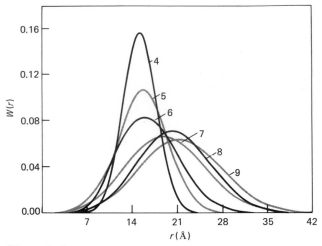

Figure 5–9
Radial distribution function of the distances between naphthalene and dansyl groups attached to the ends of peptides of 4 to 9 residues of N-*hydroxyethyl-Gln. (From E. Haas, et al., Proc. Natl. Acad. Sci. U.S.A. 72:1807–1811, 1975.)*

where Φ is a constant, independent of the nature of the polymer. For very long polymers, it has the value 2.1×10^{23} when the dimensions of the polymer are given in centimeters and $[\eta]$ is in units of cm^3/g; with relatively short polypeptides, however, a somewhat lower value applies.

A related measure is the **frictional coefficient,** f, which is generally evaluated from the sedimentation coefficient (see Eq. 1–54, p. 29) or from the diffusion coefficient, D (Eq. 1–55). For a random coil, its value should be given by

$$f = 5.1 \, \eta_0 \, \langle r^2 \rangle^{1/2} \tag{5–15}$$

where η_0 is the viscosity of the solvent. The frictional coefficient is generally expressed as the frictional ratio of the observed value of f to that expected for a solid rigid sphere of radius r, f_0:

$$f_0 = 6\pi\eta_0 r = 6\pi\eta_0 \left(\frac{3 \, M \, \bar{v}}{4\pi \, N_A}\right)^{1/3} \tag{5–16}$$

This ratio, f/f_0, gives a measure of the departure of the overall shape from a sphere.

Most uses of the viscosity and frictional coefficients have been to measure molecular weights of synthetic polymers of known conformation or to infer the conformation from the dependence on the molecular weight, which is usually different for most overall shapes. Viscosity measurements are useful to follow conformational changes, as random polypeptides have higher viscosities than folded, globular proteins, whereas those of long helices are even higher (see Figure 5–8).

Spectral Properties   The distances between specific groups of polypeptides may be estimated spectrally if the one group is a fluorescent energy donor and the other a suitable energy acceptor. If the absorption spectrum of the acceptor overlaps the emission spectrum of the donor, fluorescent light emitted by the donor will be absorbed by the acceptor. The efficiency of this process depends upon the sixth power of the distance between them, most usefully within a range of 10 to 60 Å. Oligomers of poly(Pro) with 1 to 12 residues separating a naphthalene donor and a dansyl acceptor by distances ranging from 12 to 46 Å gave the expected variation of energy transfer with distance. The same acceptor and donor were attached to the ends of flexible oligomers of N-(hydroxyethyl)glutamine:

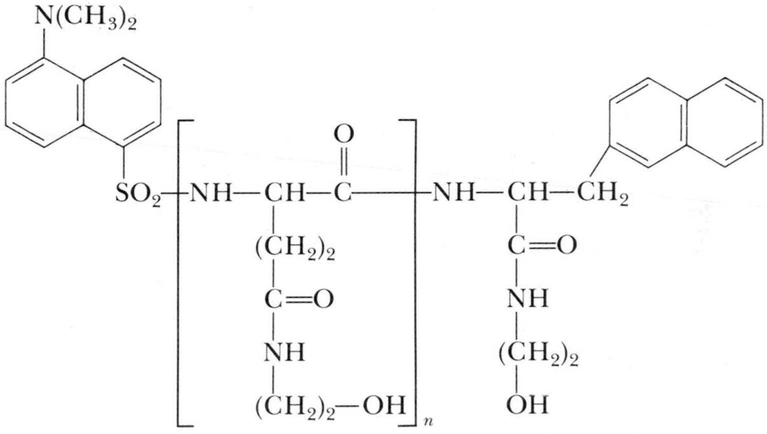

$$(5\text{--}17)$$

with $n = 4$ through 9. The observed efficiencies of energy transfer could be used to reconstruct the distribution function of the distances between the groups (Figure 5–9).

Other spectral properties of polypeptides may be directly sensitive to the conformations of the polypeptide chain. Of greatest use is the optical activity of polypeptides due to the asymmetric centers of the L-amino acids and to their asymmetric conformations. This asymmetry causes them to interact differently with right- and left-circularly polarized light. If the two beams consequently travel at different speeds through the molecule, polarized light is rotated; **optical rotatory dispersion (ORD)** is the dependence of this rotation upon wavelength, λ. In a wavelength region where the molecule does not absorb light, the rotation varies gradually with wavelength, but in an absorbance region, the rotation first increases sharply in one direction, falls to zero at the absorption maximum, and then rises sharply in the opposite direction; this is known as a **Cotton effect.** There will also be unequal absorption of left- and right-circularly polarized light, which is known as **circular dichroism.** Both phenomena have the same cause and consequently are related, as illustrated in Figure 5–10.

The polypeptide backbone absorbs and is optically active in the far UV region, below approximately 240 nm. The magnitude of the absorb-

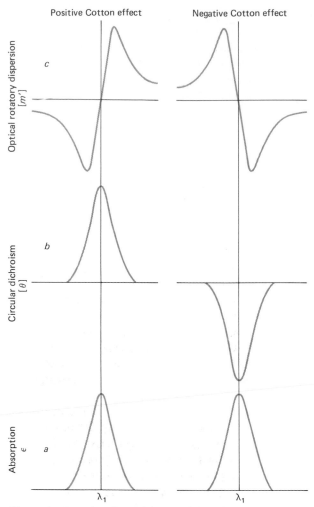

Positive Cotton effect   Negative Cotton effect

Optical rotatory dispersion [m']   c

Circular dichroism [θ]   b

Absorption ε   a

λ_1 λ_1

Figure 5–10
A typical electronic absorption band a *with its two types of circular dichroism* b *and optical rotatory dispersion* c *curves. (From A. J. Adler, et al.: Methods Enzymol. 27:675–735, 1973, © Academic Press Inc.)*

ance is somewhat dependent upon conformation, but this is most apparent in the ORD and circular dichroism spectra. Early ORD studies used the gradual variation of rotation, $[m']$, with wavelength at long wavelengths, above the Cotton effect, and fitted the curve to the Moffitt equation

$$[m']_\lambda = \frac{a_0\lambda_0^2}{\lambda^2 - \lambda_0^2} + \frac{b_0\lambda_0^4}{(\lambda^2 - \lambda_0^2)^2} \qquad (5-18)$$

where a_0, b_0, and λ_0 are constants. When λ_0 was taken to be 212 nm with polypeptides, the b_0 value was found to be proportional to the α-helix content, being 0 with no helix and -630 for totally α-helical conforma-

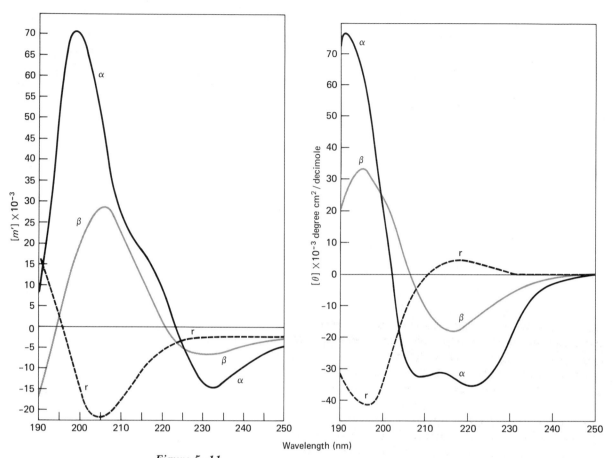

Figure 5–11
Optical rotatory dispersion (left) and circular dichroism (right) spectra of poly(Lys) in the α-helical (α), β-sheet (β), and random coil (r) conformations. (From N. J. Greenfield, et al.: Biochemistry 6:1630, 1967; 8:4108, 1969, © American Chemical Society.)

tions. At lower wavelengths, the Cotton effect was more sensitive to conformation, including β-sheet (Figure 5–11, *left*). However, circular dichroism has come to displace ORD, as it has the advantage of giving discrete spectral bands, which may be positive or negative; normal spectra composed of multiple bands are consequently easier to resolve.

Examples of the circular dichroism spectra of poly(Lys) in random, α-helical, and β-sheet forms are shown in Figure 5–11, *right*. This polyamino acid is disordered at neutral and acidic pH values, where the side chains are ionized, owing to electrostatic repulsions between them. At high pH values, the nonionized form adopts the α-helical conformation, but by gently heating, e.g., at 52° C for 15 minutes, it may be converted to aggregated antiparallel β-sheets. Similar spectra are generally obtained for other polyamino acids in these conformations, so they reflect primarily the conformation of the backbone. To the extent that these spectra are general, observed spectra of other polypeptides may be deconvoluted to determine the relative proportions of the three conformations present. Problems are often encountered, possibly because of spectral contributions by some side chains or because of the presence of other conformations, but circular dichroism remains the spectral probe most sensitive to polypeptide backbone conformation.

Circular dichroism and optical rotatory dispersion of proteins and polypeptides. A. J. Adler, et al. Methods Enzymol. 27:675–735, 1973.

Distribution of end-to-end distances of oligopeptides in solution as estimated by energy transfer. E. Haas, et al. Proc. Natl. Acad. Sci. U.S.A. 72:1807–1811, 1975.

Fluorescence energy transfer as a spectroscopic ruler. L. Stryer. Ann. Rev. Biochem. 47:819–846, 1978.

Determination of protein secondary structure in solution by vacuum ultraviolet circular dichroism. S. Brahms and J. Brahms. J. Mol. Biol. 138:149–178, 1980.

Diffusion-enhanced fluorescence energy transfer. L. Stryer, et al. Ann. Rev. Biophys. Bioeng. 11:203–222, 1982.

DYNAMIC BEHAVIOR IN SOLUTION

Molecules in solution are not static at normal temperatures but undergo a variety of movements, ranging from translation and rotation of the entire molecule, through rotations about single bonds, to bond stretching and bond angle vibrations. The greater the size of the molecule, the greater the degree of complexity of its dynamic properties.

Dynamic processes within a molecule—that is, the movement of parts relative to each other—are generally expressed as the **relaxation time.** This is the average time required for a population of molecules to change from their original positions to $1/e$ of their equilibrium positions. If a rotation is involved, the average value is 68°, that for which the cosine has the value $1/e$. Many resonance techniques use **correlation times,** which are one third of the relaxation times, or **rotational diffusion constants,** which are the reciprocals of twice the relaxation time.

Rates of conformational transitions in biological macromolecules and their analogs. H. Morawetz. Adv. Protein Chem. 26:242–277, 1972.

Kinetics of cooperative conformational transitions of linear biopolymers. G. Schwarz and J. Engel. Angew. Chem. Int. Ed. 11:568–575, 1972.

Low frequency vibrations and the dynamics of proteins and polypeptides. W. L. Peticolas. Methods Enzymol. 61:425–458, 1978.

Overall Translation and Rotation

Molecules undergo rotation and translation because of Brownian motion, in which they are subjected to multiple collisions with the surrounding atoms of their environment. **Translational movement** is characterized by the diffusion coefficient, D, which is classically defined as and measured by the decrease with time (t) of a concentration gradient (dc/dx) of the substance:

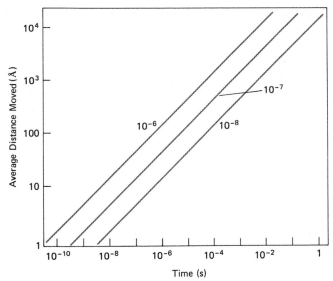

Figure 5–12
Average distance moved as a function of time by molecules with typical translational diffusion coefficients of 10^{-6}, 10^{-7} and 10^{-8} cm²/sec. Values were calculated with Equation 5–20.

$$\frac{\partial c}{\partial t} = D\,\frac{\partial^2 c}{\partial x^2} \qquad (5-19)$$

Diffusion arises from the kinetic movement of the molecules, which will follow a random flight path. The diffusion coefficient is the average square distance traveled per molecule, $\bar{x}^2$, per unit of time:

$$D = \frac{\bar{x}^2}{2t} \qquad (5-20)$$

Owing to the "random walk" nature of diffusion, the average distance traveled is proportional to the square root of time. Theoretical values of the average distances traveled as a function of time are given in Figure 5–12 for molecules with diffusion coefficients of 10^{-6}, 10^{-7}, and 10^{-8} cm²/sec, the range observed for proteins and related molecules (Table 5–3).

The translational diffusion coefficient depends upon both the size and the shape of a molecule, generally expressed by the frictional coefficient (see Eqs. 1–54 and 1–55, pp. 29 and 30), but the rate of rotation is most sensitive to the shape. Spherical molecules of a given size rotate most rapidly, with a rotational relaxation time, τ_R, predicted by the equation:

$$\tau_R = \frac{3V\,\eta_0}{k_B\,T} \qquad (5-21)$$

where V is the volume of the sphere and η_0 is the viscosity of the solvent.

Table 5–3   Examples of Translational and Rotational Diffusion Rates

MOLECULE	TRANSLATIONAL DIFFUSION COEFFICIENT (10^{-7} cm²/sec)	ROTATIONAL RELAXATION TIME
H_2O	200	10^{-2} nsec
Glycine	106[a]	
Alanine	91[a]	
Ala-Gly	72[a]	
Tryptophan		8.7 nsec[d]
Globular proteins		
Myoglobin		30 nsec[d]
Ribonuclease A	12.6[b]	22 nsec[i]
Lysozyme	10.6[b]	30 nsec[h]
Chymotrypsin		45 nsec[h]
Immunoglobulin G	3.8[b]	504 nsec[c]
Serum albumin	6.7[a]	125 nsec[c]
Unfolded proteins		
Serum albumin	1.9[f]	
Pepsinogen	2.5[f]	
Chymotrypsinogen	3.2[f]	
Tropomyosin	2.2[a]	
Fibrinogen	2.0[e]	3.5 msec[e]
Myosin	0.84[b]	
Collagen		0.5 msec[e]
Poly(benzyl-Glu) (M = 3.4×10^5)		
α-Helix	0.85[g]	
Random coil	1.30[g]	
Tobacco mosaic virus	0.3 to 0.4[b]	1.2 to 1.6 msec[b]

[a]From L. J. Gosting, Adv. Protein Chem. 11:429–554, 1956.

[b]From F. D. Carlson, Ann. Rev. Biophys. Bioeng. 4:243–264, 1975.

[c]From J. Yguerabide, et al., J. Mol. Biol. 51:573–590, 1970.

[d]From S. R. Anderson, et al., Biochemistry 9:4723–4729, 1970.

[e]From V. A. Bloomfield, Ann. Rev. Biophys. Bioeng. 10:421–450, 1981.

[f]From M. E. McDonnell and A. M. Jamieson, Biopolymers 15:1283–1299, 1976.

[g]From N. C. Ford, et al., J. Chem. Phys. 50:3098–3100, 1969.

[h]From D. R. Bauer, et al., J. Am. Chem. Soc. 99:2580–2582, 1975.

[i]From S. Krause and C. T. O'Konski, Biopolymers 1:503–515, 1963.

For spheres of radii of a few up to 50 Å in water at room temperature, this time is expected to be in the range of 10^{-9} to 10^{-7} sec. Folded, globular proteins rotate at rates close to those expected. Rodlike molecules, such as α-helices, rotate about their helical axis at comparable rates, but they rotate at right angles much more slowly, with relaxation times of about 10^{-4} sec. In the case of the rod-shaped tobacco mosaic virus, 180 Å in diameter and 3000 Å long, this time is about 10^{-3} sec (Table 5–3). Which rotation is measured depends upon the technique used.

Rotation of proteins has primarily been measured by **fluorescence depolarization.** A molecule absorbs light with maximum efficiency when the electric vector of the impinging light wave vibrates along certain specific directions in the molecule, depending upon its electronic structure. Fluorescent light emitted from such molecules is similarly polarized. If the exciting light is plane-polarized, with the vectors of all waves the same, only those molecules in a solution with the appropriate orientation will be excited. Fluorescent light emitted from such molecules will also be polarized, but the polarization will be decreased by the extent to which the molecules have rotated during the time between absorbing and emitting light; this is generally of the order of 10^{-9} to 10^{-8} sec. Consequently, the decrease in polarization measures the rotation of the molecules during the lifetime of the excited state.

The fluorescence depolarization may be measured in the steady state, with constant excitation and emission; measurements are made as a function of temperature and viscosity of the solvent, which directly affect the rate of rotation. Alternatively, the decrease in polarization with time may be measured directly on the nanosecond time scale, following a very short pulse of polarized exciting light.

Other techniques that may be used to give information about both translational and rotational motions of macromolecules are flow birefringence, NMR, and quasi-elastic light scattering, also known as intensity fluctuation spectroscopy.

Translations and rotations of the entire molecule are relatively simple only with a completely rigid molecule. In other cases, rotations of individual groups will also occur on faster time scales; for example, the side chains of the residues of an α-helix rotate about their single bonds at a rapid rate relatively independently of movement of the overall molecule. The situation is most complex with random coils, where many parts of the molecule can move relatively independently.

Rotational Brownian motion and polarization of the fluorescence of solutions. G. Weber. Adv. Protein Chem. 8:415–459, 1953.

Measurement and interpretation of diffusion coefficients of proteins. L. J. Gosting. Adv. Protein Chem. 11:429–554, 1956.

The application of intensity fluctuation spectroscopy to molecular biology. F. D. Carlson. Ann. Rev. Biophys. Bioeng. 4:243–264, 1975.

Hydrodynamics in biophysical chemistry. V. A. Bloomfield. Ann. Rev. Phys. Chem. 28:233–259, 1977.

Rates of Conformational Changes in Disordered Polypeptides

Movements within polymers, by rotation, stretching, or bending of the individual bonds, are complicated and not entirely understood. Each conformational parameter of a random coil is independent of all others at equilibrium, but this cannot be the case in fluctuations on a short time scale. If one bond near the middle of the chain were to rotate by 180°, with no other rotations, the ends of the chain would undergo extremely large movements. Such a process is implausible in a viscous solution; it seems intuitively obvious, therefore, that the rotations of all bonds must be coordinated in such a way as to produce more plausible types of movements. This complexity, along with the conformational complexity of the random coil, makes it impossible to describe in detail the movement of a disordered polypeptide, but it *is* possible to describe the average rates at

Table 5–4   Rotational Relaxation Times in a Random Polypeptide Chain

CARBON ATOM		RELAXATION TIME (10^{-9} sec)
Ala	C^β	0.21
Thr	C^β	1.56
	C^γ	0.18
Lys	C^β	0.81
	C^γ	0.54
	C^δ	0.60
	C^ϵ	0.27
Peptide	C^α	1.4 to 2.6

The values were measured at 45° C on performic acid–oxidized ribonuclease A by ^{13}C nuclear magnetic resonance.
From V. Glushko, et al. J. Biol. Chem. 247:3176–3185, 1972.

which individual bonds sample all their possible conformations by measurement of their relaxation time.

Such measurements are best made with ^{13}C nuclear magnetic resonance (see Chapter 7); values for various groups in performic acid–oxidized ribonuclease, which approximates a random coil, are given in Table 5–4. Terminal groups of the side chains have relaxation times of 0.2 to 0.3 nanosecond, similar to those observed in small molecules; these very rapid rotations reflect the small intrinsic barriers to rotation of the bonds as well as any necessary rearrangement of the solvent. The necessity for coordination of movements is shown by the increasing relaxation times observed for atoms of Lys side chains further from the ends and closer to the polypeptide backbone. The C^α atoms of the backbone have the largest relaxation times of 1.4 to 2.6 nsec, about 100 times longer than those observed with small peptides.

Another way of envisaging the complex movements of a disordered polypeptide is to consider the rate at which the ends of the polypeptide chain move with respect to each other, which may be expressed as a diffusion constant. Such measurements have been made with the oligomers of N-(hydroxyethyl)glutamine (Eq. 5–17) described earlier, using the dynamics of energy transfer between fluorescent donor and acceptor on the ends of the chains. Diffusion coefficients of from 3×10^{-8} to 8.5×10^{-8} cm²/sec were found for chains of four to nine residues. These values are an order of magnitude smaller than those expected for the free chromophores, indicating that the chain possesses appreciable internal friction that resists motion. Resistance to movement is greatest in the smaller oligomers and probably arises from the need to coordinate rotations about individual bonds. Nevertheless, both these data and those of Table 5–4 indicate that such internal friction slows transitions only by a factor of about 10, so individual parts of a disordered polypeptide chain separated by 50 to 100 residues will tend to diffuse over distances comparable to their average rms separation (Figure 5–8) in 10^{-5} to 10^{-6} sec (Figure 5–12). Different groups of a disordered polypeptide chain will therefore come into contact about 10^5 to 10^6 times per second.

Conformational states of bovine pancreatic ribonuclease A observed by normal and partially relaxed carbon-13 nuclear magnetic resonance. V. Glushko, et al. J. Biol. Chem. 247:3176–3185, 1972.

Carbon 13 nuclear magnetic resonance of pentapeptides of glycine containing central residues of aliphatic amino acids. P. Keim, et al. J. Biol. Chem. 248:6104–6113, 1973.

Determination of rotational mobilities of backbone and side-chain carbons of poly(γ-benzyl L-glutamate) in the helical and random-coil states from measurements of carbon-13 relaxation times and nuclear Overhauser enhancements. A. Allerhand and E. Oldfield. Biochemistry 12:3428–3433, 1973.

Brownian motion of the ends of oligopeptide chains in solution as estimated by energy transfer between the chain ends. E. Haas, et al. Biopolymers 17:11–31, 1978.

Activation energy insensitivity to barrier-crossing correlations in long-chain molecules. D. Perchak and J. H. Weiner. Macromolecules 14:785–792, 1981.

α-Helix → Coil Transitions

The formation of an α-helix from a disordered polypeptide is one of the most thoroughly studied transitions in synthetic polypeptides. Interconversion of the two conformations can occur upon changing the temperature or the solvent, or by altering the pH when the side chain can ionize. Where the equilibrium can be shifted from totally random state to totally helical, the transition between the two is very abrupt, indicative of a cooperative system. Moreover, the rate of forming a helix from the random coil state is very fast, generally occurring within 10^{-5} to 10^{-7} sec, and is independent of the length of the polypeptide chain. In contrast, the rate of unraveling is strongly size-dependent.

These general types of observation have been interpreted in terms of a "nucleation–zippering" model, like those proposed by Zimm and Bragg and by Lifson and Roig. Initiation of the helix in a random coil is the slowest and energetically the most unfavored step, whereas subsequent growth of the helix nucleus is rapid and relatively favored. Initiation of a helix can occur anywhere in the random coil, but growth and unraveling occur only at the ends of helices. This may be expressed simply by interconversion of the coil, C, and helical states, H_i, where i is the number of hydrogen bonds involved in the helical conformation:

$$C \underset{k_b}{\overset{\sigma k_f}{\rightleftarrows}} H_1 \underset{k_b}{\overset{k_f}{\rightleftarrows}} H_2 \underset{k_b}{\overset{k_f}{\rightleftarrows}} \cdots \underset{k_b}{\overset{k_f}{\rightleftarrows}} H_{n-1} \underset{k_b}{\overset{k_f}{\rightleftarrows}} H_n$$

$$\frac{k_f}{k_b} = s \qquad (5\text{--}22)$$

where k_f is the rate constant for adding a residue to the helix, and k_b is that for removing a residue; both have very similar values in the region of 10^8 to 10^{11}/sec, depending upon the amino acid. The value of the equilibrium constant for each residue added to the helix, s, has a value in the region of unity. The nucleation factor σ reflects the difficulty in initiating a helix and generally has a value of 10^{-5} to 10^{-2}.

Why is initiating a helix much more unfavorable than adding an additional residue? An α-helix is defined by hydrogen bonds between residues four apart in the polypeptide chain (Figure 5–4). Consequently, both the participating residues and the three intervening residues must be appropriately fixed in space before the nucleus can be formed. Subsequently adding each additional residue requires that only one residue be fixed; this residue would be in reasonable proximity owing to its position as the next one along in the chain. In other words, the effective concentrations (see p. 154) of two residues four apart in the random coil are much lower than when one is part of the helix and the other is the next residue to the end of the helix.

Other factors may also play a role. For example, the α-helix has all the peptide bonds oriented in the same direction, so the dipoles of neighboring residues are aligned unfavorably. Once the helix is formed, this is compensated for by head-to-tail interactions between the dipoles. Only the unfavorable interactions would occur in the nucleus, but the favorable interactions would assist helix growth.

Owing to the difficulty of nucleation, individual polypeptide molecules tend to be either entirely helical (H_n) or entirely disordered (C). The equilibrium constant between the two, K_n, is given by

$$K_n = \frac{[H_n]}{[C]} = \sigma\, s^n \tag{5–23}$$

With σ values of 10^{-3} or 10^{-4} and s values not much greater than unity, large values of n are required to give values of K_n greater than unity, so only long polypeptides are helical under such conditions. At the midpoint of the transition in very long chains, the average length of a helix is $\sigma^{-\frac{1}{2}}$. Therefore, with $\sigma = 10^{-4}$, the average helical segment is 100 residues long, followed by an average of 100 coil residues. The observed rate constants for helix formation (k_1) and unraveling (k_{-1}) are given by

$$k_1 = \sigma\, k_f\, \frac{s-1}{s}$$

$$k_{-1} = k_b\, \frac{s-1}{s}\left(\frac{1}{s}\right)^{n-1} \tag{5–24}$$

The term $(s-1)/s$ gives the probability that helix formation will be completed once a nucleus has been formed.

Although apparently rigid structures (Figure 5–4), α-helices are usually dynamic systems in solution, being rapidly formed and unfolded, perhaps 10^5 to 10^7 times per second.

The various amino acids differ in their tendencies to form α-helices in water, as measured in synthetic polypeptides. Where the polyamino acid is not sufficiently soluble, the amino acid is incorporated to varying extents as a "guest" in a water-soluble "host" polypeptide. The values of σ and s determined in this way for 18 amino acids by Scheraga are given in Table 5–5. The values of s range from 0.60 to 1.17; presumably Pro would have a lower value, but it has not been measured.

In spite of the great amount of work carried out, the variation in

Table 5–5   Helix Formation Parameters for Amino Acid Residues

AMINO ACID	$\sigma\ (\times\ 10^4)$	s
Gly	0.1	0.60
Asp$^-$	70	0.66
His$^+$	0.1	0.68
Ser	0.1	0.77
Asn	0.1	0.79
His	210	0.80
Thr	0.1	0.83
Lys$^+$	1	0.94
Val	1	0.96
Gln	33	0.96
Glu$^-$	6	0.96
Tyr	66	0.99
Arg$^+$	0.1	1.02
Ala	8	1.06
Trp	77	1.08
Phe	18	1.08
Ile	55	1.12
Leu	33	1.14
Met	54	1.17

σ is the equilibrium constant for initiating helical conformation in a random coil, while s is the equilibrium constant for adding a residue of helical conformation to the end of a helix (see the text). The ionization state of the residue is indicated by the charge on the side-chain.

The data were measured in water at 25° C and were kindly provided by H. A. Scheraga.

tendencies of the amino acids to form α-helices is not understood, nor are the energetic contributions of hydrogen bonds, van der Waals interactions, dipole interactions, and so on. Some indication of the complexity is given in Figure 5–13, which describes the complex temperature-dependence of the value of s for some of the amino acids.

Theory of the phase transition between helix and random coil in polypeptide chains. B. H. Zimm and J. R. Bragg. J. Chem. Phys. 31:526–535, 1959.

On the theory of the helix–coil transition in polypeptides. S. Lifson and A. Roig. J. Chem. Phys. 34:1963–1974, 1961.

On the kinetics of the helix–coil transition of polypeptides in solution. G. Schwarz. J. Mol. Biol. 11:64–77, 1965.

A general treatment of the helix–coil equilibria in macromolecular systems. P. J. Flory and W. G. Miller. J. Mol. Biol. 15:284–297, 1966.

Theory of Helix–Coil Transitions in Biopolymers. D. Poland and H. A. Scheraga. New York, Academic Press, 1970.

Use of random copolymers to determine helix–coil stability constants of the naturally occurring amino acids. H. A. Scheraga. Pure Appl. Chem. 50:315–324, 1978.

Correlated helix–coil transitions in polypeptides. M. R. Pear, et al. Biopolymers 20:629–632, 1981.

Poly(Pro) I and II Interconversion

The right-handed helix of poly(Pro) I, with *cis* peptide bonds, is stable in propanol or butanol, whereas the left-handed helix of form II, with *trans* peptide bonds, is stable in water, acetic acid, and benzyl alcohol. The interconversion between the two forms may be followed after altering the

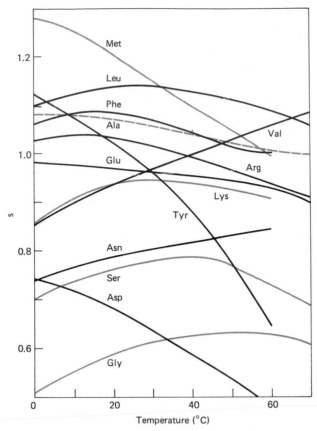

Figure 5–13
Variation of the helix–coil transition parameter s *of several amino acids as a function of temperature. (From H. A. Scheraga, Pure Appl. Chem. 50:315–324, 1978.)*

solvent. The observed rate is nearly independent of the concentration of the initial form (i.e., zero-order), suggesting that the interconversion occurs by a "zipper" mechanism: Change of conformation starts at one end and progresses sequentially along the polypeptide chain. The interconversion is also very slow, occurring on the hour time scale, and the half-time is greater for longer chains.

This mechanism contrasts with that for the transition from coil to α-helix, where nucleation of helix formation occurs anywhere in the chain and is rate-limiting. With poly(Pro), perpetuation of the conformational chain is rate-limiting. This is due to the involvement of *cis–trans* interconversion of the peptide bond—an intrinsically slow process, which occurs on the minute time scale in small model compounds. The half-time for an irreversible zipper mechanism with a polypeptide chain would then be the product of number of residues per chain and the intrinsic half-time. The interconversion is so slow that a proline-specific peptidase able to cleave only amino-terminal *trans* Pro-Pro bonds was used to show that I→II interconversion starts at the amino end of the chain, whereas the reverse interconversion starts at the carboxyl end.

Consideration of the possibility that the slow step in protein denaturation reactions is due to *cis–trans* isomerism of proline residues. J. F. Brandts, et al. Biochemistry 14:4953–4963, 1975.

Kinetic mechanism for conformational transitions between poly-L-prolines I and II: a study utilizing the *cis–trans* specificity of a proline-specific protease. L.-N. Lin and J. F. Brandts. Biochemistry 19:3055–3059, 1980.

β-Sheet Transitions

The mechanism of forming β-sheets is not known, owing primarily to the absence of a good model system with which to study the process. Synthetic polypeptides that form β-sheets generally produce large, insoluble structures comprising many polypeptide chains. Where the β-sheets are soluble, intramolecular structures are formed only at very high dilutions. Under these circumstances, the kinetics of forming the β-structure are simple first-order, but the rates are quite slow, generally occurring on the second-to-minute time scale. Until the mechanism is understood, the significance of this slow rate will remain obscure.

Beta poly(L-lysine): a model for biological self-assembly. R. Hartman, et al. J. Mol. Biol. 90:415–429, 1974.

Kinetics of the disordered chain-to-β transformation of poly(L-tyrosine) in aqueous solution. H. E. Auer and E. Patton. Biophys. Chem. 4:15–21, 1976.

FIBROUS PROTEINS

The class of fibrous proteins contains a variety of structural proteins with relatively simple, regular linear structures. They are related to the regular secondary structures just described, with the polypeptide chain extended. This class of fibrous proteins does not include elongated structures such as microtubules, actin filaments, or flagella, which are long aggregates of globular protein molecules.

Because of their regular, linear conformations, the basis for the regularity of this conformation can be sought in regularities in the amino acid sequence. Characterization of protein structure is rather more straightforward in these relatively simple cases than in that of the globular proteins, which are the subject of the remainder of this volume. Nevertheless, there are still many questions to be answered about their structures, and interest in all these structural proteins has increased recently with the realization that some are involved in the cytoskeletons of cells.

Conformation in fibrous proteins and related synthetic polypeptides. R. D. B. Fraser and T. P. MacRae. New York, Academic Press, 1972.

Intermediate filaments as mechanical integrators of cellular space. E. Lazarides. Nature 283:249–256, 1978.

Intermediate filaments: a chemically heterogeneous, developmentally regulated class of proteins. E. Lazarides. Ann. Rev. Biochem. 51:219–250, 1982.

Silk Fibroin

Silk fibroin is a structural protein synthesized as polypeptide chains of molecular weights of 350,000 to 415,000; it consists primarily of antiparallel β-pleated sheets (Figure 5–6), with interruptions of 100 to 200 residues in other, irregular conformations. The segments in the β-sheet conformation consist of about 50 repeated -(Gly-Ala)$_2$-Gly-Ser-Gly-Ala-Ala-Gly-(Ser-Gly-Ala-Gly-Ala-Gly)$_8$-Tyr- sequences. Neighboring residues in β-sheets have their side chains extending on opposite sides of the sheet, so silk sheets probably have Gly residues on one side, with Ser and Ala residues on the other. Sheets are then stacked on top of each other, those with faces of Gly side chains being in contact, alternating with faces of Ala and Ser in contact. This arrangement gives alternating distances between sheets of 3.5 Å and 5.7 Å. Consequently, silk is not very extensible, since the polypeptide chain is already nearly fully extended, but it is strong, owing to the resistance provided directly by the covalent bonds of the polypeptide backbone. Silk is also flexible, since the sheets are held together only by relatively weak interactions between the individual β-sheets.

It is probably the irregular stretches of the protein that determine the elasticity of the silk; the proportion varies with different species. The irregular conformations arise from the presence of amino acids other than Gly, Ala, and Ser, since larger side chains cannot be accommodated in the normal close-packed silk sheets.

An investigation of the structure of silk fibroin. R. E. March, et al. Biochim. Biophys. Acta 16:1–34, 1955.

The amino acid sequence in a fraction of the fibroin of *Bombyx mori*. F. Lucas, et al. Biochem. J. 66:468–479, 1957.

The silk fibroins. F. Lucas, et al. Adv. Protein Chem. 13:107–242, 1958.

Comparative studies of fibroins. II. The crystal structures of various fibroins. J. O. Warwicker. J. Mol. Biol. 2:350–362, 1960.

A novel structural model for silk fibroin: $\alpha_L\alpha_R$ β-structure. V. I. Lim and S. V. Steinberg. FEBS Letters 131:203–207, 1981.

The silkworm, a model for molecular and cellular biologists. J.-P. Garel. Trends Biochem. Sci. 7:105–108, 1982.

Coiled Coils

Structural proteins such as α-keratins, myosin, tropomyosin, and fibrinogen have coiled coils of two or three α-helices wound around each other, forming a left-handed "super-helix" with a repeat distance of about 140 Å (Figure 5–14). Because of the super-coiling, the ϕ, ψ angles of the backbone deviate slightly from those of a normal α-helix; there are 3.5 residues per turn of helix rather than the usual 3.6. The interactions between the helices are governed by the packing of the side chains, those of one helix fitting into gaps between those of the other. Residues likely to be interacting between helices are nonpolar, whereas the others are usually polar and on the outside. They are on one side of the helix and occur at every third to fourth residue. The α-helix is repeated every seven residues; this repeating sequence of nonpolar amino acids, therefore, is observed throughout the primary structure.

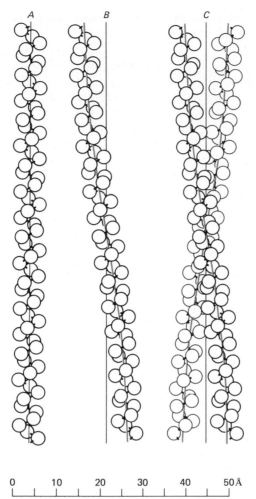

0   10   20   30   40   50 Å

Figure 5–14

Forming a two-stranded coiled-coil structure, as probably occurs in myosin, tropomyosin, and keratin. A normal right-handed α-helix is shown in a, *with a sphere centered on the C^α atom for each residue. This helix is distorted in* b *into a coiled coil; two such molecules are intertwined in* c. *(From R. D. B. Fraser, et al., Keratins, Their Composition, Structure and Biosynthetis. Springfield, Ill., Charles C. Thomas, 1972.)*

The packing of α-helices: simple coiled coils. F. H. C. Crick. Acta Cryst. 6:687–688, 1953.

Amino acid sequence of rabbit skeletal tropomyosin and its coiled-coil structure. J. Sodek, et al. Proc. Natl. Acad. Sci. U.S.A. 69:3800–3804, 1972.

The structure and chemistry of keratin fibers. J. H. Bradbury. Adv. Protein Chem. 27:111–211, 1973.

Sequence repeats in α-tropomyosin. A. D. McLachlan, et al. J. Mol. Biol. 98:281–291, 1975.

The 14-fold periodicity in α-tropomyosin and the interaction with actin. A. D. McLachlan and M. D. Stewart. J. Mol. Biol. 103:271–298, 1976.

Structure of α-keratin: structural implications of the amino acid sequences of the type I and type II chain segments. D. A. D. Parry, et al. J. Mol. Biol. 113:449–454, 1977.

Structure of the three-chain unit of the bovine epidermal keratin filament. P. M. Steinert. J. Mol. Biol. 123:48–70, 1978.

Coiled-coil formation and sequence regularities in the helical regions of α-keratin. A. D. McLachlan. J. Mol. Biol. 124:297–304, 1978.

Designation of sequences involved in the "coiled-coil" interdomainal connections in fibrinogen: construction of an atomic scale model. R. F. Doolittle, et al. J. Mol. Biol. 120:311–325, 1978.

Crystal structure and molecular interactions of tropomyosin. G. N. Phillips, et al. Nature 278:413–417, 1979.

Periodicity of α-helical potential in tropomyosin sequence correlates with alternating actin binding sites. L. B. Smillie, et al. J. Mol. Biol. 136:199–202, 1980.

Tropomyosin: a model protein for studying coiled-coil and α-helix stabilization. J. A. Talbot and R. S. Hodges. Acc. Chem. Res. 15:224–230, 1982.

Periodic features in the amino acid sequence of nematode myosin rod. A. D. McLachlan and J. Karn. J. Mol. Biol. 164:605–626, 1983.

Collagen

Collagen is the main constituent of animal frameworks—the bones, tendons, skin, ligaments, blood vessels, and supporting membranous tissue. In spite of this great diversity of roles, there are only several distinct, but closely related, collagen polypeptide chains in any individual, each with lengths of just over 1000 residues, and all have very similar overall conformations. Except for a few residues at the ends, the polypeptides have a repetitive sequence in which every third residue is Gly, i.e., (Gly-X-Y)$_n$, and about one third of the X and Y residues are Pro. Many of the Pro and Lys residues in the Y position are hydroxylated by posttranslational modification (p. 79).

The reason for the repeating sequence of the collagen chains is apparent from its conformation. Each chain has a slightly twisted, left-handed threefold helical conformation like that of poly(Pro) II and poly(Gly) II (Table 5–2 and Figure 5–7). Three such chains are coiled about each other, linked by hydrogen bonds between the NH of Gly residues and the carbonyl groups of the backbone of residues in position X, to give a right-handed super-helix (Figure 5–15). This regular conformation is adopted by all but the ends of each polypeptide chain to give a molecule 14 Å in diameter and 3000 Å in length. Every third residue of all three chains must be Gly, as they come into close proximity with the other two chains in the center of the coiled coil, too close to permit a side chain. This helical conformation is one of the few accessible to Pro residues, so their presence imparts rigidity to the structure. The hydroxyl groups on hydroxy-Pro residues are involved in hydrogen bonding between chains and are important for stabilizing the triple helix.

Upon disruption of the helical conformation of collagen, for example, by heating, it is converted to gelatin, where the individual polypeptides are unraveled, although still linked by any covalent cross-links. The chains are disordered at high temperatures but tend to assume a collagen-like conformation upon cooling by combining rather randomly with other chains

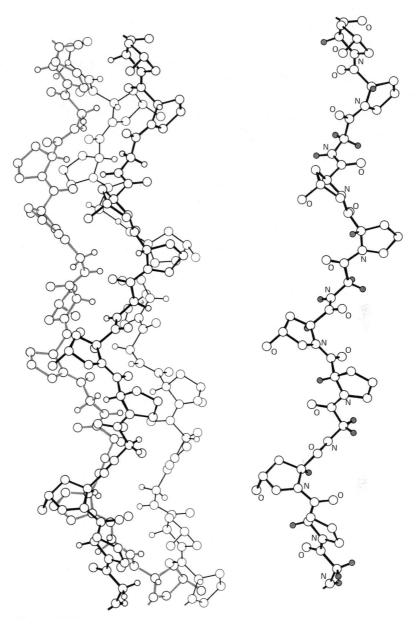

Figure 5–15
Current model for the structure of the collagen molecule, represented as the repeating sequence -Gly-Pro-γOH Pro-. The three-stranded coiled coil is on the left; a single strand is shown on the right. (From R. D. B. Fraser, et al., J. Mol. Biol. 129:463–481, © 1979, London, Academic Press Inc.)

to form short stretches of triple helices. A more regular collagen-like structure will re-form only slowly upon carefully controlled cooling, but the ends of the polypeptide chains tend not to be in register. The assembly in vivo of the regular collagen helix, with the chains in register, is accomplished with nonhelical, globular extensions of just over 100 and 300 residues at the amino and carboxyl ends, respectively, a form known as procollagen. These extensions are later removed by specific proteases. They undoubtedly help to align the three chains, and they are often cross-linked by disulfides, but the posttranslational hydroxylation of Pro and Lys, as well as glyosylation of some of the latter modified residues, occurs before the triple helix is assembled. Because procollagen is secreted it is also synthesized as a precursor, pre-procollagen, with an initial signal sequence rapidly cleaved after secretion.

Within the rather regular repeating amino acid sequence of collagen, there is still much scope for diversity, by varying the residues other than Gly and Pro. This variety is undoubtedly important for determining further aspects of its structure and interactions with other molecules. The next higher order of structure is the aggregation of triple helices into microfibrils. This structure is not entirely understood, but is generated by side-by-side interactions between triple helices to produce arrays of parallel molecules, but with a characteristic axial stagger of about 670 Å between them, corresponding to 234 residues. The dimensions of these arrays vary widely with different collagen molecules and their site of assembly. Once assembled, they are stabilized by a wide variety of covalent cross-links between the triple helices, involving primarily the hydroxy-Lys side chains.

These collagen fibrils are used to generate the wide variety of connective tissues, often by combining with other types of molecules. Tendon is almost pure collagen, with the fibrils arranged in parallel to give it great tensile strength. The fibrils in skin are cross-woven into sheets that can be stretched. In cartilage, the fibrils are embedded in a polysaccharide matrix. In bone, collagen provides a matrix that is then cemented into a rigid structure by deposits of crystals similar to calcium hydroxyapatite, $Ca_{10}(PO_4)_6(OH)_2$. All of these phenomena are no doubt governed by the detailed structure of the collagen polypeptide chain, but this level of structure is not understood.

The molecular structure of collagen. A. Rich and F. H. C. Crick. J. Mol. Biol. 3:483–506, 1961.

Polymers of tripeptides as collagen models. A. Yonath and W. Traub. J. Mol. Biol. 43:461–477, 1969.

The chemistry and structure of collagen. W. Traub and K. A. Piez. Adv. Protein Chem. 25:243–352, 1971.

Procollagen. G. R. Martin, et al. Adv. Enzymol. 42:167–191, 1975.

Biochemistry of Collagen. G. N. Ramachandran and A. H. Reddi (eds.). New York, Plenum Press, 1976.

Chain conformation in the collagen molecule. R. D. B. Fraser, et al. J. Mol. Biol. 129:463–481, 1979.

The chemistry and biology of collagen. P. Bornstein and W. Traub. In The Proteins, 3rd ed. H. Neurath and R. L. Hill (eds.). Vol. 4, pp. 411–632. New York, Academic Press, 1979.

Structurally distinct collagen types. P. Bornstein and H. Sage. Ann. Rev. Biochem. 49:957–1003, 1980.

Collagen: molecular diversity in the body's protein scaffold. D. R. Eyre. Science 207:1315–1322, 1980.

Crystal and molecular structure of a collagen-like polypeptide (Pro-Pro-Gly)$_{10}$. K. Okuyama, et al. J. Mol. Biol. 152:427–443, 1981.

Collagen evolution. E. Solomon and K. S. E. Cheah. Nature 291:450–451, 1981.

A novel structural model for collagen: water–carbonyl helix. V. I. Lim. FEBS Letters 132:1–5, 1981.

Molecular packing in collagen fibrils. A. Miller. Trends Biochem. Sci. 7:13–18, 1982.

Molecular conformation and packing in collagen fibrils. R. D. B. Fraser, et al. J. Mol. Biol. 167:497–521, 1983.

6

THE FOLDED CONFORMATIONS OF GLOBULAR PROTEINS

The great majority of proteins of biological origin differ dramatically from the synthetic polypeptides of random or simple repetitive conformation and from the structural proteins described in the previous chapter. They have much smaller dimensions in solution and are of nearly spherical shape; hence they are generally referred to as **globular.** Moreover, these physical properties do not change in a continuous manner as the environment is altered—for example, by changing the temperature, pH, or pressure—as do random polypeptides. Instead, they generally exhibit little or no change until a point is reached where there is a dramatic change in physical properties and, invariably, a loss of biological function. This phenomenon is known as **denaturation,** and the denatured polypeptide chain is much more like a random polypeptide than like the original protein.

It became obvious from many such observations that denaturation did not require rupture of covalent bonds, but that it involved disruption of some folded conformation of the polypeptide chain, upon which its biological properties are critically dependent. The nature of this folded conformation was uncertain until the structures of the proteins hemoglobin and myoglobin were determined by Perutz and Kendrew, just over two decades ago. Since that time, about 200 protein structures have been determined, depending somewhat on what is considered a unique structure, and a wealth of information about protein structure has become available. A list of the various proteins determined in this way, with references to the most recent or detailed description of it, is given in Table 6–1. These protein structures provide the basis for much of the remainder of this volume.

Table 6–1 *Proteins of Known High-Resolution Structure*

FUNCTIONAL CLASSIFICATION	PROTEIN	MOST COMPREHENSIVE OR RECENT REFERENCE
Oxygen-binding proteins		
Globin family	Hemoglobins	
	Mammalian	G. Fermi and M. F. Perutz, Haemoglobin and Myoglobin, Oxford, Clarendon Press, 1981
	Lamprey	W. A. Hendrickson, et al., J. Mol. Biol. 74:331–361, 1973
	Annelid	E. Padlan and W. E. Love, J. Biol. Chem. 249:4067–4078, 1974
	Myoglobin	S. E. V. Phillips, J. Mol. Biol. 142:531–554, 1980
	Erythrocruorin	W. Steigemann and E. Weber, J. Mol. Biol. 127:309–338, 1979
	Leghemoglobin, root nodule	B. K. Vainshtein, et al., Dokl. Akad. Nauk. S.S.R. 223:238–241, 1977
Hemerythrins	Monomer	W. A. Hendrickson, et al., Proc. Natl. Acad. Sci. U.S.A. 72:2160–2164, 1975
	Trimer	J. L. Smith, et al., Nature 303:86–88, 1983
	Octamer	R. E. Stenkamp, et al., Biochemistry 17:2499–2504, 1978
Oxidation–reduction proteins		
Cytochrome c family	Cytochrome c	T. Takano and R. E. Dickerson, J. Mol. Biol. 153:79–94, 1981
	Cytochrome c_2	F. R. Salemme, et al., J. Biol. Chem. 248:3910–3921, 1973
	Cytochrome c_{550}	R. Timkovich and R. E. Dickerson, J. Biol. Chem. 251:4033–4046, 1976
	Cytochrome c_{551}	Y. Matsuura, et al., J. Mol. Biol. 156:389–409, 1982
	Cytochrome c_{555}	Z. R. Korszun and F. R. Salemme, Proc. Natl. Acad. Sci. U.S.A. 74:5244–5247, 1977
Other cytochromes	Cytochrome b_5	P. Argos and F. S. Mathews, J. Biol. Chem. 250:747–751, 1975
	Cytochrome b_{562}	F. Lederer, et al., J. Mol. Biol. 148:427–448, 1981
	Cytochrome c'	P. C. Weber, et al., J. Mol. Biol. 153:399–424, 1981
	Cytochrome c_3	M. Pierrot, et al., J. Biol. Chem. 257:14341–14348, 1982
Non-heme	Azurin	G. E. Norris, et al., J. Mol. Biol. 165:501–521, 1983
	Ferredoxins	
	2-Iron	K. Fukuyama, et al., Nature 286:522–524, 1980
	7-Iron	J. B. Howard, et al. J. Biol. Chem. 258:508–522, 1983
	8-Iron	E. T. Adman, et al., J. Biol. Chem. 251:3801–3806, 1976
	Rubredoxin	K. D. Watenpaugh, et al., J. Mol. Biol. 131:509–522, 1979
	High-potential iron protein	C. W. Carter, et al., J. Biol. Chem. 249:4212–4225, 1974
	Plastocyanin	J. M. Guss and H. C. Freeman, J. Mol. Biol. 169:521–563, 1983
	Thioredoxin	
	Bacterial	A. Holmgren, et al., Proc. Natl. Acad. Sci. U.S.A. 72:2305–2309, 1975
	Bacteriophage	B.-O. Söderberg, et al., Proc. Natl. Acad. Sci. U.S.A. 75:5827–5830, 1978
Proteases		
Serine proteases		
Subtilisin	Subtilisin	J. D. Robertus, et al., Biochem. Biophys. Res. Commun. 42:334–339, 1971
Trypsin family	Chymotrypsinogen	S. T. Freer, et al., Biochemistry 9:1997–2007, 1970
	α-Chymotrypsin	J. J. Birktoft and D. M. Blow, J. Mol. Biol. 68:187–240, 1972
	γ-Chymotrypsin	G. H. Cohen, et al., J. Mol. Biol. 148:449–479, 1981
	Elastase	L. Sawyer, et al., J. Mol. Biol. 118:137–208, 1978
	Kallikrein A	W. Bode, et al., J. Mol. Biol. 164:237–282, 1983
	α-Lytic protease	G. D. Brayer, et al., J. Mol. Biol. 131:743–775, 1979
	Streptomyces griseus proteases	
	A	A. R. Sielecki, et al., J. Mol. Biol. 134:781–804, 1979
	B	L. T. J. Delbaere, et al., Can. J. Biochem. 57:135–144, 1979

Table 6–1   *(Continued)*

FUNCTIONAL CLASSIFICATION	PROTEIN	MOST COMPREHENSIVE OR RECENT REFERENCE
	Trypsinogen	H. Fehlhammer, et al., J. Mol. Biol. 111:415–438, 1977
	Trypsin	W. Bode and P. Schwager, J. Mol. Biol. 98:693–717, 1975
Thiol proteases	Papain	J. Drenth, et al., Adv. Protein Chem. 25:79–115, 1971
	Actinidin	E. N. Baker, J. Mol. Biol. 141:441–484, 1980
Carboxyl proteases	*Endothia parasitica* pepsin	C.-H. Wong, et al., Biochemistry 18:1638–1640, 1979
	Penicillopepsin	M. N. G. James and A. R. Sielecki, J. Mol. Biol. 163:299–361, 1983
	Porcine pepsin	N. S. Andreeva, et al., Molek. Biol. 12:922–936, 1978
	Rhizopus chinensis pepsin	R. Bott, et al., Biochemistry 21:6956–6962, 1982
Metalloproteases	Carboxypeptidase	
	A	D. C. Rees, et al., J. Mol. Biol. 168:367–387, 1983
	B	M. F. Schmid and J. R. Herriott, J. Mol. Biol. 103:175–190, 1976
	Bacterial	O. Dideberg, et al., Nature 299:469–470, 1982
	Thermolysin	M. A. Holmes and B. W. Matthews, J. Mol. Biol. 160:623–639, 1972
Dehydrogenases	Alcohol dehydrogenase	H. Eklund, et al., J. Mol. Biol. 102:27–59, 1976
	Glyceraldehyde-P dehydrogenase	
	Lobster	M. Buehner, et al., J. Mol. Biol. 90:25–49, 1974
	Bacterial	A. G. W. Leslie and A. J. Wonacott, J. Mol. Biol. 165:375–391, 1983
	Lactate dehydrogenase	J. J. Holbrook, et al., in The Enzymes, 3rd ed. P. D. Boyer (ed.). Vol. 11, pp. 191–292. New York, Academic Press, 1975
	Malate dehydrogenase	E. Hill, et al., J. Mol. Biol. 72:577–591, 1972
	6-Phosphogluconate dehydrogenase	M. J. Adams, et al. EMBO J. 2:1009–1014, 1983
Kinases	Adenylate kinase	G. E. Schulz, et al., Nature 250:120–123, 1974
	Hexokinase	C. M. Anderson, et al., J. Mol. Biol. 123:1–13, 1978
	Phosphoglycerate kinase	
	Yeast	H. C. Watson, et al., EMBO J. 1:1635–1640, 1982
	Horse	R. D. Banks, et al., Nature 279:773–777, 1979
	Phosphofructokinase	P. R. Evans and P. J. Hudson, Nature 279:500–504, 1979
	Pyruvate kinase	D. I. Stuart, et al., J. Mol. Biol. 134:109–142, 1979
Other enzymes	Aldolase, KDPG	I. M. Mavridis, et al., J. Mol. Biol. 162:419–444, 1982
	Alkaline phosphatase	J. M. Sowadski, et al., J. Mol. Biol. 150:245–272, 1981
	Aspartate aminotransferase	G. C. Ford, et al., Proc. Natl. Acad. Sci. U.S.A. 77:2559–2563, 1980
	Aspartate transcarbamylase	R. B. Honzatko, et al., J. Mol. Biol. 160:219–263, 1982
	Carbonic anhydrase	S. Lindskog, et al., in The Enzymes, 3rd ed. P. D. Boyer (ed.). Vol. 5, pp. 587–665. New York, Academic Press, 1971
	Catalase	
	Bovine	M. R. N. Murthy, et al., J. Mol. Biol. 152:465–499, 1981
	Fungal	B. K. Vainshtein, et al., Nature 293:411–412, 1981
	Citrate synthetase	S. Remington, et al., J. Mol. Biol. 158:111–152, 1982
	Cytochrome *c* peroxidase	T. L. Poulos, et al., J. Biol. Chem. 255:575–580, 1980
	Dihydrofolate reductase	J. T. Bolin, et al., J. Biol. Chem. 257:13650–13662, 1982
	Ferredoxin reductase	S. Sheriff and J. R. Herriott, J. Mol. Biol. 145:441–451, 1981
	Flavodoxin	W. W. Smith, et al., J. Mol. Biol. 165:737–755, 1983
	Glucose-6-P isomerase	P. J. Shaw and H. Muirhead, J. Mol. Biol. 109:475–485, 1977
	Glutathione peroxidase	O. Epp, et al., Eur. J. Biochem. 133:51–69, 1983
	Glutathione reductase	R. Thieme, et al., J. Mol. Biol. 152:763–782, 1981
	Glycolate oxidase	Y. Lindqvist and C.-I. Brändén, J. Mol. Biol. 143:201–211, 1980
	p-Hydroxybenzoate hydroxylase	R. K. Wierenga, et al., J. Mol. Biol. 131:55–73, 1979

Table 6–1   (Continued)

FUNCTIONAL CLASSIFICATION	PROTEIN	MOST COMPREHENSIVE OR RECENT REFERENCE
	Lysozyme	
	Hen	T. Imoto, et al., in The Enzymes, 3rd ed. P. D. Boyer (ed.). Vol. 7, pp. 665–868. New York, Academic Press, 1972
	Human	P. J. Artymiuk and C. C. F. Blake, J. Mol. Biol. 152:737–762, 1981
	Bacteriophage	S. J. Remington, et al., J. Mol. Biol. 118:81–98, 1978
	Streptomyces	S. Harada, et al., J. Biol. Chem. 256:11600–11602, 1981
	Phosphoglycerate mutase	J. W. Campbell, et al., Nature 250:301–303, 1974
	Phospholipase A_2	
	Pancreatic	B. W. Dijkstra, ct al., J. Mol. Biol. 147:97–123, 1981
	Snake venom	C. Keith, et al., J. Biol. Chem. 256:8602–8607, 1981
	Phosphorylase	
	a	S. Sprang and R. J. Fletterick, J. Mol. Biol. 131:523–551, 1979
	b	L. N. Johnson, et al., J. Mol. Biol. 140:565–580, 1980
	Rhodanese	J. H. Ploegman, et al., J. Mol. Biol. 123:557–594, 1978
	Ribonucleases	
	Bovine	
	A	A. Wlodawer and L. Sjölin, Biochemistry 22:2720–2728, 1983
	S	F. M. Richards and H. W. Wyckoff, Ribonuclease-S. Oxford, Clarendon Press, 1973
	Microbial	
	Barnase	Y. Mauguen, et al., Nature 297:162–164, 1982
	St	K. T. Nakamura, et al., Nature 299:564–566, 1982
	T_1	U. Heinemann and W. Saenger, Nature 299:27–31, 1982
	Staphylococcal nuclease	F. A. Cotton and E. E. Hazen, in The Enzymes, 3rd ed. P. D. Boyer (ed.). Vol. 4, pp. 153–175. New York, Academic Press, 1971
	Superoxide dismutase	
	Cu, Zn	J. A. Tainer, et al., J. Mol. Biol. 160:181–217, 1982
	Fe	D. Ringe, et al., Proc. Natl. Acad. Sci. U.S.A. 80:3884–3888, 1983
	Taka-amylase A	Y. Matsuura, et al., J. Biochem. 87:1555–1558, 1980
	Triose phosphate isomerase	D. C. Phillips, et al., J. Mol. Biol. 119:319–351, 1978
	tRNA synthetases	
	Methionyl	C. Zelwer, et al., J. Mol. Biol. 155:63–81, 1982
	Tyrosyl	T. N. Bhat, et al., J. Mol. Biol. 158:699–709, 1982
Immunoglobulins		
Bence Jones proteins		
L chain dimer	Mcg	A. B. Edmundson, et al., Biochemistry 18:3953–3961, 1975
V_L dimers	Rei	O. Epp, et al., Eur. J. Biochem. 45:513–524, 1974
	Au	H. Fehlhammer, et al., Biophys. Struct. Mech. 1:139–146, 1975
	Rhe	W. Furey, Jr., et al., J. Mol. Biol. 167:661–692, 1983
	Roy	P. M. Colman, et al., J. Mol. Biol. 116:73–79, 1977
F_{ab} fragments	New	F. A. Saul, et al., J. Biol. Chem. 253:585–597, 1978
	M603	D. M. Segal, et al., Proc. Natl. Acad. Sci. U.S.A. 71:4298–4302, 1974
	Kol	M. Matsushima, et al., J. Mol. Biol. 121:441–459, 1978
F_c fragment		J. Deisenhofer, et al., Z. Physiol. Chem. 357:1421–1434, 1976
Intact IgG	Kol	M. Marquart, et al., J. Mol. Biol. 141:369–391, 1980
	Dob	E. W. Silverton, et al., Proc. Natl. Acad. Sci. U.S.A. 74:5140–5144, 1977
Lectins	Concanavalin A	G. N. Reeke, et al., J. Biol. Chem. 250:1525–1547, 1975
	Wheat germ agglutinin	C. S. Wright, J. Mol. Biol. 111:439–457, 1976
Inhibitors and toxins	Bovine pancreatic trypsin inhibitor	J. Deisenhofer and W. Steigemann, Acta Cryst. B31:238–250, 1975

Table 6–1 *(Continued)*

FUNCTIONAL CLASSIFICATION	PROTEIN	MOST COMPREHENSIVE OR RECENT REFERENCE
	Secretory trypsin inhibitor	M. Bolognesi, et al., J. Mol. Biol. 162:839–868, 1983
	Carboxypeptidase A inhibitor	D. C. Rees and W. N. Lipscomb, J. Mol. Biol. 160:475–498, 1982
	Mellitin	T. C. Terwilliger and D. Eisenberg, J. Biol. Chem. 257:6016–6022, 1982
	Ovomucoid fragment	E. Papamokos, et al., J. Mol. Biol. 158:515–537, 1982
	Scorpion neurotoxin	J. C. Fontecilla-Camps, et al., Proc. Natl. Acad. Sci. U.S.A. 77:6496–6500, 1980
	Snake neurotoxin	
	Short	B. W. Low, et al., Proc. Natl. Acad. Sci. U.S.A. 73:2991–2994, 1976; D. Tsernoglou and G. A. Petsko, FEBS Letters 68:1–4, 1976
	Long	M. D. Walkinshaw, et al., Proc. Natl. Acad. Sci. U.S.A. 77:2400–2404, 1980
	Soybean trypsin inhibitor	R. M. Sweet, et al., Biochemistry 13:4212–4228, 1974
	Subtilisin inhibitor	Y. Mitsui, et al., J. Mol. Biol. 131:697–724, 1979
Hormones	Glucagon	K. Sasaki, et al., Nature 257:751–757, 1975
	Insulin	T. Blundell, et al., Adv. Protein Chem. 26:279–402, 1972
	Pancreatic polypeptide	T. L. Blundell, et al., Proc. Natl. Acad. Sci. U.S.A. 78:4175–4179, 1981
Viruses	Influenza virus	
	Hemagglutinin	I. A. Wilson, et al., Nature 289:366–373, 1981
	Neuraminidase	J. N. Varghase, et al., Nature 303:35–40, 1983
	Satellite tobacco necrosis virus	L. Liljas, et al., J. Mol. Biol. 159:93–108, 1982
	Southern bean mosaic virus	M. G. Rossmann, et al., J. Mol. Biol. 166:37–83, 1983
	Tobacco mosaic virus coat protein	A. C. Bloomer, et al., Nature 276:362–373, 1978
	Tomato bushy stunt virus	S. C. Harrison, et al., Nature 276:368–373, 1978
Other proteins	Anaphylatoxin C3a	R. Huber, et al., Z. Physiol. Chem. 361:1389–1399, 1980
	Arabinose-binding protein	G. L. Gilliland and F. A. Quiocho, J. Mol. Biol. 146:341–362, 1981
	Bacteriochlorophyll *a* protein	B. W. Matthews, et al., J. Mol. Biol. 131:259–285, 1979
	Canavalin	A. McPherson, J. Biol. Chem. 255:10472–10478, 1980
	Catabolite gene activator	D. B. McKay, et al., J. Biol. Chem. 257:9518–9524, 1982
	Crambin	W. A. Hendrickson and M. M. Teeter, Nature 290:107–113, 1981
	Cro repressor	W. F. Anderson, et al., Nature 290:754–758, 1981
	Crystallin	T. Blundell, et al., Nature 289:771–777, 1981
	Elongation factor Tu	K. Morikawa, et al., J. Mol. Biol. 125:325–338, 1978
	Ferritin	S. H. Banyard, et al., Nature 271:282–284, 1978
	Galactose-binding protein	S. L. Mowbray and G. A. Petsko, J. Biol. Chem. 258:7991–7997, 1983
	Gene 5 product, bacteriophage *fd*	G. D. Brayer and A. McPherson, J. Mol. Biol. 169:565–596, 1983
	λ repressor	C. O. Pabo and M. Lewis, Nature 298:443–445, 1982
	Parvalbumin	P. C. Moews and R. H. Kretsinger, J. Mol. Biol. 91:201–228, 1975
	Prealbumin	C. C. F. Blake, et al., J. Mol. Biol. 121:339–356, 1978
	Protein A from *Staphylococcus aureus*	J. Deisenhofer, Biochemistry 20:2361–2370, 1981
	Ribosomal protein L7/L12 fragment	M. Leijonmarck, et al., Nature 286:824–826, 1980
	Uteroglobulin	J. P. Mornon, et al., J. Mol. Biol. 137:415–429, 1980
	Vitamin D–dependent Ca^{2+}-binding protein	D. M. E. Szebenyi, et al., Nature 294:327–332, 1981

Knowledge of the three-dimensional structure of a protein is a prerequisite to understanding its physical, chemical, and biological properties. In turn, knowledge of the structure is almost totally dependent upon the technique of x-ray diffraction analysis of crystalline proteins. Appreciation and interpretation of the protein structure requires some basic understanding of the technique—of its power and of its limitations. The first part of this chapter describes the technique very briefly and simply, concentrating on those aspects important for the non-crystallographer interested in proteins; then the general properties of protein structure are summarized, followed by some consideration of the physical basis of this structure.

The use of X-ray diffraction in the study of protein and nucleic acid structure. K. C. Holmes and D. M. Blow. Methods Biochem. Anal. 13:113–239, 1965.

X-ray studies of crystalline proteins. A. C. T. North and D. C. Phillips. Prog. Biophys. Mol. Biol. 16:1–32, 1969.

X-ray crystallography and enzyme structure. D. Eisenberg. In The Enzymes, 3rd ed. P. D. Boyer (ed.). Vol. 1, pp. 1–89. New York, Academic Press, 1970.

X-ray studies of crystalline proteins. C. C. F. Blake. Prog. Biophys. Mol. Biol. 25:85–130, 1972.

Protein Crystallography. T. L. Blundell and L. N. Johnson. New York, Academic Press, 1976.

X-ray crystallographic studies of proteins. B. W. Matthews. Ann. Rev. Phys. Chem. 27:493–523, 1976.

X-ray structure of proteins. B. W. Matthews. In The Proteins, 3rd ed. H. Neurath and R. L. Hill (eds.). Vol. 3, pp. 403–590. New York, Academic Press, 1977.

Three-dimensional structure determination by electron microscopy of two-dimensional crystals. L. A. Amos, et al. Prog. Biophys. Mol. Biol. 39:183–231, 1982.

Preparation and Analysis of Protein Crystals. A. McPherson. New York, John Wiley & Sons, 1982.

VISUALIZING THREE-DIMENSIONAL STRUCTURES

Basic Principles of Diffraction

A microscope works in two stages: In the first, the object under examination scatters the light in all directions, forming a diffraction pattern. This pattern arises because the lightwaves scattered from different parts of the object combine to produce a wave of large or small amplitude in any direction, depending upon whether the light waves are in or out of step (i.e., phase) with each other. In the second stage of the microscope, the objective lens collects the diffracted waves and recombines them to form an image of the object.

The fineness of detail (i.e., the resolution) with which objects may be seen in a microscope is inherently limited by the wavelength of the light used; the smallest objects that can be resolved are those with dimensions comparable to the wavelength of the light. Visualization of molecules at atomic resolution then requires "light" with wavelengths of about 1 Å; x rays, electrons, or neutrons are most generally used. All three forms of radiation can be used to give diffraction patterns, but only electrons, being charged, may be focused by an objective lens and therefore are used in a microscope. Microscopes using x rays or neutrons are not practical. With these two types of radiation it is necessary to separate the two stages of microscopy: to record the diffraction pattern and then to simulate the second stage of reconstruction of the image by mathematical computation.

Atomic structure analysis is also limited by the very low contrast possible with individual molecules. In particular, the H, C, O, N, and S atoms that make up proteins do not have sufficient mass to interact substantially with radiation, making it impossible to detect individual molecules. This difficulty is generally overcome in electron microscopy by staining the structure with a very dense material and then looking at the holes left in the stain by the specimen; however, this greatly limits the resolution that will be apparent in the image, and the stain may change the specimen. The problem is overcome in crystallography by observing not just one molecule but the approximately 10^{15} molecules that make up a crystal. With a crystalline form, all the molecules are held in regular positions so that their diffraction patterns may add up; the crystal amplifies the diffraction pattern.

A crystal has all the molecules arranged in specific positions and orientations on a three-dimensional lattice. The basic unit of a crystal is the **unit cell,** which is the smallest parallelepiped within the crystal that, when repeated by translations parallel to its edges in all three directions, without rotating the unit, gives the **crystal lattice** (Figure 6–1). The three directions in which the unit cell is translated in the crystal lattice define the three crystal axes. However, in some instances the definitions are relaxed somewhat to allow orthogonal axes (at right angles to each other), to give unit cells described as "face-centered" or "body-centered."

In many crystal lattices, the unit cell has internal symmetry in that two or more identical structures are related by axes or planes of symmetry. For example, two identical molecules may occur in a unit cell, having different orientations with respect to the crystal axes. The symmetry axes or planes often coincide with those of the crystal axes; such symmetry within the unit cell is designated "crystallographic." The individual unit that, when repeated an integral number of times by such crystallographic operations, yields the entire unit cell is known as the **asymmetric unit.** There are 65 possible types of crystal lattices (i.e., space groups) in which asymmetric units may crystallize. Non-crystallographic symmetry is also possible, in which two or more identical units are related by symmetry axes different from those of the crystal lattice; the asymmetric unit, therefore, may consist of more than one molecule.

Not all the space groups are possible with proteins, as they consist of L-amino acids and have a unique handedness, so mirror planes are not possible. Furthermore, a single polypeptide chain of L-amino acids may have no exact structural symmetry; therefore, an asymmetric unit of a protein crystal must have at least one molecule of a complete polypeptide chain. On the other hand, more than one molecule of a polypeptide chain may be contained in one asymmetric unit.

For the remainder of this discussion, the radiation will be assumed to

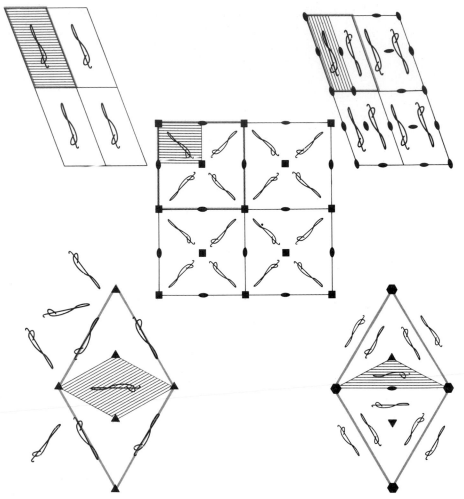

Figure 6–1
Some examples of unit cells and asymmetric units. A single unit cell is outlined heavily; the asymmetric unit is shaded. The following represent the rotational symmetry elements: ●, *twofold;* ▲, *threefold;* ■, *fourfold;* ⬢, *sixfold.*

be x rays, as they have been used in most of the work on protein structure determination. For technical reasons, the x rays used have nearly always been those of wavelength 1.542 Å, the so-called K_α x rays emitted upon electron bombardment of copper, although the increasing use of the x-ray irradiation of synchrotrons, with a continuous spectrum of wavelengths, will probably result in future use of x rays of other wavelengths. The limited work in neutron and electron diffraction will be described briefly later. The important feature of x rays is that they are scattered primarily by the electrons of the object; the scattering by each atom is then proportional to its atomic number (Table 6–2), and the structure determined by x-ray diffraction analysis is that of the electron density.

Table 6–2   Amplitudes of X-ray Waves and Neutron Waves Scattered by the Atoms Comprising Proteins

ELEMENT OR ISOTOPE	NUMBER OF ELECTRONS	X-RAY WAVE	NEUTRON WAVE
H	1	0.28	−0.37
^{2}H (deuterium)	1		0.67
C	6	1.69	0.67
N	7	1.97	0.94
O	8	2.25	0.58
S	16	4.5	0.28

All amplitudes are of positive sign unless otherwise noted and are in units of 10^{-12} cm/atom. Only coherent scattering is considered, which is the case with the great majority of these atoms.

From B. P. Schoenborn and A. C. Nunes, Ann. Rev. Biophys. Bioeng. 1:529–552, 1972.

The diffraction of x rays by atoms is analogous to the scattering of visible light on a larger scale, and the general principles of diffraction may be illustrated in two dimensions using optical transforms, which are prepared by illuminating masks drilled with holes to represent the atoms of the molecule, as shown in Figure 6–2. An example of a diffraction pattern of a molecule with six atoms is illustrated in *A*. The important point is that it is a continuous function in two dimensions of varying intensity that is determined by the structure of the molecule. However, when the molecule is repeated in a crystal lattice, the lattice acts as a diffraction grating, so that the scattered waves are confined to a number of discrete directions— those in which the waves scattered by all the molecules of the lattice are in phase. The continuous diffraction pattern of the individual molecule is then sampled at only a few points, as in *B* and *C* in Figure 6–2, even though the scattered radiation is increased by the inclusion of more molecules. The directions of such scattered x rays, generally designated **reflections,** are dependent only upon the crystal lattice, and not upon the structure of the molecule; see *C* through *F* in the figure. A three-dimensional crystal lattice then determines a three-dimensional lattice of scattered x rays in what is termed "reciprocal space," since the reciprocal lattice has parameters that are the inverse of the crystal lattice; compare *D* and *E* in the figure. This is a result of the Bragg law of diffraction:

$$d \sin \theta_n = n\lambda \qquad (6–1)$$

where d is the spacing between molecules in the crystal, θ_n is the angle of diffraction for the nth diffraction order, and λ is the wavelength of the radiation. For larger values of d, the scattering angle is decreased, so the spacing is decreased in the reciprocal lattice. The directions of the three axes of the reciprocal lattice are also related in a reciprocal manner to the directions of the crystal axes. By measuring the spacings and the axes of the reciprocal lattice produced by diffraction of a crystal with radiation of a known wavelength, the dimensions and axes of the crystal lattice in real space may be determined. The reciprocal lattice of x-ray crystallography

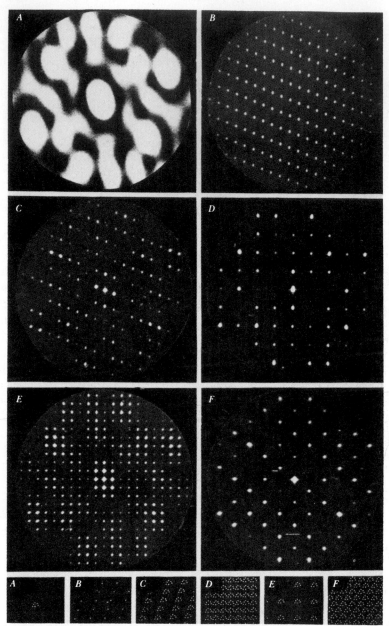

Figure 6–2
*The effect of the crystal lattice on the diffraction pattern, illustrated with optical transforms.
The objects used to generate the diffraction patterns are shown at the bottom, which are the
masks used to generate the optical transforms at the top. The continuous transform of a single
"six-atom molecule" is shown in* A. *The pattern from a lattice only is shown in* B, *giving
just a series of reciprocal lattice points. The consequences of incorporating the "molecule" of*
A *into various lattices in* C *through* F *is to sample the continuous transforms of* A *at points
of a reciprocal lattice. (From C. A. Taylor and H. Lipson, Optical Transforms. London, Bell
& Hyman Ltd., 1964.)*

is visualized directly in precession photographs, which record a two-dimensional layer of the three-dimensional reciprocal lattice.

The individual reflections of the reciprocal lattice are generally designated by their **Miller indices,** h, k, l, which are integers (either positive or negative) giving the number of spacings from the origin ($h = 0$, $k = 0$, $l = 0$) in the reciprocal lattice at which the reflection occurs. A specific reflection is then referred to as (h, k, l), giving the index values, e.g., (1, 2, 1), (4, $\bar{3}$, 2), and so on. Two-dimensional layers of the reciprocal lattice as recorded in precession photographs are designated as, for example, ($h0l$) for the layer of all reflections for which $k = 0$.

The structural information about the contents of the unit cell is contained in the intensities of the individual reflections (compare B and C in Figure 6–2). In the first instance, any crystallographic symmetry within the unit cell will be reflected in a corresponding, but inverse, symmetry of the reciprocal lattice. Any such symmetry is in addition to one element that is always present, even with a nonsymmetric unit cell, in that the intensity of reflection (h, k, l) is always the same as that of ($\bar{h}$, $\bar{k}$, $\bar{l}$); the latter designation is for negative values of h, k, and l. This relationship holds for all values of h, k, and l, except for the small effects of anomalous scattering (see p. 212). Crystallographic symmetry of the unit cell may also result in systematic absences of some reflections; e.g., ($0k0$) reflections will be present only if k is an even integer in some space groups.

From the diffraction pattern of a crystal, it is usually possible to determine the dimensions of the unit cell and the space group of the lattice. This gives the number of asymmetric units in the unit cell and the volume of each. The molecular weight of the protein in the asymmetric unit may also be estimated if the relative volumes of protein and solvent are determined. This is usually accomplished by varying the density of the solvent and then measuring the crystal density, usually by finding where it floats in a density gradient of liquids not miscible with the crystal solvent. Such measurements need not be very accurate, since the mass of protein within the asymmetric unit must consist of an integral number of protein polypeptide chains; the value of the integer, therefore, is usually apparent. It is often adequate to assume that the solvent content within a protein crystal is in the usual range of 40 to 60 per cent, although extremes of 30 and 78 per cent have been observed.

Phase Determination

Reconstruction of the detailed structure of the asymmetric unit from the diffraction pattern is the goal of diffraction analysis. The diffraction pattern breaks down the structure into discrete sine waves, as occurs in Fourier analysis (Figure 6–3). Any shape may be represented as the sum of sine waves of varying amplitudes and phases in three dimensions. The individual reflections of a diffraction pattern represent such waves, having wavelength components in the three dimensions inversely proportional to the value of h, k, and l, respectively. The original object can be reconstructed by recombining the individual sine waves, as occurs in the objective lens

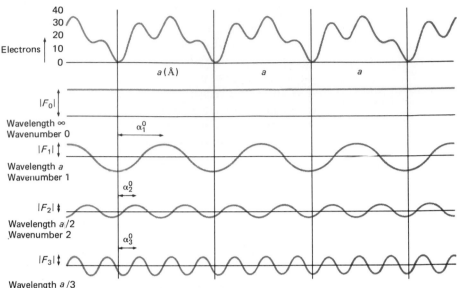

Figure 6–3
Fourier synthesis of a one-dimensional electron density profile (top) using four terms, F_0 through F_3. The unit cell dimension is a. $|F_i|$ *and* α_i *are respectively the amplitude and the phase of the ith term. The sum of these four terms gives the electron density at the top. (Adapted from C. C. F. Blake, Essays Biochem. 11:37, 1975.)*

of a microscope. However, when the x-ray diffraction pattern is recorded, only the intensities of the reflections are recorded, not their relative phases, so it is impossible to recombine them properly. This is the well-known **phase problem** of crystallography.

The phase problem was overcome in work with proteins by the method of **isomorphous replacement.** This technique depends upon the preparation of protein crystals into which additional heavy atoms that scatter x rays very strongly, such as uranium, platinum, or mercury, have been introduced at a few specific positions, without otherwise affecting the crystal structure. The modified crystals must have the same crystal lattice, i.e., be isomorphous, so that the diffraction pattern is sampled at the same points in the reciprocal lattice. The heavy atoms must contribute sufficiently to the diffraction pattern to alter significantly the intensities of the reflections, but they must cause minimal alterations of the protein, so that the diffraction pattern is the crystallographic sum of the original protein and of the heavy atoms. The heavy atoms must also be in only one or a few positions, so that their positions within the unit cell may be deduced from the way they alter the protein diffraction pattern.

The preparation of heavy atom derivatives generally relies on trial-and-error testing of many different compounds, rather than on rational design. The need for specific binding at only one or two sites on a protein makes it impossible to use reagents that react with all the protein groups

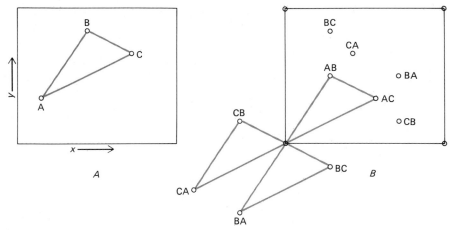

Figure 6–4

Illustration of a two-dimensional Patterson map. A, A hypothetical molecule with three atoms (A, B, and C) in its unit cell in the crystal. B, The theoretical Patterson map, with the positions of the peaks indicated. Each is designated as arising from the vector from the first atom to the second. The map may be generated by placing in turn each of the three atoms at the origin and marking the positions of the other two atoms when the molecule is in the correct orientation; the three representations of the molecule are indicated by the lines joining the appropriate peaks. The Patterson map is repetitive, just as is the crystal, and one unit cell is indicated. Large peaks occur at each of the unit cell origins, owing to the self-vectors (AA, BB, and CC).

of one type. Instead, selective binding must occur at a few sites owing to specificity produced by the protein structure; the nature of this specificity is not known, however, when the heavy atom derivatives must be prepared. Considerable experience has yielded a collection of different compounds of uranium, lead, platinum, mercury, gold, silver, and the lanthanides that have been successful in the past and that are generally tested. After the protein structure has been determined, the observed mode of binding may usually be rationalized, although surprises still occur.

Difference Patterson maps are usually used initially to determine the positions of the heavy atoms in a protein crystal. The Patterson map is calculated using the differences in intensities produced by the heavy atom; thus in the ideal case it is equivalent to determining the structure of only the heavy atoms. Only the intensities of the reflections, not their phases, are used in the calculation; such maps show the vectors between all atoms of the unit cell, superimposed and emanating from the origin of the map (Figure 6–4). There are $n(n - 1)$ such vectors from a crystal containing n atoms in the unit cell; consequently, Patterson maps of only very simple structures may be interpreted.

Once the positions of the heavy atoms are determined, both the intensities and the phases of their diffraction pattern may be calculated. How the intensity of each reflection of the protein crystal is affected by the heavy atom contribution depends upon the relative phases of the two. With knowledge of the latter, the phase of the protein contribution can be determined, except that two solutions are usually possible. It is then necessary to have a second, different heavy atom derivative in order to resolve the ambiguity. Even with two derivatives, the phases are usually not accurately determined, owing to experimental errors in the measurement of relatively small differences in intensities produced by the heavy atoms and imperfect isomorphism, so additional derivatives are generally desirable. The phases determined by isomorphous replacement are generally accurate only within about 40°, out of a possible range of 180°.

Other methods may be used to complement phase determination by heavy atom derivatives. Anomalous scattering of x rays by the inner electrons of many heavy atoms will cause small differences between the otherwise identical intensities of reflections (hkl) and $(\bar{h}\bar{k}\bar{l})$. These differences provide information about the phase of the reflection if the positions of the anomalous scattering atoms in the unit cell are known. In favorable cases, one isomorphous heavy atom derivative with anomalous scattering may be sufficient. In the case of crambin, anomalous scattering by the sulfur atoms alone was used for initial phase determination.

When the asymmetric unit of a crystal contains two or more identical molecules, related by non-crystallographic symmetry, there are restrictions on the possible values of the phases. The higher the degree of symmetry, the greater the restrictions; an extreme example is the tobacco mosaic virus coat protein disk, which has 17-fold non-crystallographic symmetry. This aspect is generally used to refine the initial phases to improve the electron density map.

Even in the absence of non-crystallographic symmetry, the information present in only the intensities of a complete set of reflections to high resolution is redundant, in that there are generally 10 reflections per atom to be fixed in space. The positions of these atoms are also not independent, being linked by the known covalent structure. In addition, the electron density can never be negative. The intensities alone therefore provide a considerable amount of information, and the initial phases need not be particularly accurate to define the structure. The initial phases may then be refined, using the measured intensities and the constraints of the covalent structure and nonnegative electron density to give an improved structure.

Calculation of the Electron Density Map

Following determination of the phases of each of the reflections, the structure of the unit cell of the crystal is reconstructed by recombining mathematically the individual reflections of the diffraction pattern, a computation known as a **Fourier synthesis.** The electron density, ρ, at a point (x, y, z) in the unit cell, where x, y, and z are expressed as fractions of the unit cell dimensions a, b, and c, is given by

$$\rho(x, y, z) = \frac{1}{V} \sum_h \sum_k \sum_l F(h, k, l) e^{i\alpha(h,\ k,\ l)} e^{-2\pi i(hx + ky + lz)} \qquad (6\text{--}2)$$

where V is the volume of the unit cell, $F(h, k, l)$ is the amplitude (the square root of the intensity) of the reflection with indices h, k, and l, and α (h, k, l) is its phase. The only reason for presenting this equation here is to note that the calculation of the electron density at each point (x, y, z) includes the amplitudes and phases of *all* the reflections of the diffraction pattern. Consequently, every reflection contains information about all parts of the unit cell, just as each atom of the crystal contributes to each reflection. A portion of the diffraction pattern does not correspond to a part of the crystal. Determination of a crystal structure is therefore an all-or-none process in that all of the structure is determined simultaneously because the data obtained are equally relevant to all parts of the crystal.

However, the quality factor of a crystal structure depends upon the number of reflections included in the summation of the Fourier synthesis of Equation 6–2 above. The Fourier synthesis consists of adding sinusoidal waves (representing electron density) with varying wavelengths; the smaller the absolute values of h, k, and l, the longer the wavelength (Figure 6–3). The waves of decreasing wavelength (i.e., increasing values of h, k, and l) confer primarily increasing detail to the electron density map. Consequently, it is possible to terminate a Fourier synthesis, using only reflections included in a sphere about the origin of reciprocal lattice; some consequent errors in the calculated map are inevitable, but the primary effect is to limit the detail. This can be illustrated visually, as in Figures 6–5 and 6–6, using optical transforms in which the optical diffraction pattern is converted directly into the image, as in a microscope. The clarity of the reconstructed image is clearly dependent upon the extent of the diffraction pattern included in the reconstruction.

Likewise, the detail visible in electron density maps depends upon the extent of the three-dimensional diffraction pattern included in the Fourier synthesis. The nominal resolution of such a structure determination is taken to be the minimum interatomic spacing (d of Eq. 6–1) that gives rise to reflections included in the Fourier synthesis. Crystallographic structure determinations consequently often proceed in stages of increasing resolution, at each step including higher-order reflections in the calculation of the electron density map. Because the crystal lattice and the reciprocal lattice are three-dimensional, doubling the resolution requires the inclusion of eight times as many reflections.

Of course, the quality of the electron density map depends not only upon the nominal resolution but also upon the accuracy of the data—the amplitudes and phases of the reflections. The degree of accuracy depends upon both the skill of the crystallographer and the regularity of the crystals used. Ultimately, the latter is the limiting factor, because the resulting electron density map is an average over all the unit cells of all the crystals used and over the substantial period of time during which the data were collected, during which x rays may be damaging the protein. Any disorder present is seen as a smearing of the electron density. It also limits the resolution attainable, as the high-order reflections fade out most rapidly

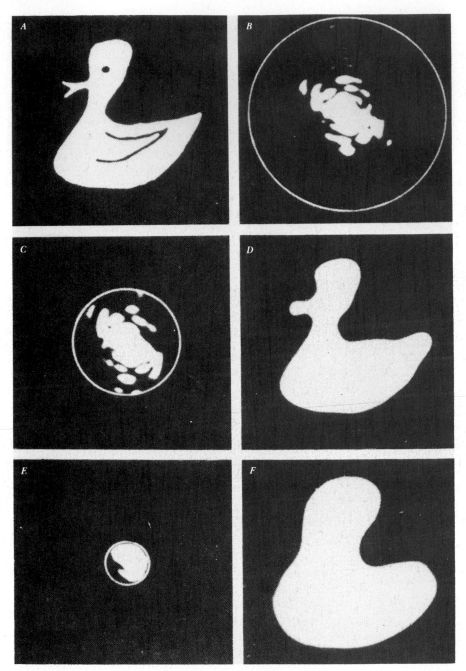

Figure 6–5
The diffraction pattern of the irregular object in A is shown in B. Only portions of the pattern were used in C and E to regenerate the original image in D and F, respectively, with correspondingly lower resolutions. (From C. A. Taylor and H. Lipson, Optical Transforms. London, G. Bell & Hyman Ltd., 1964.)

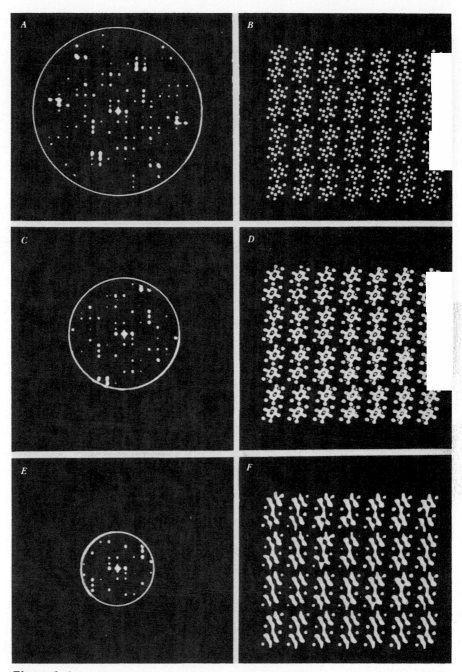

Figure 6–6
The effects of using fewer data of lower resolution in structure analysis, illustrated with optical transforms. The diffraction patterns enclosed within circles in A, C, and E were used to regenerate the original crystal lattice in B, D, and F, respectively; the clarity is correspondingly decreased upon using less of the diffraction pattern. (From C. A. Taylor and H. Lipson, Optical Transforms. London, G. Bell & Hyman Ltd., 1964.)

as a result of disorder. Protein crystals rarely diffract to a resolution better than 1.5 Å, although some do to < 1.0 Å. Unfortunately, proteins have generally been selected during evolution on the basis of properties other than their ability to crystallize.

In spite of the technical difficulties and limitations of the x-ray diffraction technique, it can produce perhaps surprisingly accurate and beautiful electron density maps, owing primarily to the large number of reflections generally included in the data. For example, in the case of the tobacco mosaic virus coat protein disk, 200,000 independent reflections were included in determining the structure at 2.8-Å resolution.

Interpretation of the Electron Density Map

The all-or-nothing aspect of crystallography requires an investment of a very substantial amount of work in protein crystallization, collection of x-ray diffraction data, search for isomorphous heavy-atom derivatives, calculation and deconvolution of difference Patterson maps, and calculation of the phase of each reflection, with virtually no relevent information about the protein obtained in the process. Yet upon calculating the electron density map, the crystallographer is overwhelmed with structural information about the molecule. Interpreting that information is not a simple task.

The protein molecule is apparent on the map as higher electron density than that of the surrounding solvent. The clarity of the electron density map is dependent upon both the accuracy and the resolution of the data. Covalent bonds are usually less than 1.5 Å long, so individual atoms are not resolved at lower resolutions. With resolution of better than 3 Å, the peptide group and the general shape of the side chains can usually be discerned, and the covalent structure may usually be traced by a continuous ribbon of electron density. With lower resolution, the polypeptide chain may not be traceable throughout the structure, except where there are dense, regular features, such as helices or β-sheets.

The clarity depends not only upon the accuracy and resolution of the data but also upon the degree of order of the protein in the crystal, since the electron density map is an average over all the unit cells of all the crystals used. Portions of the protein that are mobile, or that can adopt alternative conformations, have their electron density smeared out accordingly. Groups that attain four or more very different orientations may be completely invisible, but two alternative conformations are usually apparent.

Owing to the nature of the electron density map, its interpretation requires knowledge of the primary structure of the protein. Some side chains may be tentatively identified from their shape, but this is seldom unambiguous. The entire primary structure of no protein has been determined crystallographically, but errors in the primary structure determined in the normal manner have been detected.

In the past, the electron density map was usually displayed as contours on thin sections. A mechanical model of the protein, using standard bond lengths and angles, was usually built to fit best the electron density, often in a "Richards' box." The electron density map was displayed as a stack of transparent sheets behind a half-silvered mirror; the skeletal model of the protein was constructed in front of the mirror. An image of the model in the mirror was seen superimposed on the electron density map, enabling adjustment of the model to achieve the best possible fit.

The current practice is to "build" the model in a computer, using a graphics display system to display both the model and the electron density map simultaneously. The great advantage of this system is that the model is stored directly as coordinates of the atoms. Previously, these had to be measured from the mechanical model, with a considerable loss of accuracy.

Interpreting electron density maps is a rather subjective procedure, requiring considerable judgment and skill. It has not been possible to automate the procedure. Consequently, mistakes in interpreting electron density maps do occur, so it is permissible to question details of a protein model determined crystallographically.

The matching of physical models to three-dimensional electron-density maps: a simple optical device. F. M. Richards. J. Mol. Biol. 37:225–230, 1968.

Real-space refinement of the structure of hen egg-white lysozyme. R. Diamond. J. Mol. Biol. 82:371–391, 1974.

Real-time color graphics in studies of molecular interactions. R. Langridge, et al. Science 211:661–666, 1981.

Refinement of the Model

Following determination of the best fit of the protein's covalent structure to the electron density map, it is necessary to check the consistency of the model with the experimental diffraction data and to improve it by refinement. The amplitudes of the x-ray reflections expected with the protein model may be determined by computing its Fourier transform when in the crystal lattice. The agreement between the values calculated (F_{calc}) and those observed (F_{obs}) is generally measured by the R value:

$$R = \frac{\sum_{h,k,l} |F_{obs}(h,k,l) - F_{calc}(h,k,l)|}{\sum_{h,k,l} F_{obs}(h,k,l)} \tag{6–3}$$

where the summations are over all reflections. The smaller the R value, the better the agreement; a random assemblage of atoms in a model would be expected to give an R value of 0.59.

The R values computed for initial protein models have generally fallen in the range of 0.4 to 0.5. One reason for this seemingly large discrepancy is the common omission of the solvent area in the model. The solvent generally has a rather substantial uniform electron density, indicative of disorder of the individual molecules; increasing order, however, is usually apparent closer to the protein molecules. Including solvent molecules at these positions in the model generally decreases significantly the R value.

The extent of disorder of each atom affects its contribution to the diffraction pattern; this is taken into account by weighting its contribution by the factor

$$\exp\left(-B_i\,\frac{\sin^2\theta}{\lambda^2}\right) \qquad (6\text{--}4)$$

where θ is the angle of each reflection and λ is the wavelength of the x rays (Eq. 6–1). B_i is the temperature factor for atom i; the greater its value, the less localized the atom in the crystal. It is related to the mean square amplitude of vibration, U_i^2, in the ideal case of purely harmonic thermal vibration:

$$B_i = 8\pi^2 U_i^2 = 79\ U_i^2 \qquad (6\text{--}5)$$

A value of 79 Å^2 for B_i implies an rms vibration of 1 Å. Although termed a temperature factor, it includes not only temperature-dependent vibrations of the protein but also all kinds of static disorder within and between different unit cells.

By varying the coordinates and the temperature factor of each atom, the initial model may be refined to minimize the disagreement between F_{calc} and F_{obs}. Since the phases of each reflection have not been determined directly, and are not very accurate when determined by isomorphous replacement, their agreement with the calculated values is usually not determined. Instead, a new electron density map may be calculated using the observed amplitude of each reflection (F_{obs}) and the phase value calculated from the initial model (α_{calc}). The model is then adjusted to give the best fit to this electron density map. The new model may then be used to calculate new phases, and the process repeated. By a combination of refinement procedures, the agreement between observed and calculated amplitudes may be improved to give R values of as low as 0.11. However, they are still large compared to the values of 0.01 obtained with small molecules.

Caution must be exercised in interpreting such refined structures, since the initial model used in the refinement must be retained to some extent in the electron density map, as it determines the phases of the reflections. Fortunately, it is possible to check whether any particular feature of such a map is authentic, by omitting it from the initial model. In a map calculated with these values of F_{obs} and α_{calc}, a genuine feature should reappear, although with lower prominence, since every part of the crystal structure contributes to both the amplitude and the phase of all the reflections.

With full use of such refinement procedures, relatively accurate models of proteins may be derived. Examples of how an electron density map may be improved by refinement, and the agreement with the model obtained, are illustrated in Figure 6–7. Comparison of independent models of the same protein show that the coordinates of such models are generally accurate to within 0.15 Å, even though data to a nominal resolution of only 1.5 to 2.0 Å were used. The reason that the accuracy exceeds the resolution

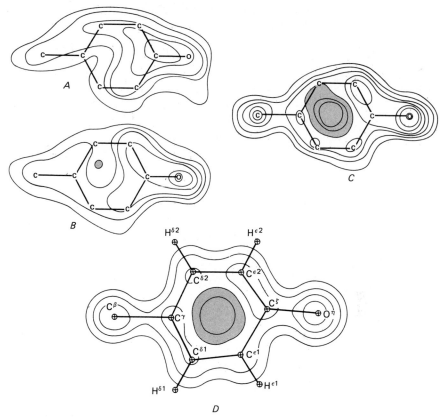

Figure 6–7

Illustration of the effect of increased resolution and phase refinement on the electron density map. One section through the plane of the side chain of Tyr 11 of rubredoxin is shown, with the skeleton structure of the model superimposed. The inner, shaded contours are of decreasing electron density. The resolution was 2.0 Å in both A and B, using phases determined by isomorphous replacement in A and phases calculated from the model in B. In C the phases were calculated, but the resolution was 1.5 Å. The fully refined map at 1.2-Å resolution is shown in D, with the optimum positions of the atoms indicated; the ideal geometry of a Tyr side chain was not imposed during the refinement. (From K. D. Watenpaugh, et al., Acta Cryst. B29:944–956, 1973; J. Mol. Biol. 138:615–633, © 1980, London, Academic Press Inc.)

is that a resolution of 2 Å means only that all reflections (i.e., distances) less than 2 Å are excluded. However, much information about the shorter distances is contained in the lower resolution reflections—hence, the apparent discrepancy.

Protein model refinement based on X-ray data. L. H. Jensen. Ann. Rev. Biophys. Bioeng. 3:81–93, 1974.

The accuracy of refined protein structures: comparison of two independently refined models of bovine trypsin. J. L. Chambers and R. M. Stroud. Acta Cryst. B35:1861–1874, 1979.

Neutron Diffraction X rays distinguish between the different atoms only to the extent that they differ in number of electrons. Of those present in proteins, sulfur atoms are the most effective scatterers of x rays, having twice as many electrons as the others (see Table 6–2). C, N, and O atoms hardly differ and are not distinguishable in practice. With only a single electron, H atoms only barely scatter x rays and are visible only in very-high-resolution maps. However, the positions of H atoms can be extremely important for the functional properties of proteins.

In such instances, diffraction by neutrons rather than by x rays is very useful. Neutrons interact in a complex manner with the nuclei of atoms, so they have widely varying scattering amplitudes (see Table 6–2). In particular, H atoms have a negative scattering factor, whereas the deuterium isotope has a large positive factor, as do most of the other atoms present in proteins. Consequently, hydrogen atoms are discernible as negative peaks with neutron diffraction and may be identified readily. The exchange of H by ^{2}H of the solvent may also be detected in this way.

However, neutron diffraction has technical difficulties, such as low beam fluxes of neutrons; consequently, long times and large crystals are required for measurements of data. Nevertheless, the importance of the information obtained with this technique has encouraged a considerable number of recent studies.

Neutron scattering. B. P. Schoenborn and A. C. Nunes. Ann. Rev. Biophys. Bioeng. 1:529–552, 1972.

Neutron Scattering for the Analysis of Biological Structures. B. P. Schoenborn (ed.). Upton, N.Y., Brookhaven National Laboratory, 1976.

Real space refinement of neutron diffraction data from sperm whale carbonmonoxymyoglobin. J. C. Hanson and B. P. Schoenborn. J. Mol. Biol. 153:117–146, 1981.

Orientation of histidine residues in RNase A: neutron diffraction study. A. Wlodawer and L. Sjölin. Proc. Natl. Acad. Sci. U.S.A. 78:2853–2855, 1981.

Neutron diffraction of crystalline proteins. A. Wlodawer. Prog. Biophys. Mol. Biol. 40:115–159, 1982.

Structure of ribonuclease A: results of joint neutron and x-ray refinement at 2.0 Å resolution. A. Wlodawer and L. Sjölin. Biochemistry 22:2720–2728, 1983.

THE GENERAL PROPERTIES OF PROTEIN STRUCTURES It has become customary to discuss protein structure in terms of four levels. The **primary structure** is the amino acid sequence, which has been described in the first three chapters. **Secondary structure** is any regular local structure of a linear segment of polypeptide chain, such as a helix or an extended strand, like those observed in polypeptides of regular primary structure described in Chapter 5. **Tertiary structure** is the overall topology of the folded polypeptide chain; **quaternary structure** is the aggregation of the polypeptides by specific interactions. In some respects, a crystal lattice is an extreme example of quaternary structure, since it is built up by interactions between protein molecules. Fortunately, protein crystals generally contain 40 to 60 per cent solvent, and the crystal lattice interactions are not very strong (see p. 266); consequently, the lattice contacts between protein molecules need not be extensive. Such contacts are observed not to be substantial, and often involve fixed solvent molecules, when the protein molecules are known not to interact significantly in solution. Where there are more extensive interactions directly between protein molecules, correspondingly greater interactions are generally observed in solution, ranging from weak, reversible interactions to essentially irreversible aggregation. In the latter cases, association of two or more polypeptide chains may be so intimate that the quaternary structure is an integral part of the tertiary structure. Consequently, the crystal structure of a protein gives information about all four levels of protein structure, although independent knowledge of the primary structure is generally required and interactions between protein molecules must be measured quantitatively in solution. Of course, all the levels of protein structure are related, but the division into the four levels is generally worthwhile.

The most striking feature of the folded conformation of a protein as determined by x-ray diffraction analysis is its complexity. This often makes it difficult to comprehend a structure and to communicate it to the reader through two-dimensional media. Simplified representations are often useful for concentrating on the gross or regular features of a structure, but there is no substitute for examination of a detailed three-dimensional model. Better still is to construct such a model; especially recommended are the Nicholson model components supplied by Labquip.*

Illustration and perception of protein structure is easier with small proteins, so many of the general properties of proteins are illustrated here with one of the smallest and simplest protein structures determined, that of bovine pancreatic trypsin inhibitor, generally abbreviated as BPTI. It consists of only 58 amino acid residues, and its three-dimensional structure has been determined to very high resolution and accuracy by Huber and his colleagues (Table 6–1), although more accurate structures have emerged subsequently. The amino acid sequence of this protein and the values of the dihedral angles of the backbone and side chains in the crystal structure are given in Table 6–3; this information is sufficient to define fully the folded conformation of the polypeptide chain, assuming standard geometry of bond lengths and angles (see Table 1–2). Its small size is illustrated in Figure 6–8 by a comparison of its model with that on the same scale of an immunoglobulin F_{ab} fragment; the latter is only about one third of an antibody molecule (see p. 372). In spite of its small size, the complexity of BPTI is apparent from more detailed photographs of the model; see Figure 6–9.

Although protein structures are extremely diverse in their molecular architecture, they have some common properties. It must be emphasized

---

*Labquip, Ashridgewood Place, Forest Road, Wokingham, Reading RG11 5RA, England.

Table 6–3 Conformational Angles of Bovine Pancreatic Trypsin Inhibitor

RESIDUE	ϕ	ψ	ω	χ^1	χ^2	χ^3	χ^4	χ^5
Arg 1		137	178	82	176	57	76	5
Pro 2	−60	149	180	177	171	−165		
Asp 3	−58	−33	179	−80	−10			
Phe 4	−63	−16	177	66	75			
Cys 5	−70	−19	177	−65	−84	−81		
Leu 6	−89	−3	178	−53	−179			
Glu 7	−78	150	177	−51	168	162		
Pro 8	−72	158	−177	−174	−172	172		
Pro 9	−67	143	−176	180	−174	168		
Tyr 10	−121	115	−177	177	82			
Thr 11	−77	−36	175	−71				
Gly 12	90	176	−176					
Pro 13	−83	−10	−177	−162	−172	165		
Cys 14	−92	159	−170	−68	95	96		
Lys 15	−119	32	169	−59	−74	174	−178	
Ala 16	−77	173	−175					
Arg 17	−130	81	−175	−66	179	−69	−94	−9
Ile 18	−109	111	−178	−61	178			
Ile 19	−78	120	176	−52	−58			
Arg 20	−122	179	172	−68	−72	180	80	8
Tyr 21	−114	147	168	−74	81			
Phe 22	−130	160	173	69	101			
Tyr 23	−87	127	176	180	−107			
Asn 24	−100	97	−172	177	−6			
Ala 25	−59	−28	173					
Lys 26	−70	−35	−174	−72	−179	174	62	
Ala 27	−90	−20	−179					
Gly 28	79	13	−178					
Leu 29	−156	171	−173	51	65			
Cys 30	−95	146	−174	−66	−119	−88		
Gln 31	−131	161	179	−65	175	160		
Thr 32	−89	156	175	50				
Phe 33	−151	164	178	75	114			
Val 34	−97	113	174	179				
Tyr 35	−86	135	180	174	44			
Gly 36	−79	−9	171					
Gly 37	99	7	172					
Cys 38	−146	157	172	71	−117	96		
Arg 39	62	36	−176	−47	−55	−171	−171	−3
Ala 40	−56	156	164					
Lys 41	−104	172	−168	−79	−170	171	163	
Arg 42	−73	−26	179	−76	−159	88	172	7
Asn 43	−82	74	−168	−169	8			
Asn 44	−163	102	−169	−177	−44			
Phe 45	−123	155	179	−56	87			
Lys 46	−86	−8	177	−72	169	−150	−176	
Ser 47	−151	158	174	73				
Ala 48	−65	−30	180					
Glu 49	−71	−39	175	−69	90	170		
Asp 50	−70	−38	179	−85	3			
Cys 51	−62	−45	179	176	−91	−88		
Met 52	−72	−35	−179	−74	61	−71		
Arg 53	−62	−38	180	178	149	39	−123	
Thr 54	−82	−42	−166	−55				
Cys 55	−111	−8	−167	−71	−65	−81		
Gly 56	−76	−6	164					
Gly 57	84	168	179					
Ala 58	−65							

Data kindly provided by J. Deisenhofer.

Figure 6–8
Comparison of the sizes of a small protein, bovine pancreatic trypsin inhibitor on the left, and the F_{ab} fragment of an immunoglobulin, on the right; the latter is only about one third of a normal immunoglobulin molecule (see Figures 2–1, 8–14, and 8–15). Note the four domains of the immunoglobulin F_{ab}.

that these general properties present a "consensus picture" of a protein structure, derived from the proteins of known structure in Table 6–1, to which there are many exceptions. The common properties illustrate the general rules of protein architecture, but each protein is unique and generally attains its functional properties by incorporating specific exceptions to these generalities. This chapter will describe briefly the common properties, while the remainder of this volume will be concerned with the exceptions.

Atlas of Macromolecular Structure on Microfiche. R. J. Feldman. Rockville, Md., Tracor-Jitco, 1977.
Principles of Protein Structure. G. E. Schulz and R. H. Schirmer. New York, Springer-Verlag, 1979.
Teaching Aids for Macromolecular Structure. R. J. Feldman and D. H. Bing. Bethesda, Md., National Institutes of Health.
Protein folding. M. G. Rossmann and P. Argos. Ann. Rev. Biochem. 50:497–532, 1981.
The anatomy and taxonomy of protein structure. J. S. Richardson. Adv. Protein Chem. 34:167–339, 1981.

The Tertiary Structure

The overall folded structures of small proteins are generally roughly spherical in overall shape, but with a very irregular surface, and remarkably compact. The BPTI molecule is not as spherical as most, being rather pear-shaped, with maximum dimensions of about 19 × 29 Å. Yet the fully extended polypeptide chain of 58 residues would be 211 Å long, and the

Figure 6–9
Model of bovine pancreatic trypsin inhibitor (BPTI) structure as determined crystallograph-
ically. The polypeptide chain backbone is white, the other groups darker. Hydrogen bonds are
the dark thin bonds. Hydroxyl and carboxyl groups appear black. The carboxyl end of the
chain is at the bottom, far right; the amino end is just to the left of it. The disulfide bond
linking Cys residues 14 and 38 is vertical, at the far left, center; that between Cys residues
5 and 55 is toward the bottom right. The α-helix between residues 47 to 56 is to the right,
running nearly vertically. Three of the four buried water molecules are visible just to the left
of center (see Figure 6–22); the fourth is hidden just to the right of the 14–38 disulfide
bond. (Model constructed from the crystallographic data of J. Deisenhofer and W. Steige-
mann, Acta Cryst. B31:238–250, 1975.)

radius of gyration of the random polypeptide chain, by Equations 5–10
and 5–12, is about 35 Å.

However, where a protein consists of more than about 200 residues,
the structure usually appears to consist of two or three rather spherical
structural units, generally referred to as **domains.** The individual domains
are associated, but they interact less extensively than do portions within
the domains. Often there is a single segment of polypeptide chain linking
domains, so that each domain consists of a single stretch of polypeptide
chain. This segregation of domains along the chain is not always the case;
in phosphofructokinase, pyruvate kinase, and arabinose-binding protein,
for example, there are two or three connections. In some cases, the ends
of the polypeptide chain interact with the other domain, appearing to serve
as straps holding the domains together.

The definition of a domain is not rigorous; in fact, the division of a

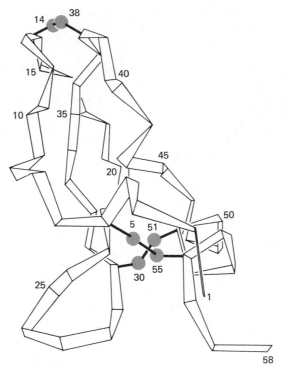

Figure 6–10
Ribbon diagram of bovine pancreatic trypsin inhibitor, showing the polypeptide backbone and the three disulfide bonds. The view is from the bottom of Figure 6–9. (From T. E. Creighton, J. Mol. Biol. 95:167–199, © 1975, London, Academic Press Inc.)

structure into domains is a very subjective process, which often is done in very different ways by different authors. Other terms and subdivisions, such as *subdomain* and *folding unit,* are also often encountered in the literature. Nevertheless, there are many instances in which the presence of domains is clear to all observers—for example, the immunoglobulins, where the heavy chain is composed of four domains and the light chain two (Figures 6–8 and 8–14).

Within a domain, the course of the polypeptide backbone is rather irregular, but it generally pursues a moderately straight course across the entire domain and then turns and continues in a more-or-less direct path to the other side. The impression is of segments of somewhat stiff polypeptide chain interspersed with relatively tight turns or bends, which are almost always on the surface of the protein. This is illustrated by the backbone topology of BPTI depicted in Figure 6–10, in that there are four moderately straight portions of polypeptide (residues 1 to 14, 15 to 26, 27 to 38, and 39 to 58), which turn or end at the top or bottom of the pear-shaped molecule. This general type of structure has been compared to the behavior of a fire hose. It may be contrasted with other possible limiting situations: one more irregular, such as that obtained upon collapsing a flexible string, and the other more curved, illustrated by a ball of string.

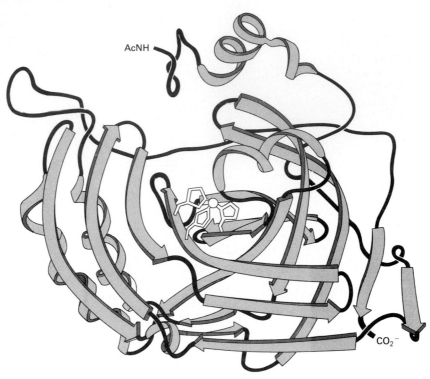

Figure 6–11
Schematic drawing of the polypeptide backbone of carbonic anhydrase. Note that the carboxyl-terminal segment is inserted through the plane of other segments of the polypeptide chain, so that a knot would result if it were pulled downward and the acetylated amino terminus upward.

The white ball in the middle is a zinc ion bound to the protein through the imidazole side chains of three His residues.

(Kindly provided by T. A. Jones and A. Liljas.)

The polypeptide backbone has never been observed to form a well-defined knot; that is, if the polypeptide chain were grasped at each end and pulled out straight, a linear chain would always result. This restriction on topology probably arises from the process by which the linear unfolded polypeptide chain attains the folded conformation. One apparent exception occurs in carbonic anhydrase; as shown in Figure 6–11, the two ends of the chain are somewhat entwined, so that if they were pulled toward the top and bottom, a knot would result. This exception is probably possible because only the very ends of the polypeptide chain are involved; one end of the chain might be tucked into the knotted topology at a very late stage of folding.

Knotted topologies would be more likely on a random basis if disulfide bonds are considered as part of the topology; yet such knots have not been observed. The most irregular topologies observed are two instances of "threaded" disulfide loops: one in BPTI (illustrated in Figure 6–12), the

Figure 6–12

The threaded topology of the polypeptide backbone of bovine pancreatic trypsin inhibitor, BPTI, relative to the covalent loop defined by the 30–51 disulfide bond. The polypeptide backbone between residues 15 and 56 is shown in projection, plus the side chains of Cys residues 30 and 51. The chain between residues 15 through 26 is shaded, as it passes through the covalent loop defined by the 30–51 disulfide bond and the intervening polypeptide chain.

The hydrogen bonds of the twisted β-sheet are shown as dashed lines. The α-helix of residues 47 through 56 is shown nearly end-on at the lower left. The direction of view is from the top of Figure 6–9.

(From T. E. Creighton, Prog. Biophys. Mol. Biol. 33:231–297, © 1978, Oxford, Pergamon Press.)

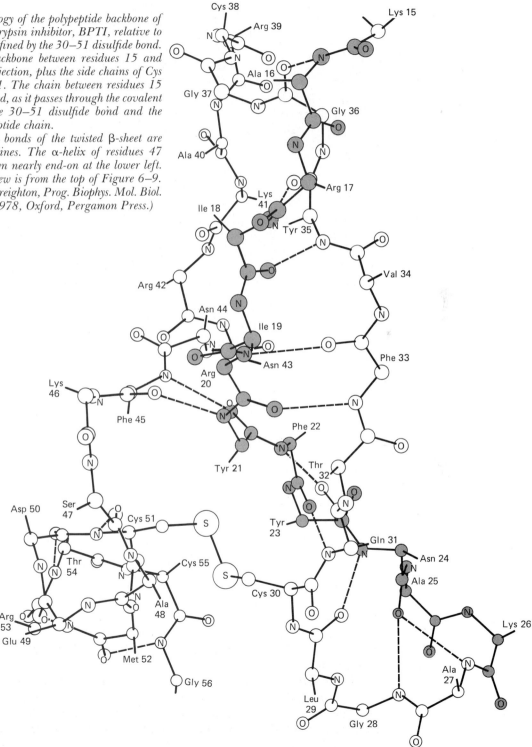

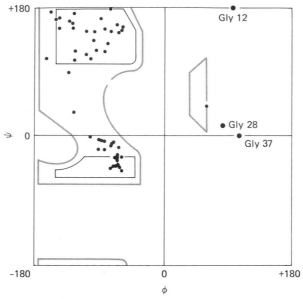

Figure 6–13
Ramachandran plot of the values of the torsion angles φ and ψ observed in bovine pancreatic trypsin inhibitor, BPTI. All the values are within the allowed regions for amino acids with side chains, except for those of three Gly residues, which are within regions allowed for such residues (see Figure 5–2A). (From T. E. Creighton, Prog. Biophys. Mol. Biol. 33:231–297, © 1978, Oxford, Pergamon Press.)

other in carboxypeptidase A. The topology in BPTI approximates a "slip-knot," in that the amino-terminal portion of the polypeptide chain passes through the covalent loop defined by the disulfide bond, and the intervening peptide chain, between cysteines 30 and 51. This topology could be imagined to be "un-threaded" by pulling the amino-terminal portion of the polypeptide chain (at about residue 15) away from the covalent loop, which would pull the threaded portion of the polypeptide chain through the loop and rotate the 30–51 disulfide bond.

Rotations about the individual bonds of both the backbone and the side chain are generally close to one of the conformations favored in the isolated structural unit. Consequently, the peptide bonds of the backbone are nearly always planar and are the *trans* isomer ($\omega = 180°$) unless the next residue is Pro, when *cis* peptide bonds ($\omega = 0$) are expected, and are found much more frequently. Only four examples of *cis* peptide bonds not involving Pro residues have been reported reliably, three in carboxypeptidase A and one in dihydrofolate reductase; others may have been overlooked, however, owing to general expectation of the *trans* form. *Cis* peptide bonds adjacent to proline residues appear to be incorporated primarily at tight bends of the polypeptide backbone.

The dihedral angles φ and ψ of the polypeptide backbone also generally lie within the limits deduced for the isolated peptide unit, as is illustrated for BPTI in Figure 6–13. Similarly, rotations about the bonds of

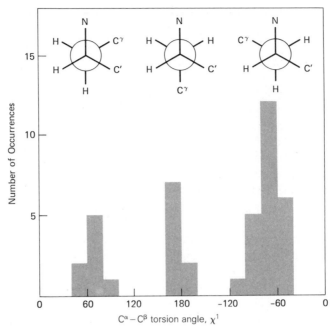

Figure 6–14

Histogram of the values of the C^α–C^β torsion angle, χ^1, observed in bovine pancreatic trypsin inhibitor, BPTI. Note that the staggered conformations are preferred, especially where the C' and C^γ atoms are trans. *(From T. E. Creighton, Prog. Biophys. Mol. Biol. 33:231–297, © 1978, Oxford, Pergamon Press.)*

the side chains are generally close to one of the three configurations in which the adjacent atoms are staggered, with that giving the greatest separation of the bulkiest groups favored (Figure 6–14). Unfavorable stereochemistry appears to be used in proteins only when required for their functional properties.

Many proteins also contain various ligands: prosthetic groups, coenzymes, metals, and so forth. The roles of these extra groups in the structures will be discussed in Chapter 8.

Studies on the conformation of amino acids. XI. Analysis of the observed side group conformations in proteins. R. Chandrasekaran and G. N. Ramachandran. Int. J. Protein Res. 2:223–233, 1970.

Topology of globular proteins. G. M. Crippen. J. Theor. Biol. 45:327–338, 1974; 51:495–500, 1975.

Directional structural features of globular proteins. G. M. Crippen and I. D. Kuntz. J. Theor. Biol. 66:47–61, 1977.

Conformation of amino acid side-chains in proteins. J. Janin, et al. J. Mol. Biol. 125:357–386, 1978.

Zinc environment and *cis* peptide bonds in carboxypeptidase A at 1.75 Å resolution. D. C. Rees, et al. Proc. Natl. Acad. Sci. U.S.A. 78:3408–3412, 1981.

Side chain characteristic main chain conformations of amino acid residues. A. S. Kolaskar and V. Ramabrahman. Int. J. Pept. Protein Res. 19:1–9, 1982.

Secondary Structure   The extended segments of the polypeptide chain very often have rather regular conformations like those observed in model polypeptides (Chapter 5). Most prominent are the right-handed α-helices and the extended β-strands, which generally associate side by side into parallel or antiparallel β-sheets. Approximately 31 per cent of the residues in known proteins occur in α-helices; 28 per cent, in β-structures.

Other regular conformations are much less frequent. A poly(Pro) helix is generally observed with only a few residues including two or more proline residues, as with residues 7 to 10 of BPTI, where residues 8 and 9 are Pro. A short segment of a left-handed, collagen-like helix, with Pro as every third residue, is found in cytochrome c_{551}. Short three-stranded, keratin-like coiled coils are found in Southern bean mosaic virus and in the hemagglutinin of influenza virus.

Regular secondary structure is a very useful focal point in attempts to comprehend a complex protein structure, so many studies of it have been made, and most schematic drawings of proteins emphasize it, often in a very idealized manner.

The segments of α-helices and β-sheets are generally rather short, being limited to the diameter of the protein globule. The length of an α-helix is generally 10 to 15 residues, while that of a β-strand is 3 to 10 residues. Consequently, it is often difficult to define exactly which residues are part of the secondary structure. For example, at both ends of an ideal α-helix there are four residues that participate in only one hydrogen bond each within the helix, whereas all other interior residues participate in two. Moreover, the ends are often irregular in conformation. Which residues should be counted as part of the helix, therefore, is often not clear. Crystallographers have used various criteria for defining secondary structure, thereby producing variable assignments, depending upon whether hydrogen bonding, conformational angles ϕ and ψ, or distances between C^{α} atoms are used.

Secondary structure in proteins is generally somewhat distorted. In α-helices, the plane of the peptide bond is often rotated so that the carbonyl group is pointing outward from the helix axis, and the terminal residues often form a local 3_{10} helix. The single α-helix of BPTI is rather regular and consists of about three turns, involving residues 47 to 56.

β-Sheets are generally twisted, rather than planar, with a right-handed twist of from 0° to 30° between strands (see Figure 6–11). The conformational parameters can also deviate considerably from ideality. For example, in the rather distorted antiparallel β-sheet of BPTI, comprising residues 17 to 24 and 29 to 35, the mean value of ϕ is $-114°$, with a standard deviation of 24°, and that of ψ is $139° \pm 26°$. The standard values for a planar sheet should be $\phi = -139°$ and $\psi = 135°$ (Table 5–2), whereas somewhat more positive values of both are generally observed with twisted sheets. Further distortions occur in β-sheets consisting of both parallel and antiparallel strands, since the ideal backbone conformations for the two differ. An extra residue is often present in a strand at the edge of a sheet, interrupting the hydrogen bond pattern and producing a "β-bulge."

β-Sheets can consist entirely of parallel or antiparallel strands or can have a mixture of the two. Purely parallel sheets are least frequent; purely antiparallel sheets are most common. Antiparallel sheets often consist of just two or three strands, whereas parallel sheets always have at least 4 and may have up to 8. Mixed sheets may consist of 3 to 15 strands. Purely parallel or antiparallel sheets of 6 or 8 strands are said often to curve around to close up the sheet into a continuous "β-barrel," although they are probably described more accurately as two β-sheets packed against each other. Most individual β-strands consist of 3 to 10 residues, with the average length being 6.5.

The strands adjacent in a sheet tend to consist of segments that are adjacent in the primary structure. This correlation is greatest for antiparallel strands and least for parallel strands.

Secondary structure is most apparent in the larger proteins, where most of the interior is composed of such regular structure. At least one important property of secondary structure is that it provides an efficient way of pairing in hydrogen bonds the internal polar groups of the polypeptide backbone; such pairing is a prerequisite for stability of the folded conformation. Since regular hydrogen bonding is a major characteristic of secondary structure, it is usually conspicuous in diagrams of hydrogen bonds in proteins, as illustrated in Figure 6–15 for BPTI.

The secondary structure is usually indicated schematically in simplified depictions of protein folded conformations. Helices are often indicated by cylinders or coiled ribbons, and extended strands of β-sheets by broad arrows, indicating the amino-to-carboxyl polarity of the polypeptide backbone. Some simplified representations of several proteins are shown in Figure 6–16.

Two-dimensional plots of the distance between the C^α atoms of residues i and j are very useful descriptions of both the secondary structure and the overall folding of the polypeptide chain, especially for comparing different proteins. In such plots, called **contact** or **distance maps,** the distances between all pairs of C^α atoms in the protein, in order of the amino acid sequence, are represented as contours, or else only those within a certain distance, usually 10 Å, are marked. The contact map for BPTI is illustrated in Figure 6–17.

In such distance plots, α-helices are evident by a greater spread of close contacts along the diagonal, i.e., between residues nearby in the amino acid sequence, since C_i^α is in close proximity to C_{i-4}^α, C_{i-3}^α, C_{i+3}^α, and C_{i+4}^α, where the subscript is the number of the residue in the polypeptide chain. Within a parallel β-sheet, in which the first two residues of two adjacent extended strands are i and j, C_i^α is next to C_j^α, C_{i+1}^α is adjacent to C_{j+1}^α, and so on; this gives rise to a series of close contacts on a diagonal line *parallel* to the main diagonal, but offset from it by $(i - j)$ residues. With two strands of antiparallel β-sheet, where residues i and j are respectively the first and last residues hydrogen-bonded, C_i^α is next to C_j^α, C_{i+1}^α is next to C_{j-1}^α, and so forth. This gives rise to a series of contacts that define a diagonal line *perpendicular* to the main diagonal. In the dis-

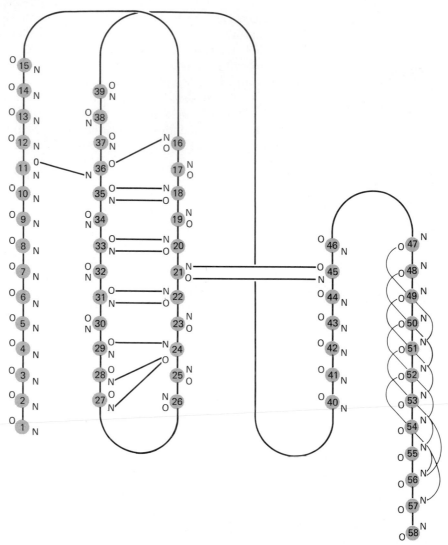

Figure 6–15
Diagram of the hydrogen bonds between polypeptide backbone atoms of bovine pancreatic trypsin inhibitor, BPTI. (From R. Huber, et al., in Proteinase Inhibitors. H. Fritz and H. Tschesche (eds.). Pp. 56–65. Berlin, Walter de Gruyter, 1971.)

Figure 6–16
Schematic drawings of the polypeptide backbones of four proteins. α-Helices are depicted as helical ribbons; β-strands, as arrows. The hemoglobin β subunit has a heme group bound; hemerythrin has two iron atoms. Lactate dehydrogenase domain 1 has a molecule of the coenzyme NAD illustrated schematically. (From J. S. Richardson, Adv. Protein Chem. 34:167–339, 1981.)

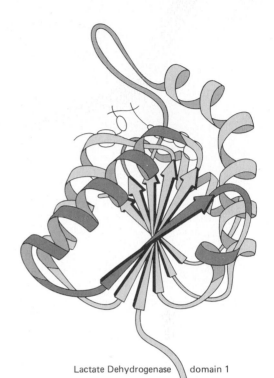

Hemoglobin β subunit

Lactate Dehydrogenase domain 1

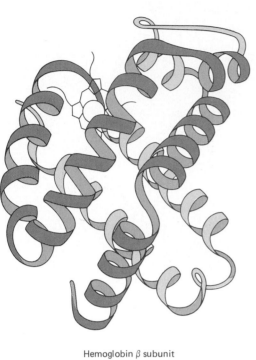

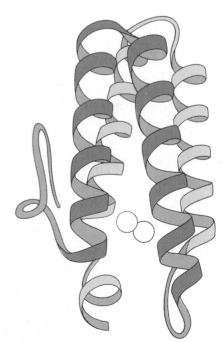

Triose Phosphate Isomerase

Hemerythrin

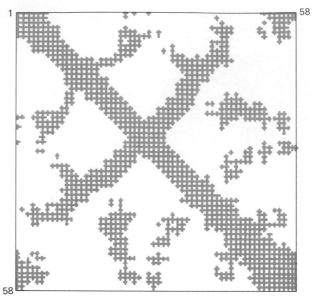

Figure 6–17
Contact map for native bovine pancreatic trypsin inhibitor, BPTI. A cross is present whenever two residues are within 10 Å of each other. (From M. Levitt, J. Mol. Biol. 104:59–107, © 1976, London, Academic Press Inc.)

tance plot for BPTI (Figure 6–17), the antiparallel β-sheet between residues 18 and 35 is apparent by the strong diagonal near the middle of the diagram perpendicular to the main diagonal. The α-helix at the carboxyl end of the chain is also apparent from the broadening of the main diagonal in the lower right-hand corner. Structural domains segregated along the polypeptide backbone are often apparent as segregated areas of contacts on the distance plots (see Figure 6–29 for the two-domain proteins chymotrypsin and elastase).

On the basis of their secondary structure, protein structures have been classified into four classes: **(α),** having only α-helices; **(β),** with primarily β-sheet structure; **(α + β),** having both helices and sheets, but in separate parts of the structure; and **(α/β),** where both helices and sheets interact and often alternate along the polypeptide chain. In (α) proteins, about 60 per cent of the residues are in helices; these helices are usually in contact. In (β) proteins, there are always two β-sheets, usually antiparallel, that pack against each other; with β-barrels, the two sheets are joined to give a continuous cylindrical sheet. In the (α + β) proteins, there may be a single β-sheet, usually antiparallel; the helices often cluster together at one or both ends of the β-sheet. The (α/β) proteins have one major β-sheet of primarily parallel strands; a helix usually occurs in each of the segments of polypeptide chain connecting the β-strands, probably owing to the necessarily long lengths of these connections. The helices pack on both sides of the sheet.

Table 6–4   Normalized Frequencies of Occurrence of Amino Acid Residues in Secondary Structure in Globular Proteins

AMINO ACID	α-HELIX (P_α)	β-SHEET (P_β)	REVERSE TURN (P_t)
Ala	1.29	0.90	0.78
Cys	1.11	0.74	0.80
Leu	1.30	1.02	0.59
Met	1.47	0.97	0.39
Glu	1.44	0.75	1.00
Gln	1.27	0.80	0.97
His	1.22	1.08	0.69
Lys	1.23	0.77	0.96
Val	0.91	1.49	0.47
Ile	0.97	1.45	0.51
Phe	1.07	1.32	0.58
Tyr	0.72	1.25	1.05
Trp	0.99	1.14	0.75
Thr	0.82	1.21	1.03
Gly	0.56	0.92	1.64
Ser	0.82	0.95	1.33
Asp	1.04	0.72	1.41
Asn	0.90	0.76	1.28
Pro	0.52	0.64	1.91
Arg	0.96	0.99	0.88

The normalized frequencies were calculated from the fraction of residues of each amino acid that occurred in each of the three conformations, divided by this fraction for all residues. Random occurrence would give a normalized frequency of 1; more frequent than random occurrences give values greater than 1.

The values were calculated from 66 protein structures, comprising 31 different conformations. Where M related proteins occur with the same conformation, they were each weighted by the factor $1/M$. This takes into account the different amino acid sequences that can give the same conformation, which have been determined a number of times, e.g., for the globins.

Adapted from M. Levitt, Biochemistry 17:4277–4285, 1978.

Many studies have been made of the occurrence of the 20 amino acids in the various structural elements of proteins, as part of attempts to understand the basis of protein structure. There are small, but significant, differences for virtually all the amino acids, as shown by the most extensive compilation (Table 6–4). Only Arg shows no tendency to occur preferentially in any particular secondary structure. Amino acids with a branched or bulky side chain (Val, Ile, and Thr) or an aromatic ring (Phe, Tyr, and Trp) occur most frequently in β-sheets. All the rest occur most often in α-helices, except for those with short polar (Ser, Asp, and Asn) or special side chains (Gly and Pro), which occur most often in reverse turns.

Conformation of twisted β-pleated sheets in proteins. C. Chothia. J. Mol. Biol. 75:295–302, 1973.

Structural patterns in globular proteins. M. Levitt and C. Chothia. Nature 261:552–558, 1976.

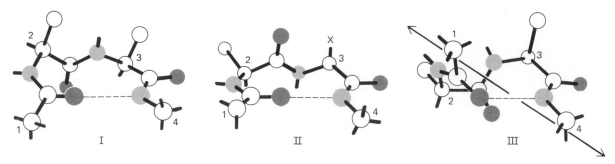

Figure 6–18
Classical types of reverse turns. Type III is similar to a 3_{10} helix; the arrow indicates where the long axis of the helix would be. The X in type II indicates that no side chain may be present on the third residue. Hydrogen bonds are represented by dashed lines; carbon atoms, white spheres; nitrogen, light gray spheres; oxygen, dark gray spheres. (Adapted from J. Birktoft and D. M. Blow, J. Mol. Biol. 68:187–240, 1972.)

Automatic identification of secondary structure in globular proteins. M. Levitt and
    J. Greer. J. Mol. Biol. 114:181–239, 1977.
On the conformation of proteins: an analysis of β-pleated sheets. M. J. E. Sternberg
    and J. M. Thornton. J. Mol. Biol. 110:285–296, 1977.
β-Sheet topology and the relatedness of proteins. J. S. Richardson. Nature 268:
    495–500, 1977.
Packing of α-helices onto β-pleated sheets and the anatomy of α/β proteins. J.
    Janin and C. Chothia. J. Mol. Biol. 143:95–128, 1980.
Relative orientation of close-packed β-pleated sheets in proteins. C. Chothia and
    J. Janin. Proc. Natl. Acad. Sci. U.S.A. 78:4146–4150, 1981.
Analysis of the tertiary structure of protein β-sheet sandwiches. F. E. Cohen, et al.
    J. Mol. Biol. 148:253–272, 1981.

Reverse Turns

Proteins have rather spherical structures, in spite of being composed of straight segments of polypeptide chain, because the polypeptide chain generally makes rather sharp bends at the surface, thereby reversing the direction of the polypeptide chain. It is these **reverse turns** that give proteins their globularity, and a substantial fraction of residues of every protein are involved in them. They are also known as *hairpin bends,* β-*bends,* and β-*turns,* because they often connect antiparallel β-strands.

The polypeptide chain is generally rather extended. Sharp chain reversals, therefore, require rather special geometry; it is perhaps not surprising that the special amino acids Gly and Pro are often involved.

Four consecutive residues in the polypeptide chain are generally considered to comprise a reverse turn, although it is only the torsion angles of the second and third residues that are critical. The first and fourth residues are usually included because a hydrogen bond between their backbone groups was originally considered necessary. Three such ideal reverse turns, generally designated I, II, and III (Figure 6–18), were predicted by Venkatachalam on the basis of allowed geometry, with planar *trans* peptide bonds. Mirror images of the backbone may occur in variants I′, II′, and III′. Type I is compatible with any amino acid residue at positions 1 through

Table 6–5 *Bend Types: Classification and Frequency of Occurrence in 26 Proteins*

BEND TYPE	DIHEDRAL ANGLES OF TWO CENTRAL RESIDUES (°)[a]				NUMBER OF OBSERVED BENDS			
	ϕ_2	ψ_2	ϕ_3	ψ_3	Ideal bends[b]	Nonideal bends	Total bends	H-bonded bends[c]
I	−60	−30	−90	0	130	46	176	99
I′	60	30	90	0	8	5	13	10
II	−60	120	80	0	41	23	64	43
II′	60	−120	−80	0	15	5	20	16
III	−60	−30	−60	−30	66	11	77	45
III′	60	30	60	30	11	2	13	7
IV	A bend with two or more angles differing by at least 40° from those given above				0	35	35	5
V	−80	80	80	−80	1	2	3	0
V′	80	−80	−80	80	0	4	4	2
VI	A *cis* Pro at position 3				8	0	8	6
VII	A kink in the protein chain created by $\psi_2 \approx 180°$ and $|\phi_3| < 60°$ or $|\psi_2| < 60°$ and $\phi_3 \approx 180°$				8	0	8	1
Total					288	133	421	234

[a]The two central residues of a tetrapeptide β-turn are the $i + 1$ and $i + 2$ or 2nd and 3rd residues characterized by $(\phi, \psi)_2$ and $(\phi, \psi)_3$ given above.

[b]Bends that do not have any angle differing by more than 50° from the $(\phi, \psi)_2$ and $(\phi, \psi)_3$ for a particular bend type are considered as ideal.

[c]Bends with $O_{(1)}$ to $N_{(4)}$ distances < 3.5 Å were considered as hydrogen-bonded.

Adapted from P. Y. Chou and G. D. Fasman, J. Mol. Biol. 115:135–175, 1977.

4, except that Pro cannot occur at position 3. In contrast, type I′ requires Gly at both positions 2 and 3. Types II and II′ require Gly at positions 3 and 2, respectively. Type III is a portion of a 3_{10} helix, and any amino acids are permissible; type III′ requires Gly at positions 2 and 3.

Many examples of turns are observed in proteins, but the conformational angles are not ideal, and a hydrogen bond is not always formed between residues 1 and 4. Other additional types of turns have been observed, generally designated types IV, V, V′, VI, and VII (Table 6–5). Type VI has Pro at position 3 and a *cis* peptide bond preceding it.

Stereochemical criteria for polypeptides and proteins. V.Conformation of a system of three linked peptide units. C. M. Venkatachalam. Biopolymers 6:1425–1436, 1968.

Chain reversals in proteins. P. N. Lewis, et al. Biochim. Biophys. Acta 303:211–229, 1973.

The reverse turn as a polypeptide conformation in globular proteins. J. L. Crawford, W. N. Lipscomb, and C. G. Schellman. Proc. Natl. Acad. Sci. U.S.A. 70:538–542, 1973.

β-Turns in proteins. P. Y. Chou and G. D. Fasman. J. Mol. Biol. 115:135–175, 1977.

Reverse turns in peptides and proteins. J. A. Smith and L. G. Pease. Crit. Rev. Biochem. 8:315–399, 1980.

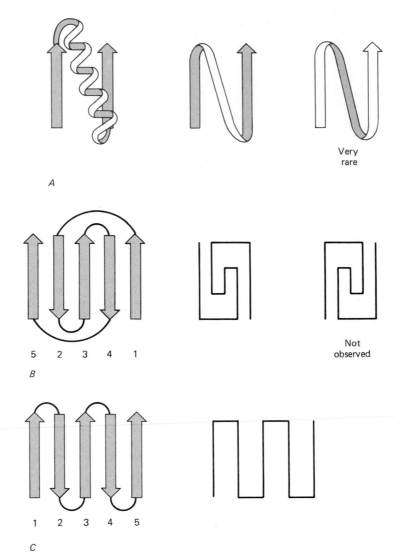

Figure 6–19
Super-secondary structures observed in proteins. A, A β–α–β unit; the segment joining the two β-strands is almost always above their plane, not below. B, The Greek-key motif of antiparallel β-sheets, where it is always of one "handedness." C, A β-meander.

Super-secondary Structures

Certain assemblies of a number of secondary structure elements, including the connecting segments of polypeptide chain, have been observed a sufficient number of times that they are becoming recognized as a further level of structure, termed **super-secondary structures.** These structures are at a higher level than is secondary structure but do not constitute entire structural domains.

One element is often designated **β–α–β,** where the two β-strands are parallel in a β-sheet, not necessarily adjacent. The α-helix occurs in the connecting segment, which must be roughly antiparallel to the β-strands (Figure 6–19A). In virtually all instances, the connecting segment lies on the same side of the β-sheet, defined as right-handed, even when there is no α-helix in the connecting strand. This unique topology may arise from the tendency of the β-strands to twist in that direction, as may be visualized by twisting a ribbon, then bringing the ends together.

Within antiparallel β-sheets, a common occurrence is the **"Greek key"** topology (Figure 6–19B), named after a common pattern found on Greek pottery. Once again, the topology almost always occurs with only one of the two possible orientations. The basis for this choice is not certain.

Finally, antiparallel β-sheets often have those strands that are sequential in the primary structure also adjacent in the β-sheet. They are connected by relatively tight reverse turns, to produce a **"β-meander"** (Figure 6–19C). Here there is no asymmetry and all connections are equivalent.

The architecture of proteins appears to be made up of segments of secondary structure packing together, so current efforts are directed at understanding the basis for the interactions between helices, between β-sheets, and between helices with sheets. Because of their basic architecture, the interactions between them are governed primarily by the amino acid side chains on their surfaces. For example, interdigitation of the side chains of two ideal α-helices would be expected only when their axes cross at angles of $-82°$, $-60°$, or $+19°$ (a rotation is positive when the lower helix is rotated clockwise relative to the top one). Helices are expected to pack onto β-sheets with their axes nearly parallel to those of the β-strands. Two normal twisted β-sheets should pack together face to face with the top sheet rotated clockwise between 30° and 80°. Proteins are found to observe these rules to a first approximation, but substantial variation is produced by the nonideality of their secondary structure and by the variability of the amino acid side chains involved in the contacts. In the great majority of cases, only nonpolar amino acid side chains are involved in these interactions.

Comparison of super-secondary structures in protein. S. T. Rao and M. G. Rossmann. J. Mol. Biol. 76:241–256, 1973.

Handedness of crossover connections in β sheets. J. S. Richardson. Proc. Natl. Acad. Sci. U.S.A. 73:2619–2623, 1976.

On the conformation of proteins: the handedness of the connection between parallel β-strands. M. J. E. Sternberg and J. M. Thornton. J. Mol. Biol. 110:269–283, 1977.

Structure of proteins: packing of α-helices and pleated sheets. C. Chothia, et al. Proc. Natl. Acad. Sci. U.S.A. 74:4130–4134, 1977.

Interiors and Exteriors

Detailed analyses of the very complex surfaces of folded protein and of the internal packing of atoms have used the procedure of Lee and Richards (Figure 6–20A). Each atom of the protein is depicted as a sphere of its van der Waals radius; overlapping spheres, where the atoms are covalently

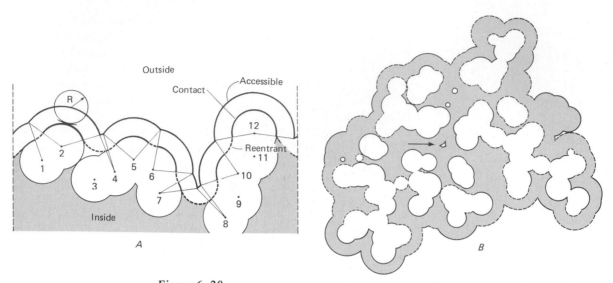

Figure 6–20

Analysis of protein surfaces. The various surfaces are defined in A *for a two-dimensional slice through a hypothetical protein composed of atoms 1 through 10, using a solvent probe of radius* R. *The van der Waals and accessible surfaces of a section through ribonuclease S are shown in* B. *The solid lines are for carbon or sulfur atoms; the dashed lines, for nitrogen or oxygen. In places, the accessible surface is controlled by atoms above or below the section shown. The arrow indicates a cavity inside the protein large enough to accommodate a molecule of radius 1.4 Å, that used for the probe. (Adapted from F. M. Richards, Ann. Rev. Biophys. Bioeng. 6:151–176, 1977; and B. Lee and F. M. Richards, J. Mol. Biol. 55:379–400, 1971.)*

bonded, are truncated. The complex surface that results is referred to as the **van der Waals surface;** it has a strictly defined surface area and encloses a definite volume. However, no chemical procedure ever measures directly this area or volume, as any chemical probe has significant dimensions. The surface accessible to any chemical probe, such as the water molecules of solvent, is explored by imagining a spherical probe of appropriate radius, *R,* to roll on the outside of the molecule, maintaining contact with the van der Waals surface. In the hypothetical protein of Figure 6–20*A* the probe will never contact atom 3, 9, or 11; such atoms are considered to be *not* part of the surface of the molecule but interior atoms. Those parts of the van der Waals surface that make contact with the surface of the probe are designated the **contact surface;** it is a series of disconnected patches. When the probe is simultaneously in contact with more than one atom, its interior surface defines the **reentrant surface.** The contact surface and the reentrant surface together define a continuous surface, known as the **molecular surface.** The surface that is discussed most often is the **accessible surface,** which is defined by the center of the probe. The probe is generally considered to be a water molecule with a radius of 1.4 Å.

The van der Waals surface and the accessible surface of a section through the crystallographic structure of ribonuclease S are illustrated in

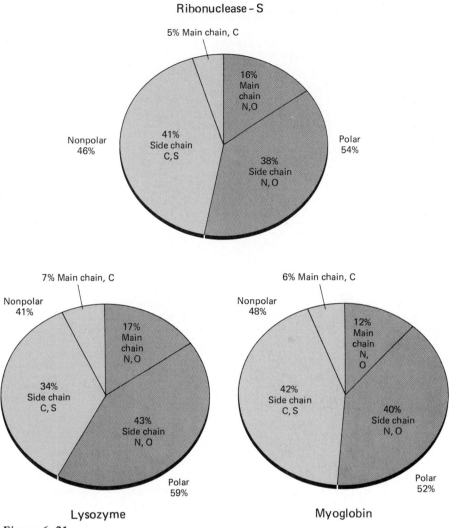

Figure 6–21

Contributions made to the total accessible surface area by different types of atoms. Those of polar atoms are shaded. The values are only approximate, as they depend upon the values for the van der Waals radii used. (From B. Lee and F. M. Richards, J. Mol. Biol. 55:379–400, © 1971, London, Academic Press Inc.)

Figure 6–20*B*. The protein atoms that define the accessible surfaces of different proteins are quite similar, consisting primarily of the atoms of side chains (Figure 6–21).

The total accessible surface area of proteins is approximately proportional to the two-thirds power of the molecular weight, as would be expected for approximately spherical objects. However, the accessible surface area is nearly two times greater than that expected for a sphere of the same size, which is some measure of the roughness of the surface. The

Table 6–6   The Packing of Residues in the Interior of Proteins

AMINO ACID	AVERAGE VOLUME (Å^3) OF BURIED RESIDUES[a]	VOLUME IN PROTEIN/ VOLUME OF RESIDUE[e]	FRACTION OF RESIDUES AT LEAST 95% BURIED[f]
Gly	66.4	1.10	0.36
Ala	91.5	1.29	0.38
Val	141.7	1.01	0.54
Leu	167.9	1.01	0.45
Ile	168.8	1.01	0.60
Ser	99.1	1.11	0.22
Thr	122.1	1.05	0.23
Asp	124.5	1.12	0.15
Asn	124.5	1.06	0.12
Glu	155.1	1.12	0.18
Gln	161.1	1.12	0.07
Lys	171.3	1.02	0.03
Arg	—[b]		0.01
His	167.3	1.09	0.17
Phe	203.4	1.07	0.50
Tyr	203.6	1.05	0.15
Trp	237.6	1.04	0.27
Cys	105.6[c]		0.40
	117.7[d]	1.08	0.50
Met	170.8	1.05	0.40
Pro	129.3	1.05	0.18

[a]From C. Chothia, Nature 254:304–308, 1975.

[b]No residues were found to be totally buried in nine proteins.

[c]When in disulfide form.

[d]When in thiol form.

[e]Calculated from the observed volume in proteins in the first column and from the volumes of the isolated amino acid residues in Table 1–1.

[f]Average for 12 proteins; from C. Chothia, J. Mol. Biol. 105:1–14, 1976.

accessible surface area (A_s) in Å^2 of a typical small protein may be estimated from its molecular weight (M) by the approximate relationship

$$A_s = 11.12\ M^{2/3} \tag{6–6}$$

Its volume, V, in Å^3, would be given by

$$V = 1.27\ M \tag{6–7}$$

Where a large protein is divided into conspicuous domains, the value of A_s becomes directly proportional to M.

The interiors of proteins are found to be very closely packed, with most adjacent atoms in van der Waals contact. About 75 per cent of the interior is filled with atoms, defined by their van der Waals radii. This is close to the value of 74 per cent possible with close packing of identical spheres and is within the range of 70 to 78 per cent found with crystals of small organic molecules (Table 6–6). It contrasts with the lower values observed with liquids, such as water (58 per cent) and cyclohexane (44 per cent). The packing density does vary somewhat throughout the interior, generally being highest in areas where the polypeptide topology is most regular. In relatively few instances are there unfilled cavities of sufficient size to accommodate other molecules; one is shown in Figure 6–20*B*. The general close-packing of protein interiors is most impressive in light of the fact that it must be compatible with the covalent connectivity of the polypeptide chain.

Water molecules are generally excluded from protein interiors. When present, however, they appear to be integral parts of the protein structure: They are well-fixed in internal cavities of the protein, isolated from the bulk solvent; they invariably form hydrogen bonds to polar groups of the protein; and they are conserved in homologous structures. Some water molecules occur singly, some in pairs, and others in extended networks. Most make two or three hydrogen bonds to polar groups or other water molecules, although some do make four, involving the oxygen atom as hydrogen acceptor in two and the two hydrogen atoms as donors in two further hydrogen bonds. Many bind to the NH and CO groups of the main-chain backbone, although some are attached to polar side chains.

Such buried water molecules appear to be important for filling holes and, probably the more important, pairing with internal polar groups of the protein in hydrogen bonds. Even in the small protein BPTI, there are four internal H_2O molecules (Figure 6–22).

Virtually all ionized groups on a protein are on the surface, exposed to the solvent. Uncharged polar groups of side chains are usually on the surface; when buried in the interior, they are generally involved in hydrogen bonds to other polar groups. Nonpolar groups tend to be in the interior, but many do occur on the surface. It is inevitable that the polar groups of much of the backbone be buried in the interior; these also are paired in hydrogen bonds, most often in secondary structure. As a result of these differences between internal and external residues, an α-helix near the surface of a protein tends to have nonpolar side chains on one side and mainly polar side chains on the other. Consequently, there is often a repeating pattern in this segment of the primary structure, with every third or fourth residue tending to be of the same class. Likewise, β-pleated sheets near the surface tend to have polar side chains on the exterior side, nonpolar on the other, which is reflected in an alternating pattern of polar and nonpolar residues in the amino acid sequence. Where helices or β-sheets are buried, all the side chains tend to be nonpolar; the central strands of a sheet tend to have the most hydrophobic amino acid residues.

Because of the size and complexity of many of the amino acid side chains, it is difficult to classify them as being simply buried or exposed. For example, in the long side chains of Lys and Arg, the ionized terminal groups are almost invariably exposed to the solvent, but the remaining hydrophobic methylene carbons are often buried in the interior; consequently, the C^α atoms of these residues may be rather far from the surface, and most of the side chain may be buried. Different attempts to analyze the roles of the various amino acids have used the fraction buried, the

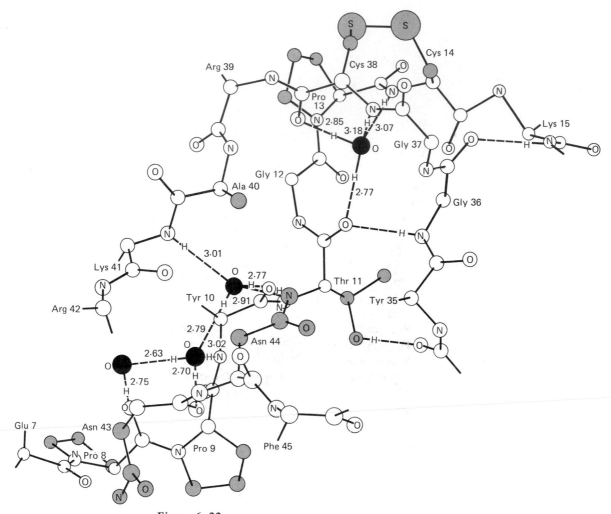

Figure 6–22
The four internal water molecules of bovine pancreatic trypsin inhibitor, BPTI. The oxygen atoms observed crystallographically are shown in black, with the presumed positions of the hydrogen atoms indicated. The lengths in angstroms of the hydrogen bonds between N and O atoms are indicated. See also Figure 6–9, where the three-dimensional positions of these water molecules are visible. (From T. E. Creighton, Prog. Biophys. Mol. Biol. 33:231–297, © 1978, Oxford, Pergamon Press.)

radial position of the C^α atom or that at the other end of the side chain, or the radial orientation of the side chain, whether pointing inward or outward. The criterion that seems most meaningful is the fraction of the residues having at least 95 per cent of the side chain inaccessible to the solvent. As shown in Table 6–6, this ranges from 0.01 and 0.03 for the very basic Arg and Lys residues to 0.54 and 0.60 for the hydrophobic Val and Ile, respectively.

The environments of the amino acid side chains within the interior of proteins and on their exteriors have been analyzed extensively, but there appear to be no conspicuous tendencies for the 20 different amino acid side chains to be adjacent preferentially to each other or to the peptide backbone, other than the general tendencies just described for hydrophobic and polar groups as classes. There are, then, apparently no simple structural rules to relate conformation to the amino acid sequence. This point will become even more obvious in the following section, in the discussion of the variety of amino acid sequences that result in the same folded conformation.

In summary, the general properties of protein structures seem to be compatible with sound physical principles, as outlined in Chapter 4, with very few unfavorable features. The nearly spherical globule and the hydrophobic side chains on the interior tend to minimize the contact of nonpolar surfaces with water. Ionized and polar groups are not removed from contact with the solvent without pairing with oppositely charged groups or in hydrogen bonds. Where ion pairs are formed, they tend to be on the surface, probably so as not to disrupt their hydration shell. The bond angles, lengths, and rotations are also close to the most favorable values.

How these features can account for the net stability of the folded state, and why any particular folded state is assumed, is the subject of Chapter 7.

The interpretation of protein structures: estimation of static accessibility. B. Lee and F. M. Richards. J. Mol. Biol. 55:379–400, 1971.

On the nature of the protein interior. M. H. Klapper. Biochim. Biophys. Acta 229:557–566, 1971.

The interpretation of protein structures: total volume, group volume distributions and packing density. F. M. Richards. J. Mol. Biol. 82:1–14, 1974.

Structural invariants in protein folding. C. Chothia. Nature 254:304–308, 1975.

The nature of the accessible and buried surfaces in proteins. C. Chothia. J. Mol. Biol. 105:1–14, 1976.

Areas, volumes, packing, and protein structure. F. M. Richards. Ann. Rev. Biophys. Bioeng. 6:151–175, 1977.

On the conformation of proteins: hydrophobic ordering of strands in β-pleated sheets. M. J. E. Sternberg and J. M. Thornton. J. Mol. Biol. 115:1–17, 1977.

Correlation of sequence and tertiary structure in globular proteins. G. M. Crippen. Biopolymers 16:2189–2201, 1977.

A study of the preferred environment of amino acid residues in globular proteins. P. Manavalan and P. K. Ponnuswarmy. Arch. Biochem. Biophys. 184:476–487, 1977.

Hydrophobicity, hydrophilicity, and the radial and orientational distributions of residues in native proteins. S. Rackovsky and H. A. Scheraga. Proc. Natl. Acad. Sci. U.S.A. 74:5248–5251, 1977.

Quaternary Structure

Many proteins are observed in solution to exist as aggregates of two or more polypeptide chains, either identical or different. (Different polypeptide chains are designated by letters; for example, hemoglobin is normally $\alpha_2\beta_2$, aspartate transcarbamylase is r_6c_6, and very many proteins are di-

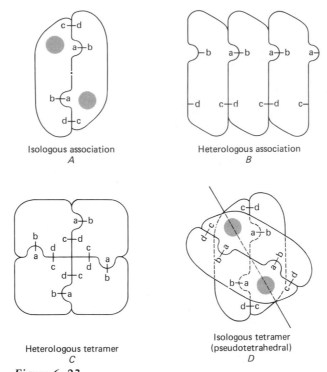

Figure 6–23
*Schematic illustrations of isologous and heterologous association between protein subunits. A,
Isologous association to form a dimer with a twofold symmetry axis perpendicular to the plane
of the figure. B, Heterologous association leading to infinitely long polymers. C, Heterologous
association to form a closed, finite structure, in this case a tetramer. D, Forming a tetramer
by isologous association, using two different bonding interactions: a–b, c–d, plus the cross-
hatched circles of A. (Adapted from J. Monod, et al., J. Mol. Biol. 12:88–118, 1965.)*

mers, trimers, tetramers, or even higher-order aggregates of identical polypeptide chains.) In many instances, such quaternary structure is vital for
the structure and function of the protein. Because it is invariably observed
in the crystal structure, the detailed nature of these interactions is known.

Each subunit is generally folded into an apparently independent globular conformation, which then interacts with other subunits. The interfaces
between different subunits are generally very similar to the interiors of the
individual molecules. They are closely packed and involve both hydrophobic and polar interactions. Consequently, the interacting surfaces of the
individual molecules are highly complementary, both in shape and in pairing of polar groups.

Two fundamental types of interactions between identical subunits are
possible, designated **isologous** and **heterologous** (Figure 6–23). In the former, the interacting surfaces are identical, which gives rise to a closed,
dimeric structure with a twofold axis of symmetry. Further interactions,
to give a tetramer, must use different sets of interfaces. Many tetramers

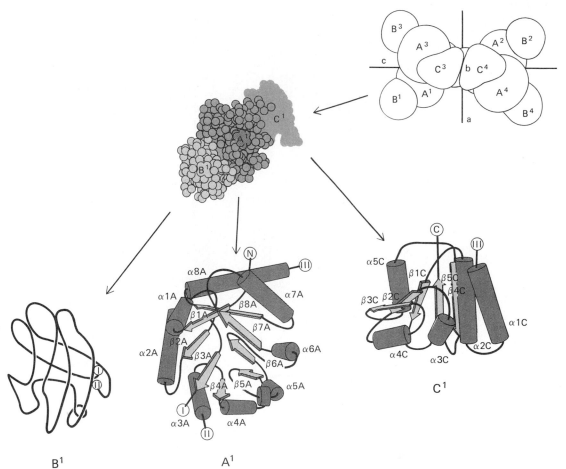

Figure 6–24

Schematic representation of the domain and subunit structure of muscle pyruvate kinase. The full tetramer of four identical polypeptide chains is shown in outline at the top right, with the three twofold axes a, b, and c. The three domains of each monomer are labeled A, B, and C, with superscripts to indicate the different polypeptide chains. Subunit 1 is shown at the center, with a sphere representing each amino acid residue. The three domains are dissected at the bottom. A has the amino terminus of the polypeptide chain, N, and is depicted as cylindrical α-helices and ribbon β-strands. I and II mark the connections of the polypeptide chain to domain B, which is depicted only by the polypeptide backbone and has two antiparallel β-sheets. III is the polypeptide connection between the A and C domains; the latter has the carboxyl terminus, marked C. (From D. I. Stuart, et al., J. Mol. Biol. 134:109–142, © 1979, London, Academic Press Inc.)

are formed by two sets of isologous interactions, to give three twofold axes of symmetry (Figure 6–24). For example, prealbumin and concanavalin A associate to dimers by associating the β-sheets of different molecules; then two such dimers pack together to give a tetramer (Figure 6–25).

In heterologous association the interfaces that interact are not identical. The surfaces must be complementary but need not be symmetric.

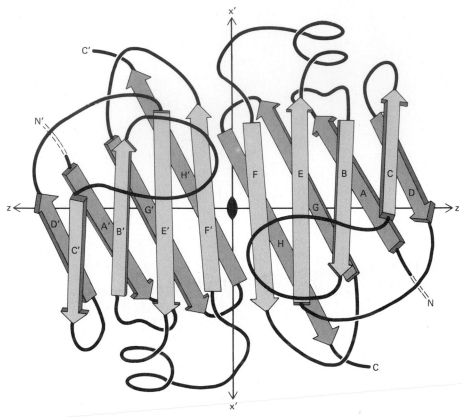

Figure 6–25
Schematic drawing of the polypeptide backbone in the prealbumin dimer. The arrows depict the β-strands, labeled A to H in one monomer, A′ to H′ in the other. Note that the two monomers associate to continue the two β-sheets, by strand F interacting with F′ and H with H′. A tetramer is formed by isologous interactions between the side chains protruding outward from sheet D′ A′ G′ H′ H G A D in both dimers, which are packed together at nearly right angles to each other. (From C. C. F. Blake, et al., J. Mol. Biol. 88:1–12, © 1974, London, Academic Press Inc.)

Such association is open-ended and would give rise to long assemblies of varying length, unless the geometry of the interaction is such as to produce closed, cyclic structures (Figure 6–23). The latter situation usually applies, and trimers of bacteriochlorophyll protein and of aspartate transcarbamylase and tetramers of hemerythrin and of neuraminidase are produced in this way.

A special case arises when the two heterologous surfaces overlap. Only a dimer may be formed by such interactions, since the second surface is sterically inaccessible, but the dimer does not have a twofold axis of symmetry and the two molecules are not structurally equivalent. Such association is observed in hexokinase. Other unexplained instances of nonequivalence are found in insulin and α-chymotrypsin.

With higher-order structures, such as spherical viruses, more complex interactions often occur, and the protein molecule is designed to be more flexible so that structural "quasi-equivalence" is observed, as in tomato bushy stunt and Southern bean mosaic viruses.

Association of nonidentical molecules requires only spatial and physical complementarity of the interacting surfaces. This subject will be discussed in Chapter 8.

Structure and symmetry of oligomeric enzymes. B. W. Matthews and S. A. Bernhard. Ann. Rev. Biophys. Bioeng. 2:257–317, 1973.

Principles of protein–protein recognition. C. Chothia and J. Janin. Nature 256: 705–708, 1975.

The symmetry of self-complementary surfaces. R. S. Morgan, et al. J. Mol. Biol. 127:31–39, 1979.

Enzymes with asymmetrically arranged subunits. Y. Degani and C. Degani. Trends Biochem. Sci. 5:337–341, 1980.

Flexibility Detected Crystallographically

The protein models resulting from a crystallographic structure determination are invariably static, but this should not be taken to imply that the molecule is equally as static, even in the crystal lattice. The electron density map provides direct evidence for flexibility in protein molecules, in the extent to which the electron density is smeared out, usually expressed as the temperature factor, B (Eq. 6–5). Proteins have considerably larger B values than those determined for crystals of small molecules, where the value of B is seldom greater than 10 $Å^2$. Atoms in proteins seldom have lower values; for interior atoms of well-ordered crystals, they are in the range of 10 to 20 $Å^2$. Such B values imply rms amplitudes of vibrations of 0.3 to 0.5 Å; instantaneous deviations of up to 1.0 to 1.5 Å, therefore, would be expected frequently.

The value of B varies throughout the structure, with the largest values for those atoms on the surface of the molecule. In many instances, surface side chains or small portions of the polypeptide chain may be invisible, presumably owing to a large possible number of different conformations. For example, no significant electron density was observed in the electron density map of BPTI for most of the side chains of Glu 7, Lys 15, Lys 26, and Lys 41. In other instances, two alternative conformations may be apparent, as was the case for $C^γ$ of Glu 7; the terminal carboxyl group of this side chain was invisible. In the case of trypsinogen and of IgG(Kol), entire sections of the proteins were not visible.

Unfortunately, the value of the temperature factor is affected not only by flexibility but also by any disorder within the crystal lattice, so interpretation of B values in terms of flexibility within a protein molecule is not straightforward. It is necessary to distinguish between lattice disorder and internal flexibility—for example, by comparing different parts of the electron density map to determine which variation can be explained by slightly different packing of the rigid molecule within the unit cell; by comparing maps of the same molecule in different crystal lattices or in different environments within the same lattice (when there is non-crystallographic sym-

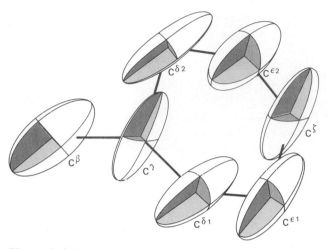

Figure 6–26
Thermal ellipsoids for the Phe 30 side chain of rubredoxin. (From K. D. Watenpaugh, et al., J. Mol. Biol. 138:615–638, © 1980, London, Academic Press Inc.)

metry); or by varying the temperature, to see if the flexible portions of the molecule can be frozen in fixed conformations at low temperatures.

It seems clear that intramolecular flexibility of proteins accounts for much of the smearing of the electron density, but detailed interpretations are still being developed. Illustrations of the degree of flexibility and disorder indicated by the electron maps are often given by thermal ellipsoids, where the variation in position of each atom in three dimensions is determined. The resulting ellipsoid represents that volume of space in which the atom was present with 50 per cent probability. An example for one relatively rigid, internal, side chain of the extensively refined protein rubredoxin is shown in Figure 6–26; that the major axes of the ellipsoids are aligned indicates that the variation is due to movement of the rigid aromatic ring, not to simple experimental errors, although the movement could be of the entire molecule or of just this particular side chain.

Temperature-dependent X-ray diffraction as a probe of protein structural dynamics. H. Frauenfelder, G. A. Petsko, and D. Tsernoglou. Nature 280:558–563, 1979.

Crystallographic studies of the dynamic properties of lysozyme. P. J. Artymiuk, et al. Nature 280:563–568, 1979.

Dynamic information from protein crystallography. An analysis of temperature factors from refinement of the hen egg-white lysozyme structure. M. J. E. Sternberg, et al. J. Mol. Biol. 130:231–253, 1979.

Conformational flexibility and its functional significance in some protein molecules. R. Huber. Trends Biochem. Sci. 4:271–276, 1979.

Low temperature protein crystallography. Effect on flexibility, temperature factor, mosaic spread, extinction and diffuse scattering in two examples: bovine trypsinogen and Fc fragment. T. P. Singh, et al. Acta Cryst. B36:621–627, 1980.

Conformational substates in a protein: structure and dynamics of metmyoglobin at 80° K. H. Hartmann, et al. Proc. Natl. Acad. Sci. U.S.A. 79:4967–4971, 1982.

The Solvent

The structure of the aqueous solvent surrounding protein molecules is one of the most important aspects of protein structure, but it is also one of the most difficult to determine. The structure determined crystallographically is averaged over both the extensive time required to collect the crystallographic data and over all the many molecules of the crystal lattice. The molecules of the liquid solvent are particularly mobile and by themselves would give only an extremely uniform, average electron density throughout the nonprotein areas of the crystal.

Nonetheless, distinct electron density peaks corresponding to relatively fixed solvent molecules are observed on the surfaces of proteins. For example, 43 solvent molecules could be localized around each molecule of BPTI, which represents about one third of the total crystal solvent content. In well-refined and well-ordered crystal structures, the number of fixed water molecules is about double the number of amino acid residues. Of course, the electron density map does not identify the molecules, and crystal solvents usually contain high concentrations of salts or other agents to induce crystallization, but most such fixed molecules appear to be water molecules; this is generally assumed in the absence of indications to the contrary.

The degree of order of the solvent is generally proportional to its proximity to the protein surface and to its participation in hydrogen bonding. Water molecules are most highly ordered, with B values as low as 13 Å^2, when extensively hydrogen-bonded in crevices on the protein surface and when bridging between different molecules in the crystal lattice. Almost invariably, they are anchored by hydrogen bonding to fixed polar groups of the protein surface, and one or more firmly anchored molecules can apparently serve to fix adjacent water molecules in a hydrogen-bonded lattice. Beyond this, there is a continuum of degree of order of solvent molecules, with B values increasing to the point where it is no longer deemed worthwhile to include them in the structure.

Ordered water structures are generally not observed around nonpolar surfaces, as might be expected from the hydrophobic interaction (Figure 4–4). However, the ordered networks might be only transient, or they might readily "slip" along the nonpolar surface and thus be smeared out in the electron density map.

The solvent within protein crystals has also been examined with other physical methods, with results broadly in line with the crystallographic observations. Three types of water molecules have been distinguished: The first is like bulk water, with a rotational relaxation time of only 10^{-11} sec, and constitutes the largest fraction. The second type is bound water, with a longer relaxation time of about 10^{-9} sec and an altered freezing point, presumably produced by interactions with the protein surface. The amount of this water type is approximately 0.3 g per gram of protein, or roughly 2 water molecules per average amino acid residue—about the same number as observed crystallographically. A monolayer of water molecules around an average protein molecule would have about the same mass as the protein, so only one third of the solvent surrounding the protein is relatively fixed. The third type of water molecule comprises those bound very tightly, not rotating significantly with respect to the protein. This type represents only a small fraction of the total solvent and presumably corresponds to those molecules observed crystallographically to be most firmly fixed.

Very similar observations are made of the solvent surrounding proteins in solution, so the environment of a protein is not expected to be very different when in solution, except for the absence of the ordered networks produced by adjacent protein molecules in the crystal lattice.

The properties of water in biological systems. R. Cooke and I. D. Kuntz. Ann. Rev. Biophys. Bioeng. 3:95–126, 1974.

Hydration of proteins and polypeptides. I. D. Kuntz and W. Kauzmann. Adv. Protein Chem. 28:239–345, 1974.

Water structure in a protein crystal: rubredoxin at 1.2 Å resolution. K. D. Watenpaugh, et al. J. Mol. Biol. 122:175–190, 1978.

Computer simulation of the solvent structure around biological macromolecules. A. T. Hagler and J. Moult. Nature 272:222–226, 1978.

Protein–water interactions. Heat capacity of the lysozyme–water system. P. H. Yang and J. A Rupley. Biochemistry 18:2654–2661, 1979.

Correlation of IR spectroscopic, heat capacity, diamagnetic susceptibility and enzymatic measurements on lysozyme powder. G. Careri, et al. Nature 284:572–573, 1980.

X-ray studies of water in crystals of lysozyme. C. C. F. Blake, et al. J. Mol. Biol. 167:693–723, 1983.

PROTEINS WITH SIMILAR FOLDED CONFORMATIONS

With the very wide diversity in amino acid sequences observed in proteins, described in Chapters 1 and 3, it might be expected that there would be a corresponding diversity in folded conformations. In contrast, many similarities between protein structures have been observed, and the tertiary structures of proteins appear to have been much more conserved during evolution than have their primary structures. The degree of this evolutionary conservation is a matter of intense debate and speculation, as similarities in tertiary structures are being found in cases where there was no evolutionary or functional reason to expect them. These findings raise fundamental questions about the evolution and the physical basis of protein structure that are still far from being answered.

Evolutionarily Related Proteins

Proteins with substantial similarity in their primary structures are almost certain to have arisen from a common ancestor during evolution (Chapter 3), and they have invariably been found to have very similar folded conformations. For example, the three-dimensional structures of horse and human hemoglobin are virtually identical, even though they differ in 43 of the 287 residues of the α and β chains. Similarly, the cytochromes c of horse and tuna have very similar structures, even though they differ at 17 of the 104 amino acid residues. Such conservation of conformation is understandable, in that the related proteins probably serve the same function in the different species, with very similar conformational requirements.

That more distantly related proteins also have very similar conformations was shown by the very first two proteins whose structures were determined, sperm whale myoglobin and horse hemoglobin. The single polypeptide chain of myoglobin and the α, β, and γ chains of hemoglobin are remarkably similar in the general topology of the polypeptide backbone (Figure 6–27). These proteins have amino acid sequences that are sufficiently similar to indicate that they are related evolutionarily (Figure 3–11). They also have similar functions in reversibly binding oxygen at a bound heme group. Other oxygen-binding, heme-containing proteins from a wide variety of vertebrates, a marine annelid worm, a larval insect, and lupine root nodules have very similar conformations (Figure 6–27). These proteins are usually grouped together as the globin family (Table 6–1).

In some of these instances, the primary structures are not detectably homologous. In all, the globins have only two amino acid residues in common. Nevertheless, their closely similar conformations support the current assumption that they diverged from a common evolutionary ancestor.

Another family of proteins with similar conformations is that of the cytochrome c–type protein. Conformation of the closely similar cytochromes c found in the mitochondria of vertebrates (Figure 3–5) resembles that in a variety of functionally related cytochromes from bacteria (Figure 6–28). The bacterial cytochromes had been recognized from their properties to be similar to cytochrome c but were sufficiently different in primary structure and physical and functional properties to be given names such as c_2, c_{550}, c_{551}, and c_{555}. The polypeptide chains of these proteins range in length from 82 to 135 residues and have only 5 residues in common.

The other well-characterized family of homologous proteins is that of the trypsin-like serine proteases, so named because they all have an important serine residue at the active site and similar catalytic mechanisms (Chapter 9). This family contains the related proteins of higher organisms—trypsin, chymotrypsin, elastase, and thrombin—as well as similar enzymes from microbial sources (Table 6–1). In the case of bovine α-chymotrypsin and elastase, 39 per cent of their residues are identical, and their conformations can be superimposed so that on average their polypeptide backbone atoms differ in relative position by only 1.80 Å. This similarity in topology is immediately obvious in their distance plots (Figure 6–29). Trypsin and α-chymotrypsin may be similarly juxtaposed to within 0.75 Å, even though only 44 per cent of their residues are identical. The bacterial serine proteases have diverged more in both sequence and overall conformation; they have fewer than 20 per cent of their residues identical with those of any of the mammalian proteins, with numerous insertions

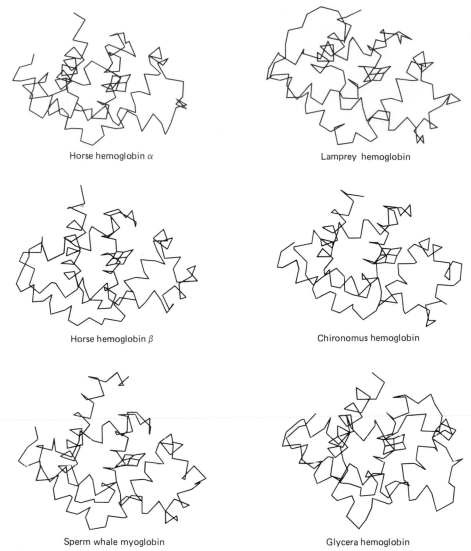

Horse hemoglobin α

Lamprey hemoglobin

Horse hemoglobin β

Chironomus hemoglobin

Sperm whale myoglobin

Glycera hemoglobin

Figure 6–27
Comparison of the similar polypeptide backbone foldings in the globins. (Adapted from
W. Love, Cold Spring Harbor Symp. Quant. Biol. 36:349, 1971.)

and deletions, not much more similar than random sequences would be
expected to be. Yet their three-dimensional conformations are more obviously similar than are the primary structures; 55 to 64 per cent of the
residues of the mammalian and bacterial proteins have been determined
to be topologically equivalent, in that their C^α atoms occupy the same
relative overall position. With the greater divergence of amino acid sequence, there is a somewhat greater change in structure, although the
structure is the more conserved.

The first question to be asked about these observations is how such wide variation in amino acid sequences is compatible with such similar folded conformations. The greatest variation in amino acids occurs at positions in the polypeptide chain where the side chain is on the surface. Here the chemical nature of the side chain has often been changed; however, there is a tendency for the residues at reverse turns to have one of the short, polar side chains, or to be Gly or Pro—those residues found most frequently at these positions. Interior residues are changed least frequently, and the nonpolar nature of the side chain is most conserved. Changes in the volume of the side chains often tend to compensate, so that the total volume of the interior tends to be conserved.

The most highly conserved residues are those involved directly in the functional properties of the protein. Examples are the invariant cytochrome c Cys residues 14 and 17, to which the heme group is attached, and the ligands to the heme iron atoms, His 18 and Met 80 (Figures 3–6 and 6–28). Likewise, the His side chains of hemoglobin that bind the heme iron (Figure 8–23) are conserved, as are the catalytically important residues in the active sites of the serine proteases (see Figure 9–8).

Insertions and deletions within the polypeptide chain occur at reverse turns on the surface, with little perturbation of the interior. Significant changes in this respect are apparent in the different c-type cytochromes (Figure. 6–26); some external loops are lengthened, whereas others are shortened.

Disulfide bridges are often added or deleted. For example, in the serine protease family, trypsin has 6, chymotrypsin 5, elastase 4, α-lytic protease 3, and proteases A and B of *Streptomyces griseus* 2 each. However, the Cys residues have never been observed to change their pairings. If a disulfide is deleted, both Cys residues are generally changed to other amino acid residues, but their relative three-dimensional positions in the folded structure remain the same. Therefore, it may be concluded that disulfide bonds do not determine the conformation, nor are they absolutely necessary.

Elements of secondary structure can change in length or even disappear altogether, but a helix is not replaced by a β-sheet, nor vice versa. In particular, neither the orientation of individual strands in a β-sheet—i.e., parallel or antiparallel—nor their order has ever been observed to differ in proteins with homologous primary structures. The relative positions of α-helices are retained, but often within rather wide limits. For example, in some of the widely differing globins, corresponding pairs of α-helices may differ in proximity by 7 Å and in relative orientation by 30°.

In summary, the three-dimensional structures of evolutionarily related proteins appear to have been remarkably conserved during evolution, suggesting that the three-dimensional structure was well-defined at the early stage of evolution in the common ancestor and that it is crucial for the function of the protein. Changes in the primary structure due to genetic mutations have accumulated during evolutionary divergence, but

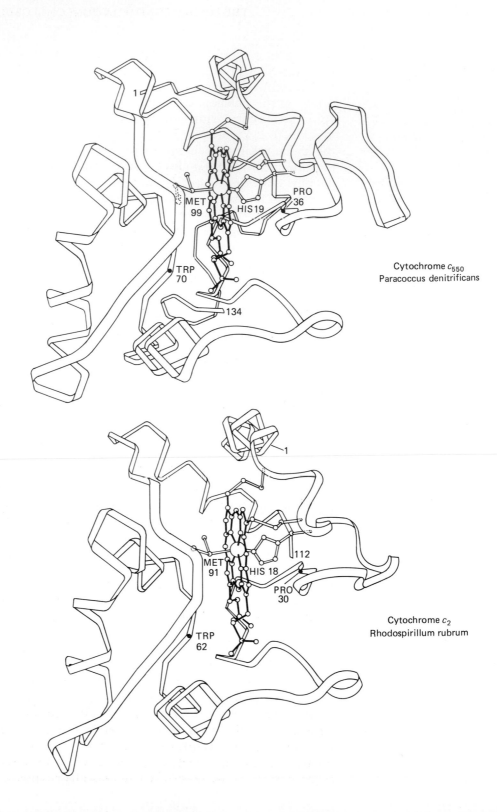

Cytochrome c_{550}
Paracoccus denitrificans

Cytochrome c_2
Rhodospirillum rubrum

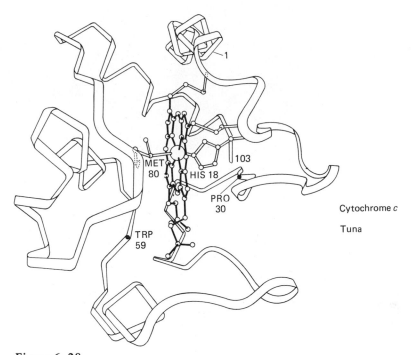

Figure 6–28
The similar folded conformations of distantly related cytochromes. (Adapted from F. R. Sal-emme, Ann. Rev. Biochem. 46:299–329, 1977.)

only those that were compatible with the folded conformation and with its function have been retained. These are the factors apparently responsible for the constraints on evolutionary change of the primary structure noted in Chapter 3. The structural reasons for the conservation of some of the residues in cytochrome *c* noted in Figure 3–6 are given in Table 6–7. The degree to which such accumulated mutations have been independent, or whether change at one position favored a compensating change at a second position, is not yet certain. Also not clear is whether there have been further restraints on the change permitted during evolution, for example, owing to importance in the process of folding to the final conformation. Nevertheless, the great variety of amino acid sequences that produce the same folded conformation indicates considerable flexibility in the rules of protein folding.

Knowledge of the three-dimensional structures of evolutionarily related proteins is also very useful for inferring the genetic basis of the evolutionary divergence, in that it identifies which residues of the primary structure should be aligned. It is generally assumed that structurally equivalent residues are also those evolutionarily related. Alignment of the sequences of distantly related proteins is often uncertain, especially when there have been numerous deletions and insertions. The usual experi-

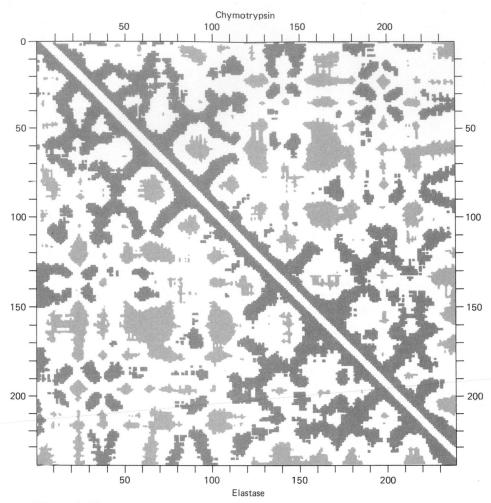

Figure 6–29
Similar contact maps for the homologous structures of α-chymotrypsin (top right) *and elastase* (lower left). *Distances between residues greater than 30 Å are indicated by light gray; distances less than 15 Å are indicated by dark gray areas. (From L. Sawyer, et al., J. Mol. Biol. 118:137–208, © 1978, London, Academic Press Inc.)*

mental approach involving minimization of the number of genetic mutations has often given alignments inconsistent with the structural homology.

Structure and function of haemoglobin. II. Some relations between polypeptide chain configuration and amino acid sequence. M. F. Perutz, et al. J. Mol. Biol. 13:669–678, 1965.

Sequence and structure homologies in bacterial and mammalian-type cytochromes. R. E. Dickerson. J. Mol. Biol. 57:1–15, 1971.

Invariant features of globin primary structure and coding of their secondary structure. O. B. Ptitsyn. J. Mol. Biol. 88:287–300, 1974.

Table 6–7   Roles of Evolutionarily Invariant Residues of Cytochrome c

RESIDUE	APPARENT REASON FOR INVARIANCE
Gly 1	Stabilizes C helix
Gly 6	No room for side chain
Phe 10	Good shape for nonpolar close-packing
Cys 17	Covalently attached to heme
His 18	Fifth ligand to heme iron
Pro 30	Maintain His 18 orientation by hydrogen bond
Leu 32	Provide hydrophobic pocket for heme
Gly 34	No room for side chain
Arg 38	Hydrogen bonds to heme propionyl and carbonyl of residue 33
Gly 41	No room for side chain
Tyr 48	Hydrogen bonds to heme propionic group
Trp 59	Hydrogen bonds to heme propionic group
Leu 68	Provide hydrophobic pocket for heme
Asn 70	Free in solvent, may interact with electron donor or acceptor proteins
Pro 71	Participate in reverse turn of polypeptide chain
Lys 73	Free in solvent, may interact with electron donor or acceptor proteins
Pro 76	Participate in sharp bend of polypeptide chain
Gly 77	No room for side chain
Thr 78	Hydrogen bonds to heme propionic group and H_2O molecule, may affect redox potential
Lys 79	Stabilizes both sides of heme crevice
Met 80	Sixth ligand to heme iron
Phe 82	Nonpolar environment for heme
Gly 84	No room for side chain
Arg 91	Stabilizes conformation of residues 80 to 90 by hydrogen-bonding to carbonyl of residue 85

Adapted from T. Takano and R. E. Dickerson, J. Mol. Biol. 153:79–94, 1981.

Tertiary structural differences between microbial serine proteases and pancreatic serine enzymes. L. T. J. Delbaere, et al. Nature 257:758–763, 1975.

The cytochrome fold and the evolution of bacterial energy metabolism. R. E. Dickerson, et al. J. Mol. Biol. 100:473–491, 1976.

A comparison of the structures of electron transfer proteins. E. T. Adnam. Biochim. Biophys. Acta 549:107–144, 1979.

How different amino acid sequences determine similar protein structures: the structure and evolutionary dynamics of the globins. A. M. Lesk and C. Chothia. J. Mol. Biol. 136:227–270, 1980.

Structural Homology without Apparent Sequence Homology

Has the folded conformation of a protein been so conserved during evolution that a similar structure, even in the absence of detectable homology in primary structure, is indicative of a common evolutionary ancestor? In this case, what degree of structural homology is significant? At what point do similarities in structure reflect simply the general principles of protein structure? Does a common backbone folding and a similar function require a common ancestor? Alternatively, can a similar conformation arise as a result of evolutionary convergence to a common function and the structural requirements for such a function? These are just a few of the questions raised by observations of similar folded conformations where none was expected on the basis of the primary structures.

Examples of the conformationally related proteins with similar functions in extremely different species occur in the thioredoxins of *E. coli* and of bacteriophage T4 and the lysozymes of chicken and of bacteriophage T4. The thioredoxins are quite similar in conformation, except that the bacterial protein has an additional helix and a β-strand at one end of the polypeptide chain. The chicken and bacteriophage lysozymes appeared initially to be unrelated, but subsequent closer comparison revealed apparently significant similarities. Likewise, cytochrome b_5 and the globins have some similarities in conformation; they have similar functions only in that both use heme groups: the globins, to reversibly bind oxygen; the cytochrome, to bind electrons.

It is tempting to conclude that these similarities reflect very ancient common origins, so distant that the primary structures are no longer detectably similar and that varying degrees of change of the conformations have also occurred. However, it does seem clear that similar functions need not imply a common ancestry, even when the common function reflects similar structures at the active sites. The bacterial protease subtilisin and the trypsin proteases have very similar geometries of residues at their active sites, but no other structural similarities. Likewise, the bacterial protease thermolysin and bovine carboxypeptidase have another catalytic mechanism with similar active sites, but no other detectable similarities. It could be argued that these pairs had common evolutionary origins and that both the sequences and the conformations have diverged drastically, except for that around the active sites. However, the equivalent residues within the active sites do not occur in comparable positions in the primary structure. Changes in the overall topology by divergence would require either drastic changes in the folding process or genetic rearrangements within the structural gene. Although these processes cannot be ruled out, it seems most likely that these are instances of evolutionary convergence to similar active sites and catalytic mechanisms, but not to similar folded conformations.

A more complex and intriguing case of structural homology involves the dehydrogenases and the binding of nucleotides. These enzymes carry out similar oxidation–reduction of different small molecules using the coenzyme nicotinamide adenine dinucleotide (NAD). The lactate and malate dehydrogenases have quite similar tertiary structures, even though they normally exist as tetramer and dimer, respectively. Each polypeptide chain is folded into two structural domains; one binds the coenzyme NAD, whereas the other binds the substrate and provides many of the residues involved in catalysis. Both domains are similar in conformation in the two dehydrogenases. Two other enzymes, alcohol dehydrogenase and glyceraldehyde phosphate dehydrogenase, also have two domains, and the NAD-binding domains of both are closely similar in conformation to those of lactate and malate dehydrogenases. However, the second domains are very different in conformation and also pack differently against the NAD-binding domains. The amino acid sequences of the NAD-binding domains are detectably homologous, suggesting a common ancestor, but the other domains either must have diverged considerably in both primary and tertiary

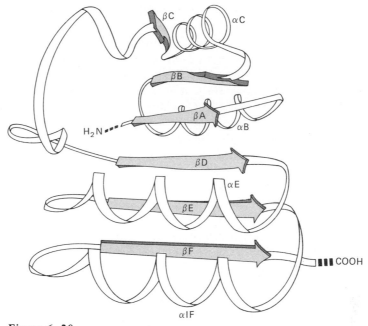

Figure 6–30
The NAD-binding domain of the dehydrogenases, which is composed of two nucleotide-binding units (β–α–β–α–β). Nucleotides are generally bound on the right, near the carboxyl ends of the β-strands (see Figure 6–16). (From I. Ohlsson, et al., J. Mol. Biol. 89:339–354, © 1974, London, Academic Press Inc.)

structures or must have arisen independently and been joined to the NAD domains by a process of gene fusion. (See Figure 8–13 and pp. 368–369.)

The structure of the NAD-binding domains has also been observed in many proteins that bind other nucleotides such as flavin adenine dinucleotide or AMP, ADP, or ATP—moieties that represent half the structure of NAD (which is AMP and nicotinamide nucleotide joined in a phosphodiester bond). The dehydrogenase NAD-binding domain consists of a six-stranded parallel pleated sheet and four helices, with a unique topology (Figure 6–30). The structural half comprising the first three β-strands (β–A, β–B, and β–C) plus two helices (α–B and α–C) binds the AMP portion of NAD. The other half, also consisting of three β-strands and two α-helices, similar to the first half, binds the nicotinamide portion of NAD. These two similar halves are known as **mononucleotide-binding domains,** as they are found individually in other proteins that bind nucleotides. Most retain the same topology and polypeptide connections between the β-strands and helices, except for dihydrofolate reductase, where substantial differences are reported. As there is no significant amino acid sequence homology in most of these cases, it is an open question at the present time whether these segments arose from a common ancestor or whether this is an example of evolutionary convergence for functional or structural reasons.

Similar uncertainty exists with the very similar β-sheet structures found in proteins as diverse as the domains of immunoglobulins and the enzyme superoxide dismutase. Both have very similar β-barrels of eight strands in which the orders of strands within the barrel are similar. Individual domains of seemingly unrelated proteins are also found to be very similar. For example, the triose phosphate isomerase structure (Figure 6–16) is also found in KDPG aldolase and in one domain of pyruvate kinase (A in Figure 6–24), although in the latter a second domain (B) is inserted into the polypeptide chain at one of the reverse turns. A third domain of pyruvate kinase (C) has the mononucleotide-binding fold.

Interpretation of these observations would be most straightforward if further evolutionary data were available, such as documentation of other protein structures with which to construct, or to rule out, an evolutionary phylogeny. In the absence of such information, the evolutionary origin of these contemporary proteins can only be guessed. Whether evolutionary divergence from a common ancestor or convergence to similar structures is considered the more likely depends upon how unique a particular protein folded conformation is considered to be; i.e., how much it may be altered gradually by divergence and how likely it is to occur by independent events during convergence. At the moment it seems impossible to decide, but since there is no documented example of convergence to similar folded conformations, the well-established evolutionary divergence is considered the more probable.

Similarity of three-dimensional structure between the immunoglobulin domain and the copper, zinc superoxide dismutase subunit. J. S. Richardson, et al. J. Mol. Biol. 102:221–235, 1976.

Exploring structural homology of proteins. M. G. Rossmann and P. Argos. J. Mol. Biol. 105:75–96, 1976.

The taxonomy of protein structure. M. G. Rossmann and P. Argos. J. Mol. Biol. 109:99–129, 1977.

Convergence of active site geometries. R. M. Garavito, et al. Biochemistry 16:5065–5071, 1977.

Comparison of the structures of carboxypeptidase A and thermolysin. W. R. Kester and B. W. Matthews. J. Biol. Chem. 252:7704–7710, 1977.

Recognition of phylogenetic relationships from polypeptide chain fold similarities. G. E. Schulz. J. Mol. Evol. 9:339–342, 1977.

A general method to assess similarity of protein structures, with applications to T4 bacteriophage lysozyme. S. J. Remington and B. W. Matthews. Proc. Natl. Acad. Sci. U.S.A. 75:2180–2184, 1978.

Similarities of protein topologies: evolutionary divergence, functional convergence or principles of folding? O. B. Ptitsyn and A. V. Finkelstein. Quart. Rev. Biophys. 13:339–386, 1980.

Relation between structure and function of α/β proteins. C.-I. Brändén. Quart. Rev. Biophys. 13:317–338, 1980.

Common precursor of lysozymes of hen egg-white and bacteriophage T4. B. W. Matthews, et al. Nature 290:334–335, 1981.

Comparison of the folding of KDPG aldolase, triose phosphate isomerase and pyruvate kinase. Implications in molecular evolution. L. Lebioda, et al. J. Mol. Biol. 162:445–458, 1982.

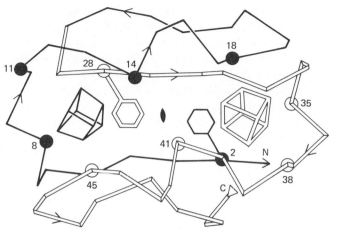

Figure 6–31
Internal symmetry within ferredoxin, resulting from internal homology within its primary structure. The sequence homology between the two halves of the polypeptide chain is shown in Figure 3–15; the amino-terminal half of the polypeptide backbone is dark, the other half light. The distorted cubes are the iron–sulfur clusters used to bind electrons. Circles indicate Cys residues, which bind the iron–sulfur clusters. (From A. D. McLachlan, J. Mol. Biol. 128:49–79, © 1979, London, Academic Press Inc.)

Structural Homology within a Polypeptide Chain

Amino acid sequence homology within a polypeptide chain indicates elongation of the gene (and the protein) by gene duplication (Chapter 3). All the structures of such duplicated proteins, such as ferredoxin (Figure 3–15), parvalbumin, and the immunoglobulins, have also indicated internal homology of the protein at the level of tertiary structure (Figure 6–31).

Other instances of structural homology within a protein, with a corresponding symmetry of the total structure, have been observed in cases where there is no apparent homology within the primary structure. For example, within the two domains of rhodanese, each of about 140 residues, 117 C^α atoms of each may be superimposed with an average deviation of only 1.95 Å. Other proteins with two similar halves, but no sequence homology, are tobacco mosiac virus coat protein, cytochrome b_5, arabinose-binding protein, the NAD-binding domains (Figure 6–30), trypsin proteases, acid proteases, glutathione reductase, hexokinase, γ-crystallin, and rubredoxin. A threefold repeat is observed in soybean trypsin inhibitor.

As with homology between different proteins in the absence of sequence homology, it is difficult to determine whether these instances arose by gene duplication. However, in all these intramolecular cases, the similar conformations are related by symmetry axes, very close to those expected for oligomeric proteins (Figure 6–23). With twofold repeats, the two parts are related by rotations extremely close to 180°; with the threefold repeat it is very close to 120°. These findings suggest that each of these proteins were originally oligomers and that the gene was duplicated to give a single fused polypeptide chain; otherwise there is no apparent reason for having such symmetry axes. The original amino acid homology would have disappeared by divergence; yet both the structures and the packing of the original monomers have been retained.

The structure of ferredoxin demonstrates the twofold symmetry (Figure 6–31) and further strengthens the argument for a dimer precursor during evolution. The two halves of the primary structure do not form individual domains, but each half of the tertiary structure is composed of residues from both halves of the polypeptide. Therefore, the two original polypeptide chains probably were adapted to fit together before they were fused, which would imply a dimeric structure.

Internal symmetry within proteins thus provides further evidence for the role of evolutionary divergence and the rather strict maintenance of the folded conformation.

Structural evidence for gene duplication in the evolution of the acid proteases. J. Tang, et al.: Nature 271:618–621, 1978.

Gene duplications in the structural evolution of chymotrypsin. A. D. McLachlan. J. Mol. Biol. 128:49–79, 1979.

Gene duplication in glutathione reductase. G. E. Schulz. J. Mol. Biol. 138:335–347, 1980.

An examination of the expected degree of sequence similarity that might arise in proteins that have converged to similar conformational states. The impact of such expectations on the search for homology between the structurally similar domains of rhodanese. P. Keim, et al. J. Mol. Biol. 151:179–197, 1981.

7

PROTEINS IN SOLUTION

The folded conformations of proteins described in the previous chapter give them physical, chemical, and biological properties quite unlike those of unfolded polypeptides or of the sum of their constituent amino acids. Owing to their compactness, they diffuse and rotate relatively rapidly, and they are relatively resistant to proteases. When peptide bonds in mobile surface loops are cleaved by proteases, the folded structure often remains intact. Groups are placed in unique environments, with often marked effects on their physical and chemical properties. Functional groups are often held in proximity, making their effective concentrations relative to each other extremely high, so that reactions may occur between them that would be negligible between the same two groups on individual molecules.

Because many of these properties are evident only with proteins in solution, outside the crystal lattice, when they also have enhanced degrees of flexibility not apparent in the crystal structure, it is necessary to describe their properties under these conditions. The following discussion emphasizes globular proteins in dilute aqueous solution, even though many proteins normally exist in supermolecular assemblies, such as filaments, tubules, ribosomes, or viruses. Proteins embedded in membranes are not considered here simply because so little information is available about them. Proteins in large aggregates can probably be described by extrapolation from small soluble proteins, but those in membranes are thought to have unique structures and properties, which are only now being studied. Future books on proteins will probably devote a large proportion of space to membrane proteins, but extensive discussion here would be largely speculative in nature.

Much to the collective relief of protein crystallographers, it is now clear from different types of evidence that a protein in solution generally retains the conformation determined within a crystal lattice. The many properties of a protein in solution to be described here are usually explicable in terms of its crystal structure; our understanding of these properties, however, is far from complete, so this is not concrete evidence. It is more convincing that a protein crystallized in a number of different crystal lattices, often produced by very different crystallization procedures, invariably has been found to have essentially the same structure in each. The same is true of related proteins; indeed, there have been many instances of surprising similarities in protein crystal structures, whereas there have been no instances of finding different structures when they were expected to be similar. The specific interactions between adjacent molecules that determine the crystal lattice appear to be relatively weak and unable to perturb the general structure of a stable folded conformation. Fortunately, these intermolecular interactions are usually similar to the intramolecular ones that specify the folded conformation; crystallization conditions that favor the crystal lattice, therefore, also favor the folded structure.

This assertion of the identity of crystal and solution conformations must be qualified somewhat: The crystal lattice interactions undoubtedly do affect local aspects of the conformation near the surface of the protein. Most often, side chains may be held in specific orientations owing to interactions with neighboring molecules, whereas they would be mobile in solution or perhaps engaged in other, intramolecular interactions. For example, the Lys 41 and Tyr 10 side chains of BPTI are involved in intermolecular interactions with other groups in the crystal, whereas NMR data indicate they interact with each other intramolecularly in solution. Specific aspects of the crystallization conditions may also affect local aspects of the conformation that do not apply under different conditions in solution. For example, the terminal amino and carboxyl groups of BPTI appear to interact in a salt bridge in solution at neutral pH, whereas they are 8 Å apart in the crystal structure. Disruption of this salt bridge in the crystal is probably due to the high pH (10.5) used for crystallization, which deprotonates the α-amino group, even in solution. Changes within protein interiors are much less extensive; interactions with neighboring atoms, especially hydrogen bonds, are maintained to a very high degree.

Very small proteins are the most likely exceptions to the generalization of retention of the crystal structure in solution, as they often are unable to maintain a single conformation in solution; these are probably designated more appropriately as peptides. The 29-residue hormone glucagon is the most notable example, as it approximates a random coil in dilute solution. Only in concentrated solution does it tend to acquire a trimeric helical structure, which is probably like that observed in the crystal lattice. Because there appears to be a lower limit of roughly 50 amino acid residues required for maintenance of a single conformation in solution, it is necessary to study the solution properties of all small proteins (which may turn out to be simply large peptides) to determine the relevance of their crystal structures.

Aside from the exceptions of very small proteins and of local aspects of the protein surface, the available evidence indicates that a folded protein structure is unique. No protein has been observed to adopt two substantially different conformations. Nevertheless, the literature is full of references to conformational changes in proteins; in fact, protein chemists almost invariably suggest such changes to explain any sudden change in properties of a protein. Most of these so-called conformational changes probably involve rather localized alterations in conformation or changes in the degree of flexibility. In all cases where they have been determined crystallographically, substantial changes in conformation involve primarily rearrangement of subunits or structural domains relative to each other. The individual subunits or domains maintain their overall conformation, so it is these individual structural units for which the architecture is apparently unique.

Conformational nature of monomeric glucagon. B. Panijpan and W. B. Gratzer. Eur. J. Biochem. 45:547–554, 1974.
Two conformations of crystalline adenylate kinase. W. Sachsenheimer and G. E. Schulz. J. Mol. Biol. 114:23–36, 1977.
[1]H-nuclear magnetic resonance studies of the molecular conformation of monomeric glucagon in aqueous solution. C. Boesch, et al. Eur. J. Biochem. 91:209–214, 1978.
The influence of a single salt bridge on static and dynamic features of the globular solution conformation of the basic pancreatic trypsin inhibitor. L. R. Brown, et al. Eur. J. Biochem. 88:87–95, 1978.
Spurious conformational transitions in proteins? A. Cooper. Proc. Natl. Acad. Sci. U.S.A. 78:3551–3553, 1981.
Transmission of conformational change in insulin. C. Chothia, et al. Nature 302:500–505, 1983.

PHYSICAL AND CHEMICAL PROPERTIES

Proteins have been studied extensively by virtually every physical and chemical technique available. The most useful methods were described in Chapter 5, where they were applied to unfolded polypeptides. The same techniques give very different results with folded proteins.

Some physical probes of enzyme structure in solution. S. N. Timasheff. In The Enzymes, 3rd ed. P. D. Boyer (ed.). Vol. 2, pp. 371–443. New York, Academic Press, 1970.

Hydrodynamic Properties

The most obvious difference between a folded protein and an unfolded polypeptide chain is the compactness of the former. The consequent lower resistance to movement through the solvent results in a much lower viscosity and in greater rates of both translational and rotational diffusion (Table 5–3), and of sedimentation. Hydrodynamic parameters for a number of folded proteins of known structure are given in Table 7–1.

Table 7–1 Hydrodynamic Properties of Proteins of Known Structure

PROTEIN (SOURCE)	HYDRODYNAMIC DATA			MOLECULAR WEIGHT		f/f_0 [f]	DIMENSIONS[g] (Å)
	$s^0_{20,w}$ [a] (S)	$D^0_{20,w}$ [b] (10^{-7} cm²/sec)	$\bar{v}$ [c] (ml/g)	Structure[d]	Measured[e]		
Pancreatic trypsin inhibitor (bovine)	1.0	12.9	0.718	6,520	6,670	1.321	29 × 19 × 19
Cytochrome c (equine)	1.83	13.0	0.715	12,310	11,990	1.116	25 × 25 × 37
Ribonuclease A (bovine)	1.78	10.7	0.703	13,690	13,600	1.290	38 × 28 × 22
Lysozyme (hen)	1.91	11.3	0.703	14,320	13,800	1.240	45 × 30 × 30
Myoglobin (sperm whale)	1.97	11.3	0.745	17,800	16,600	1.170	44 × 44 × 25
Adenylate kinase (porcine)	2.30	10.2	0.74	21,640	21,030	1.167	40 × 40 × 30
Trypsin (bovine)	2.50	9.3	0.727	23,200	23,890	1.187	50 × 40 × 40
Bence Jones REI[h] (human)	2.6	10.0	0.726	23,500	23,020	1.156	40 × 43 × 28
Chymotrypsinogen (bovine)	2.58	9.48	0.721	25,670	23,660	1.262	50 × 40 × 40
Elastase (porcine)	2.6	9.5	0.73	25,900	24,600	1.214	55 × 40 × 38
Subtilisin novo (B. amyloliq.)	2.77	9.04	0.731	27,530	27,630	1.181	48 × 44 × 40
Carbonic anhydrase (human)	3.23	10.7	0.729	28,800	27,020	1.053	47 × 41 × 41
Superoxide dismutase (bovine)	3.35	8.92	0.729	33,900	33,600	1.132	72 × 40 × 38
Carboxypeptidase A (bovine)	3.55	9.2	0.733	34,500	35,040	1.063	50 × 42 × 38
Phosphoglycerate kinase (yeast)	3.09	6.38	0.749	45,800	46,800	1.377	70 × 45 × 35
Concanavalin A	3.8	6.34	0.732	51,260	54,240	1.299	80 × 45 × 30
Hemoglobin, oxy[i] (equine)	4.22	6.02	0.750	64,610	67,980	1.263	70 × 55 × 55
Malate dehydrogenase[h] (porcine)	4.53	5.76	0.742	74,900	73,900	1.344	64 × 64 × 45
Alcohol dehydrogenase[h] (equine)	5.08	6.23	0.750	79,870	79,070	1.208	45 × 55 × 110
Lactate dehydrogenase[j] (dogfish)	7.54	4.99	0.74	146,200	141,000	1.273	74 × 74 × 84

[a]Sedimentation coefficient at 20° C in water, extrapolated to zero protein concentration, in Svedberg units.

[b]Translational diffusion coefficient at 20° C in water, extrapolated to zero protein concentration.

[c]Partial specific volume.

[d]Calculated from covalent structure.

[e]Calculated from hydrodynamic data (see Eq. 1–56).

[f]Frictional ratio from experimental hydrodynamic data.

[g]From crystal structure.

[h]Dimer of identical polypeptide chains.

[i]$\alpha_2\beta_2$ tetramer.

[j]Tetramer of identical polypeptide chains.

From P. G. Squire and M. E. Himmel, Arch. Biochem. Biophys. 196:165–177, 1979.

Although many proteins are nearly spherical, their rates of sedimentation and diffusion are not as great as those expected for a sphere of the same size (p. 178). In other words, the frictional ratio (f/f_0) is greater than unity; values of 1.05 to 1.38 are observed with a sample of proteins (Table 7–1). Two factors other than departure from spherical shape are responsible for the observed difference: the roughness of the surface and the bound solvent. The former increases the resistance to movement through the solvent, whereas the tightly bound solvent effectively increases the mass of the particle. Attempts to separate the contributions of shape and solvent, primarily to measure the latter, were inadequate in the past, because the shape could be represented only as a smooth ellipsoid, neglecting the roughness of the surface. Consequently, the contribution of hydration was overestimated. Only recently has it become possible to calculate the shape factor from the known crystal structures of proteins. The inclusion of water molecules bound tightly to polar groups on the surface, of the type seen in crystal structures, gives reasonable agreement between calculated and measured frictional coefficients.

The translational friction coefficient of proteins. D. C. Teller, et al. Methods Enzymol. 61:103–124, 1979.

Hydrodynamic properties of complex, rigid, biological macromolecules: theory and applications. J. Garcia de la Torre and V. A. Bloomfield. Quart. Rev. Biophys. 14:81–139, 1981.

Spectral Properties

The variety of environments of the chromophores of a folded protein and the unique stereochemistry of the polypeptide chain result in various spectral effects, which are readily used to characterize, as well as to follow changes in, the folded conformation in solution.

Absorbance of UV light is not very sensitive to conformation or environment, except for that by the aromatic rings of Phe, Tyr, and Trp residues (Figure 1–3). Their absorbance spectra are shifted somewhat to longer wavelengths (red-shifted) in a nonpolar environment, such as the interior of a protein. Also, when so buried, their absorbance is not sensitive to changes in the solvent, whereas that of aromatic groups on the surface may be perturbed significantly by addition of reagents such as glycerol, ethylene glycol, and sucrose. Such solvent perturbation may be used to measure the fraction of aromatic residues exposed to the solvent. However, proteins have a wide variety of environments of aromatic side chains that are not readily classified as buried or exposed; the value obtained reflects the average exposure of all the aromatic rings.

The degrees of solvent exposure measured from the absorbance spectrum of the protein and from its perturbation by solvent are complementary. They can give similar results, as indicated in Table 7–2 for a series of conformational states of bovine pancreatic trypsin inhibitor (BPTI), ranging from the unfolded reduced form, with at least 85 per cent exposure of the Tyr rings (relative to that in a small model peptide), to the

Table 7–2   Exposure of Tyrosine Residues of Various Conformational States of Bovine Pancreatic Trypsin Inhibitor (BPTI)

| FORM OF BPTI[a] | FRACTIONAL EXPOSURE OF TYROSINES | | | |
| | Compared to R | | Compared to Gly-Tyr-Gly | |
	Comparison Spectra[b]	Perturbation Spectra[c]	Comparison Spectra	Perturbation Spectra
R	100	100	84	86
(5-30)	73	80	59	69
(30–51)	64	67	51	57
(30-51, 5-14) + (30-51, 5-38)	60	63	47	53
(30-51, 14-38)	49	49	37	42
(30-51, 5-55)	27	41	16	35
Refolded	36	37	25	32
Native	36	35	25	30

[a]The various forms of BPTI were isolated as trapped intermediates in refolding of the reduced protein, R (see Figures 7–17 and 7–18). They are designated by the numbers of the Cys residues paired in disulfide bonds; native and refolded BPTI are (30-51, 5-55, 14-38). BPTI has four Phe and four Tyr residues, but the latter dominate the absorbance.

[b]The "red-shift" of the absorbance spectrum was used to quantitate the fraction of residues buried.

[c]The perturbant was 20 per cent ethylene glycol.

From P. Kosen, et al.: Biochemistry 19:4936–4944, 1980.

native form with 25 to 36 per cent exposure. The crystal structure indicates 31 per cent exposure, primarily of Tyr 10 and Tyr 21 (Figure 6–9).

Fluorescence by the aromatic groups (Table 1–3) is much more sensitive to their environment, but in a wide variety of ways. The quantum yield may be either increased or decreased; a folded protein, therefore, may have higher or lower fluorescence than that of the unfolded form. Thus, the magnitude of the fluorescence is not very informative in itself, but it can serve as a very sensitive probe of any perturbations of the folded state. The close proximity of the aromatic groups in a folded protein usually results in very efficient energy transfer between them (p. 179). Consequently, light absorbed by one chromophore may be transferred to another absorbing at a longer wavelength, which may then emit it as fluorescence. Since the absorption wavelengths of the aromatic amino acids are in the order Phe < Tyr < Trp (Figure 1–3), proteins containing all three residues generally emit fluorescent light typical of Trp; Tyr fluorescence is generally observed only in the absence of Trp; that of Phe, only in the absence of both Tyr and Trp.

The circular dichroism (CD) and optical rotatory dispersion (ORD) spectra of a protein are very sensitive to its conformation. In the far-UV region (below 250 nm), these spectral characteristics are primarily determined by the polypeptide backbone conformation, especially secondary structure. Such a spectrum of a protein of known structure is usually close to that expected from the average of the spectra of α-helices, β-sheet, and irregular conformations of model polypeptides (Figure 5–11), weighted according to the fraction of the polypeptide chain in each conformation. Consequently, CD spectra of proteins of unknown structure may be used to determine the relative proportions of the various types of secondary structure. However, the agreement is not always satisfactory, and it is clear that other chromophores can contribute significantly to the far-UV spectrum, so care must be used in interpreting CD and ORD spectra. For example, the CD spectrum of BPTI differs substantially from that expected on the basis of its structure.

Folded proteins generally also have significant optical activity in the near-UV region (250–300 nm), owing primarily to the presence of the aromatic side chains in asymmetric environments and to the chirality of any disulfide bonds. The theoretical basis of near-UV optical activity is much better understood than that at lower wavelengths, and the spectra of small proteins may be calculated from their known structures.

Nuclear magnetic resonance (NMR), using especially the ^1H atoms, but also ^{13}C (either that naturally present or the isotope introduced by covalent modification), is the other spectral technique that has proved to be of great use in studying protein structure in solution. The physical basis of this technique is too complex to be described here, but it is sensitive to very small differences in the environments of individual nuclei, giving highly resolved spectra. Individual atoms may give rise to distinct peaks in the spectrum, in contrast to the previously described spectral methods, which give poorly resolved spectra averaged over the entire molecule. Although NMR spectra are very detailed, they are not readily interpreted, because no single parameter determines the position of a resonance peak in the spectrum. Merely assigning the many peaks to the many individual hydrogen atoms of a protein has been difficult; in fact, mistakes were made in some early studies—most notably in the initial assignment of those of the four His residues of ribonuclease A. The most reliable assignments come from correlating spectral changes with specific covalent modifications of individual groups, but this is a daunting task even with a small protein. Recently developed techniques have used the effects of irradiating one hydrogen atom on the resonances of atoms nearby in the covalent structure; with the primary structure and one starting point, the resonances of most hydrogen atoms may be assigned sequentially along the polypeptide chain. BPTI has been studied most extensively in this way (Figure 7–4); results of these studies have been used to develop the NMR technology.

The magnitude of the spectral interaction between hydrogen atoms depends upon the distance between them; this information has been used to obtain semiquantitative estimates of distances within proteins. Between hydrogen atoms along the polypeptide backbone, the different types of secondary structure can be detected. Because spectral interactions also occur between atoms close in space, but not covalently bonded, this provides semiquantitative information about the folded conformation. These NMR techniques hold great promise for the determination, in some detail, of the folded conformations of proteins in solution.

Applications of nuclear magnetic resonance spectroscopy to the study of macromolecules. O. Jardetzky and N. G. Wade-Jardetzky. Ann. Rev. Biochem. 40:605–634, 1971.

Long-range non-radiative transfer of electronic excitation energy in proteins and polypeptides. I. Z. Steinberg. Ann. Rev. Biochem. 40:83–114, 1971.

Uses of fluorescence in biophysics: some recent developments. G. Weber. Ann. Rev. Biophys. Bioeng. 1:553–570, 1972.

Ultraviolet difference spectroscopy—new techniques and applications. J. W. Donovan. Methods Enzymol. 27:497–525, 1973.

Correlation proton magnetic resonance studies at 250 MHz of bovine pancreatic ribonuclease. I. Reinvestigation of the histidine peak assignments. J. L. Markley. Biochemistry 14:3546–3554, 1975.

Nuclear Magnetic Resonance in Biological Research: Peptides and Proteins. K. Wüthrich. Amsterdam, Elsevier/North Holland, 1976.

Carbon-13 nuclear magnetic resonance studies of proteins. W. Egan, H. Shindo, and J. S. Cohen. Ann. Rev. Biophys. Bioeng. 6:383–417, 1977.

Enzyme Structure, Part G. C. H. W. Hirs and S. N. Timasheff (eds.). Methods Enzymol., Vol. 49, 1978.

Circular polarization of luminescence: biochemical and biophysical applications. I. Z. Steinberg. Ann. Rev. Biophys. Bioeng. 7:113–137, 1978.

The application of high resolution nuclear magnetic resonance to biological systems. I. D. Campbell and C. M. Dobson. Methods Biochem. Anal. 25:1–133, 1979.

Carbon-13 nuclear magnetic resonance: new techniques. A. Allerhand. Methods Enzymol. 61:458–549, 1979.

The interpretation of near-ultraviolet circular dichroism. P. C. Kahn. Methods Enzymol. 61:339–378, 1979.

Magnetic circular dichroism of biological molecules. J. C. Sutherland and B. Holmquist. Ann. Rev. Biophys. Bioeng. 9:293–326, 1980.

Systematic application of two-dimensional ^{1}H-nuclear-magnetic-resonance techniques for studies of proteins. K. Nagayama, et al. Eur. J. Biochem. 114:365–374, 375–384, 1981.

Sequential resonance assignments as a basis for determination of spatial protein structures by high resolution proton nuclear magnetic resonance. K. Wüthrich, et al. J. Mol. Biol. 155:311–388, 1982.

Chemical Properties

The unique environments of reactive groups in folded proteins can have very substantial effects on their chemical properties. Surveys of the chemical reactivities of individual groups of a protein with various reagents, like those described in Chapter 1 (pp. 4–24, 37–42), and of the consequent effect on its biological activity, were the preferred method for inferring the architecture and functional design of a protein before characterization of crystal structure became possible. Such studies uncovered a bewildering range of reactivities, from total unreactivity to hyperreactivity, where the reaction was much more rapid than with a normal, accessible group. Two factors were believed to be usually involved: the effect of the environment on the electronic state of a group, i.e., its intrinsic reactivity, or steric effects on accessibility to the modifying reagent. However, explanation of such results in the light of a protein's crystal structure has very often not been satisfactory. Groups on the surface often are unreactive, whereas those buried in the interior may be most reactive, as in the case of cytochrome *c,* where the two internal Tyr residues (48 and 67) are much more reactive toward tetranitromethane than are the two at the surface of the molecule (residues 74 and 97). In contrast, iodination modifies primarily residues 67 and 74.

That the relative reactivity of a residue often depends upon the nature of the reagent suggests that it is the local concentration of the reagent, determined by its interaction with the neighboring parts of the protein, that is often the crucial factor. Repulsions may reduce the local concentration to far below that of the bulk solvent, leading to unreactivity with a group, whereas tight binding of a reagent can produce extremely high local concentrations, up to the equivalent of about 10^{10} M (see Chapter 9), leading to apparent hyperreactivity of a nearby group in appropriate proximity.

One of the classic examples of hyperreactivity involves the Ser residue characteristic of the so-called serine proteases (trypsin, chymotrypsin, elastase, and so on; see Table 6–1). These related enzymes were observed to be inactivated upon treatment with acylating reagents, such as diisopropyl fluorophosphate (DFP), which reacted with a unique Ser residue (number 195 in the usual numbering system based upon chymotrypsinogen):

$$[(CH_3)_2CH-O-]_2 \overset{\overset{\displaystyle O}{\|}}{P}-F \ + \ HOCH_2- \ \longrightarrow$$

$$\underset{\text{DFP}}{} \qquad\qquad \underset{\text{Ser 195}}{}$$

$$[(CH_3)_2CH-O-]_2 \overset{\overset{\displaystyle O}{\|}}{P}-OCH_2- \ + \ F^- \ + \ H^+ \qquad (7\text{--}1)$$

Because a Ser hydroxyl group normally does not react with such acylating reagents, and Ser 195 does not when the native conformation is disrupted, this reaction was attributed to greatly enhanced nucleophilicity of this particular group in the native conformation. However, it is now clear that reagents such as DFP bind in the active sites of serine proteases, much like their substrates, leading to very high reactivities (see Chapter 9).

Another well-known example of hyperreactivity of specific residues is the reaction of His 12 and His 119 of ribonuclease A with iodoacetate, iodoacetamide, and other alkyl halides (Table 7–3). Normal His residues will react with these reagents, but much less rapidly. The relative reaction rates of both His 12 and His 119 depend upon the nature of the reagent, indicating interactions between it and the surrounding protein to be important. The orientation of the His residues in the folded conformation is also important, as His 12 invariably is alkylated on the $N^{\epsilon 2}$ atom, and His 119 on the $N^{\delta 1}$ atom.

The close proximities of other groups in a folded protein can also lead to unexpected reactions. In hen lysozyme, iodine unexpectedly attacks preferentially one of the least accessible Trp residues (Trp 108) presumably owing to favorable interactions with the reactive species, probably the

Table 7–3   Relative Rates of Alkylation of Histidine and of Two His Residues of Ribonuclease A

| ALKYLATING REAGENT | SECOND-ORDER RATE CONSTANT $(10^{-4}\,sec^{-1}M^{-1})$ | | |
	L-Histidine	Ribonuclease[a] His 12	His 119
Iodoacetate		7.3	51.1
Iodoacetamide	0.012	1.1	0
Bromoacetate	0.086	20.5	184.5
L-α-Bromopropionate	0.0027	0.19	0.66
D-α-Bromopropionate	0.0028	4.16	1.84
D-α-Bromo-*n*-butyrate		3.60	1.11
β-Bromopyruvate		0	911
β-Bromopropionate	0.0229	0	6.33

[a]His 12 is always alkylated at atom $N^{\epsilon 2}$, His 119 at $N^{\delta 1}$; reaction of one atom inhibits reaction at the other.

From R. L. Heinrickson, et al., J. Biol. Chem. 140:2921–2934, 1965; and R. G. Fruchter and A. M. Crestfield, J. Biol. Chem. 242:5807–5812, 1967. Reactions were carried out at 25° C and pH 5.3 to 5.5.

I^{+} ion; other reagents tend to react with other Trp residues. In solution, oxidation of an indole ring is believed to occur by attack by the reagent, followed by displacement of HI by hydroxyl ion or water, to yield a residue of oxindole alanine:

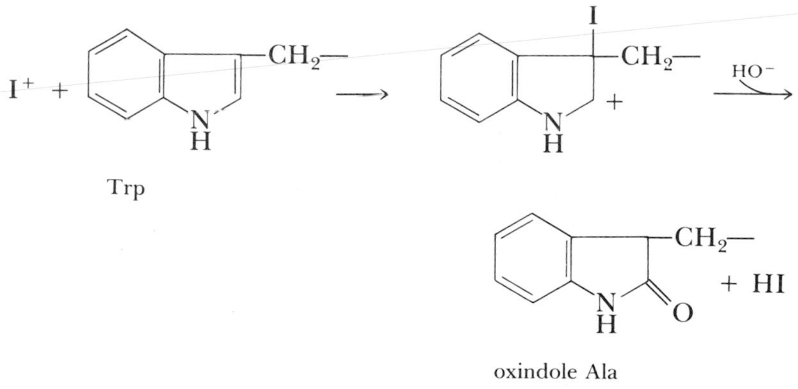

$$(7\text{–}2)$$

oxindole Ala

However, in lysozyme the attack by a solvent hydroxyl does not take place, but attack by a neighboring Glu 35 carboxyl group does, to give an internal ester (Figure 7–1). If the cross-linked protein is unfolded, the internal ester is hydrolyzed to regenerate Glu 35 and to convert Trp 108 to the oxindole alanine derivative.

Altered intrinsic reactivities of functional groups are often apparent from perturbations of their pK_a values. They may be affected by a wide variety of environmental (Figure 4–1 and Table 4–1) and electrostatic effects. Many charged groups are brought into close proximity on the surface of the folded protein, so ionization of groups that would increase

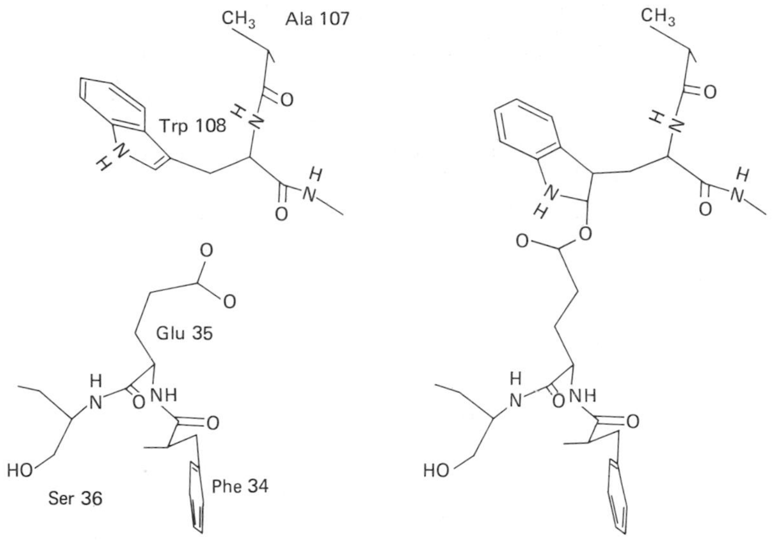

Figure 7–1
Covalent cross-link between Glu 35 and Trp 108 of hen lysozyme produced by iodine treatment. The positions of these two residues in native lysozyme are shown on the left. Iodine presumably reacts with Trp 108, but then Glu 35, rather than water, reacts preferentially with the adduct owing to the proximity of its side chain. The structure of the cross-linked protein is shown on the right. (Adapted from C. R. Beddell, et al., J. Mol. Biol. 97:643–654, 1975.)

the net charge is hindered. This general electrostatic effect alters all the groups. Specific interactions also occur, such as hydrogen bonding or salt bridges, affecting primarily ionization of the particular group. The titration behavior of many proteins have been extensively studied; more recently, the ionization of individual residues have been followed by NMR. The pK_a values of different residues have been observed to vary widely, often over a range of 3 to 4 pH units, presumably owing to their different environments. These electrostatic effects are only partly understood; in particular, the dielectric constant of the interior or surface of a protein is not known, and the concept of a macroscopic dielectric constant is of uncertain validity on the atomic scale. In the only recent attempts to simulate electrostatic effects in proteins, all groups of a particular type were assumed to have the same intrinsic pK_a, irrespective of their environments (but see Figure 4–1 and Table 4–1), but the actual pK_a value is modulated by the effects of other ionized groups on the protein. The effect of each ionized group is weighted by the extent to which it is inaccessible to the solvent.

Electrostatic interactions between the various groups can make their ionization behavior quite complex, with unusual titration curves, since the pK_a value of a group will depend upon the ionization of its neighbors.

Role of a buried acid group in the mechanism of chymotrypsin. D. M. Blow, et al. Nature 221:337–340, 1969.

Chemical modification as a probe of structure and function. L. A. Cohen. In The Enzymes, 3rd ed. P. D. Boyer (ed.). Vol. 1, pp. 147–211. New York, Academic Press, 1970.

Competitive labelling, a method for determining the reactivity of individual groups in proteins. H. Kaplan, et al. Biochem. J. 124:289–299, 1971.

Interpretation of protein titration curves. Application to lysozyme. C. Tanford and R. Roxby. Biochemistry 11:2192–2198, 1972.

Oxidation of lysozyme by iodine. T. Imoto, et al. J. Mol. Biol. 80:637–648, 648–656, 657–667, 1973.

The charge relay system in chymotrypsin and chymotrypsinogen. A. R. Fersht and J. Sperling. J. Mol. Biol. 74:137–149, 1973.

Electrostatic effects in myoglobin. Hydrogen ion equilibria in sperm whale ferrimyoglobin. S. J. Shire, et al. Biochemistry 13:2967–2979, 1974.

Observation of histidine residues in proteins by means of nuclear magnetic resonance spectroscopy. J. L. Markley. Acc. Chem. Res. 8:70–80, 1975.

An X-ray study of the structure and binding properties of iodine-inactivated lysozyme. C. R. Beddell, et al. J. Mol. Biol. 97:643–654, 1975.

The chemical modification of proteins by group-specific and site-specific reagents. A. N. Glazer. In The Proteins, 3rd ed. H. Neurath and R. L. Hill (eds.). Vol. 2, pp. 1–103. New York, Academic Press, 1976.

The intrinsic pK_a values of functional groups in enzymes: improper deductions from the pH-dependence of steady-state parameters. J. R. Knowles. Crit. Rev. Biochem. 4:165–173, 1976.

Proton nuclear magnetic resonance study of histidine ionizations in myoglobins of various species. Comparison of observed and computed pK values. L. H. Botello, et al. Biochemistry 17:5197–5205, 1978.

Experimental evaluation of the effective dielectric constant of proteins. D. C. Rees. J. Mol. Biol. 141:323–326, 1980.

FLEXIBILITY OF PROTEIN STRUCTURE

The crystal structures of proteins demonstrate varying degrees of flexibility, in that the electron density of an atom is spread over space to varying extents, reflecting in part significant populations of alternative conformations (pp. 248–251). Upon removal from any constraints of the crystal lattice, even greater flexibility would be expected in solutions; indeed, it is a thermodynamic requirement that molecules the size of proteins have substantial transient fluctuations. Moreover, as the following section will show, the folded conformations are only marginally stable and therefore must spontaneously undergo transient, but complete, unfolding with a frequency of somewhere between 10^{-4} to 10^{-12} per second, depending upon the protein and the conditions. Since the more rapid vibrations occur on a time scale of 10^{14}/sec, protein flexibility involves movements of widely varying magnitudes on a time scale spanning perhaps 26 orders of magnitude. Presumably, the greater the movement, the lower the frequency, but rather little is known about flexibility.

The most common and best-understood movements of atoms are the small-scale vibrations of bond lengths and angles detectable by infrared and Raman spectroscopy techniques. These vibrations in proteins are similar to those observed in small molecules and occur at frequencies between

6×10^{12} and 10^{14}/sec. Lower-frequency vibrations also occur in proteins and undoubtedly involve larger-scale movements, presumably of groups of atoms, owing to the close-packing of the interior. Because the nature of the less frequent, more extensive movements has not been determined directly by experimental methods, computer simulations of molecular dynamics have been used on time scales as long as 10^{-10} sec. The classical equations of motion for the protein atoms are solved using expressions for the energy as a function of the conformation. The validity of the molecular dynamics simulations depends upon that of the energy parameters used, and it is usually necessary to omit the solvent. Nevertheless, the mean vibrations of atoms were comparable to those observed crystallographically. Such studies of protein flexibility are being actively pursued at the present time, in order to understand how concerted movements of groups of atoms occur.

At the other end of the spectrum are movements, relative to each other, of individual domains of large proteins, brought about by relatively flexible "hinge" regions linking them. In antibody molecules, the domains rotate relative to each other on a time scale of 10^{-8} to 10^{-7} sec.

The phenomenon of protein flexibility is consequently immensely complicated, with a very wide variety of motions occurring over extremely wide ranges of time. Here we will consider the few aspects that have been studied experimentally.

Segmental flexibility in an antibody molecule. Y. Yguerabide, et al. J. Mol. Biol. 51:573–590, 1970.
Thermodynamic fluctuations in protein molecules. A. Cooper. Proc. Natl. Acad. Sci. U.S.A. 73:2740–2741, 1976.
Raman spectroscopy: a conformational probe in biochemistry. N. T. Yu. Crit. Rev. Biochem. 4:229–280, 1977.
Laser Raman scattering as a probe of protein structure. T. G. Spiro and B. P. Gaber. Ann. Rev. Biochem. 46:553–572, 1977.
Motions in proteins. F. R. N. Gurd and T. M. Rothgeb. Adv. Protein Chem. 33:73–165, 1979.
Low frequency vibrations and the dynamics of proteins and polypeptides. W. L. Peticolas. Methods Enzymol. 61:425–458, 1979.
Protein structural fluctuations during a period of 100 ps. M. Karplus and J. A. McCammon. Nature 227:578, 1979.
Simulation of protein dynamics. J. A. McCammon and M. Karplus. Ann. Rev. Phys. Chem. 31:28–45, 1980.
The internal dynamics of globular proteins. M. Karplus and J. A. McCammon. Crit. Rev. Biochem. 9:293–349, 1981.
Effect of constraints, solvent and crystal environment on protein dynamics. W. F. van Gunsteren and M. Karplus. Nature 293:677–678, 1981.
Characterization of the distribution of internal motions in the basic pancreatic trypsin inhibitor using a large number of internal NMR probes. G. Wagner. Quart. Rev. Biophys. 16:1–57, 1983.

Hydrogen Exchange

The most direct, convincing, and well-established evidence for motility of protein structure is that internal groups of proteins do react at a finite rate with appropriate reagents, even if only very slowly. Either the normally buried group must occasionally be at the surface, accessible to the reagent, or the reagent must permeate the protein interior; both would require disruption of the normal protein conformation. Such studies can provide information about even the slowest, or least frequent, structural motility, limited only by the stability of the protein or the patience of the investigator. However, they are often complicated by the tendencies of reagents to bind specifically to the protein, reacting rapidly with nearby groups or perhaps perturbing the conformation; the most useful reagent, therefore, is one that is normally present—namely, water. It may be used in its isotopic forms (^1H_2O, ^2H_2O, ^3H_2O) to measure the tendencies of the various hydrogen atoms of the protein to exchange with the solvent.

A protein in which the hydrogen atoms are of one isotope is placed in water of another isotope, and the exchange between the two is measured. Hydrogen atoms attached to various atoms in model compounds exchange with solvent at different rates. Two mechanisms of exchange are involved: The first requires temporary abstraction of the hydrogen atom by either H_2O or ^-OH upon hydrogen bonding; the hydrogen atom is then replaced by another molecule of the solvent. In the second mechanism, the molecule is transiently protonated by the solvent; exchange takes place when a different hydrogen atom is lost upon deprotonation. The rates at which these two processes occur depend upon the pH of the solvent and upon the tendency of the group to ionize (Figure 7–2). Hydrogen atoms on O, N, or S atoms exchange rather rapidly, whereas those attached to C atoms hardly do so at all.

Amide hydrogens are most often involved in exchange measurements with proteins, as they exchange on a convenient time scale and are often buried in proteins, further slowing exchange. Both mechanisms of exchange in model amides are significant, so the process is both acid- and base-catalyzed. The slowest rate occurs at pH 3, where the acid- and base-catalyzed processes are equal in rate; the rate increases tenfold for each unit change in pH away from the minimum value (Figure 7–2). The intrinsic rate of exchange is temperature-dependent, generally increasing about threefold with a 10° increase in temperature, which corresponds to an activation energy of 17 to 20 kcal/mole.

However, the rate of exchange is also affected by the environment and by inductive and charge effects on the amide, so it is more complex in large molecules like proteins. Within an unfolded protein, such as performic acid–oxidized ribonuclease, the rate of exchange of individual amide hydrogens varies 100-fold, apparently owing to inductive effects of the neighboring amino acid side chains, and is not as pH-dependent as would be expected from model compounds.

Classical methods of hydrogen exchange measurement in proteins give only the average number of hydrogen atoms exchanged as a function of time. Many hydrogen atoms are observed to exchange rapidly, approximately at the rate observed with model amides, suggesting that the groups

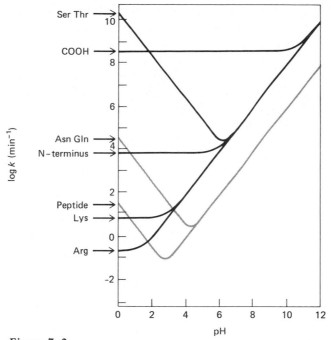

Figure 7–2
Dependence on pH of hydrogen exchange rates of model groups. (From K. Wüthrich and G. Wagner, J. Mol. Biol. 130:1–18, © 1979, London, Academic Press Inc.)

involved are on the surface of the protein. Other hydrogens exchange more slowly, with a broad distribution of rates, presumably reflecting that the groups are buried and that conformational motility is required for exchange to occur. The number of slowly exchanging hydrogens and their rates of exchange vary with the protein and the conditions. A rather rapidly exchanging, presumably flexible protein (insulin) is compared with one slowly exchanging and inflexible (BPTI) in Figure 7–3. In the latter case, the most slowly exchanging hydrogen atoms do so at about 10^{-8} the rate observed in model amides.

Such methods could not identify the atoms changing at the various rates, but if the exchange process can be made sufficiently slow, the protein can be cleaved and the distribution of hydrogen isotopes determined chemically. This is barely feasible with amide exchange, and only ribonuclease has been studied in this way, but the $C^{\epsilon 1}$ hydrogen atom of His and the $C^{\delta 1}$ of Trp residues exchange sufficiently slowly that the rates of exchange of individual residues in proteins have been measured.

The greatest advance has been the ability to follow exchange of individual protons by means of ^1H nuclear magnetic resonance (NMR), owing to its high resolution and discrimination between ^1H and ^2H atoms. BPTI has been most thoroughly studied; its slowly exchanging amide hy-

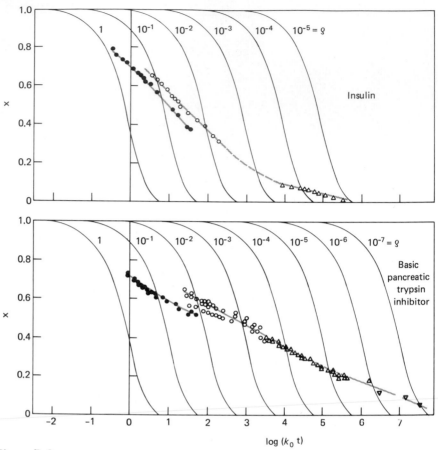

Figure 7–3

Exchange-rate curves of insulin and bovine pancreatic trypsin inhibitor at 25° C. X is the fraction of unexchanged peptide hydrogen atoms, k_0 is the exchange rate of an average solvent-exposed peptide group, and t is time. The measurements were made in the time interval from about 5 minutes to 10 hours after dissolution of the protein and at the following values of pH: insulin, (●) pH 3.0, (○) pH 4.1, (△) pH 7.7; bovine pancreatic trypsin inhibitor, (●) pH 3.4, (○) pH 5.1, (△) pH 7.3, (▽) pH 9.2. The curves are calculated for the values of ϱ indicated on the figure; ϱ is the probability of exposure to solvent of the peptide groups. (Adapted from A. Hvidt and E. J. Pedersen, Eur. J. Biochem. 48:333–338, 1974.)

drogens have been identified (Figure 7–4). Those atoms that exchange most readily are in the interior of the molecule and are involved in hydrogen bonding within the β-sheet (see Figures 6–9 and 6–15). Exchange is slower in the center of the β-strands than at either end; the interior side of the α-helix exchanges more slowly than the exterior side. All exchange rates are both acid- and base-catalyzed, but not always to the theoretical extent, and all increase in rate at higher temperatures.

Neutron diffraction readily distinguishes between ^1H and ^2H atoms (p. 220), permitting measurement of exchange within the crystalline state,

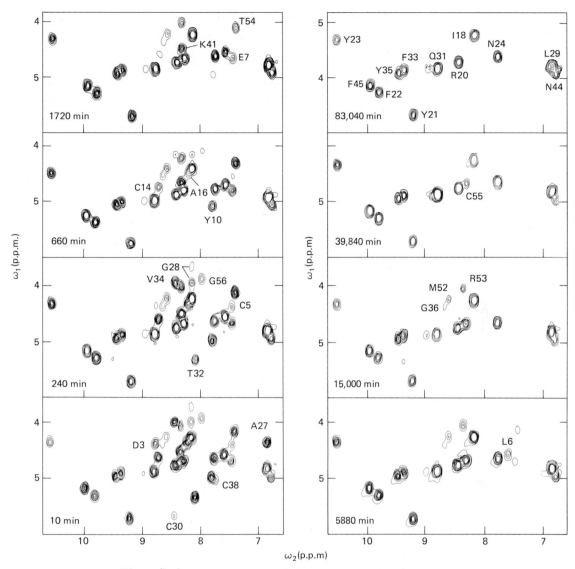

Figure 7–4
Hydrogen exchange of individual backbone amide protons in BPTI followed by two-dimensional correlated nuclear magnetic resonance spectroscopy. The protein was dissolved in ^2H_2O and kept at 36° C for the indicated periods of time before the spectra were measured. The resonances from the amide protons that disappear are identified on the last spectrum in which they were apparent, using the one-letter abbreviation for the amino acid (Figure 1–1) followed by the residue number. The assignments of the most slowly exchanging amides are given on the last spectrum. (From G. Wagner and K. Wüthrich, J. Mol. Biol. 160:343–361, © 1982, London, Academic Press Inc.)

although crystallographic measurements are not very amenable to rate measurements. The few studies have demonstrated primarily that exchange of many interior sites occurs even within the crystal lattice.

In spite of the very many studies of hydrogen exchange with many proteins, there is no consensus regarding how it occurs in proteins and what dynamic fluctuations are responsible for exchange of interior groups. The wide variation in exchange rates of different atoms within a protein implies that it does not occur solely by rare, transient unfolding of the entire protein; furthermore, even the slowest rates are too fast. Exchange from the unfolded state does become significant if conditions approach those that produce unfolding, and the rates of different sites converge to comparable values; however, exchange from the fully unfolded state is usually insignificant.

Only local unfolding, or "breathing," may be invoked to explain exchange of interior hydrogen atoms. This classical interpretation of hydrogen exchange, due initially to Linderstrom-Lang, invoked a variety of transient "open" conformations of the protein, with the exchanging groups exposed to the solvent:

$$\text{folded} \underset{k_{-1}}{\overset{k_1}{\rightleftharpoons}} \text{``open''} \xrightarrow{k_{\text{ex}}} \text{hydrogen exchanged} \qquad (7\text{--}3)$$

The hypothetical "open" form would be unstable and transient; consequently k_1 must be much less than k_{-1}, so that the equilibrium constant for its formation, $K_1 (\equiv k_1/k_{-1})$, is much less than unity. A variety of open states need to be postulated to account for the different rates of exchange of different sites.

The rates of the process are generally considered in terms of whether the rate-limiting step is opening of the structure, the **EX$_1$ mechanism,** or the exchange reaction, the **EX$_2$ mechanism.** In only the latter case should the rate be sensitive to factors known to affect the intrinsic exchange reaction in model compounds, especially pH. An EX$_1$ mechanism requires that $k_{\text{ex}} > k_{-1}$, so that the rate of exchange gives directly the value of k_1, the rate of opening of the protein structure. An EX$_2$ mechanism will apply if $k_{\text{ex}} < k_{-1}$, so that the observed rate is $K_1 k_{\text{ex}}$ and gives only the putative equilibrium constant for the breathing process. Proteins demonstrate EX$_2$ types of exchange under most conditions; the data in Figure 7–3 are of this type and are interpreted there as giving the equilibrium constant for opening of the proteins. Only in a few cases, especially at high pH where exchange is very rapid, is the EX$_1$ mechanism observed.

An alternative, extreme hypothesis suggests that hydrogen exchange occurs within the protein interior upon rare instances of diffusion of solvent molecules to the various sites. Quenchers of fluorescence are believed to be able to diffuse rapidly into protein interiors (see following section); therefore, it is not implausible for H_2O, H_3O^+ and ^-OH molecules to do so. The rate of exchange for each site would depend, in a complex manner, upon the general flexibility of the protein to create the necessary channels, on the probability of diffusion of solvent molecules to the appropriate location, and on the stability of the hydrogen bond in which the amide hydrogen atom will usually be involved. Interpretation of hydrogen exchange rates in terms of protein flexibility is not straightforward with this mechanism.

The available evidence is not sufficient to determine the mechanism of hydrogen exchange with any protein. That most indicative of the classical breathing process is the observation that substantial effects on the rate of exchange of *all* slowly exchanging atoms in BPTI are produced by localized covalent modification of the protein, roughly proportional to the decrease in stability of the folded conformation. However, such effects might also be explained by the solvent permeation mechanism if protein structure is sufficiently cooperative that localized modifications affect the overall flexibility of the protein required for solvent channels and the stabilities of all the hydrogen bonds. Other observations suggesting that exchange occurs within the protein interior by solvent permeation are that (1) it occurs within the crystalline state, where local unfolding would be expected to be greatly diminished, (2) the rate is not increased by low concentrations of denaturing agents, such as urea, that would be expected to favor local unfolding, and (3) the rate is markedly pressure-dependent, suggesting a need to create channels.

On the other hand, it is most likely that hydrogen exchange occurs by a large number of different processes, including both the above extremes, varying with the protein and the conditions.

Hydrogen exchange in proteins. A. Hvidt and S. O. Nielsen. Adv. Protein Chem. 21:287–386, 1966.

Hydrogen exchange. S. W. Englander, et al. Ann. Rev. Biochem. 41:903–924, 1972.

Primary structure effects on peptide group hydrogen exchange. R. S. Molday, et al. Biochemistry 11:150–158, 1972.

Determination of pK_a's of individual histidine residues in pancreatic ribonuclease by hydrogen–tritium exchange. M. Ohe, et al. J. Biochem. 75:1197–1200, 1974.

The solvent dependence of hydrogen exchange kinetics of folded proteins. C. K. Woodward, et al. J. Biol. Chem. 250:440–444, 1975.

Hydrogen–tritium exchange. S. W. Englander and J. J. Englander. Methods Enzymol. 49:24–39, 1978.

Correlation between the amide proton exchange rates and the denaturation temperatures in globular proteins related to the basic pancreatic trypsin inhibitor. G. Wagner and K. Wüthrich. J. Mol. Biol. 130:31–37, 1979.

Hydrogen exchange kinetics and internal motions in proteins and nucleic acids. C. K. Woodward and B. D. Hilton. Ann. Rev. Biophys. Bioeng. 8:99–127, 1979.

Hydrogen exchange from identified regions of the S-protein component of ribonuclease as a function of temperature, pH, and the binding of S-peptide. J. J. Rosa and F. M. Richards. J. Mol. Biol. 145:835–851, 1981.

Influence of charge on the rate of amide proton exchange. P. S. Kim and R. L. Baldwin. Biochemistry 21:1–4, 1982.

Hydrogen exchange in RNase A: neutron diffraction study. A. Wlodawer and L. Sjolin. Proc. Natl. Acad. Sci. U.S.A. 79:1418–1422, 1982.

Protein dynamics investigated by the neutron diffraction–hydrogen exchange technique. A. A. Kossiakoff. Nature 296:713–721, 1982.

Amide proton exchange and surface conformation of the basic pancreatic trypsin inhibitor in solution. G. Wagner and K. Wüthrich. J. Mol. Biol. 160:343–361, 1982.

Water catalysis of peptide hydrogen isotope exchange. R. B. Gregory, et al. Biochemistry 22:910–917, 1983.

The pH dependence of hydrogen exchange in proteins. J. B. Matthew and F. M. Richards. J. Biol. Chem. 258:3039–3044, 1983.

Fluorescence Quenching

The fluorescence of aromatic groups is instantly "quenched" by close physical encounter of the excited molecule with some small molecules, such as O_2, I^-, and acrylamide. As expected, aromatic side chains on the surface of a protein are quenched by diffusion-controlled encounters with such small molecules. Somewhat more surprising, many internal residues are quenched only slightly less efficiently by neutral molecules, e.g., O_2 and acrylamide, suggesting that they can diffuse through the interiors of proteins at rates 20 to 50 per cent of those in water, within the lifetime of the excited state. Charged quenchers, such as I^-, are an order of magnitude less efficient; this notable difference presumably reflects the inhospitable nature of the nonpolar protein interior to charged molecules. However, detailed interpretations of such studies are not possible because they are complicated by energy transfer between different groups within the protein, by their varying quantum yields, by possible perturbations of the protein structure upon excitation of the aromatic groups, and by possible binding of the quenchers to sites on or within the protein. Nevertheless, such studies suggest that small molecules may diffuse rapidly through protein interiors; this mechanism is very pertinent to interpretation of hydrogen exchange.

Quenching of protein fluorescence by oxygen. Detection of structural fluctuations in proteins on the nanosecond time scale. J. R. Lakowicz and G. Weber. Biochemistry 12:4171–4179, 1973.

Fast relaxation processes in a protein revealed by the decay kinetics of tryptophan fluorescence. A. Grinvald and I. Z. Steinberg. Biochemistry 13:5170–5178, 1974.

Room temperature phosphorescence and the dynamic aspects of protein structure. M. L. Saviotti and W. C. Galley. Proc. Natl. Acad. Sci. U.S.A. 71:4154–4158, 1974.

Dynamics of a protein matrix revealed by fluorescence quenching. M. R. Eftink and C. A. Ghiron. Proc. Natl. Acad. Sci. U.S.A. 72:3290–3294, 1975.

Fluorescence quenching studies with proteins. M. R. Eftink and C. A. Ghiron. Anal. Biochem. 114:199–227, 1981.

Oxygen quenching and fluorescence depolarization of tyrosine residues in proteins. J. R. Lakowicz and B. P. Maliwal. J. Biol. Chem. 258:4794–4801, 1983.

Rotations of Side Chains

Side chains on the surfaces of protein, as well as terminal methyl groups in the interior, are observed to have mobilities comparable to those in unfolded proteins or in small-molecule analogues, rotating on a time scale of 10^{-11} to 10^{-8} sec. Such rapid motions are observed by fluorescence depolarization (p. 184) or by ^{13}C–NMR (p. 186), but the slower motions

Table 7–4   Rotation of Aromatic Rings in BPTI

RESIDUE	FREQUENCY OF 180-DEGREE ROTATIONS (sec^{-1}) AT TEMPERATURE OF			ACTIVATION PARAMETERS Enthalpy: $\Delta H\ddagger$ (kcal/mole)	Entropy: $\Delta S\ddagger$ (cal/mole/°)
	4° C	40° C	80° C		
Tyr 10	Rotating rapidly at all temperatures[a]				
Tyr 21	Rotating rapidly at all temperatures[a]				
Tyr 23	< 5	3×10^2	5×10^4	26	35
Tyr 35	< 1	50	5×10^4	37	68
Phe 4	Rotating rapidly at all temperatures[a]				
Phe 22	< 5	< 5	5		
Phe 33	Rotating rapidly, but rates not known				
Phe 45	30	1.7×10^3	5×10^4	17	11

[a]Assuming that equivalence of both pairs of C$^\delta$ and C$^\epsilon$ hydrogen atom resonances result from rapid rotations, not equivalent fixed environments ($> 5 \times 10^4$ sec^{-1}).

Adapted from G. Wagner, et al. Biophys. Struct. Mech. 2:139–158, 1976.

undergone by internal groups are masked in many techniques by the over-all rotation of the active molecule. Nevertheless, that such slower motions do occur can be inferred from ^{1}H–NMR spectra of aromatic rings. The two *ortho* (C$^{\delta 1}$ and C$^{\delta 2}$) and *meta* (C$^{\epsilon 1}$ and C$^{\epsilon 2}$) hydrogen atoms of Tyr and Phe residues are in unique environments in folded proteins and would be expected to give individual resonances in ^{1}H–NMR spectra. However, if the rings are flipping by 180-degree rotations about the C$^\beta$—C$^\gamma$ bond rapidly (on the NMR time scale), each of the pairs of atoms spend equal amounts of time in both environments, and only a single, averaged resonance will be observed for each pair. The time scale of NMR measurements depends upon the spectral region but lies within the range of 10^{-5} to 1 sec.

Most proteins give averaged spectra for Phe and Tyr residues, sug-gesting that the rings are rotating by 180-degree flips at least 10^4 times per second, even when the rings are fully buried. Of course, such appar-ently averaged ^{1}H–NMR spectra could arise by accidental coincidence of individual resonances, but this is unlikely in general. It has been shown not to be the case in BPTI, where the two Tyr and the three Phe residues that are most buried in the interior give immobilized spectra at low tem-peratures. As the temperature is increased, averaged spectra are obtained, indicating that rotations occur more frequently (Table 7–4). Even rapidly rotating rings appear fixed on the ^{13}C–NMR time scale, indicating that the flips occur with frequencies no greater than 5×10^7/sec.

Flipping of aromatic rings within the close-packed interior of a protein requires movement of the surrounding atoms, but perhaps not as much as might be expected with their usual representation as flat planar rings, since Tyr and Phe rings are actually more like flattened spheres, or oblate ellipsoids, with thickness 3.4 Å and diameter 6.8 Å. The perturbations re-quired for flipping have been studied extensively with BPTI. Theoretical simulations indicate that adjacent atoms need move by only small bond rotations, no greater than 17 degrees.

It is likely that flipping simply is an infrequent process but occurs rapidly once initiated, since the half-rotated transient state should be quite unstable. The frequencies of flipping of Phe 45 and Tyr 35 are decreased by elevated pressures, indicating that the transition states of the protein have substantially increased volumes. On the other hand, the frequencies are not affected by local alterations that alter the overall stability of the folded state and the rates of hydrogen exchange of interior amides, indicating that different types of flexibility are involved.

In contrast to Tyr and Phe, buried Trp side chains do not detectably undergo such ring flipping, undoubtedly owing to the greater size of the indole ring and to the absence of symmetry, so that 360-degree full rotations would be required. However, it generally does undergo more restricted vibrations about its mean position on the 10^{-8} sec time scale.

Side chain torsional potentials and motion of amino acids in proteins: bovine pancreatic trypsin inhibitor. B. Gelin and M. Karplus. Proc. Natl. Acad. Sci. U.S.A. 72:2002–2006, 1975.

Dynamics of the aromatic amino acid residues in the globular conformation of the basic pancreatic trypsin inhibitor (BPTI). ^{1}H NMR studies. G. Wagner, et al. Biophys. Struct. Mech. 2:139–158, 1976.

Subnanosecond motions of tryptophan residues in proteins. I. Munro, et al. Proc. Natl. Acad. Sci. U.S.A. 76:56–60, 1979.

The influence of localized chemical modifications of the basic pancreatic trypsin inhibitor on static and dynamic aspects of the molecular conformation in solution. G. Wagner, et al. Eur. J. Biochem. 95:239–248, 1979.

Carbon-13 nuclear magnetic resonance relaxation studies of internal mobility of the polypeptide chain in basic pancreatic trypsin inhibitor and a selectively reduced analogue. R. Richarz, et al. Biochemistry 19:5189–5196, 1980.

Activation volumes for the rotational motion of interior aromatic rings in globular proteins determined by high resolution ^{1}H NMR at variable pressure. G. Wagner. FEBS Letters 112:280–284, 1980.

Side-chain rotational isomerization in proteins: a mechanism involving gating and transient packing defects. J. A. McCammon, et al. J. Am. Chem. Soc. 105:2232–2237, 1983.

UNFOLDING AND REFOLDING

The folded states of proteins are only marginally stable under the best of conditions and may be disrupted (i.e., the protein denatured) merely by environmental change, such as a rise in temperature, variation of pH, addition of a variety of denaturants, or increase in pressure. Extrapolations of such unfolding transitions to conditions optimal for stability show that even under these conditions proteins spontaneously unfold completely—surely the ultimate expression of protein flexibility—at a finite frequency. The transiently unfolded protein must then refold even more rapidly; this may be demonstrated directly by placing a fully unfolded protein under conditions where the folded state is stable.

Folding of a protein appears generally to be a self-assembly process, in that all the information required is present in the amino acid sequence, and it occurs spontaneously under the appropriate conditions, probably also in vivo after biosynthesis of the linear polypeptide chain. Proteins that do not refold in vitro are generally assumed to have undergone some interfering covalent modification or to have lost some cofactor required for the folded state (Chapter 8).

That proteins can fold spontaneously should be somewhat surprising. Polypeptide chains were shown in Chapter 5 to be capable of adopting an astronomically large number of different conformations—a number so large that it is not feasible to sample all of them on a practical time scale. For example, if a polypeptide chain of 100 residues has 10^{100} possible conformations (i.e., an average of 10 conformations per residue) and the time to convert one conformation into another is the shortest possible (i.e., about 10^{-13} sec), the average time required to sample all of them would be 10^{77} years (10^{85} sec). The magnitude of this length of time is incomprehensible, but even more conservative estimates are still impractical. Since proteins are observed to fold on time scales of 10^{-1} to 10^3 sec, the inescapable conclusion is that proteins do not fold by simply sampling randomly all possible conformations until that with the lowest free energy is encountered. Instead, the process must be directed in some way. Of the many hypotheses suggesting how this might occur, most have envisaged some nucleation step, after which further folding is very rapid, as in the helix–coil transition (p. 187). Others picture unstable fragments of conformation coalescing into the final folded state.

If kinetic factors are so important, the possibility arises that the observed folded conformation is not that with the lowest possible free energy, but simply the most stable of those kinetically accessible; other unobserved, and probably unobservable, conformations might have even lower free energies. This very basic question is almost philosophical in nature, in that answering it experimentally may be unfeasible.

Are there pathways for protein folding? C. Levinthal. J. Chim. Phys. 65:44–45, 1968.

Principles that govern the folding of protein chains. C. B. Anfinsen. Science 181:223–230, 1973.

Acquisition of three-dimensional structure of proteins. D. B. Wetlaufer and S. Ristow. Ann. Rev. Biochem. 42:135–158, 1973.

Nucleation, rapid folding, and globular intrachain regions in proteins. D. B. Wetlaufer. Proc. Natl. Acad. Sci. U.S.A. 70:697–701, 1973.

Intermediates in protein folding reactions and the mechanism of protein folding. R. L. Baldwin. Ann. Rev. Biochem. 44:453–475, 1975.

Experimental and theoretical aspects of protein folding. C. B. Anfinsen and H. A. Scheraga. Adv. Protein Chem. 29:205–300, 1975.

Protein folding dynamics. M. Karplus and D. L. Weaver. Nature 260:404–406, 1976.

Experimental studies of protein folding and unfolding. T. E. Creighton. Prog. Biophys. Mol. Biol. 33:231–297, 1978.

Protein Folding. R. Jaenicke (ed.). Amsterdam, Elsevier/North Holland, 1980.

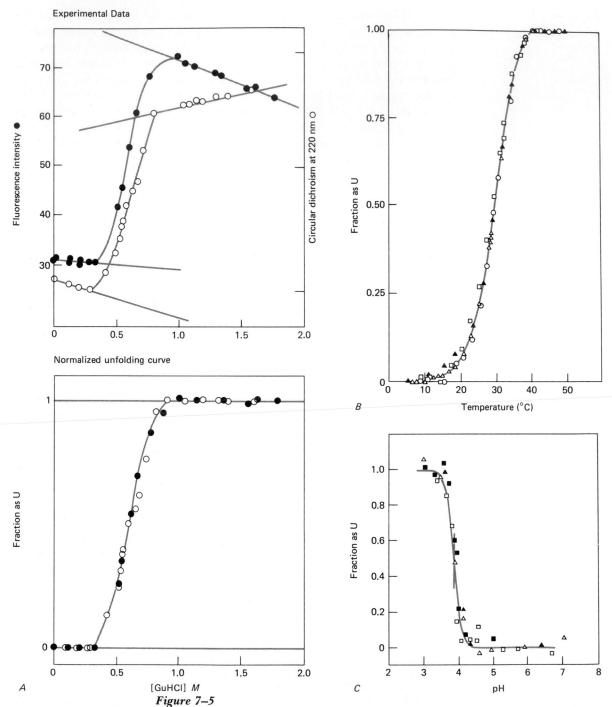

Experimental Data

Normalized unfolding curve

Figure 7–5
Equilibrium measurements of unfolding transitions. A, Guanidinium chloride (GuHCl) un-
folding of phosphoglycerate kinase. The experimental observations of unfolding using circular
dichroism and fluorescence are illustrated in the upper half. The straight lines show the effect

Equilibrium Unfolding Transitions

As the environment is gradually altered toward conditions favorable for protein unfolding, the folded conformation initially changes very little, if at all. There undoubtedly are increases in flexibility, and there may be localized changes, but the average structure is unchanged. The protein then becomes fully unfolded within a rather limited range of conditions. For example, staphylococcal nuclease undergoes acid-induced unfolding over a pH range of only 0.3 pH unit (see Figure 7–5C). Such an abrupt transition cannot be caused by the ionization of a single group, which should require 2 pH units to go from 90 per cent to 10 per cent ionization. It is indicative of a very cooperative transition, for example, where each of a number of groups can be ionized only if all of them do so simultaneously.

When the degree of unfolding is measured by a number of different methods, all generally give the same curve (Figure 7–5). This suggests that unfolding is a two-state phenomenon, with only the fully folded (N) and unfolded (U) states present: That is, at 50 per cent unfolding, 50 per cent of the molecules are N and 50 per cent are U, rather than all, or any significant fraction, of the molecules being half-folded. This also indicates that folding is a cooperative phenomenon, in which disruption of any significant portion of the folded structure leads to unfolding of all the rest. In other words, the stability of each part of the structure is dependent upon the stability of all the other parts; partially folded structures must then be unstable relative to U or N.

Unfolding transitions may be followed by many methods, but they are visualized most readily by electrophoresis in polyacrylamide gel slabs in which there is a gradient of urea perpendicular to the direction of migration (Figure 7–6). The compact folded protein migrates more rapidly than does the unfolded form, owing to its lower frictional coefficient. The continuous smooth band of protein through the unfolding transition, with a single inflexion point, is indicative of a rapidly equilibrating two-state unfolding transition.

---

of GuHCl on the spectral properties of the folded state, at low concentrations, and on the unfolded state at high concentrations. The same effects are assumed to apply throughout the transition region. The fraction of unfolding indicated by the spectral measurements is plotted in the lower half, illustrating that the two methods give the same smooth unfolding curve, consistent with a two-state unfolding transition. (Adapted from H. Nojima, et al., J. Mol. Biol. 116:429–442, 1977.)

B, The temperature-induced unfolding of bovine ribonuclease A in HCl-KCl at pH 2.1 and 0.019 ionic strength, measured by the increase in viscosity (□) and the decreases in optical rotation in 365 nm (○) and UV absorbance at 287 nm (△—△). The filled triangles show measurements of a second melting after cooling from 41°C for 16 h, indicating slight irreversibility at low temperatures. (Adapted from A. Ginsburg and W. R. Carroll, Biochemistry 4:2159–2174, 1965.)

C, The acid-induced unfolding of staphylococcal nuclease A, measured by viscosity (□) and circular dichroism at 220 nm (△), The open symbols were measurements made during acidification, the closed symbols during raising the ph. (Adapted from C. B. Anfinsen. Biochem. J. 128:737–749, 1972.)

Horse
Ferricytochrome *c*

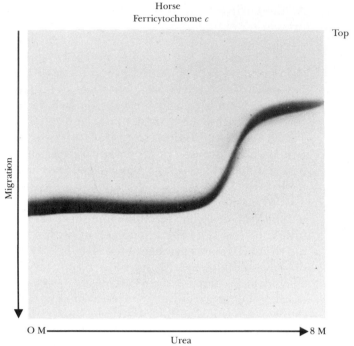

Top

O M————————————————→ 8 M
Urea

Figure 7–6

Urea-gradient electrophoresis of cytochrome c. *The folded protein was layered on the top of the polyacrylamide gel, which contained a linear gradient of urea from left to right. Electrophoresis at pH 4.0 was from top to bottom for 3.5 hours. At low urea concentrations, the protein remains folded and migrates rapidly; at high urea concentrations it is unfolded and migrates more slowly. The same pattern is obtained starting with unfolded protein, and the continuous band of protein through the abrupt unfolding transition also indicates that unfolding and refolding were rapid, so the equilibrium fraction of unfolding determined the rate of migration. The smooth shape of the transition, with a single apparent inflexion point, indicates that only two conformational states with different electrophoretic mobilities were present to significant extents. (From T. E. Creighton, J. Mol. Biol. 129:235–264, © 1979, London, Academic Press Inc.)*

Many apparent exceptions to two-state transitions have been reported, but in most cases with small single-domain proteins they appear to be incorrect. Noncoincidence of unfolding transitions can result from incorrect extrapolation of the properties of the N and U states (Figure 7–5A), heterogeneity or precipitation of the protein, or effects of the experimental measurements on the folding equilibrium, which are especially likely if they involve antibodies or proteolytic enzymes, which bind to either U or N. Spurious conformational transitions can also arise by dynamic processes, within either the unfolded or the folded states, that affect the parameter measured.

Nevertheless, well-established complex transitions are known. They usually occur with large multidomain proteins, and it appears that individual domains may undergo independent unfolding transitions. This need not always be the case, for phosphoglycerate kinase exhibits an apparently two-state transition (Figure 7–5), even though it is folded into two prominent structural domains; either its two domains have closely similar stabilities, or they do not unfold independently.

For a two-state transition, the equilibrium constant between the N and U states may be measured directly from the average degree of unfolding (α) within the transition region:

$$K = \frac{[N]}{[U]} = \frac{1 - \alpha}{\alpha} \tag{7–4}$$

The relative stabilities of the two states are known over this limited range of conditions, since at each point

$$\Delta G^\circ = G_N^\circ - G_U^\circ = -RT \ln K \tag{7–5}$$

However, the most pertinent values of ΔG° are those under normal conditions, for example, in the absence of denaturant. It is likely that increasing concentrations of denaturants such as urea and guanidinium chloride affect ΔG° rather uniformly (see following section); consequently the values of ΔG° may be extrapolated from the transition region to absence of the denaturant (Figure 7–7). There is a high degree of uncertainty in such extrapolations, but they usually give optimal values of ΔG° in the region of -5 to -15 kcal/mole, indicating maximum values of K between 10^3 and 10^{11}. Therefore, such proteins have probabilities of between 10^{-3} and 10^{-11} of being fully unfolded, even under optimal conditions.

Physical Basis for Protein Denaturation

A great many studies have been made of the effects on protein stability of varying the environment, as the results were expected to give insight into the physical basis for stability of the folded state, but the complexity of the process has usually prevented simple, unequivocal conclusions. Nevertheless, some generalities are possible, although unfolding of any protein is probably caused by a number of factors.

Extremes of pH usually cause unfolding because the folded state has groups buried in nonionized form that can ionize only after unfolding. Most prevalent are His and Tyr residues, which can cause unfolding at acid and alkaline pH values, respectively. Salt bridges between ionized groups can also be disrupted by pH changes. It might also be expected that, when a protein has substantial net charge, the general electrostatic repulsion between the ionized groups might cause unfolding, since such repulsions would be minimized in the unfolded state. There is little evidence for such repulsions being sufficient to produce unfolding, although proteins tend to be most stable near their isoelectric point.

A vast majority of compounds have been studied for their effects on protein stability, and the variety of effects observed is quite staggering. It is impossible to summarize all these observations; one can point out only that the net observed effect is often a subtle interplay among effects on both the folded and unfolded states of a protein, which can vary with different proteins, and on the properties of water, which affect protein stability via the hydrophobic effect (Chapter 4).

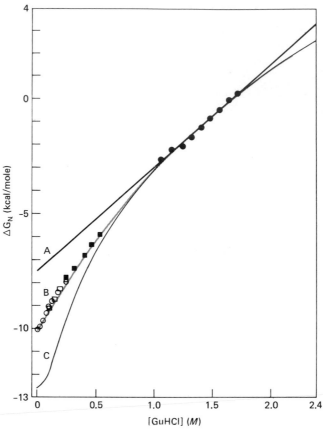

Figure 7–7
Example of estimation of the net stability of a folded protein from an equilibrium unfolding transition by the extrapolation to the absence of the denaturant. The difference between the free energies of the folded and unfolded states of horse metmyoglobin (ΔG_N) throughout the guanidinium chloride (GuHCl)–induced unfolding transition was measured at pH 6.0 (●), 4.9 (■), 4.7 (□), and 4.6 (○). The unfolding transition is very pH-dependent; the measurements were corrected to pH 7.0 by assuming that the pH-dependence is a result of only six groups with different pK_a values in the folded and unfolded states.

The three curves are different ways of extrapolating through the pH 6.0 data. Curve A assumes that the free energy difference is linearly proportional to the denaturant concentration. Curves B and C are those expected if denaturation is produced by preferential binding of GuHCl to the unfolded state; they use different but plausible affinities and numbers of bound GuHCl molecules.

The midpoint of myoglobin unfolding at pH 7 is at 1.6 M GuHCl, so only a modest extrapolation to 0 M is required; yet there is considerable uncertainty about the extrapolation. (Adapted from C. N. Pace and K. E. Vandenburg, Biochemistry 18:288–292, 1979.)

Even in the case of the most useful denaturants, urea and guanidinium chloride (often called guanidinine hydrochloride),

$$\underset{\text{urea}}{H_2N-\overset{\overset{\displaystyle O}{\|}}{C}-NH_2} \qquad \underset{\text{guanidinium chloride}}{H_2N-\overset{\overset{\displaystyle NH_2}{|}}{C}=NH_2{}^{+}Cl^{-}} \qquad (7\text{–}6)$$

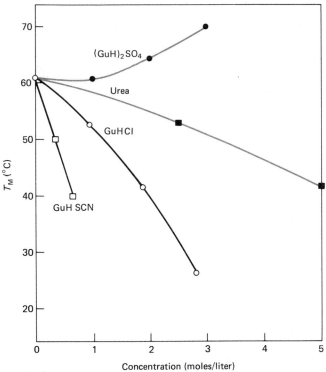

Figure 7–8
Thermal stability of ribonuclease A as a function of the concentration of urea and various guanidinium (GuH⁺) salts. The temperatures at the midpoint of the thermal unfolding transition, T$_m$, were measured at pH 7.0. (Adapted from P. H. Von Hippel and K.-Y. Wong, J. Biol. Chem. 240:3909–3923, 1965.)

there is no simple, generally accepted explanation for their effectiveness as denaturants. Initial rationalizations focused upon their obvious potential for hydrogen bonding, and they were considered to break protein hydrogen bonds. However, further reflection, plus some experimental data, indicated that they were no more potent in this respect than water. Model compound studies demonstrated that both increase the solubilities of nonpolar molecules, including those of the amino acid side chains, in proportion to their accessible surface area (Figure 4–6), diminishing the magnitude of the hydrophobic effect by up to one third. This probably is an indirect result of their effects upon the structural properties of water, resulting from their comparable hydrogen-bonding capabilities, but with different geometries. In confirmation, the denaturant potencies of guanidinium salts are affected by the nature of the anion according to the Hofmeister series (p. 150); guanidinium thiocyanate is more potent than the chloride, whereas guanidinium sulfate actually stabilizes proteins (Figure 7–8).

Diminishing the hydrophobic interaction by one third should be ample to produce unfolding of proteins, but the observed effects are usually considerably less than those predicted; therefore, there must be additional factors, probably direct interactions with the protein. Many theories of unfolding have concentrated on specific interactions between denaturants and the protein, explaining unfolding by the possibility of more interactions with the unfolded state. However, it is not clear how specific such interactions could be if concentrated solutions like 6 M guanidinium chloride are required, where the denaturant occupies half the volume of the solvent. Nevertheless, interactions with the protein are apparent by a variety of techniques, and crystallographic studies of α-chymotrypsin in the presence of guanidinium chloride and urea show a variety of interactions with the folded protein. Urea molecules, but not the charged guanidinium ion, even permeate the interior, occupying small cavities and perturbing somewhat the close-packed interior. Consequently, it is most likely that denaturants such as urea and guanidinium act indirectly by diminishing the hydrophobic interaction in a uniform, predictable manner, and that they directly interact with both the folded and unfolded states to produce a wide range of effects, depending upon the local geometry of the interacting groups in the protein.

Although most proteins are unfolded by 6 M guanidinium chloride (approaching its solubility limit), not all are. Proteins vary widely in their susceptibility to denaturation; there is no simple explanation for such differences, but undoubtedly they are caused by direct interactions between the denaturant and the protein, especially when folded. For example, the existence of a tight binding site on the folded protein for the denaturant can make it resistant to unfolding; ligand binding to one conformational state increases its relative stability (Chapter 8), even if the ligand is a denaturant. One example is the enzyme urease, which hydrolyzes urea and consequently has a specific substrate binding site for it; it is not denatured by urea.

The thermodynamics of protein denaturation. J. F. Brandts. J. Am. Chem. Soc. 86:4291–4301, 4302–4314, 1964.

Isothermal unfolding of globular proteins in aqueous urea solution. C. Tanford. J. Am. Chem. Soc. 86:2050–2059, 1964.

Methods for the study of reversible denaturation of proteins and interpretation of data. J. Hermans. Methods Biochem. Anal. 13:81–111, 1965.

Protein denaturation. C. Tanford. Adv. Protein Chem. 23:121–282, 1968; 24:1–95, 1970.

Thermodynamics of protein denaturation. Effect of pressure on the denaturation of ribonuclease A. J. F. Brandts, et al. Biochemistry 9:1038–1047, 1970.

The stability of globular proteins. C. N. Pace. Crit. Rev. Biochem. 3:1–43, 1975.

Interactions of urea and other polar compounds in water. M. Roseman and W. P. Jencks. J. Am. Chem. Soc. 97:631–640, 1975.

The role of solvent interactions in protein conformation. F. Franks and D. Eagland. Crit. Rev. Biochem. 3:165–219, 1975.

Solvent denaturation. J. A. Schellman. Biopolymers 17:1305–1322, 1978.

Expression of functionality of α-chymotrypsin. Effects of guanidine hydrochloride and urea in the onset of denaturation. L. S. Hibbard and A. Tulinsky. Biochemistry 17:5460–5468, 1978.

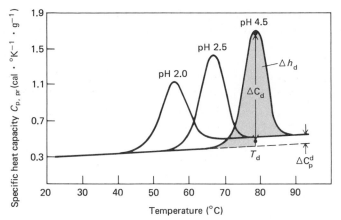

Figure 7–9

Calorimetric measurement of thermal unfolding of lysozyme at three different pH values. The calorimeter measures directly the amount of heat required to raise the temperature of the protein solution by a specified amount (i.e., the heat capacity), relative to the solvent alone. At low temperatures this gives the heat capacity of the folded protein; at high temperatures, of the unfolded protein. Both are found to be independent of pH; the difference between the two is ΔC_p^d. The heat absorbed during unfolding, when integrated over the entire transition (the shaded area at pH 4.5), gives the specific enthalpy change upon unfolding Δh_d. (Adapted from P. L. Privalov and N. N. Khechinashvili, J. Mol. Biol. 86:665–684, 1974.)

Conformation of a stable intermediate on the folding pathway of *Staphylococcus aureus* penicillinase. E. A. Carrey and R. H. Pain. Biochim. Biophys. Acta 533:12–22, 1978.

Electrophoretic analysis of the unfolding of proteins by urea. T. E. Creighton. J. Mol. Biol. 129:235–264, 1979.

The stabilization of proteins by sucrose. J. C. Lee and S. N. Timasheff. J. Biol. Chem. 256:7193–7201, 1981.

Thermodynamics of Unfolding

The temperature-dependence of protein unfolding (with any disulfide bonds kept intact) has been studied especially thoroughly, because of its intrinsic thermodynamic importance. Direct measurements of heat effects with a sensitive microcalorimeter have been made by Privalov and coworkers on a number of small proteins. The results suggest some general properties of proteins in solution, particularly when expressed per gram of protein, rather than mole, to compensate for differences in size.

Calorimetric traces of the effects of heating solutions of lysozyme at three different pH values are illustrated in Figure 7–9. The quantity measured is the heat capacity of the solution at constant pressure, C_p, which is the energy required to raise its temperature. The initial value at low temperatures, 0.32 ± 0.02 cal $°K^{-1}g^{-1}$, is similar for all folded proteins and independent of the pH. As the temperature at which unfolding occurs is approached, there is a large absorption of heat. At higher temperatures, the heat capacity becomes more constant; it is similar for all unfolded proteins and higher than that of the folded state.

The heat capacity measures the temperature-dependence of both the enthalpy and entropy, since at constant pressure and the transition midpoint

$$C_p = \frac{\partial H^\circ}{\partial T} = T \frac{\partial S^\circ}{\partial T} \qquad (7\text{--}7)$$

Consequently, integrating the area under the unfolding curve gives the changes in enthalpy and entropy upon unfolding. The changes with temperature in enthalpies and entropies of both the folded and unfolded states are obtained from their heat capacities. All such values are generally expressed relative to the enthalpy of the folded state under standard conditions, usually taken as 25° C and pH 7. The relationship between the enthalpy and entropy is measured at the midpoint of the unfolding transition, T_m, since at this temperature

$$T_m \, \Delta S_m^\circ = \Delta H_m^\circ \qquad (7\text{--}8)$$

because the folded and unfolded states have equal free energies. At other temperatures, free energies are obtained from the enthalpy and entropy, since

$$G^\circ = H^\circ - T \, S^\circ \qquad (7\text{--}9)$$

These calorimetric measurements then permit a complete thermodynamic description of the folded and unfolded states of proteins in solution.

A most important result of these measurements with small proteins is that the enthalpy change upon unfolding measured calorimetrically is very nearly the same as that measured from the temperature-dependence of the equilibrium constant, the van't Hoff enthalpy change:

$$\Delta H_{VH} = -R \frac{\partial \ln K}{\partial (1/T)} = RT^2 \frac{\partial \ln K}{\partial T} \qquad (7\text{--}10)$$

This must be the case for a simple reaction and confirms that the unfolding transitions of these proteins are cooperative and two-state. The ratio of the two values averages 1.05 ± 0.03, indicating that only 5 per cent of the molecules at the transition midpoint could be in intermediate conformations, with enthalpies different from both fully folded and fully unfolded molecules. In contrast, multidomain proteins such as papain and a dimer of immunoglobulin light chains (a Bence Jones protein), both with two domains per polypeptide, gave values of 1.80 and 1.90, respectively, indicating that the individual domains tend to unfold independently. Owing to their simplicity, thermodynamic measurements have concentrated on the two-state transitions of single-domain proteins.

These measurements are of the entire solution of protein plus solvent, not of just the protein. Because aqueous solutions can give complex thermodynamic behavior (see Chapter 4), explanation of such measurements is not straightforward, but some correlations are observed that can be rationalized.

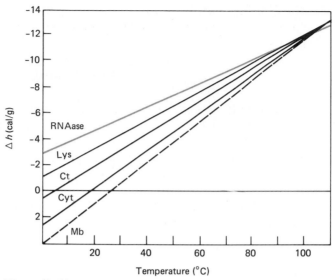

Figure 7–10
The specific enthalpy difference, Δh (per gram of protein), between the folded and unfolded states of five proteins. RNAase, ribonuclease A; Lys, hen lysozyme; Ct, bovine α-chymotrypsin; Cyt, cytochrome c; Mb, metmyoglobin. The pH of each solution is that at which the protein is most stable. (Adapted from P. L. Privalov and N. N. Khechinashvili, J. Mol. Biol. 86:665–684, 1974.)

    The substantial heat capacities of both the folded and unfolded states imply that their enthalpies are very temperature-dependent. The difference between specific enthalpies of the two states is also temperature-dependent, varying somewhat with the protein, but extrapolating to a common value at 110° C (Figure 7–10). The slopes of such curves (i.e., the heat capacity differences) correlate with the density of hydrophobic contacts measured in the proteins' crystal structures, suggesting that what is being measured is the extent of nonpolar surface exposed to solvent upon unfolding. Large heat capacity changes were described in Chapter 4 as symptomatic of the hydrophobic effect, reflecting the increased ordering of water molecules around exposed nonpolar surfaces; therefore, this correlation is satisfying. Convergence to a common value at 110° C has been explained by postulating that the hydrophobic interaction becomes negligible at this temperature, owing to the loss of structural order of water. All of these proteins have similar densities of hydrogen bonds, but other proteins with lower densities extrapolate to lower specific enthalpies at 110° C. Consequently, it appears likely that the difference in enthalpies between folded and unfolded states at 110° C is proportional to the content of hydrogen bonds, whereas the temperature-dependence reflects the magnitude of the hydrophobic interaction.

    The complete thermodynamic characterization of lysozyme illustrates the complexity of the physical basis of protein structure, since the observed net stability is a very small difference between very large, but compensat-

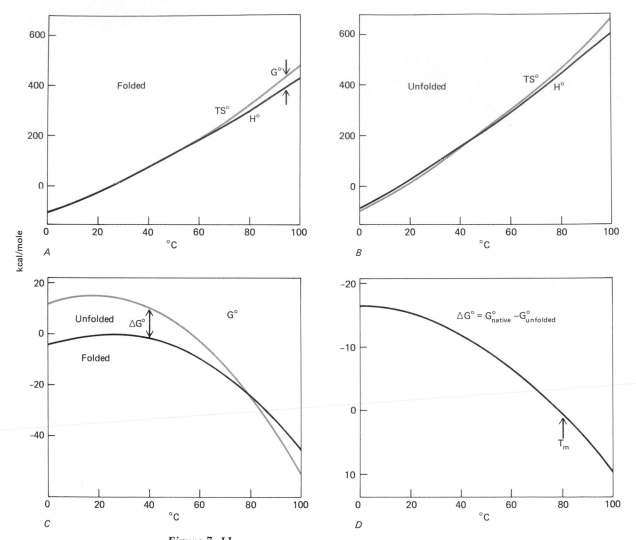

Figure 7–11

Thermodynamic parameters for the folded and unfolded forms of lysozyme at pH 7.0 and various temperatures in ° C. The enthalpic (H°) and entropic (T S°) contributions to the free energies of the folded and unfolded states are plotted as a function of temperature in A and B, respectively. The Gibbs free energies (G°) of both states are the differences in the enthalpic and entropic contributions, as indicated in A. These values for both the folded and unfolded states are plotted in C as a function of temperature. The net stability of the folded state is the difference between the free energies of the two states, as indicated in C; this value is plotted in D. Note the change in energy scale from A and B to C and then to D; the final ΔG° is a very small difference between the individual enthalpy and entropy contributions. (Data from W. Pfeil and P. L. Privalov, Biophys. Chem. 4:41–50, 1976.)

ing, individual factors. For example, both the enthalpies and entropies of both the folded and unfolded states are very temperature-dependent, and their energy contributions vary by up to 700 kcal/mole over the range 0° to 100° C (Figure 7–11A, B). Yet the enthalpy and entropy vary similarly

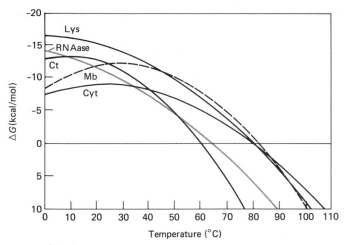

Figure 7–12
Temperature dependence of the difference in free energy between folded and unfolded states of proteins (per mole of protein). Lys, lysozyme; RNAase, ribonuclease A; Mb, metmyoglobin; Ct, α-chymotrypsin; Cyt, cytochrome c. The pH of the solution of each protein was that where it is most stable. (Adapted from P. L. Privalov and N. N. Khechinashvili, J. Mol. Biol. 86:665–684, 1974.)

and compensate each other, so that the free energy is a relatively small difference between the two, varying by only one-tenth the magnitude of the individual factors (Figure 7–11C). The free energies of the folded and unfolded states also vary similarly; thus the difference between them, the **observed net stability,** is no greater than 16 kcal/mole (Figure 7–11D). Consequently, accounting for the net stabilities of proteins in terms of the primary interactions is a hazardous accounting procedure, in which the net result is miniscule relative to the individual terms. An error of greater than 2 per cent in a term of 700 kcal/mole would obliterate the net difference of only 16 kcal/mole. In spite of this, the proteins studied have rather similar marginal net stabilities in their folded structures (Figure 7–12).

A thermodynamic approach to the problem of stabilization of globular protein structure: a calorimetric study. P. L. Privalov and N. N. Khechinashvili. J. Mol. Biol. 86:665–684, 1974.

Thermodynamic investigations of proteins. W. Pfeil and P. L. Privalov. Biophys. Chem. 4:23–32, 33–40, 41–50, 1976.

Stability of proteins. Small globular proteins. P. L. Privalov. Adv. Protein Chem. 33:167–241, 1979.

Micro- and macro-stabilities of globular proteins. P. L. Privalov and T. N. Tsalkova. Nature 280:693–696, 1979.

Thermodynamics of α-lactalbumin unfolding. W. Pfeil. Biophys. Chem. 13:181–186, 1981.

The problem of the stability of globular proteins. W. Pfeil. Mol. Cell Biochem. 40:3–28, 1981.

Nature of the Unfolded State

The unfolded states of proteins appear to approach the random-coil state described in Chapter 5 for disordered polypeptides. This is best established for proteins unfolded by urea and guanidinium chloride, where their hydrodynamic, spectral, and chemical properties are close to those expected. Less certain are acid- and heat-denatured states, where evidence for nonrandom conformation may usually be found, although this is disputed by calorimetric experiments. In any case, it would be surprising to find denaturing conditions that could abolish simultaneously all the possible types of intramolecular interactions possible in a chemically diverse polypeptide, as is required for a truly random-coil state. Nevertheless, small unfolded proteins are certainly grossly disordered, and the general flexibility of disordered polypeptides (Chapter 5) makes it unlikely that any "residual structure" from the folded state will be retained, although certain conformations may be slightly favored, owing to favorable intramolecular interactions.

The unfolded form of a protein is considered a discrete state, but it is important to remember that this form represents an enormous collection of different conformations. At any instant of time, each molecule in a population has a unique conformation, but it is not static and is constantly being altered by rapid rotation about single bonds. This is particularly relevant for the kinetics of unfolding and refolding.

Evidence for residual structure in acid- and heat-denatured proteins. K. C. Aune, et al. J. Biol. Chem. 242:4486–4489, 1967.
Influence of temperature on the intrinsic viscosities of proteins in random coil conformation. F. Ahmad and A. Salahuddin. Biochemistry 13:245–249, 1974.
Reactivities of the cysteine residues of the reduced pancreatic trypsin inhibitor. T. E. Creighton. J. Mol. Biol. 96:777–782, 1975.
Calorimetric study of lysozyme denaturation by guanidinine hydrochloride. W. Pfeil and P. L. Privalov. Biophys. Chem. 4:33–40, 1976.

Kinetic Pathways

The two-state nature of protein unfolding transitions at equilibrium is convenient for thermodynamic analysis but makes it difficult to decipher how it occurs, i.e., what pathway the polypeptide chain follows. The pathway of unfolding and refolding is defined by what intermediate conformational states (I) the molecule passes through.

$$U \rightleftharpoons I_1 \rightleftharpoons I_2 \rightleftharpoons \cdots \rightleftharpoons I_{n-1} \rightleftharpoons I_n \rightleftharpoons N \quad (7\text{–}11)$$

For a complex process like protein folding, there are bound to be a large number of intermediate states, although only a very small fraction of those theoretically possible.

The two-state nature of equilibrium folding transitions indicates that all the intermediate states are less stable than either U or N under all conditions, so that they are not populated by a significant fraction of molecules at equilibrium. Such unstable intermediates might accumulate to detectable levels as transient kinetic intermediates in refolding, upon suddenly placing the unfolded protein under conditions where it will fold, or vice versa in unfolding. However, intermediates need not accumulate sub-

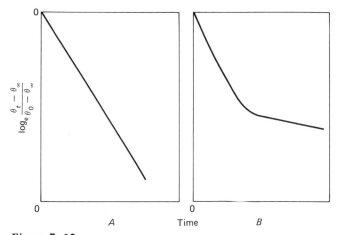

Figure 7–13
Examples of one-step (A) *and two-step, biphasic* (B) *reaction kinetics. A parameter of the extent of the reaction,* θ_t, *is plotted versus time;* θ_0 *is the initial value of the parameter,* θ_∞ *the final value. The slope of each curve gives the value of the apparent rate constant; if the reverse of a reaction is significant, the apparent rate constant is the sum of those of the forward and reverse steps.*

stantially, and only those preceding the rate-limiting step in the pathway may do so. For example, if the conversion of $I_2 \rightarrow I_3$ is the slowest step in the scheme just described, species I_3, I_4 . . . I_n will not accumulate during refolding. Intermediates I_1 and I_2 could accumulate to significant levels, but only if the rate constant for their formation is greater than both rate constants for their disappearance, forward to the next intermediate and backward to the preceding one. Consequently, crucial intermediates need not be observed kinetically.

With a simple, one-step reaction, the kinetics should be characterized by a single rate constant, k:

$$A \xrightarrow{k} B$$

$$\frac{-da}{dt} = \frac{db}{dt} = ka \tag{7–12}$$

where a and b represent the concentrations of reactant and product, respectively, and t is time. This differential equation may be readily solved:

$$\frac{da}{a} = -k\,dt$$

$$\ln \frac{a}{a_0} = -kt \tag{7–13}$$

where a_0 is the initial concentration.

A logarithmic plot of the time-dependent change in a (or b) relative to a_0 should be linear, with the slope giving the value of the rate constant (Figure 7–13A). If the observed curve is not linear but has multiple phases, a correspondingly greater number of reaction steps (and species) must be involved (Figure 7–13*B*).

Many kinetic studies of protein refolding (with disulfides kept intact) have detected complexities, with two or more phases, whereas the kinetics of unfolding are usually simple. The kinetic complexities in refolding were initially interpreted in terms of various types of intermediates, either on or off the main production pathway:

$$ U \rightleftharpoons I \longrightarrow N \quad \text{or} \quad \begin{matrix} I \\ \upharpoonleft\!\!\downharpoonright \\ U \end{matrix} \longrightarrow N \qquad (7–14) $$

However, thorough kinetic studies by Baldwin and colleagues showed that the fully refolded protein was the product of both slow and faster transitions; if the complexities were due to kinetic intermediates on the pathway, fully folded protein should be formed only at the rate of the slowest step. It is now clear that most such kinetic complexities are due to two or more populations of unfolded protein molecules that differ in their rates of refolding. The simplest kinetic scheme with fast (U_F) and slow (U_S) refolding species is

$$ U_S \underset{\text{slow}}{\overset{\text{slow}}{\rightleftharpoons}} U_F \overset{\text{fast}}{\longrightarrow} N \qquad (7–15) $$

If the initially folded protein, N, is rapidly unfolded and then immediately refolded, all the molecules refold rapidly, because only U_F was formed. However, if refolding is delayed sufficiently for $U_F \rightarrow U_S$, a fraction of the molecules refold slowly. Consequently, the interconversion of U_F and U_S is slow in both directions; at equilibrium, 80 per cent of the molecules of unfolded ribonuclease (with the four disulfides intact) refold slowly.

This is a consequence of the conformational heterogeneity of the unfolded protein (p. 30) and indicates that not all the conformations possible with the unfolded protein are rapidly interconverted. In this case, the slow transition appears to involve *cis–trans* isomerization of the peptide bonds adjacent to proline residues (see p. 163). Folded proteins have each such bond either *cis* or *trans* in all the molecules, but in the unfolded protein there is an approximately 80 per cent probability that each bond will be *trans*, 20 per cent *cis*. This interconversion is intrinsically slow, with a half-time of about one minute under normal conditions. Upon unfolding of the N protein, the unique collection of isomers slowly equilibrates to the equilibrium mixture. Only those molecules with all such bonds in an isomeric state compatible with the folded state (U_F) can refold rapidly, whereas the others (U_S) must change the incorrect isomers. The latter process appears not to be limited solely by *cis–trans* isomerization; rather, it is a complex process in which conformational forces within the protein participate. Molecules with certain incorrect isomers may fold to a native-like conformation, and then the incorrect isomer is corrected.

Slowly interconverted forms of unfolded proteins may be demonstrated directly by rapid urea-gradient electrophoresis (Figure 7–14). In

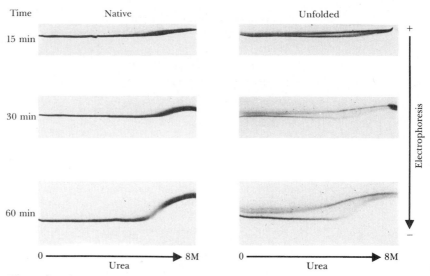

Figure 7–14

Slow- and fast-refolding forms of β-lactoglobulin demonstrated by urea-gradient electrophoresis. Folded (left) *and unfolded* (right) *forms of protein were subjected to electrophoresis (see Figure 7–6) rapidly at 2° C for the indicated periods of time. Starting with folded β-lactoglobulin, only rapid equilibration between N and U_F (fast-refolding) states is observed. Starting with unfolded protein, similar rapid equilibration of N and U_F forms is observed; however, a large fraction of molecules, U_S, do not refold but only equilibrate with compact forms at low urea concentrations. The difference between the U_F and U_S forms is not known in this case. (Adapted from T. E. Creighton, J. Mol. Biol. 137:61–80, 1980.)*

virtually all cases where a slow refolding form of a protein was apparent, it did not remain fully unfolded at low urea concentrations; instead, it rapidly equilibrated with more compact, but nonnative, conformations. Hydrogen atoms of the backbone are also protected from exchange with the solvent, indicating a nonrandom conformation. Moreover, folding occurs more rapidly under these conditions, suggesting the compact states have roles in refolding.

In contrast, the kinetic process of unfolding is not usually observed to be preceded by any significant transient partial unfolding; rather, it is an all-or-none type of process.

These observations suggest that the rate-limiting step is at a very late stage in refolding, and at a very early stage of unfolding, when the degree of folding is measured by the compactness of the molecule. The unfolded protein equilibrates rapidly with partially folded conformations when placed under refolding conditions. Rapid equilibration suggests that only a very small fraction of the many possible conformations are present. Such partially folded conformations are metastable and transient; consequently their natures are not known.

The highest energy transition state, which represents the rate-limiting step in both folding and unfolding, is close to the fully folded state and

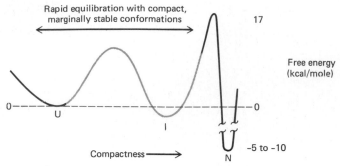

Rapid equilibration with compact, marginally stable conformations

17

Free energy (kcal/mole)

Compactness ⟶

-5 to -10

Figure 7–15
Proposed general scheme for the energetics of protein unfolding and refolding, under conditions favoring the folded state, and ignoring complications such as slow cis–trans isomerization of peptide bonds. The horizontal reaction coordinate is measured by the compactness of the polypeptide chain. The deep, narrow energy well of the folded state, N, of a depth in the range usually measured experimentally for small proteins, is envisaged to be that limiting its flexibility. The height of the energy barrier is that of the transition state for a protein that refolds with a rate of 1 sec⁻¹ (see text, p. 399, for how to calculate the energy of a transition state).

U is the unfolded state, which is a very large set of conformations of comparable energies, while I represents the compact, partially folded states observed to be rapidly equilibrated with U under refolding conditions. This energy surface is indicated by a thin line, as it will be very complex, but was drawn to indicate that all the observed conformational transitions have relatively low energy barriers.

probably is a perturbed variant of it. Its high energy probably reflects the cooperativity of the folded conformation, as its disruption requires simultaneous disruption of many stabilizing interactions.

Under optimal conditions, the U_F forms of small proteins can refold with half-times as short as 0.1 sec. Larger proteins may require a few minutes. An average rate constant for refolding might be 1/sec. The rate constant for spontaneous, transient unfolding under the same conditions can be estimated, since

$$U_F \underset{k_{\text{unfolding}}}{\overset{k_{\text{refolding}}}{\rightleftharpoons}} N$$

$$K = \frac{k_{\text{refolding}}}{k_{\text{unfolding}}} \qquad (7\text{–}16)$$

The value of the equilibrium constant, K, is known to be generally in the range of 10^3 to 10^{11} (p. 291); hence the value of $k_{\text{unfolding}}$ must be in the range of 10^{-3} to 10^{-11}/sec. Therefore, even under optimal conditions, the marginal stability of a folded protein requires that it have a finite probability of completely unfolding. Of course, this probability increases as conditions are made less favorable for stability.

These general observations lead to the general energetic scheme of Figure 7–15 for protein unfolding and refolding, under conditions for optimal stability. If correct, it indicates that refolding does not occur by a rate-limiting nucleation event in the unfolded protein, after which further refolding is rapid, analogous to α-helix formation (p. 187). Any such nucleation event occurs only in a very compact conformation. Nucleation events in the random coil could lead only to rapid equilibration between the compact, nonnative conformations.

Consideration of the possibility that the slow step in protein denaturation reactions is due to *cis–trans* isomerism of proline residues. J. F. Brandts, et al. Biochemistry 14:4953–4963, 1975.

Unfolding and refolding occur much faster for a proline-free protein than for most proline-containing proteins. J. F. Brandts, et al. Proc. Natl. Acad. Sci. U.S.A. 74:4178–4181, 1977.

Role of proline isomerization in folding of ribonuclease A at low temperatures. K. H. Cook, et al. Proc. Natl. Acad. Sci. U.S.A. 76:6157–6161, 1979.

Structural intermediates trapped during the folding of ribonuclease A by amide proton exchange. P. S. Kim and R. L. Baldwin. Biochemistry 19:6124–6129, 1980.

Kinetic study of protein unfolding and refolding using urea-gradient electrophoresis. T. E. Creighton. J. Mol. Biol. 137:61–80, 1980.

Specific intermediates in the folding reactions of small proteins and the mechanism of protein folding. P. S. Kim and R. L. Baldwin. Ann. Rev. Biochem. 51:459–489, 1982.

Mechanism for the unfolding and refolding of ribonuclease A. Simulations using a simple model with no structural intermediates. L.-N. Lin and J. F. Brandts. Biochemistry 22:573–580, 1983.

Trapping Intermediates

Kinetic intermediates are usually difficult to study because they are transient and often present only at rather low concentrations. For example, with a sequential reaction of three steps of equal rates, the two intermediates will accumulate only to maximum levels of 37 per cent and 27 per cent, respectively, of the population, the remainder being either the initial or final forms (Figure 7–16). However, if they could be trapped in a stable form, they could be isolated and characterized.

Such trapping is possible in protein folding using disulfides formed between cysteine residues. The redox nature of this interaction (pp. 22–24) makes it possible to control experimentally the rates of formation and breakage of disulfides. Most useful is thiol–disulfide exchange with a disulfide reagent, RSSR:

Reduced protein	Mixed disulfide	Protein disulfide

$$(7-17)$$

The first step is simply thiol–disulfide exchange between RSSR and a protein thiol group; the second is the intramolecular step between two protein Cys residues, which must come into appropriate proximity, reflecting the

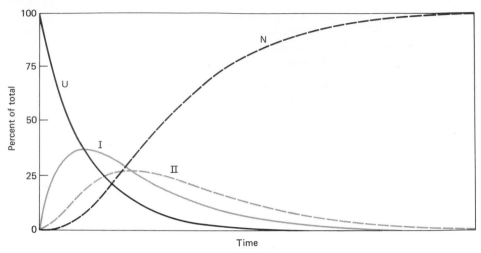

Figure 7–16
Kinetics of a three-step sequential reaction, $U \rightarrow I \rightarrow II \rightarrow N$, in which all three steps have the same rate constant. The intermediates accumulate to relatively low concentrations, and there is then always a complex mixture of species present. Only if one rate constant is much greater than the next does an intermediate accumulate to high levels.

conformational properties of the protein. The equilibrium and rates of interconversion between the reduced and disulfide forms of the protein may be controlled via the concentrations of RSH and RSSR. This interconversion may also be stopped by rapidly blocking all thiol groups present; whatever disulfides were present at that time are then trapped and are stable under appropriate conditions.

Many proteins that have disulfide bonds in their folded states, such as BPTI and ribonuclease, unfold when they are broken, even in the absence of a denaturant. If disulfide formation is then permitted, either by adding a disulfide reagent or by air oxidation, they may form the correct disulfides and refold to a native-like conformation. The process by which this occurs can be determined by trapping the protein molecules with different numbers of disulfides present at different times, separating them (Figure 7–17), and determining their disulfide bonds (see Figure 1–11) and other conformational properties.

The main pathway elucidated for BPTI in this way is shown schematically in Figure 7–18. The reduced protein initially makes the first disulfide bond among any of the possible pairs of the six Cys residues, reflecting its unfolded conformation, but the initial disulfide is rapidly rearranged by intramolecular thiol–disulfide exchange to favor the two predominant single-disulfide intermediates: one with a disulfide between Cys 30 and Cys 51, designated (30-51); the other, (5-30). The former predominates and has a disulfide that is present in native BPTI; the second most stable intermediate has a nonnative disulfide. Their favorable stabilities result from conformational interactions determined by the primary

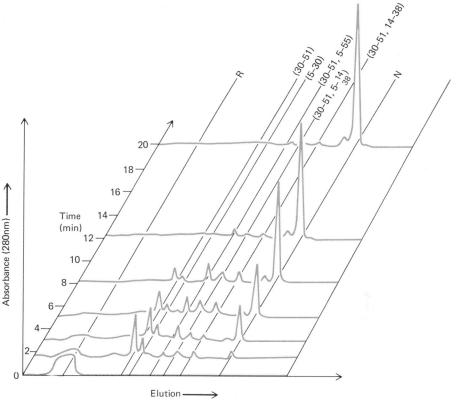

Figure 7–17

Isolation of intermediates trapped during refolding of reduced BPTI. Refolding and disulfide formation were initiated by addition of glutathione disulfide and stopped at the indicated times by addition of iodoacetate to 0.1 M, which rapidly reacts with all thiol groups, converting free Cys residues to acidic groups. Protein molecules with different numbers of disulfide bonds then differ in their net charge and are readily separated by ion-exchange chromatography, on carboxymethyl cellulose in this instance. BPTI molecules are eluted roughly in order of their content of disulfide bonds, although molecules with the same number, but different pairings, are also resolved, as the charge distribution on the protein molecule is also important in its binding to the resin.

Fully reduced BPTI, R, was present initially and has largely disappeared after 1.5 minutes, when one-disulfide intermediates are near their maximum levels. Two-disulfide intermediates accumulate more slowly, reaching their maximum levels after about 5 minutes. The major intermediates are identified by the disulfides they contain. Refolded BPTI, N, eventually predominates. The small peak preceding N contains incorrectly folded molecules, with wrong disulfide bonds. (Adapted from T. E. Creighton, J. Mol. Biol. 95:167–199, 1975.)

structure and the environment. The other two native disulfides, 14-38 and 5-55, are not present in the single-disulfide intermediates at significant concentrations. Because all further intermediates retain the 30-51 disulfide, probably only (30-51) readily forms second disulfides; 14-38, 5-14, and 5-38 are formed at comparable rates. Only the first is a native-like disulfide. None of the resulting two-disulfide species can readily form third

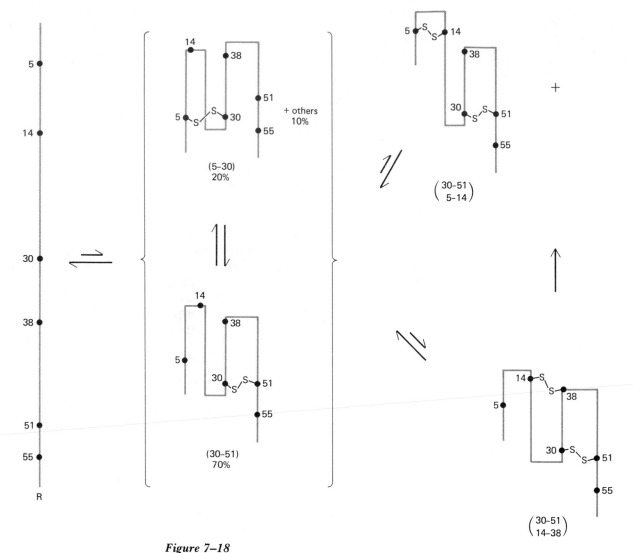

Figure 7–18

The pathway of unfolding and refolding of BPTI that accompanies disulfide bond breakage and formation. The polypeptide chain of 58 residues is shown schematically as a solid black line, with the positions of the six Cys residues indicated. Reduced BPTI, R, is unfolded and is depicted as a straight chain. The conformations of species N_S^S and N_{SH}^{SH} are both meant to represent the native conformation (Figures 6–9, 6–10, and 6–12). The residue numbers of the pairs of Cys residues involved in disulfide bonds are indicated below each of the intermediates. Because the one-disulfide intermediates are in rapid equilibrium, they are enclosed in brackets; only the two most predominant species are indicated, along with their relative proportions under standard conditions. The "+" between intermediates (30-51, 5-14) and (30-51, 5-14) indicates that they both have the kinetic role indicated. (Adapted from T. E. Creighton, J. Mol. Biol. 113:275–293, 1977.)

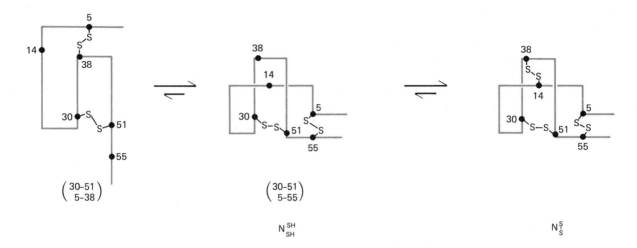

disulfides, even though (30-51, 14-38) has two native-like disulfides. Instead, they complete refolding by rearranging the second disulfides to give the intermediate (30-51, 5-55), which adopts a stable native-like folded conformation. Cys 14 and Cys 38 are now in correct proximity and can rapidly form a disulfide to complete refolding.

The pathway of unfolding upon disulfide reduction is just the reverse. The 14-38 disulfide bond is on the surface of the molecule, and its reduction neither requires nor causes any conformational changes. The resulting two-disulfide species (30-51, 5-55) is quite stable in the native-like conformation and unfolds further only by rearranging the 5-55 disulfide bond to either 5-14 or 5-38. This step occurs quite slowly under normal conditions and represents a very extreme example of protein flexibility: It must occur by thiol–disulfide exchange between the thiol group of either Cys 14 or Cys 38, at the top of the pear-shaped folded protein (see Figure 6–9), and the 5-55 disulfide at the other, bottom end of the molecule.

Therefore, BPTI does not unfold or refold by simple, sequential breakage or formation of its three disulfides; rather, it goes through complex rearrangements at the two-disulfide stage. The energetics of the folding transition provide a possible rationalization (Figure 7–19). The rate-limiting intramolecular step is the two-disulfide rearrangement; formation and breakage of the disulfides occurs much more rapidly. The rate-limiting barrier separates the native-like species from all the other, much more unfolded species, as the intermediate (30-51, 5-55) is essentially only a minor variant of the folded state. It is this energy barrier separating the fully folded state from all others that prevents intermediates (30-51) and (30-51, 14-38) from forming directly the 5-55 disulfide, even though, or, more appropriately, because, the folded state would result. This barrier results from the need to go through a high-energy, perturbed form of the native conformation, exactly analogous to the general energetic scheme of Figure 7–15.

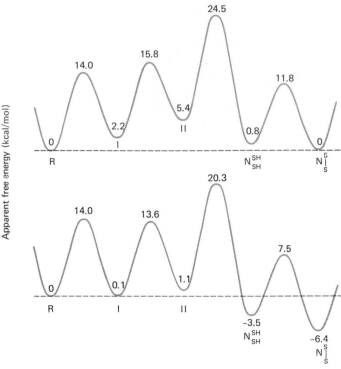

Figure 7–19
Energetics of the folding transition of BPTI. At the top, the stabilities of the three disulfide bonds of fully folded N_S^S were such as to give it the same free energy as the fully reduced state, R. This illustrates the cooperativity of the folding process, as all the partially folded intermediates have higher energies and are not populated substantially at equilibrium. At the bottom, the stability of each disulfide bond is increased by about 2.1 kcal/mole, making the fully folded state, 6.4 kcal/mole, more stable, a somewhat arbitrary value comparable to that observed with small proteins with no disulfide bonds. The energy of a disulfide bond relative to two thiol groups depends upon the redox potential of the environment and has no fixed value.

The energies of the transition states were determined (see text, p. 399) from the rate constants for the intramolecular steps in forming or rearranging disulfide bonds.

I represents all the one-disulfide intermediates; II, the two-disulfide intermediates (30-51, 5-14), (30-51, 5-38), and (30-51, 14-38), while N_{SH}^{SH} is the native-like two-disulfide intermediate (30-51, 5-55), all depicted in Figure 7–18. The placing of intermediates along the reaction coordinate is in order of their kinetic roles, but the spacing is arbitrary. If the spacing were according to compactness, as in Figure 7–15, the intermediates I and II would be approximately midway between R and N_S^S, while N_{SH}^{SH} is indistinguishable in this way from N_S^S. (Adapted from T. E. Creighton, J. Mol. Biol. 113:295–312, 1977.)

Similar considerations apply in the folding transition of ribonuclease and are likely for other proteins. Yet a detailed explanation of the BPTI pathway, which is the only one known, in terms of conformation and intramolecular interactions is not yet possible, owing to the difficulty in obtaining such information about the rather flexible trapped intermediates in solution. Nevertheless, this type of approach seems the only one capable of elucidating complex folding pathways, and further details should be forthcoming.

The reader should be cautioned that this description of protein folding is based largely on the author's own work and is not universally accepted.

The kinetics of formation of native ribonuclease during oxidation of the reduced polypeptide chain. C. B. Anfinsen, et al. Proc. Natl. Acad. Sci. U.S.A. 47:1308–1314, 1961.

Pathways of folding of reduced bovine pancreatic ribonuclease. R. R. Hantgan, et al. Biochemistry 13:3421–3431, 1974.

Intermediates in the refolding of reduced pancreatic trypsin inhibitor. T. E. Creighton. J. Mol. Biol. 87:579–602, 1974.

Conformational restrictions on the pathway of folding and unfolding of the pancreatic trypsin inhibitor. T. E. Creighton. J. Mol. Biol. 113:275–293, 1977.

Energetics of folding and unfolding of pancreatic trypsin inhibitor. T. E. Creighton. J. Mol. Biol. 113:295–312, 1977.

Kinetics of refolding of reduced ribonuclease. T. E. Creighton. J. Mol. Biol. 113:329–341, 1977.

Intermediates in the refolding of reduced ribonuclease. T. E. Creighton. J. Mol. Biol. 129:411–431, 1979.

Role of the environment in the refolding of reduced pancreatic trypsin inhibitor. T. E. Creighton. J. Mol. Biol. 144:521–550, 1980.

Folding of Complex Proteins

The preceding discussion considered only single-domain proteins; those with multiple domains or composed of subunits are clearly more complex. In some cases, individual domains have been excised proteolytically from the protein and shown to retain a folded conformation, which can be unfolded and refolded. Domains are also observed to unfold independently of the others in a protein, which produces complex unfolding transitions for such proteins. Therefore, individual domains appear to fold and then to associate. Similarly, individual subunits of oligomeric proteins are observed to fold and then to aggregate. However, the exact nature of the initial folding of a domain or a subunit alone is not clear, since some conformational adjustments occur upon association of domains or of subunits.

Analysis of the reconstitution of oligomeric enzymes by cross-linking with glutaraldehyde: kinetics of reassociation of lactic dehydrogenase. R. Hermann, et al. Biochemistry 20:5195–5201, 1981.

Folding and association of proteins. R. Jaenicke. Biophys. Struct. Mech. 8:231–256, 1982.

Stability of proteins. Proteins which do not present a single cooperative system. P. L. Privalov. Adv. Protein Chem. 35:1–104, 1982.

Biosynthetic Folding

Because proteins are synthesized on ribosomes in vivo as linear polypeptide chains, they must fold to their final conformations either during or after biosynthesis; however, virtually nothing is known directly about how or when it happens. Since the polypeptide chain is assembled unidirectionally, starting from the amino end (Chapter 2), it was attractive to envisage the initial amino-terminal fragment to fold into a nucleus of fixed conformation, which then directed folding of the remainder of the polypeptide chain as it was assembled. However, any such asymmetric folding cannot be a general kinetic requirement, since intact, complete polypeptides can refold readily. It is also unlikely that an incomplete nascent polypeptide adopts any stable conformation until that sufficient for an entire structural domain has been synthesized, because fragments of such domains are generally observed to be very disordered (see subsequent section, Cleavage of the Polypeptide Chain). Removal of just a few carboxyl-terminal residues, the last to be added to a nascent chain, is often sufficient to produce unfolding of small, single-domain proteins.

These generalizations about protein folding in vivo are extrapolations from in vitro studies and assume that folding is strictly a self-assembly process. There is no need to invoke special factors to explain folding in vivo, since intact proteins can fold by themselves in vitro, but their existence in the biosynthetic apparatus cannot be ruled out.

Additional complexities are introduced by the many instances of covalent modifications of the polypeptide chain after its biosynthesis (Chapter 2). A few of these, such as glycosylation, probably occur on the nascent chain, before folding, and might even be required for folding to occur. Many modifications occur only after folding, and the folded conformation is usually important for the specificity of the modifications. In such cases, the covalent modification of the folded protein could be imagined to affect, or even inhibit, refolding of the polypeptide chain. However, it must be remembered that all proteins studied have finite rates of spontaneous complete unfolding, which would be irreversible if any modification interfered with refolding. Such a phenomenon could be useful for turnover of protein molecules (Chapter 10), but proteins that are stable for long periods of time must be able to refold spontaneously.

One posttranslational covalent modification that is often important for folding is disulfide bond formation between Cys residues. This process has been studied extensively in vitro (see preceding section), but how it occurs normally after biosynthesis is not known, nor is even the nature of the oxidizing agent. It may occur by thiol–disulfide exchange with small-molecule disulfides, of which glutathione is the most prevalent in cells. The great majority of glutathione is present intracellularly as the reduced, thiol form, rather than the disulfide. Other such disulfides in the same environment are likely to be present in the same redox equilibrium, since thiol–disulfide exchange is rapid, and enzymes are also present to catalyze such reactions. Consequently, the cell cytoplasm is often considered to be too reducing to permit protein disulfide formation. However, this is not the case when the protein conformation brings two Cys residues into proximity; under such conditions, reduced BPTI and lysozyme will fold and form disulfides in vitro. On the other hand, where the protein conformation keeps Cys residues apart, they should remain largely in the thiol form.

Relatively few intracellular proteins are known to form disulfides; most proteins that form disulfides are normally extracellular. This state of affairs is not unreasonable, as a disulfide bond will provide more stability in a more oxidizing environment. Such proteins are normally secreted during biosynthesis; disulfide formation, therefore, probably occurs outside the cytoplasm, possibly within the endoplasmic reticulum (see Figure 2–10).

Rearrangement of disulfide bonds is often important during folding of such proteins in vitro, and there is an enzyme capable of catalyzing it found in all cells examined. When isolated, this enzyme increases the rates of all steps that involve both disulfide interchange and substantial conformational alterations of the protein. It is found in microsomes, membranes to which ribosomes are attached, synthesizing secreted proteins. This is the logical place for it to be to function in vivo, but there is no other direct information about its physiological role.

Disulfide bonds are stable under appropriate conditions, and they may stabilize a folded conformation that is no longer thermodynamically stable, owing to covalent modification after biosynthesis and folding. For example, many hormones and enzymes, such as insulin and chymotrypsin, are synthesized initially as inactive precursors that fold and form disulfide bonds and are then activated by proteolytic cleavage. The active forms are stable under most conditions, but if disulfide interchange is permitted, these proteins will rearrange their disulfides, unfold, and become inactive. In these particular instances, the original disulfides are stabilizing an inherently unstable folded conformation, although this need not always be the case.

Disulfide interchange and the three-dimensional structure of proteins. D. Givol, et al. Proc. Natl. Acad. Sci. U.S.A. 53:676–684, 1965.
Formation of an intrachain disulfide bond on nascent immunoglobulin light chains. L. W. Bergmann and W. M. Kuehl. J. Biol. Chem. 254:8869–8876, 1979.
Catalysis by protein–disulphide isomerase of the unfolding and refolding of proteins with disulphide bonds. T. E. Creighton, et al. J. Mol. Biol. 142:43–62, 1980.
A role for oligosaccharides in glycoprotein biosynthesis. R. Gibson, et al. Trends Biochem. Sci. 5:290–293, 1980.
Conformational changes associated with proteolytic processing of presecretory proteins allow glutathione-catalyzed formation of native disulfide bonds. G. Scheele and R. Jacoby. J. Biol. Chem. 257:12277–12282, 1982.

STABILITY OF THE FOLDED STATE

There are two aspects to the problem of understanding the stability of a protein folded conformation: The first is to account for its net stability relative to the unfolded state. The second is to explain why a protein adopts its particular folded conformation, and not some other. The first aspect is considered here; the second, in the last section of this chapter.

Both aspects depend upon the primary structure, and it is this relationship that will be examined. Stability relative to the unfolded state was discussed in the previous section in terms of its dependence on the environment, but variation of the protein should be a more direct way of investigating its physical basis. The covalent structures of proteins may be modified in a great number of ways (Chapters 1 and 2); in addition, nature has provided a great wealth of evolutionary and mutational variations of amino acid sequences producing very similar folded conformations (Chapter 3), but often with varying stabilities. For example, proteins produced by thermophiles, organisms that grow at high temperatures, usually have greater thermal stabilities than those of normal organisms. Proteins from halophiles, organisms that grow at high salt concentrations, usually tolerate or even require such conditions.

No comprehensive study of stability of the folded state of one protein as a function of alteration of its primary structure has been carried out, for the scope of such a study would be immense. Nevertheless, some data are available, and we will attempt to account for protein stability semi-quantitatively.

Life at high temperatures. T. D. Brock. Science 158:1012–1019, 1967.
Salt-dependent properties of proteins from extremely halophilic bacteria. J. K. Lanyi. Bacteriol. Rev. 38:272–290, 1974.

Variation of the Primary Structure

The amino acid side chains at the surfaces of proteins often appear to have relatively little role in stabilizing the folded conformation. These are the residues that change most freely during evolution, indicating fewer constraints on their roles. Surface groups may also be modified extensively, without disrupting the folded conformation or preventing refolding. There are exceptions: Some surface charged groups appear to participate in energetically important salt bridges; surface groups are also important for the solubility of the folded protein and are those responsible for the protein's usual biological role in interacting with other molecules (Chapter 8).

The stabilities of various collections of proteins to unfolding by a variety of denaturing conditions have been found to correlate with different parameters, such as the surface area buried in the folded state, the number of hydrogen bonds or salt bridges, but no simple, general rules have emerged. All these factors are undoubtedly important for stability, and no single one would be expected to predominate. Many of these comparisons have been complicated by the multiplicity of differences in primary structure; mutant forms differing at only single residues should be much more informative. In the relatively few mutant proteins studied, the mutations have produced a wide variety of effects on net stability—it being unchanged or either somewhat increased or decreased. The folded structures of these mutant proteins have not been altered substantially by the change in amino acid, but the physical basis of the change in stability has usually not been obvious. Even where there is little change in net free energy of stability, the enthalpy and entropy changes upon unfolding may both be altered substantially, but in compensating directions. The problem is clearly a complicated one.

The greatest collection of individual mutant forms of a protein is that

Table 7–5   Causes of Instability in Abnormal Hemoglobins

	NUMBER OF OCCURRENCES
Loss of nonpolar "plug" that normally seals the protein surface	16
Loss of hydrogen bonds or salt bridges	7
Introduction of interior charge or dipole	5
Introduction of interior gap	6
Introduction of wedge between helices	6
Misfit at subunit contact	4
Introduction of Pro into α helix	13
Unclear	2
Introduction of side chain at position of invariant Gly in β-bend	1
Deletions	11
Other causes	2
Total	73

From G. Fermi & M. F. Perutz, *Haemoglobin and Myoglobin,* Oxford, Clarendon Press, 1981.

of the more than 300 known human hemoglobin variants. Detailed studies of their stabilities have not been carried out, but the 40 or so that are markedly unstable are very apparent from their adverse effects on the health of the individual, since precipitation of the unfolded protein produces erythrocyte lysis. The single amino acid changes in these cases suggest that the mutant hemoglobins have lower stabilities due to any of several causes: alteration of close-packing of the interior; the loss of one group that normally participates in a hydrogen bond or salt bridge; the introduction of a charged or polar group into the interior; or the insertion into a helical region of a Pro residue, which must distort the α-helix (Table 7–5). In contrast, other, sometimes radical changes of surface groups—for example, even introduction of a nonpolar side chain—have no great effect on stability.

Mechanism of denaturation of haemoglobin by alkali. M. F. Perutz. Nature 247:341–344, 1974.

Reversible thermal unfolding of thermostable phosphoglycerate kinase. Thermostability associated with mean zero enthalpy change. H. Nojima, et al. J. Mol. Biol. 116:429–442, 1977.

Relationship of protein thermostability to accessible surface area. E. Stellwagen and H. Wilgus. Nature 275:342–343, 1978.

Reversible thermal unfolding of thermostable cytochrome *c*-552. H. Nojima, et al. J. Mol. Biol. 122:33–42, 1978.

Molecular basis of thermostability in the lysozyme from bacteriophage T4. M. G. Grütter, et al. Nature 277:667–669, 1979.

Heat stability of a tetrameric enzyme, D-glyceraldehyde-3-phosphate dehydrogenase. J. E. Walker, et al. Eur. J. Biochem. 108:581–586, 1980.

Effect of single amino acid substitutions on the thermal stability of the α subunit of tryptophan synthetase. C. R. Matthews, et al. Biochemistry 19:1290–1293, 1980.

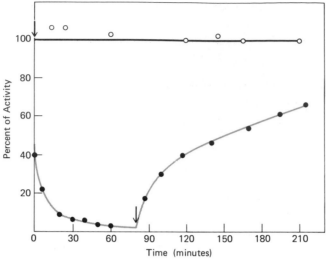

Figure 7–20

Inactivation and disulfide interchange of ribonuclease S-protein by the enzyme protein disulfide isomerase. At the times indicated by the arrows, 1.3 equivalents of S-peptide were added. The S-peptide prevents inactivation if present initially by forming the stable ribonuclease S complex, and it can reverse the disulfide rearrangement of S-protein, presumably by pulling the equilibrium toward the form with the folded conformation and the correct disulfides. (Adapted from I. Kato and C. B. Anfinsen, J. Biol. Chem. 244:1004–1007, 1969.)

Effect on protein stability of reversing the charge on amino groups. M. Hollecker and T. E. Creighton. Biochim. Biophys. Acta 701:395–404, 1982.

Mutations and protein stability. J. A. Schellman, et al. Biopolymers 20:1989–1999, 1981.

Protein thermostability. D. J. Merkler, et al. Int. J. Pept. Protein Res. 18:430–442, 1981.

Cleavage of the Polypeptide Chain

Fully folded structural domains of proteins are usually relatively resistant to proteolytic enzymes. However, the polypeptide segments linking independent structural domains are often flexible and susceptible to cleavage, providing a diagnostic test for the presence of such domains and a method for their separation and isolation. The separated domains often unfold and refold like normal small single-domain proteins.

Exposed flexible polypeptide loops on the surfaces of folded proteins may also be cleaved proteolytically; the folded conformation is often still stable. The classic example is ribonuclease S, in which the peptide bond between residues 20 and 21 is cleaved; the two fragments are held together by noncovalent forces, with only local alterations of the conformation. The cleaved protein unfolds more readily than normal, and the two fragments then dissociate.

Such fragments of single-domain proteins generally have no stable conformation individually, even under conditions favoring folding. The S-peptide fragment of ribonuclease S-residues 1 to 20 is largely unfolded, with only a limited tendency to adopt the helical conformation it has in the folded complex. The other fragment, the S-protein, has the four correct disulfide bonds and retains a folded conformation with some similarities to native ribonuclease. However, it unfolds if disulfide interchange is permitted (Figure 7–20), when the disulfides are "scrambled."

Many such pairs of fragments will reassemble to generate the original noncovalent complex, in a complicated, little-understood process involving both folding and recognition of the two parts. For example, addition of S-peptide to scrambled S-protein can regenerate ribonuclease S and the four correct disulfides (Figure 7–20). By testing various combinations of peptides, such systems have been used to investigate which portions of the primary structure are required for a stable folded conformation. Generally, virtually all the original residues must be represented at least once in the two fragments. Folded complexes can occur if a segment of residues is present in both fragments; the redundant segments can then be removed proteolytically, indicating that only one of the two repeated segments, often from either of the two fragments, is involved in the folded conformation (Figure 7–21). However, not all such pairs of fragments, even if containing all the original residues, will yield stable conformations. In these cases, cleavage greatly destabilizes the folded conformation; different covalent bonds of the polypeptide chain apparently differ widely in their energetic importance for stability.

Within the folded conformation of a cleaved polypeptide chain, the newly generated amino and carboxyl ends may still be in sufficient proximity to regenerate the original peptide bond. The equilibrium constant may be sufficiently unfavorable for bond hydrolysis that even proteolytic enzymes may be used to catalyze peptide bond formation—the reverse of the normal reaction. Such a phenomenon is utilized by protease inhibitors, such as BPTI, which bind tightly at the active site of proteolytic enzymes, without being degraded; hydrolysis of the peptide bond is readily reversible, with equilibrium constants in the region of unity. Peptide bond reformation may also be induced chemically; for example, the homoserine lactone generated at the carboxyl end upon cyanogen bromide cleavage at Met residues (p. 48) is sufficiently reactive that a newly generated amino group held in proximity will react with it. Such procedures should make possible the preparation of semisynthetic proteins, in which one fragment has been modified specifically or synthesized de novo, a short-cut to the production of specific protein variants.

In many cases, a few residues can be removed from the ends of a polypeptide chain with retention of a stable folded conformation. However, at some point, depending upon the protein, the stability disappears, and the remaining polypeptide chain is unfolded. Removal of carboxyl-terminal residues seems most deleterious with the few proteins studied in this way.

On the stabilization of ribonuclease S-protein by ribonuclease S-peptide. I. Kato and C. B. Anfinsen. J. Biol. Chem. 244:1004–1007, 1969.

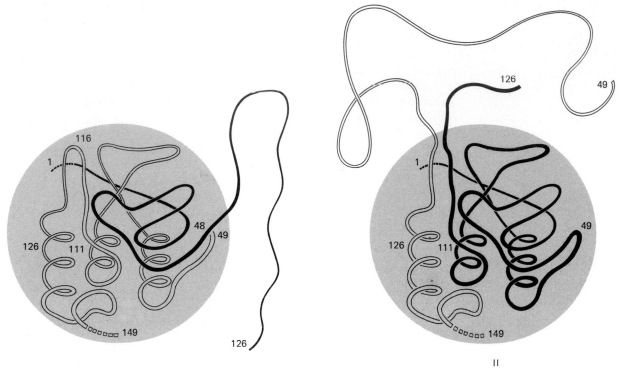

Figure 7–21
Schematic representation of the two types of complementation formed by fragments of residues 1 to 126 (dark ribbon) and 49 to 149 (light ribbon) of staphylococcal nuclease. The normal enzyme consists of 149 residues. In complex I, residues 49 to 116 of the folded structure are provided by fragment (49-149), while in II they are provided by fragment (1-126). In both cases, the redundant fragments may be cleaved off by proteases, leaving a stable complex. The conformation of the ordered complex, enclosed by the circle, is assumed to be like that of the native enzyme. (Adapted from H. Taniuchi and C. B. Anfinsen, J. Biol. Chem. 246:2291–2301, 1971.)

Formation of randomly paired disulfide bonds in des-(121-124) ribonuclease after reduction and reoxidation. H. Taniuchi. J. Biol. Chem. 245:5459–5468, 1970.

The dynamic equilibrium of folding and unfolding of nuclease T'. H. Taniuchi. J. Biol. Chem. 248:5164–5174, 1973.

Spontaneous re-formation of a broken peptide chain. D. F. Dykes, et al. Nature 247: 202–204, 1974.

Protein complementation. I. Zabin and M. R. Villarejo. Ann. Rev. Biochem. 44:295–314, 1975.

Enzymatic resynthesis of the hydrolyzed peptide bonds in ribonuclease S. G. A. Homandberg and M. Laskowski, Jr. Biochemistry 18:586–592, 1979.

Protein inhibitors of proteinases. M. Laskowski, Jr., and I. Kato. Ann. Rev. Biochem. 49:593–626, 1980.

A salt bridge stabilizes the helix formed by isolated C-peptide of RNase A. A. Bierzynski, et al. Proc. Natl. Acad. Sci. U.S.A. 79:2470–2474, 1982.

Conformational Equilibria in Polypeptide Fragments

A polypeptide chain will appear from its average properties to be a random coil simply if no conformation is populated by a substantial fraction of the molecules. Consequently, a seemingly random-coil polypeptide could have some particular conformation populated by, say, 10^{-2} of its molecules, whereas this value would be expected in a truly random coil to be populated much less, perhaps 10^{-n}, where n is the number of amino acid residues. The actual value is important for understanding the energetics of polypeptide conformation. Such conformational equilibria are usually measured by K_{conf}, the equilibrium constant between the particular, folded conformation of interest, F, and all other conformations, U:

$$U \overset{K_{conf}}{\rightleftharpoons} F$$

$$K_{conf} = \frac{[F]}{[U]} \tag{7-18}$$

Relatively small values of K_{conf} may be measured immunochemically, using the conformational specificity of antibodies against proteins. Antibodies obtained against the native protein are specific for the native conformation, at least for the unique configuration of groups on its surface that is recognized by the antibody. Antibodies directed against unfolded proteins are quite different, recognizing many linear segments of the polypeptide chain in very many different conformations. The cross-reaction between the two sets can be used to measure the probability that an unfolded polypeptide is in the native-like conformation, and therefore is bound to antibodies against that conformation, or that the folded protein is unfolded and is bound to antibodies against the unfolded protein.

Consider a polypeptide chain that possesses all the parts recognized by antibodies against the folded protein, N, but that is in equilibrium with other conformations, e.g., U. If this polypeptide is in the folded conformation, like N, a reasonable assumption is that it will bind to the anti-N antibodies with the same affinity, K_N, as that of the folded protein:

$$U \overset{K_{conf}}{\rightleftharpoons} N \underset{Ab}{\overset{K_N}{\rightleftharpoons}} Ab \cdot N$$

$$K_N = \frac{[Ab \cdot N]}{[N][Ab]} \tag{7-19}$$

where Ab is an antibody-combining site. The observed affinity of native, folded protein, essentially always in the N conformation, for Ab is K_N, while that for the polypeptide involved in the conformational equilibrium will be

$$K_{app} = \frac{[Ab \cdot N]}{([U] + [N])[Ab]} = \frac{[Ab \cdot N]}{\left(1 + \dfrac{1}{K_{conf}}\right)[N][Ab]} = \frac{K_N}{\left(1 + \dfrac{1}{K_{conf}}\right)} \tag{7-20}$$

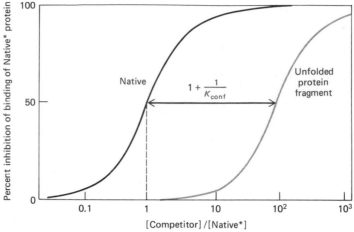

Figure 7–22

Immunochemical measurement of the equilibrium between U and N conformations in a protein fragment using antibodies recognizing only the N conformation. A constant amount of radioactive folded protein, designated Native*, *is mixed with varying amounts of the nonradioactive protein or fragment to be tested and is added to a limiting amount of antibodies recognizing the N conformation. Binding of radioactivity to the antibodies is measured; competition by the unlabeled, competing protein for the antibodies is reflected in inhibition of binding of radioactive protein.*

The black curve is that expected with unlabeled protein in the folded conformation, indistinguishable from the radioactive protein; an equivalent amount of radioactive and nonradioactive proteins should produce 50 per cent inhibition. The gray curve is that expected for a polypeptide that is in the N conformation only 0.01 per cent of the time, i.e., $K_{conf} = 10^{-2}$. In general, such a curve is offset to the right by the factor $(1 + 1/K_{conf})$.

Therefore, the affinity of the polypeptide will be lower by the factor $[1 + (1/K_{conf})]$. If K_{conf} is very small, this factor becomes $1/K_{conf}$.

The actual values of the apparent affinities may be measured, but antibodies elicited in the normal manner are very heterogeneous, with a wide range of affinities. Monoclonal antibodies minimize this problem; however, only relative values of affinities are required, not the actual values. Therefore, the competition of the folded protein and the polypeptide for a limiting concentration of antibodies is more readily measured, as in radioimmunoassay. For example, the folded protein can be radiolabeled and its displacement from the antibodies measured by varying concentrations of unlabeled polypeptide (Figure 7–22).

Values of K_{conf} measured with a few unfolded proteins or fragments have been in the region of 10^{-3} to 10^{-4}, consistent with their generally unfolded state, but considerably larger than might be expected for random occurrence of a unique conformation in a fully unfolded polypeptide chain.

An immunologic approach to the conformational equilibria of polypeptides. D. H. Sachs, et al. Proc. Natl. Acad. Sci. U.S.A. 69:3790–3794, 1972.

An immunological approach to the conformational equilibrium of staphylococcal nuclease. B. Furie, et al. J. Mol. Biol. 92:497–506, 1975.

Immunochemical analysis of the conformational properties of intermediates trapped in the folding and unfolding of bovine pancreatic trypsin inhibitor. T. E. Creighton, et al. J. Mol. Biol. 123:129–147, 1978.

Precise determination of the entire antigenic structure of lysozyme. M. Z. Atassi. Immunochemistry 15:909–936, 1978.

The antigenic structure of myoglobin and initial consequences of its precise determination. M. Z. Atassi. Crit. Rev. Biochem. 6:337–369, 1979.

Basic principles of antigen–antibody reactions. E. A. Kabat. Methods Enzymol. 70:3–49, 1980.

Stabilities of Folded Conformations

How can we understand the observed net stabilities of the folded states of proteins, their disruption by some modifications and not others, the cooperativity of the unfolding transition, and the values of K_{conf} measured with protein fragments? Because there is no single unifying theory of protein stability, a rigorous explanation is not possible. An empirical approach will be adopted here, that introduced in Chapter 4 for considering intramolecular interactions. This should suffice to demonstrate that the folded conformations of proteins result from a set of weak interactions that operate simultaneously to give a mutually reinforcing structure.

The main factor that will tend to keep a polypeptide chain unfolded is believed to be the large conformational entropy that is gained, owing to the large number of conformations possible. In addition, there are many favorable interactions with the solvent in the unfolded state, as well as the absence of stereochemically unfavorable interactions that are inevitable in a specific folded conformation. All these factors must be overcome energetically to give a stable folded state; they can be included in a term $K_{conf,U}$, which is the equilibrium constant for forming a specific conformation in the absence of all stabilizing interactions. It is expected to be a very small number for a polypeptide chain.

Consider an unfolded polypeptide chain in which there are two groups, A and B, capable of interacting favorably, as in a hydrogen bond, a salt bridge, or a nonpolar hydrophobic interaction.

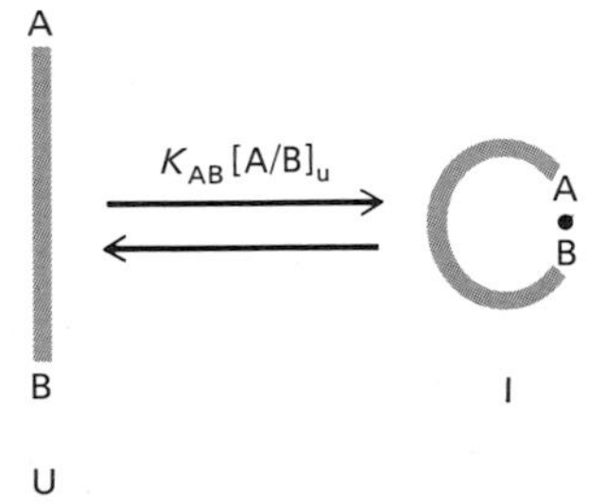

$$(7\text{–}21)$$

The interaction between two such groups on separate small molecules, but under the same conditions, is measured by an association constant, K_{AB}. For weak interactions of the type present in proteins, in aqueous solution, values of K_{AB} are in the range of 10^{-2} to 0.4 M^{-1} (Table 4–5); these values include the effects of interaction of groups A and B individually with water. When attached to the same molecule, the equilibrium constant for interaction of the two groups is

$$K_{obs,U} = K_{AB} \, [A/B]_U \qquad (7-22)$$

where $[A/B]_U$ is the effective concentration of the two groups relative to each other on the unfolded polypeptide, U. Groups attached to moderate-sized random polypeptides are expected to have effective concentrations in the region of 10^{-2} to 10^{-5} M, depending upon their relative positions (Table 5–1); thus values for $K_{obs,U}$ of between 4×10^{-3} and 10^{-7} are expected. Consequently, the presence of just a single weak interaction between two groups can give conformational equilibrium constants comparable to those measured immunochemically with protein fragments.

Multiple interactions between two or more pairs of groups attached to the same molecule are not independent, but they can assist or interfere with each other. With two pairs of groups on a polypeptide, the following equilibria are possible:

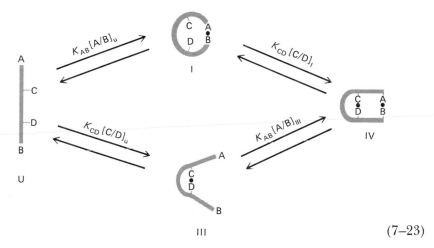

$$(7-23)$$

If both interactions are possible simultaneously, the presence of one interaction will increase the effective concentration of the other two groups, in a mutual manner, since the thermodynamics of cyclic equilibria require that the two be linked:

$$\frac{[A/B]_{III}}{[A/B]_U} = \frac{[C/D]_I}{[C/D]_U} \qquad (7-24)$$

Consequently a second simultaneous interaction is more stable than when it is present alone.

If additional groups are present on the polypeptide chain that may also interact simultaneously, the above equilibria are extended in the same way. The overall equilibrium constant between the final state, with all the interactions, and the unfolded state is the product of the individual equilibrium constants along any of the conceivable reaction paths; for example,

$$K_{net} = (K_{AB}[A/B]_U)(K_{CD}[C/D]_I)(K_{EF}[E/F]_{IV}) \cdots \qquad (7\text{--}25)$$

The value of K_{net} is independent of the reaction path, so we need not know or propose a specific "folding pathway."

The final state will be stable, i.e., populated by most of the molecules, only if the value of K_{net} is greater than unity. With weak interactions, the first will be very weak, with an equilibrium constant of between 10^{-3} and 10^{-7} (see Eq. 7–21). The second may be somewhat larger, but if it is still less than unity, the product of the two will be even smaller than the first. The net stabilities of conformations with additional weak interactions will decrease still further until the effective concentrations of additional interacting groups are increased sufficiently to make each additional equilibrium constant greater than unity. The value of K_{net} will then increase in magnitude. With sufficient weak interactions present simultaneously, the polypeptide chain will lose its flexibility, and those interactions added last will result from the high effective concentrations of their groups. The value of K_{net} will become greater than unity, and the folded conformation will be stable (Figure 7–23).

Partially folded states, with incomplete stabilizing interactions, will be unstable relative to the initial and final states, and the transition is expected to be cooperative.

In order to consider the net contribution of each interaction to stability of the final state, irrespective of the order in which it was added, the equilibria should be dissected artificially into all the factors stabilizing the unfolded state and those stabilizing the folded conformation. Considering only two interactions,

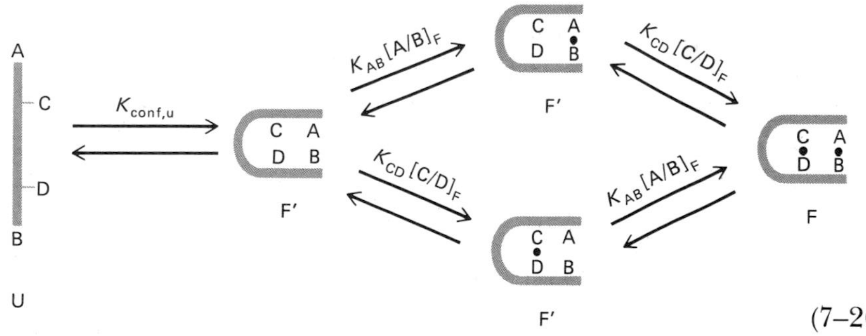

$$(7\text{--}26)$$

The folded states F′, without all the interactions, are strictly hypothetical. $K_{conf,U}$ is defined above and includes all the factors stabilizing the unfolded state; it represents the very small probability of finding the polypeptide chain in this folded state in the absence of any stabilizing interactions. $[A/B]_F$ and $[C/D]_F$ are the effective concentrations of these groups in the folded states F′. The overall equilibrium constant between the folded and unfolded states will be

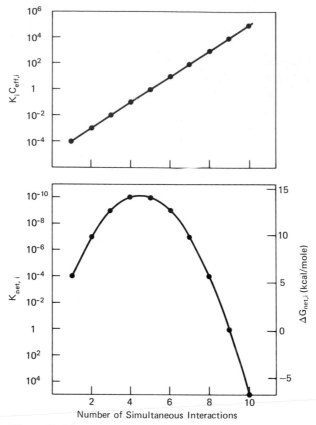

Figure 7–23

Hypothetical illustration of the cooperativity of folding produced by multiple weak interactions. Up to 10 interactions are possible simultaneously. The contribution of the ith interaction is given at the top, where the initial interaction has $K_1 C_{eff,1} = 10^{-4}$ and each additional interaction is ten times stronger than the previous one. The overall equilibrium constant $K_{net,i}$ (bottom) is the product of the contributions of the interactions present (see Eq. 7–25). Only with 10 such interactions is $K_{net,i} > 1$, implying stability of the folded state. The free energy of each state relative to U is given by $\Delta G^{\circ}_{net,i} = -RT \ln K_{net,i}$, with the scale on the right pertaining to 25° C.

$$K_{net} = K_{conf,U}(1 + K_{AB}[A/B]_F)(1 + K_{CD}[C/D]_F) \qquad (7\text{--}27)$$

Additional interactions would simply extend this equation, so that

$$K_{net} = K_{conf,U}(1 + K_{AB}[A/B]_F)$$
$$(1 + K_{CD}[C/D]_F)(1 + K_{EF}[E/F]_F) \cdots \qquad (7\text{--}28)$$

The energetic contribution of each interaction to stability of the folded state is then given by

$$\Delta G_{AB}{}^{\circ} = -RT \ln (1 + K_{AB}[A/B]_F) \qquad (7\text{--}29)$$

so that

$$\Delta G_{\text{net}}° = \Delta G_{\text{U}}° + \Delta G_{\text{AB}}° + \Delta G_{\text{CD}}° + \Delta G_{\text{EF}}° + \cdots \qquad (7\text{--}30)$$

where $\Delta G_{\text{U}}°$ is the sum of all the unfavorable entropic and enthalpic factors in the fully folded state:

$$\Delta G_{\text{U}}° = -RT \ln K_{\text{conf,U}} \qquad (7\text{--}31)$$

The final state is stable only when the sum of the stabilizing interactions is greater in magnitude than the destabilizing $\Delta G_{\text{U}}°$.

The important conclusion of this analysis is that it is the effective concentration of interacting groups in the folded, rather than the unfolded, state that is relevant to their contribution to the stability of that state, but the loss of conformational entropy and any other unfavorable aspects of the folded state must be included.

Effective concentrations of groups may be very high in intramolecular reactions, up to 10^{10} M (Table 4–7), when the two groups are held in correct proximity. Such high values are probably also possible in the folded states of proteins. For example, the effective concentration of the thiol groups of Cys 5 and Cys 55 in the interior of the folded conformation of BPTI has been measured to be about 5×10^5 M. For Cys 14 and Cys 38, which are on the surface of the molecule, so that the side chains are mobile, the effective concentration is only 2×10^2 M.

With such large effective concentrations, even weak interactions, with K_{AB} of the order of 10^{-2} M, can provide substantial net stabilization to the folded state. However, the maximum effective concentration possible depends upon the type of interaction—upon the entropy lost in the interaction; those involving covalent bonds give the highest possible values. Hydrogen bonds, which are moderately rigid, should have somewhat lower values; flexible, nonspecific van der Waals interactions should be even lower, perhaps 10^3 M.

The effective concentration of two groups in a folded protein depends upon the extent to which they are held in proximity in the folded state when not interacting; in turn, this depends upon the stability of all the surrounding interactions. All parts of a protein, therefore, are mutually dependent.

The contribution of each interaction to net stability of the folded state depends upon the effective concentration of the interacting groups. If they are on the surface, or in a flexible part of the structure, their effective concentration will be low and the interaction will provide little, if any, net stability. Breaking that interaction, which could even involve a covalent bond such as a disulfide or peptide bond, will have little effect. On the other hand, interactions between groups within relatively rigid parts of the structure will yield high effective concentrations and will provide substantial net stability. Removing or altering such an interaction would have a large effect on stability.

It is impossible at the present time to account rigorously for the net stability of the folded states of proteins, because accurate values for the individual factors are not known. Potential energy calculations on proteins have been carried out, but they primarily check the internal stereochemistry of the structure, demonstrating too-close contacts between adjacent groups, unlikely bond lengths and angles, unfavorable torsion angles, and so forth. They ignore the solvent and the conformational entropy of the molecule, essentially calculating the potential energy of a static molecule at $0°$ K in a vacuum. Such calculations are most useful for refining crystallographic models of proteins.

Calculation of the net free energy of the folded state of a protein requires estimation of the difference between free energies of the unfolded and folded states of the protein in water. The main stabilizing factor for the unfolded state, its greater conformational entropy, is a complex factor; its value is unknown. Estimates of the contribution to the free energy at room temperature have usually varied in the range of 3.3 to 10 kcal/mole per amino acid residue. Conformational strain in the unfolded state is likely to be negligible, but energy calculations of crystallographic folded structures indicate small, but significant, unfavorable interactions averaging 2 kcal/mole per residue. These two factors of between 5.3 and 12 kcal/mole per residue must be overcome by favorable interactions in the folded state; for a protein of 100 residues, this is 530 to 1200 kcal/mole (Table 7–6), and $K_{conf,U}$ should be between 10^{-390} and 10^{-882} at $25°$ C.

The hydrophobic interaction has long been considered the primary stabilizing force; it undoubtedly accounts for the heat capacity change upon folding, and its favorable contribution to the free energy change may be estimated from the accessible surface area buried in the folded state (Figure 4–5). For an average protein of 100 residues, this is likely to contribute 264 kcal/mole (Table 7–6), significantly less than the total required. However, this value was derived from transfer of nonpolar molecules to liquids, whereas the interiors of proteins are much more closely packed. In other words, the effective concentrations of nonpolar groups in the interiors of proteins are greater than in liquids; consequently the van der Waals interactions between these should be greater. This factor can be estimated from the enthalpy of fusion to the solid state of small molecules, which is generally about -30 cal/g, and it nearly doubles the net contribution of nonpolar interactions (Table 7–6).

The hydrophobic interaction alone also cannot account for the temperature-dependence of protein stability. In model systems it increases in magnitude with increasing temperature, up to about $60°$ C, whereas proteins decrease in stability over this range, having much lower temperatures of maximum stability (Figure 7–12). In other words, the enthalpy change upon unfolding is positive, rather than negative, the folded state having a lower value than the unfolded state. The increased van der Waals interactions in the folded state should contribute negatively to the enthalpy of the folded state, but the available data suggest this contribution alone is not sufficient.

Table 7–6 Estimates of the Net Contributions of Individual Factors to Stability of an Average Protein of 100 Residues

FACTOR	FREE ENERGY DIFFERENCE BETWEEN N AND U AT 25° C (kcal/mole)	CONTRIBUTION TO ENTHALPY DIFFERENCE
Conformational entropy at 25° C	$+330$ to $+1000$	negligible
Unfavorable interactions in folded state	$+200$	positive
Hydrophobic interactions[a]	-264	positive
Increased van der Waals interactions due to close-packing[b]	-227	negative
Required contribution of hydrogen bonds[c]	-49 to -719	negative
Observed net effect	-5 to -10	negative

[a]The free energy of the hydrophobic interaction is proportional to the buried accessible surface area (-24 cal/mole/Å^2; Figure 4–5). The average surface area buried in a protein of molecular weight M, in Å^2, is given by $1.449 M - 11.116 M^{2/3}$: from D. C. Teller, Nature 260:729–731, 1976.

[b]The free energy change is estimated from the average enthalpy of fusion observed with small, nonpolar model compounds, -30 cal/g, multiplied by the fraction of accessible surface of the protein involved in close-packed interactions (given by $0.595 + .0536 M \times 10^{-5}$ for an average protein: C. Chothia, Nature 254:304–308, 1975.

[c]The average number of protein hydrogen bonds would be 74 (C. Chothia, Nature 254:304–308, 1975), so an average contribution of -0.7 to -9.7 kcal/mole would be required. With $K_{AB} = 10^{-2}$ M^{-1} for hydrogen bonding in water (Table 4–5), effective concentrations of the interacting groups of 310 to 1.4×10^9 M would be required in proteins.

Hydrogen bonds within proteins have not been considered recently to contribute to the net stability of the folded state, since the unfolded protein was considered to make comparable hydrogen bonds to water. The concentration of hydrogen bonding groups in water is 110 M; with a value of K_{AB} for this interaction in water of 10^{-2} M^{-1} (Table 4–5), the equilibrium constant should be about unity in an unfolded polypeptide. However, the effective concentration of hydrogen-bonding groups in proteins is likely to be considerably greater than 110 M; intramolecular hydrogen bonds, therefore, may be significantly more stable than those with water and would then contribute to the net stability.

Hydrogen bonds within bulk water are present only half the time at room temperature, and many are distorted; those between the solvent and surface groups of proteins likewise are observed crystallographically to be only partially occupied and variable in geometry. In contrast, hydrogen bonds within proteins are usually present essentially all the time, and the groups are held in correct proximity; consequently they would be expected to be energetically more favorable. They would then contribute more negative enthalpy to the folded state, further helping to explain the observed temperature-dependence of protein stability (Table 7–6).

Other factors probably contribute to protein stability, such as electrostatic interactions, especially between dipoles and in specific salt bridges, but they are numerically and energetically less important and are omitted from this compilation. As pointed out above, the marginal net stabilities of folded proteins are small differences between large, compensating individual factors. With our limited understanding of these factors, it is possible only to account semiquantitatively for protein stability.

A discussion of pressure–volume effects in aqueous protein solutions. A. Hvidt. J. Theor. Biol. 50:245–254, 1975.

Stability of protein conformation: internal packing and enthalpy of fusion of model compounds. J. Bello. J. Theor. Biol. 68:139–142, 1977.

Tight packing of protein cores and interfaces. J. Bello. Int. J. Peptide Res. 12:38–41, 1978.

Dipoles of the α-helix and β-sheet: their role in protein folding. W. G. J. Hol, et al. Nature 294:532–536, 1981.

An empirical approach to protein conformation stability and flexibility. T. E. Creighton. Biopolymers 22:49–58, 1983.

PREDICTION OF FOLDED CONFORMATIONS

The specific folded conformation of a protein is believed to be dictated solely by its amino acid sequence, which opens the tantalizing possibility of being able to predict it given just the primary structure. However, this has not yet been possible owing to the astronomical number of conformations theoretically possible and uncertainty about how and why folding occurs. The large variation in amino acid sequence possible with essentially the same folded conformation (Chapter 6) indicates that there is substantial flexibility, or redundancy, in the rules relating primary and tertiary structures. Clearly, the particular amino acid side chain at each residue of a protein is not vital for determining the folded conformation. In particular, wide variation is observed in residues on the surface. Some internal residues may be vital and are conserved in all the related proteins, whereas in others, only the general class of side chain is preserved (e.g., aromatic, nonpolar, hydrogen-bonding, or ionic). Nevertheless, the general rules of protein structure noted in the previous chapter are always followed: The overall shape is roughly spherical; the interior is close-packed, with no large empty spaces; charged and polar groups are on the surface accessible to the solvent, unless they are paired in hydrogen bonds; and the torsion angles for bond rotations are relatively favorable, so that there is minimal conformational strain. Indeed, it is possible that any variation of a protein's amino acid sequence compatible with these principles would adopt the same conformation, although perhaps not on a finite time scale.

This is not to say that any sequence of amino acids will adopt *some* stable folded conformation; these structural principles must be compatible with the covalent structure of the polypeptide chain—a severe constraint. Only a very small fraction of polypeptides of random sequences, therefore, are likely to be capable of folding to a stable conformation.

With knowledge of a protein's primary structure, it is theoretically possible, but not feasible at the present time, to search for a conformation compatible with the general rules of protein structure. There is likely to be only one such conformation, but the number of possibilities is too great. Consequently, other approaches have been necessary to predict protein conformation. Simplified models of proteins make conformational calculations much more rapid, but predictions with these have not been generally successful. Most attention has been given to the more modest, one-dimensional problem of predicting elements of secondary structure.

Computer simulation of protein folding. M. Levitt and A. Warshel. Nature 253: 694–698, 1975.

A simplified representation of protein conformations for rapid simulation of protein folding. M. Levitt. J. Mol. Biol. 104:59–107, 1976.

Protein folding. G. Nemethy and H. A. Scheraga. Quart. Rev. Biophys. 10:239–352, 1977.

On the formation of protein tertiary structure on a computer. A. T. Hagler and B. Honig. Proc. Natl. Acad. Sci. U.S.A. 75:554–558, 1978.

Prediction of protein structure from amino acid sequence. M. J. E. Sternberg and J. M. Thornton. Nature 271:15–20, 1978.

Protein conformation, dynamics and folding by computer simulation. M. Levitt. Ann. Rev. Biophys. Bioeng. 11:251–271, 1982.

Secondary Structure   Most approaches to protein structure have concentrated on the elements of secondary structure, because these local, regular segments simplify the comprehension and description of complex structures. There is also a feeling that if the secondary structure elements could be predicted accurately, it would then be feasible to learn how to pack them together to generate the folded conformation, since 90 per cent of the residues in most proteins are involved in either α-helices (38 per cent), β-strands (20 per cent), or reverse turns (32 per cent).

The tendencies of the various amino acids to be involved in the structural elements, i.e., P_α, P_β, and P_t in a folded protein of known structure (compiled in Table 6–4), have been examined in the hope that there would be clear conformational preferences. There are indeed significant preferences in that 19 of the 20 amino acids occur preferentially in one conformation; only Arg is indifferent. However, they are rather marginal preferences. For example, the most helical-preferring amino acids (Glu and Met) occur in α-helices less than 50 per cent more frequently than would be expected on a random basis, and even Gly and Pro residues are found in helices about 50 per cent as often as expected at random, even though they are stereochemically rather incompatible with the helical conformation.

Fortunately, helices, β-sheets, and reverse turns are determined not by individual residues but by a number of them adjacent in the sequence. Consequently, a segment of a particular secondary structure is much more probable when several sequential residues tend to prefer that structure. A number of prediction schemes based upon such empirical observations have been proposed. The easiest to use, and therefore the most well-known, is that of Chou and Fasman, which classifies the amino acids as favoring, breaking, or being indifferent to each type of conformation. An α-helix is predicted if four out of six adjacent residues are helix-favoring and if the average value of P_α is greater than 1.0 and greater than P_β; it is extended along the sequence until Pro or four sequential residues with average value $P_\alpha < 1.00$ is reached. A β-strand is predicted if three out of five residues are sheet-favoring and if the average value of P_β is greater than 1.04 and greater than P_α; it is extended until a sequence of four residues with an average value of $P_\beta < 1.0$ is reached. A reverse turn is the likely conformation when sequences of four residues characteristic of reverse turns are found (see Table 6–5).

A more a priori theory based upon stereochemical considerations of the hydrophobic, hydrophilic, and electrostatic properties of the side chains in terms of the structural rules of folded proteins was proposed by Lim. Interactions between side chains separated by up to three residues in the sequence are considered in terms of their packing and interactions in either the helical or β-sheet conformations. For example, a sequence with alternating hydrophobic and hydrophilic side chains is likely to be a strand in a β-sheet, in which the orientations of the side chains alternate, with the hydrophilic side exposed to the surface and the hydrophobic side buried in the interior of the protein. A helical segment must have hydrophobic residues every three or four residues in order to give at least one side of the helix a hydrophobic surface with which to be attached to the remainder of the protein structure.

None of these procedures is described in detail here, for none gives sufficiently reliable results. In the best cases, about 80 per cent of the residues may be correctly predicted as either helical or nonhelical, and similarly for β-strands and reverse turns. However, if the accounting procedure considers whether residues are predicted correctly as α-helix, β-strand, reverse turn, or irregular conformation, the score drops considerably, to approximately 50 per cent correct. This degree of accuracy is certainly significant and is considerably better than guesswork, but it is still not adequate for predicting a folded structure. The empirical parameters (Table 6–4) have tended to change as the number of protein structures analyzed has increased, but the accuracy of the prediction methods has not improved correspondingly. It seems clear that interactions between residues distant in the amino acid sequence, which are not considered in such predictions, have substantial roles to play in determining the secondary structure of any segment.

Nevertheless, there are indications that strictly local interactions within and between neighboring residues are important. For example, Figure 7–24 illustrates that there is a correlation between the tendency of an amino acid to be helical in a protein (P_α) and its tendency to adopt the helical conformation in model polypeptides (s in the helix–coil transition; see Table 5–5). The correlation is not perfect, but it is also not clear how they should be related, since the isolated helix–coil transition in model peptides is not immediately pertinent to folded proteins. In the former case, helices are rapidly fluctuating and must consist of hundreds of resi-

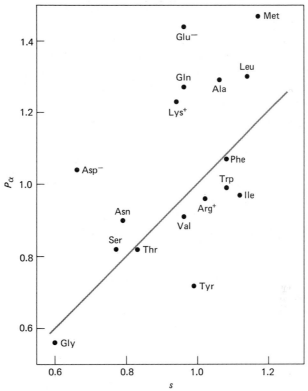

Figure 7–24
Correlation between the different amino acids' tendency to be helical in proteins (P$_\alpha$) and in synthetic polypeptides (s). The values of P$_\alpha$ are from Table 6–4; of s, from Table 5–5.

dues in order to be stable, whereas those in proteins are relatively stable and short.

On the other hand, the tendency for the amino acids to occur preferentially in one type of secondary structure need not imply simply that they intrinsically "prefer" that conformation; it could also reflect the more general rules of protein structure. For example, the large, bulky, nonpolar side chains of Val, Ile, Tyr, Phe, Trp, and Leu tend to occur most in β-sheets, which are generally internal, just where nonpolar side chains are most expected from the general principles of protein structure. Similarly, the amino acids with short hydrophilic side chains, such as Asn, Asp, and Ser, occur most often in reverse turns, which are normally at the protein surface, where such side chains must be. The frequent occurrences of Gly and Pro in reverse turns are clearly due to their unique conformational properties (see p. 236).

It is not yet clear whether secondary structure predictions will improve further and whether they will then permit prediction of the entire three-dimensional conformation from the primary structure alone. How protein conformation predictions will be accomplished is not at all obvious, but that a procedure will be found seems almost certain. The elucidation of protein amino acid sequences, particularly from the sequences of their genes, is occurring much more rapidly than three-dimensional structure determinations; consequently the demand for predictive schemes is certain to increase.

Algorithms for prediction of α-helical and β-structural regions in globular proteins. V. I. Lim. J. Mol. Biol. 88:873–894, 1974.

Comparison of predicted and experimentally determined secondary structure of adenyl kinase. G. E. Schulz, et al. Nature 250:140–142, 1974.

Comparison of the predicted and observed secondary structure of T4 phage lysozyme. B. W. Matthews. Biochim. Biophys. Acta 405:442–451, 1975.

Relationship between helix–coil transition parameters for synthetic polypeptides and helix conformation parameters for globular proteins. A simple model. E. Suzuki and B. Robson. J. Mol. Biol. 107:357–367, 1976.

Conformational preferences of amino acids in globular proteins. M. Levitt. Biochemistry 17:4277–4285, 1978.

Empirical predictions of protein conformation. P. Y. Chou and G. D. Fasman. Ann. Rev. Biochem. 47:251–276, 1978.

Prediction of the secondary structure of proteins from their amino acid sequence. P. Y. Chou and G. D. Fasman. Adv. Enzymol. 47:45–148, 1978.

The Chou-Fasman secondary structure prediction method with an extended data base. P. Argos, M. Hanei, and R. M. Garavito. FEBS Letters 93:18–24, 1978.

How good are predictions of protein secondary structure? W. Kabsch and C. Sander. FEBS Letters 155:179–182, 1983.

Homologous Structures

The one instance in which a tertiary structure can be predicted with reasonable certainty is when the amino acid sequence is homologous to that of a protein of known structure. The two proteins almost certainly have the same conformation; thus far, there are no exceptions to this rule. A model of the unknown protein may then be constructed using its amino acid sequence and the backbone conformation of the known structure. The accuracy of such a model will depend upon the degree of homology between the two primary structures, especially the number of insertions and deletions of residues.

Such a model should be consistent with the general rules of protein structure. The closer the homology, the more rigorous is the test.

Prediction of protein conformation from the primary structure alone may be possible only when a homologous structure is known. Fortunately, this may be feasible in most instances in the future, since homologous conformations are being discovered by crystallographic studies sufficiently often to suggest that there are only a limited number of folded conformations that are used by all proteins, perhaps 200 to 500, of which nearly 100 are known. To determine the conformation of a newly sequenced protein, it will then be necessary only to search for homology with that of a protein of known structure, using the general rules of protein conformation to assess the probability of conformational homology.

Structure and function of haemoglobin. II. Some relations between polypeptide chain configuration and amino acid sequence. M. F. Perutz, J. C. Kendrew, and H. C. Watson. J. Mol. Biol. 13:669–678, 1965.

A possible three-dimensional structure of bovine α-lactalbumin based on that of hen's egg-white lysozyme. W. J. Browne, et al. J. Mol. Biol. 42:65–86, 1969.

Computation of structures of homologous proteins. α-Lactalbumin from lysozyme. P. K. Warme, et al. Biochemistry 13:768–782, 1974.

Comparison of the predicted model of α-lytic protease with the X-ray structure. L. T. J. Delbaere, G. D. Brayer, and M. N. G. James. Nature 229:165–168, 1979.

Protein differentiation: emergence of novel proteins during evolution. G. E. Schulz. Angew. Chem. Int. Ed. 20:143–151, 1981.

8

INTERACTIONS WITH OTHER MOLECULES

The biological functions of proteins almost invariably depend upon their direct, physical interaction with other molecules. Virtually every substance with which a cell comes into contact is recognized and bound by some protein, and every small molecule within a cell was first bound by the enzyme that produced it or by the receptor on the cell surface that enabled it to be taken up. Proteins may bind very tightly and specifically to other proteins, generating large complexes; to nucleic acids, especially when controlling their replication and expression; to polysaccharides, especially important being those on the surfaces of cell membranes; and to lipids, often becoming incorporated within membranes. Every aspect of the structure, growth, and replication of an organism is dependent upon such interactions.

Proteins are generally classified according to the purpose and consequences of binding—for example, structural proteins, enzymes, repressors, lectins, toxins, immunoglobulins, hormones, receptors, membrane transport proteins, and proteins of motility. The physical principles of the interactions are similar in all these cases and are the subject of this chapter. (Systems involving chemical changes, as in enzyme catalysis, active transport, and mechanical movement, are left to the next chapter.) The following discussion focuses on the protein; whatever molecule it interacts with, even if it is another protein, is designated the **ligand.**

This chapter emphasizes reversible interactions, in which there is always a significant concentration of free ligand in equilibrium with the complex, or where the protein is normally found in both the free and liganded forms. Accordingly, stable supramolecular assemblies such as viruses and ribosomes, where the complex is the normal structural unit, are not considered here; likewise, little attention is given to interactions with normal components of the solvent, such as water molecules and hydrogen ions in the case of water-soluble proteins, and lipids in the case of membrane proteins. These interactions have been described earlier and are not fundamentally different, but they occur at many sites on the protein and are generally weak, occurring only because the solvent molecules are present in such high concentrations.

The interactions described here are distinguished by being specific for the appropriate ligand and occurring at a very limited number of sites on the protein, most often just one per polypeptide chain, with only a very limited number of different binding sites. The specificity of binding—that is, the discrimination between even closely related ligands—is determined by their relative binding affinities, which are discussed first. Next, the molecular structures of protein–ligand complexes are described briefly, followed by discussion of the molecular basis of the binding affinity and specificity. The final topic is the consequences of binding for the structure of the protein and how binding sites within a protein interact with each other—that is, the phenomenon of allostery.

ENERGETICS AND DYNAMICS OF BINDING

Binding Affinities   The affinity between a protein (P) and a ligand (A) is measured simply by the association constant, K_a, for the binding reaction at equilibrium:

$$\text{P} + \text{A} \xrightleftharpoons{K_a} \text{P} \cdot \text{A} \tag{8–1}$$

$$K_a = \frac{[\text{P} \cdot \text{A}]}{[\text{P}][\text{A}]} \tag{8–2}$$

All species are presumed to be present at sufficiently low concentration for thermodynamic ideality to apply. K_a is a constant under a given set of conditions and is measured experimentally by the dependence of binding on the free ligand concentration. Several commonly used graphic methods of analyzing binding data are illustrated at Figure 8–1. The ratio of bound to free protein should be directly proportional to the free ligand concentration:

$$\frac{[\text{P} \cdot \text{A}]}{[\text{P}]} = K_a[\text{A}] \tag{8–3}$$

An experimentally more useful measure of binding is the fraction (y) of protein molecules with bound ligand:

$$y = \frac{[\text{P} \cdot \text{A}]}{[\text{P}] + [\text{P} \cdot \text{A}]} = \frac{K_a[\text{A}]}{1 + K_a[\text{A}]} \tag{8–4}$$

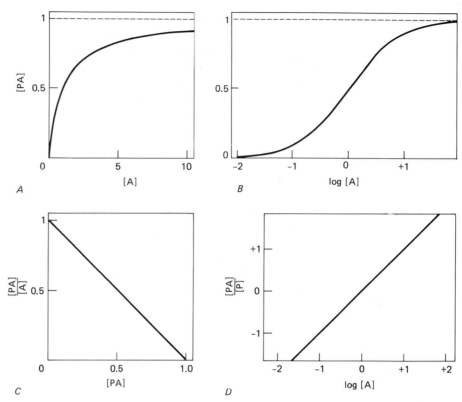

Figure 8–1

Some common methods of plotting binding data, using theoretical curves for the simple binding reaction P + A ⇌ P · A. The concentration of free ligand is expressed relative to its dissociation constant, which is that concentration which gives half maximal binding. The concentrations of free and liganded forms of the protein are relative to the concentration of total protein.

The normal hyperbolic relationship between binding and free ligand concentration is illustrated in A, demonstrating that a ligand concentration 9-fold greater than its dissociation constant produces only 90 per cent of maximal binding (indicated by the dashed line); 99-fold greater is required for 99 per cent saturation. The large range of free ligand concentrations required for a complete binding curve is emphasized when a logarithmic scale is used, as in B.

C, A Scatchard plot. The negative slope gives the value of the association constant (the reciprocal of the dissociation constant). The horizontal intercept gives the extrapolated extent of the maximal binding.

D, A Hill plot. An accurate value for the maximum binding is required for this plot, as both the liganded and free protein concentrations are required. The value of the dissociation constant is given by the value of the free ligand concentration where the vertical axis is zero, i.e., at half-maximal binding. This plot is used primarily for analyzing cooperative binding (see Figure 8–3).

The greater the value of K_a, the greater the affinity. However, the value of K_a has units of (concentration)$^{-1}$, and it is often intuitively easier to consider the dissociation constant, K_d; it is simply the reciprocal of K_a and has units of concentration. With concentrations of free ligand below

K_d, little binding occurs. With a concentration equal to K_d, half the protein molecules have ligand bound. An occupancy of 90 per cent requires a nine times' greater concentration of free ligand, whereas 99 per cent occupancy requires that it be 99 times K_d.

Specific binding of one ligand, and not another, is dependent upon their having different affinities. If two ligands are present at a concentration of 10^{-5} M, but have different values of K_d—say, 10^{-3} M and 10^{-6} M— only the latter will be bound significantly. If both are present at higher concentrations, say, 10^{-2} M, specific binding will still be possible, both would be bound to their individual sites. However, if they compete for the same site:

$$P \cdot A \underset{K_d^A}{\overset{A}{\rightleftharpoons}} P \underset{K_d^B}{\overset{B}{\rightleftharpoons}} P \cdot B \tag{8-5}$$

$$P = \frac{[P \cdot A] \, K_d^A}{[A]} = \frac{[P \cdot B] \, K_d^B}{[B]} \tag{8-6}$$

$$\frac{[P \cdot A]}{[P \cdot B]} = \frac{K_d^B \, [A]}{K_d^A \, [B]} \tag{8-7}$$

In this case, that ligand with the higher affinity would be bound to a correspondingly greater extent.

The energetics of binding are often expressed by the Gibbs free energy of binding, $\Delta G°$:

$$\Delta G° = -RT \ln K_a = RT \ln K_d \tag{8-8}$$

However, it must be kept in mind that K_a and K_d have units of concentration and that the value of $\Delta G°$ will depend upon which units are used, i.e., the standard state. If the units are moles/liter, the standard state is 1 M, and the calculated value of $\Delta G°$ will apply only under the rather arbitrary situation when the concentration of free ligand is 1 M. In many instances a "unitary" free energy of binding is used as a measure of the intrinsic affinity: the free energy that would occur with a hypothetical ligand at a concentration of 55 M, the normal concentration of water. However, this parameter is not of any special significance, except when the ligand is water, and it does not represent the free energy of interaction that would occur in a unimolecular interaction (see pp. 360–366).

The energetics of binding are defined more explicitly as the difference in free energies of the free and bound protein, $\Delta G_b°$:

$$\Delta G_b° = -RT \ln (K_a[A]) = RT \ln \left(\frac{[A]}{K_d}\right) \tag{8-9}$$

where the concentration of free ligand must be specified.

As described later, it is currently impossible to rationalize the value of K_a or K_d, but relative affinities for two related ligands (e.g., A and B) are more easily analyzed, because the ratio is dimensionless:

$$\Delta(\Delta G^\circ)_{A-B} = -RT \ln \frac{K_a{}^A}{K_a{}^B} = +RT \ln \frac{K_d{}^A}{K_d{}^B} \qquad (8-10)$$

The observed affinities of proteins for ligands vary, ranging from very high values, where dissociation is immeasurably small, to very low values, where the concentration of free ligand required for binding is so great as to cast doubt upon the nature of its effect upon the protein. Whether or not a given affinity is significant depends upon the concentration of ligand the protein is likely to encounter; no other generalizations are possible.

Where the affinity is very high, the protein is likely to be found and isolated as the complex; if such a ligand is relatively small, it is designated a **prosthetic group.** Examples are the heme groups of the globins and cytochromes, some coenzymes, and metal ions that are integral parts of the protein structure. With lower, moderate affinities, ligands originally bound are likely to be lost during isolation of the protein, unless they are added to the protein solution.

Ligand binding is quite simple in dilute solution, but proteins often function in extremely concentrated aqueous solutions, as in the cell cytosol. For example, the interior of the red blood cell is about 35 per cent by weight hemoglobin. Consequently, such solutions are very nonideal, and the pertinent equilibria must be expressed in terms of thermodynamic activities of the species, which can be very different from their concentrations. Even though a particular protein might not be present at high concentration, the presence of other molecules rather than water in the environment can lead to substantial excluded volume effects. These will favor any conformational or binding reaction that leads to a more spherical shape; consequently, binding of a ligand is often considerably greater than might otherwise be expected. It is possible that all proteins in the cytosol normally exist bound to each other, to membranes, to cytoskeleton, or to some other organized structure.

Rates of Binding and Dissociation

$$P + A \underset{k_d}{\overset{k_a}{\rightleftharpoons}} P \cdot A$$

$$K_a = \frac{k_a}{k_d} \qquad (8-11)$$

The rate constant for binding of a ligand to a protein, k_a, may vary considerably, depending upon their sizes and upon what changes must take place in both upon binding. Nevertheless, many small ligands are found to bind very rapidly, at rates approaching those expected for diffusion control, k_D. This may be estimated from the diffusion coefficients of the protein and ligand, D_P and D_A, respectively, treating them as small spherical molecules that must approach within a distance r_{PA}:

$$k_D = 4\pi N_A (D_P + D_A) r_{PA} \qquad (8-12)$$

For molecules with average diffusion coefficients (Table 5–3) under normal circumstances, values of k_D in the region of 10^9 M^{-1}sec^{-1} are expected. Larger molecules have smaller diffusion coefficients, but the value of k_D does not decrease accordingly, because r_{PA} is correspondingly larger. If the two molecules attract or repel each other at a distance, the term r_{PA} in Equation 8–12 should be replaced by a term containing the energy of interaction as a function of distance, $U(r)$:

$$k_D = \frac{4\pi N_A (D_P + D_A)}{\int_{r_{PA}}^{\infty} \dfrac{e^{-U(r)/k_B T}}{r^2}\, dr} \qquad (8\text{–}13)$$

For example, electrostatic interactions are significant over substantial distances and, when favorable, can increase rate constants for association to 10^{11} M^{-1}sec^{-1}.

Lower rates of binding imply either that the two molecules must be in defined orientations for productive binding to occur, or that changes occur during binding to produce a multistep association reaction. Both phenomena are undoubtedly important with proteins. The binding sites on proteins bind ligands usually in only defined orientations and comprise only small fractions of the protein surface. Consequently, most encounters between ligand and protein would be expected to be unproductive, and many such association reactions are relatively slow, but there are exceptions. For example, cytochromes c are believed to transfer electrons to and from other proteins through only 0.6 per cent of their surface, where the heme group is accessible (see Figure 6–28) and only when the two proteins interact in very specific orientations. On this basis, the rate of their interaction would be expected to be lower than that for diffusion-controlled encounters by a factor of at least 1000, but it is not. The reason is believed to be that asymmetric distributions of charges on the proteins orient them so that they tend to approach each other rapidly in a productive manner. For horse cytochrome c, the charge distribution indicates a large dipole moment of just over 300 debye units, and the dipole axis passes through the presumed binding site. Consequently, protein interactions are not governed simply by diffusive encounters.

The surfaces of proteins and of many of their ligands are flexible; many of these flexible portions must be fixed in a more defined manner upon binding, or parts of the protein may have to move to permit binding. Consequently, binding may occur via diffusion-controlled formation of an unstable "encounter" complex, followed by rearrangement to the final complex:

$$\text{P} + \text{A} \underset{k_{-1}}{\overset{k_1}{\rightleftharpoons}} \left\{\text{P}\cdots\text{A}\right\} \underset{k_{-2}}{\overset{k_2}{\rightleftharpoons}} \text{P}\cdot\text{A} \qquad (8\text{–}14)$$

Here, only the rate of formation of the encounter complex depends upon the concentration of the free ligand (i.e., k_1 is second-order); all the other

Table 8–1   Rate Constants for Formation of Complexes of tRNA Synthetase (P) and tRNA

$$P + tRNA \underset{k_{-1}}{\overset{k_1}{\rightleftharpoons}} P \cdot tRNA \underset{k_{-2}}{\overset{k_2}{\rightleftharpoons}} P * tRNA$$

tRNA SYNTHETASE	tRNA	k_1 ($M^{-1}sec^{-1}$)	k_{-1} (sec^{-1})	k_2 (sec^{-1})	k_{-2} (sec^{-1})
Yeast Ser[a]	Yeast Ser	2.7×10^8	220	760	330
Yeast Phe[b]	Yeast Phe	2×10^8	250	420	750
Yeast Phe[b]	*E. coli* Tyr	8×10^8	1600	—	—

The source and amino acid specificity of each tRNA synthetase and tRNA are indicated. The difference between the two complexes P · tRNA and P * tRNA is not known.

[a]Data from D. Riesner, et al., Eur. J. Biochem. 68:71–80, 1976.

[b]Data from G. Krauss, et al., Eur. J. Biochem. 68:81–93, 1976.

steps are unimolecular (i.e., first-order). For the overall rate of such a reaction to be proportional to the free ligand concentration, but slower than diffusion-controlled k_1—the usual situation—the encounter complex must be unstable; i.e., $k_{-1} > k_1$ [A]. In addition, it must dissociate more rapidly than it completes binding; i.e., $k_{-1} > k_2$.

Even where the observed rate of association is apparently diffusion-controlled, additional steps following the initial association may occur, but they will be observed only if they are made to be rate-limiting by increasing the rate of the initial bimolecular association. For example, the initial interaction between a tRNA molecule and a tRNA synthetase enzyme can occur at nearly the diffusion-controlled rate, about 10^8 $M^{-1}sec^{-1}$, but rearrangements occur subsequently on the millisecond time scale (Table 8–1).

Diffusion-limited encounters will occur with all molecules in a solution; therefore, stable and specific binding—that is, high values of K_a or low values of K^d—must be reflected in slow rates of dissociation, k_d (Eq. 8–11). Energetically favorable rearrangements of the complex after initial association have the effect of decreasing the apparent rate of dissociation. For example, binding of a yeast tRNA synthetase to a bacterial tRNA is not productive and is weaker than with the homologous pairs, but the initial association is somewhat faster. The weaker binding results from a greater rate of dissociation of the initial complex and from the apparent absence of a second step (Table 8–1).

Very little is known about what occurs structurally during the course of binding of ligands to proteins. There is considerable scope for conformational changes in both protein and ligand in most cases, but techniques for following them have yet to be devised.

Elementary steps in enzyme reactions. M. Eigen and G. G. Hammes. Adv. Enzymol. 25.1–38, 1963.

Relaxation spectrometry of biological systems. G. G. Hammes. Adv. Protein Chem. 23:1–57, 1968.

Rapid reactions and transient states. G. G. Hammes and P. R. Schimmel. In The

Enzymes, 3rd ed. P. D. Boyer (ed.). Vol. 2, pp. 67–114. New York, Academic Press, 1970.

Transients and relaxation kinetics of enzyme reactions. H. Gutfreund. Ann. Rev. Biochem. 48:315–344, 1971.

Binding of flexible ligands to macromolecules. A. S. V. Burgen, et al. Nature 253:753–755, 1975.

Excluded volume as a determinant of macromolecular structure and reactivity. A. P. Minton. Biopolymers 20:2093–2120, 1981.

The asymmetric distribution of charges on the surface of horse cytochrome c: functional implications. W. H. Koppenol and E. Margoliash. J. Biol. Chem. 257:4426–4437, 1982.

The ionic strength dependence of the rate of a reaction between two large proteins with a dipole moment. J. W. vanLeeuwen. Biochem. Biophys. Acta 743: 408–421, 1983.

Thermodynamics of protein-ligand interactions: calorimetric approaches. H.-J. Hinz. Ann. Rev. Biophys. Bioeng. 12:285–317, 1983.

Multiple Binding Sites and Interactions between Them

Association between a protein and a single ligand molecule is quite simple, as just described, but binding of multiple ligands at multiple sites can lead to complex behavior. The multiple sites may be identical and have the same intrinsic binding properties, as when each subunit of an oligomeric protein contains the same combining site, or there may be different sites with different affinities. Moreover, binding at one site may increase or decrease the affinities of the other sites.

There is often considerable confusion about the interpretation and significance of the complex binding curves that can result from such phenomena. This is partly due to the use of different methods for measuring binding, which can give different types of information. The most commonly used methods measure binding physically, as with equilibrium dialysis, where the free ligand, but not the protein or the complex, equilibrates across a semipermeable membrane. Such methods can measure the number of ligands bound per protein molecule, but averaged over all the binding sites of the protein population; they do not distinguish between the different sites on the protein. Some procedures, such as radioimmunoassay, measure a protein antigen as bound so long as at least one antibody molecule is bound, irrespective of the actual number. Spectral measurements of binding, which use spectral changes upon binding in either the protein or the ligand, have the potential to distinguish between different classes of binding sites, but they have the disadvantage of usually determining only the fraction of maximum possible binding, not stoichiometries.

Identical and Independent Sites

The binding curves are similar in shape to those with a single ligand, but the maximum stoichiometry of binding should be an integer, n, equal to the number of sites per molecule (Figure 8–2A). Similarly, the intrinsic K_d of an individual site is given by the concentration of free ligand producing

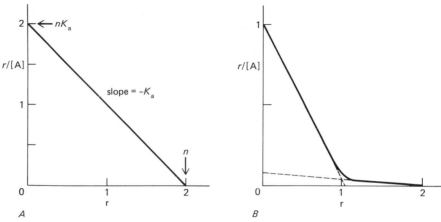

Figure 8–2

Scatchard plots of binding of ligand A to two sites on a protein. The degree of binding is expressed as r, *the moles of ligand bound per mole of protein.*

A, *Two identical and noninteracting sites, each with an intrinsic association constant,* K_a, *of 1.0, expressed in the same units as the concentration of ligand. The slope is* $-K_a$, *and the x-intercept is the extrapolated maximum number of ligand molecules bound,* n. *The y-intercept is* n K_a.

B, *Two different, independent sites with intrinsic association constants differing 50-fold:* $K_a^1 = 1.00$ *and* $K_a^2 = 0.02$. *The two linear segments of the curve correspond approximately, but not exactly, to the binding to the two sites. The y-intercept of the steep portion gives an apparent value of 1.02 for the first binding constant, which is* $K_a^1 + K_a^2$. *The extrapolated value for the second phase is 0.0196, which is* $K_a^1 K_a^2/(K_a^1 + K_a^2)$. *The negative slopes of the two phases are very close to the values of* K_a^1 *and* K_a^2. *(Adapted from S. Ferguson-Miller and W. H. Koppenol, Trends Biochem. Sci. 6:4–7, 1981.)*

half of the maximum binding. However, such binding curves are often interpreted in terms of association or dissociation constants for the 1st, 2nd, . . . *n*th ligand molecules to bind; the use of such constants introduces a statistical factor. For example, the first ligand to bind has *n* empty binding sites available, so the rate constant for association is *n* times that for an individual site, k_a. Its rate constant for dissociating is that for an individual site, k_d. The association constant for binding the first ligand, K_a^1, is then

$$K_a^1 = \frac{nk_a}{k_d} = nK_a \qquad (8\text{--}15)$$

or *n* times that of an individual site. Conversely, the *n*th ligand has only one site available for binding, but there are *n* ways of losing one ligand from the fully saturated molecule, so

$$K_a^n = \frac{k_a}{nk_d} = \frac{1}{n}K_a \qquad (8\text{--}16)$$

or $1/n$ that of an individual site. Statistical factors also apply to the other association constants.

Independent but Nonidentical Sites

The sites with highest affinity become occupied at the lowest ligand concentrations, followed by those of decreasing affinity. Unless site-specific binding is measured, the curves of total binding are complex. In particular, Scatchard plots become *concave*, the curvature depending upon the relative affinities. The curve may have two or more linear portions if the affinities of the different sites are very different, and the stoichiometries and affinities of the different classes of sites may then be estimated (Figure 8–2B).

Cooperative Binding of a Single Ligand

Binding of ligand to one site can increase or decrease the affinities of the other sites; i.e., there can be either positive or negative cooperativity.

Positive cooperativity produces a sigmoidal relationship between degree of binding and free ligand concentration (Figure 8–3A), as the affinity for successive ligands increases. A *convex* Scatchard plot results (Figure 8–3B); this plot is most sensitive to low degrees of cooperativity but is not widely used for positive cooperativity, since its interpretation is not obvious. Hill plots are most widely used (Figure 8–3C), because the slope in the middle of the curve, the **Hill coefficient,** gives a quantitative measure of the degree of cooperativity. It corresponds to the hypothetical number of ligand molecules that would need to be bound fully cooperatively, in an all-or-none fashion, to give such a slope. In this case there would be present only empty and fully occupied protein molecules:

$$P + nA \rightleftharpoons P \cdot A_n \qquad (8\text{--}17)$$

$$K_a = \frac{[P \cdot A_n]}{[P][A]^n} \qquad (8\text{--}18)$$

$$\frac{y}{1 - y} = \frac{[P \cdot A_n]}{[P]} = K_a[A]^n \qquad (8\text{--}19)$$

$$\log\left(\frac{y}{1 - y}\right) = \log K_a + n \log A \qquad (8\text{--}20)$$

so that a linear Hill plot would be expected, with a slope of n.

In reality, binding is never totally cooperative, and the curve has a slope greater than 1 only in the middle. Partially occupied protein molecules are present, but at lower concentrations than with independent binding (Figure 8–3D). The ends of the curves have slopes of unity and correspond to the binding of the first and last ligand molecules and give their association constants. The Hill coefficient is the maximum slope of the curve and may have any value, not necessarily an integer, depending on the degree of cooperativity. Its value could be as large as the number of interacting binding sites only if there were complete cooperativity.

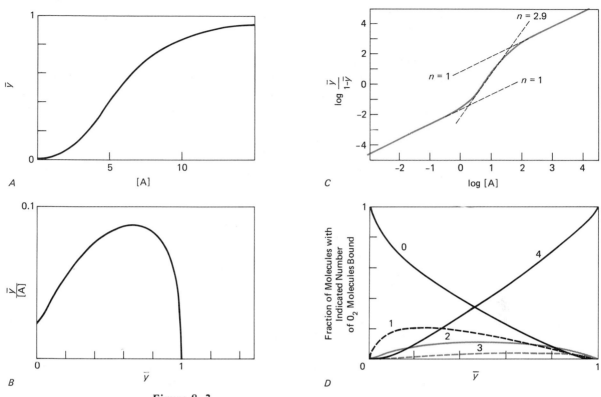

Figure 8–3
A, B, *and* C, *Positive cooperativity in ligand binding.* A, *Plot of* $\bar{y}$, *the average fraction of complete binding, versus the ligand concentration,* [A]. B, *Scatchard plot.* C, *Hill plot. The binding curve in each case is that for binding of* O_2 *by normal hemoglobin at 25° C in 0.1 M NaCl at pH 7.4, where the intrinsic association constants for the first, second, third, and fourth* O_2 *molecules bound to the four heme groups are respectively 0.024, 0.074, 0.086, and 7.4 mm^{-1}; the O_2 concentration is expressed as its partial pressure in mm of mercury. (From Y. Tyuma, et al., Biochemistry 12:1491–1498, 1973.)*

The fractions of hemoglobin molecules with zero to four O_2 molecules bound are plotted in D *as a function of the fractional saturation. As a result of the cooperativity, the molecules with zero and four bound O_2 molecules predominate at all stages of binding, with partially saturated molecules being relatively rare. In contrast, with no cooperativity, partially saturated molecules predominate from approximately $\bar{y} = 0.16$ to 0.84.*

Negative cooperativity produces Hill plots with slopes less than unity, owing to the decreased affinity for successive ligands. However, Scatchard plots are used more frequently to analyze such binding data, as they give concave curves, often biphasic, like those with sites of different affinities (Figure 8–2*B*). However, such curves resulting from negative cooperativity are not readily interpretable in terms of the affinities for the successive ligands or of the interaction between them.

Binding curves alone cannot distinguish among negative cooperativity between otherwise identical sites, that between independent but different sites on the same protein, and simple heterogeneity of the protein molecules present. Negative cooperativity can be distinguished by the effect of added ligand on the rate of dissociation of radioactive ligand from partially occupied protein molecules; the rate should be increased by added ligand only if there is negative cooperativity.

Complex binding curves can arise from other effects and need not imply cooperativity. Most binding studies measure association of only a particular ligand to the protein and assume that all sites not occupied by it are free, whereas an unidentified second ligand could be present. It is also generally assumed that ligand not bound to the protein is free and not interacting with any other component of the solution. For example, a ligand might aggregate when free and the protein might bind noncooperatively only to the least prevalent form. If this is the monomer, the binding curves will suggest negative cooperativity, whereas binding of only the rare aggregate would suggest positive cooperativity.

Nevertheless, examples of both positive and negative true cooperativity are well-characterized, such as the positive cooperative binding of four O_2 molecules by hemoglobin and the negative cooperativity in the binding of four NAD molecules by the enzyme glyceraldehyde phosphate dehydrogenase.

Interactions between Different Ligands: Linked Functions
Just as binding of one ligand molecule can affect the subsequent binding of further molecules of the same ligand, producing interactions termed **homotropic,** so it can affect the binding of different ligands to other sites. These **heterotropic** effects are easier to analyze because the concentrations of the different ligands can usually be varied independently, so that the interactions can be observed directly.

This simple situation can be used to illustrate an important feature of interactions between sites: They are **linked functions.** Consider the binding of two ligands, A and B, to different sites on a protein, P:

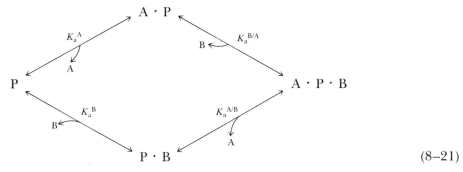

$$(8\text{--}21)$$

It is a thermodynamic necessity that the free energy change around such a cycle be zero—that is, that the products of the equilibrium constants around the cycle be unity. Therefore

$$\frac{K_a^{\,A}}{K_a^{\,A/B}} = \frac{K_a^{\,B}}{K_a^{\,B/A}} \qquad (8\text{--}22)$$

In other words, whatever effect the binding of ligand A has on the affinity for ligand B, the binding of B must have exactly the same effect on the affinity for A. For example, binding of organic phosphates, hydrogen ions, and CO_2 affect the oxygen affinity of hemoglobin; likewise, the binding of oxygen must affect the affinity of hemoglobin for these ligands. If an inducer affects the binding of a repressor molecule to the operator region of DNA, binding to the DNA must affect the repressor's affinity for the inducer.

It is not only ligand binding for which this relationship holds. Any two properties of a protein for which a cyclic equilibrium holds, such as ionization, aggregation, or conformational isomerization, are linked functions.

Linked functions and reciprocal effects in hemoglobin: a second look. J. Wyman, Jr. Adv. Protein Chem. 19:223–286, 1964.

Allosteric linkage. J. Wyman. J. Am. Chem. Soc. 89:2202–2218, 1967.

Properties of graphical representations of multiple classes of binding sites. I. M. Klotz and D. L. Hunston. Biochemistry 10:3065–3069, 1971.

Protein interactions with small molecules. I. M. Klotz. Acc. Chem. Res. 7:162–168, 1974.

The meaning of Scatchard and Hill plots. F. W. Dahlquist. Methods Enzymol. 48:270–299, 1978.

Protein affinities for small molecules: conceptions and misconceptions. I. M. Klotz and D. L. Hunston. Arch. Biochem. Biophys. 193:314–328, 1979.

Affinity Chromatography

The interaction of proteins with specific ligands makes possible one of the most powerful methods of protein purification, **affinity chromatography.** In this technique, the ligand is chemically attached to a solid support suitable for chromatography, i.e., insoluble and porous, in such a way that the pertinent protein is still able to bind to it. Obviously, the appropriate site on the ligand for attachment to the support is one not involved in binding. Also, to ensure that the ligand is sufficiently distant from the support so that binding of the protein is not physically obstructed, a spacer group is usually inserted between the ligand and the support. If the parts of the ligand required for binding to the protein are not known, different linkages must be tried until one is found to work.

The desired protein then binds to the ligand resin; other proteins do not but instead pass through the column. After washing to remove all other proteins, those bound tightly are eluted by adding soluble ligand to compete with that bound, or by changing the conditions to decrease the affinity of protein for the bound ligand; in extreme cases, the protein may need to be unfolded before it is eluted. The use of soluble ligand permits quantitative measurements of the proteins' affinities for both the free and bound ligand, which need not be the same. Multiple proteins that bind to the same ligand may often be eluted specifically by different ligands (Figure 8–4) or by different concentrations of the same ligand (Figure 8–5).

The principles of affinity chromatography are similar to those of the other types of column chromatography often used for proteins, such as

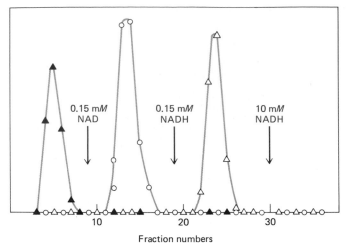

Figure 8–4
Separation of three proteins by affinity chromatography. The column matrix was Sepharose with the cofactor NAD linked covalently via a diaminohexane spacer. A mixture of bovine serum albumin, glyceraldehyde 3-phosphate dehydrogenase, and lactate dehydrogenase was applied to the column. Serum albumin (▲) does not bind NAD and consequently was not bound to the column. The two dehydrogenases do and were bound. Lactate dehydrogenase (△) binds NADH more tightly and was eluted by low concentrations of it, whereas glyceraldehyde-P dehydrogenase (○) binds NAD more tightly and was eluted by low concentrations of the free cofactor. (Adapted from K. Mosbach, et al., Biochem. J. 127:625, 1972.)

ion-exchange, hydrophobic, or reversed-phase chromatography. Affinity chromatography differs from other techniques only in the specificity of the interactions between the protein and the column resin.

Affinity chromatography. P. Cuatrecasas and C. B. Anfinsen. Ann. Rev. Biochem. 40:259–278, 1971.
Affinity chromatography of macromolecules. P. Cuatrecasas. Adv. Enzymol. 36:29–89, 1972.
Affinity techniques. W. B. Jakoby and M. Wilchek. Methods Enzymol. 34:1974.
Biospecific affinity chromatography and related methods. J. Porath and T. Kristiansen. In The Proteins, 3rd ed. H. Neurath and R. L. Hill (eds.). Vol. 1, pp. 95–178. New York, Academic Press, 1975.
The purification of biologically active compounds by affinity chromatography. M. Wilchek and C. S. Hexter. Methods Biochem. Anal. 23:345–385, 1976.

STRUCTURES OF PROTEIN COMPLEXES

Binding studies in solution provide valuable information about the stoichiometries and energetics of binding, but they do not provide structural information directly. The stoichiometries of binding are informative because very often there is one binding site for a particular ligand on each copy of a particular polypeptide chain, although there are notable exceptions: With some oligomeric proteins of apparently identical polypeptide chains, there are only half the expected number of binding sites, owing to

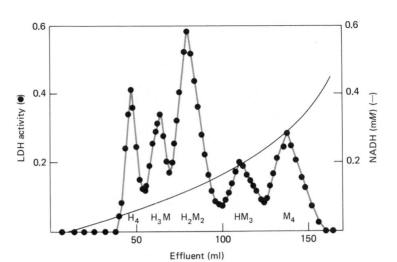

Figure 8–5

Separation of the five isozymes of lactate dehydrogenase by affinity chromatography. It is a tetramer formed from two different but related polypeptide chains, usually designated H and M. The two chains can associate nearly randomly, so that five tetramers are possible: the H_4, H_3M, H_2M_2, HM_3, and M_4 isozymes. The individual chains bind NADH independently, with H chains having a fivefold greater affinity. Consequently, H_4 is eluted from an affinity column with lower concentrations of free NADH in the buffer, the others in order of their average affinities. The affinity column was AMP, which corresponds to half an NADH molecule, bound to Sepharose. (Adapted from P. Brodelius and K. Mosbach, FEBS Letters 35:223, 1973.)

either direct or indirect interactions between the sites. Binding studies can also provide indirect information regarding the interaction between ligand and protein by the effect of altering the chemical structure of the ligand on the binding affinity. Removal or alteration of a group on the ligand that interacts directly with the protein generally decreases the affinity correspondingly, but modifications of other groups may have little or no effect on affinity. Similarly, the effect on affinity of modifications of the protein may indicate which portions of the protein are involved in binding.

Nevertheless, the most information by far about the binding interaction comes from crystallographic determination of the structure of the complex, as with the protein itself. A wide variety of complexes have been determined directly in this way in many of the protein crystal structures listed in Table 6–1. In most of these cases, the ligands are integral parts of the protein structure, either very tightly bound or known to be required free in solution for stability or activity of the protein. In many of the complexes the ligand is another protein molecule, either the same or different, as in multimeric proteins (discussed earlier in Chapter 6 under Quaternary Structure) or with some protein inhibitors of enzymes, especially proteases. Other ligands include prosthetic groups such as heme, NAD, flavins, and pyridoxal phosphate, as well as many types of ions. Some complexes with ligands of varying size are illustrated in Figure 8–6.

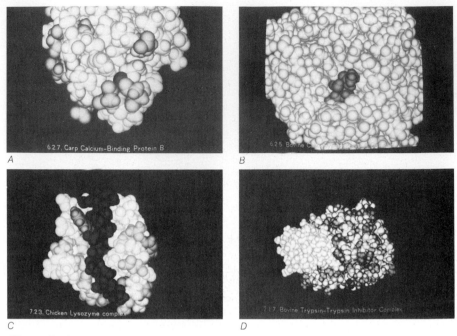

A

B

C

D

Figure 8–6
Examples of some protein–ligand complexes, illustrated as space-filling models viewed from the surface.

A, A Ca^{2+} ion (dark) bound to parvalbumin. The Asp and Glu residues are shaded.

B, An inhibitor, glycyl tyrosine (dark), bound to carboxypeptidase A. A Zn^{2+} ion also bound tightly to the protein is barely visible and is behind the inhibitor.

C, A hexasaccharide, (N-acetylglucosamine)$_6$ (black), bound to hen lysozyme. Trp, Phe, Arg, Lys, Asn, and Gln residues are shaded.

D, The complex of BPTI (white) with bovine trypsin (some atoms are shaded).

(Adapted from R. J. Feldman and D. H. Bing, Teaching Aids for Macromolecular Structure. Bethesda, Md. Division of Computer Research and Technology, National Institutes of Health, 1980.)

In most cases, there is a single unique binding site for a particular ligand on each molecule of a polypeptide chain. However, if the polypeptide chain has internal symmetry, as when it has arisen by gene duplication, each of the structural units may have a binding site. For example, gene-duplicated ferredoxins contain two similar iron–sulfur complexes related by the twofold symmetry of the molecule (see Figure 6–31); each cluster is bound by both halves of the polypeptide chain, not by just one half. This arrangement suggests that, before gene duplication and doubling of the polypeptide chain, the ancestral molecule was a dimer of two identical chains binding two iron–sulfur complexes.

In many cases, a protein that binds a number of different ligands will bind them on separate domains. These domains are often designated by their binding properties; examples are illustrated in Figures 8–7, 9–17,

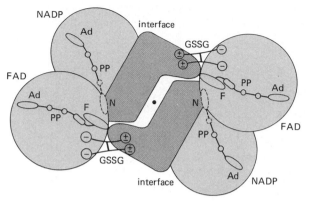

Figure 8–7
Schematic diagram of the domain structure of the dimeric enzyme glutathione reductase. The twofold axis relating the two polypeptide chains is indicated by the dot in the center. The interface domains bind to each other and determine the dimeric structure. The NADP-binding domain has the adenine (Ad), ribose (circles), diphosphate (PP), and nicotinamide (N) moieties indicated; only the nicotinamide portion is fully buried. The FAD-binding domain has the coenzyme indicated similarly, except that F is the flavin moiety; only the adenine group is exposed to the solvent. The substrate, oxidized glutathione (GSSG), binds between the two subunits. (Adapted from G. E. Schulz, et al., Nature 273:120–124, 1978.)

and 9–18. In some cases, ligands are bound between domains or between subunits, both participating in binding. Such ligands are often substrates and are altered covalently after binding, or their binding is regulated by changes elsewhere in the protein. Binding sites between domains or subunits seem to be rather dynamic locations, probably owing to the flexibility of the protein architecture.

There are some notable exceptions to the generalization of one binding site for a particular ligand per protein structural unit: A single-chain cytochrome c_3 binds four identical heme groups in unique environments (Figure 8–8). A bacteriochlorophyll protein from a green photosynthetic bacterium binds seven chlorophyll molecules, each in a unique position within a "string bag" of a 15-stranded β-sheet closed to form a flattened barrel (Figure 8–9).

Closer inspection of the interactions between proteins and ligands shows that there is steric as well as physical complementarity between the two: The interface between them is closely packed; all polar groups are paired in hydrogen bonds; and electrostatic charges are generally neutralized. They follow structural rules similar to those within the proteins themselves; indeed, the interface between two interacting protein molecules is not basically different from the interior of either protein.

The earlier description of protein structures (Chapter 6) noted their tendency to be spherical. However, binding sites for ligands represent the greatest exceptions to this generality: They are often sizable depressions on the surface. It is probably more correct to state that the complex of protein and its ligands tends to be spherical, and that the interacting sur-

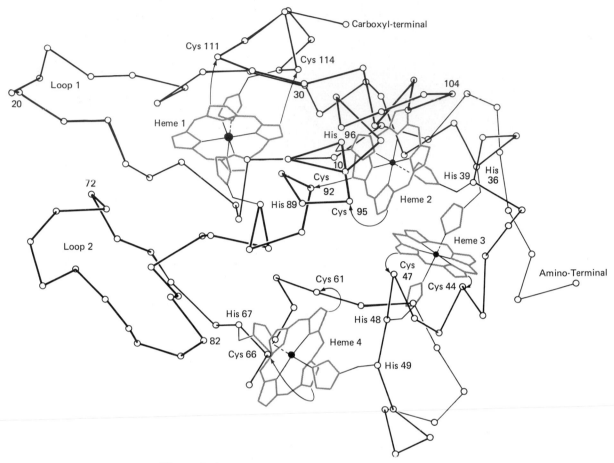

Figure 8–8

The four heme groups bound to the cytochrome c₃ from Desulfovibrio desulfuricans. *The polypeptide backbone is shown as the solid line, with the positions of the Cᵅ atoms marked. The four heme groups are indicated, along with the Cys residues, by which they are covalently attached to the protein, and the His residues, which ligate the iron atom of each heme. (Adapted from R. Haser, et al., Nature 282:806–810, 1979.)*

Figure 8–9

One subunit of the bacteriochlorophyll protein, showing the seven chlorophyll molecules inside the "string bag" made up from the polypeptide backbone, depicted as the solid line. The positions of the chlorophylls are determined by the close-packing of amino acid side chains within the protein, which are not indicated in this drawing. For clarity, only the porphyrin ring of each bacteriochlorophyll A molecule (shown in detail at the bottom) is depicted in the protein. (From B. W. Matthews and R. E. Fenna, Acc. Chem. Res. 13:308–317, 1980, © American Chemical Society.)

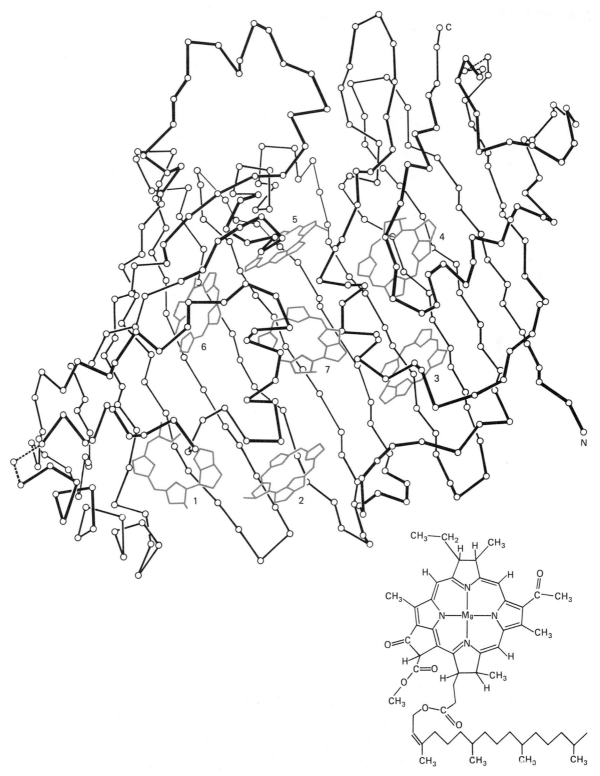

Bacteriochlorophyll A

face between them tends to be maximized. A very small ligand does not perturb the dimensions much and tends to be bound inside a rather spherical protein molecule. Somewhat larger ligands tend to bind in depressions on the surface; however, those that do not need to associate and dissociate often, such as hemes and some other prosthetic groups, are bound deep in the protein interior. Long, linear ligands, such as polysaccharides and nucleic acids, tend to be bound in clefts on the surface. If the protein and its ligand are of similar size, e.g., two associating protein molecules, their interface tends to be flat.

By using these general principles, it is often possible to guess correctly the structure of a protein–ligand complex, if the individual structures of the two components are known. Such predictions are based as well on the assumption that the protein or ligand when complexed does not adopt conformations drastically different from those it normally has individually; this appears to be the general case.

A model for the association of bovine pancreatic trypsin inhibitor with chymotrypsin and trypsin. D. M. Blow, et al. J. Mol. Biol. 69:137–144, 1972.

Structure of the complex formed by bovine trypsin and bovine pancreatic trypsin inhibitor. Crystal structure determination and stereochemistry of the contact region. A. Ruhlmann, et al. J. Mol. Biol. 77:417–436, 1973.

Principles of protein–protein recognition. C. Chothia and J. Janin. Nature 256:705–708, 1975.

Prealbumin and the thyroid hormone nuclear receptor. C. C. F. Blake. Proc. R. Soc. Lond. [Biol.] 211:413–431, 1981.

Molecular geometry of cytochrome c and its peroxidase: a model for biological electron transfer. J. Kraut. Biochem. Soc. Trans. 9:197–202, 1981.

DNA-binding proteins. Y. Takeda, et al. Science 221:1020–1026, 1983.

The Difference Fourier Technique

Once the crystal structure of a protein is known, the structures of its complexes with small ligands often can be readily determined. The ligand can be introduced into the crystal of the protein by diffusion through the large aqueous channels, or the same crystal lattice may often be obtained by crystallization of the protein and ligand together. If the crystal of the complex is isomorphous with the original crystal of the protein alone, the amplitudes and phases of each reflection will be only slightly different. The amplitudes may be measured directly, but not the phases; it is generally assumed, therefore, that the phases are unchanged by the presence of the ligand. An electron density map of the complex can then be calculated (see p. 212), using for each reflection the measured amplitude for the crystal of the complex and the phase of the normal crystal. The bound ligand should then be apparent in the map as added electron density, although not at the expected magnitude, owing to the use of somewhat incorrect phases. Nevertheless, this initial structure of the complex can be refined to produce a more accurate map, as with protein structures themselves (p. 217); this procedure is becoming more practical with advances in computing procedures.

However, the usual procedure has been to calculate a **difference Fourier map.** The Fourier calculation is exactly analogous to that in Equation 6–2 but uses the *difference* in amplitude of each reflection between the normal protein crystal and that with the ligand present. The phase is that for the normal protein crystal. The resulting difference Fourier map gives the difference in electron density between the two crystal structures. Electron density of the added ligand will be represented by positive values, and negative values will be present if something has disappeared or moved upon ligand binding. The two may compensate, as if the ligand displaces some other ligand or protein group, and no feature may appear.

The great advantage of difference Fouriers is their sensitivity to small changes that might not be apparent on comparison of the structures of the protein and the complex. If a part of the protein moves slightly upon binding the ligand, there will be a positive peak on the side to which it has moved, with a comparable negative peak on the other side. With a large movement, there will be a large negative peak in its original position and a corresponding large positive peak in its new position (Figure 8–10).

The crystallographic technique involved is analogous to that using isomorphous heavy atoms as ligands to determine the phases of the reflections of the original crystal; it differs, however, in that very small ligands may be used. The difference Fourier technique is very sensitive and accurate, permitting direct observation of ligands as small as oxygen atoms, because many of the systematic errors in Fourier maps are the same with both crystals and cancel out in the difference between them. Moreover, since a very large number of reflections from the entire protein contributes to the structure of a much smaller ligand, the changes in the intensities of the reflections need not even be apparent to give an accurate and interpretable difference Fourier map. On the other hand, the assumption that the phase of each reflection is unaltered results in values for the difference peaks of only half the expected intensity.

The mode of attachment of the azide ion to sperm whale metmyoglobin. L. Stryer, J. C. Kendrew, and H. C. Watson. J. Mol. Biol. 8:96–104, 1964.

Structure of some crystalline lysozyme-inhibitor complexes determined by X-ray analysis at 6Å resolution. L. N. Johnson and D. C. Phillips. Nature 206:761–763, 1965.

An X-ray study of azide methaemoglobin. M. F. Perutz and F. S. Mathews. J. Mol. Biol. 21:199–202, 1966.

The difference Fourier technique in protein crystallography: errors and their treatment. R. C. Henderson and J. K. Moffat. Acta Cryst. B27:1414–1420, 1971.

Crystallography of liver alcohol dehydrogenase complexed with substrates. B. P. Plapp, et al. J. Mol. Biol. 122:23–32, 1978.

Structures of product and inhibitor complexes of *Streptomyces griseus* protease A at 1.8 Å resolution. M. N. G. James, et al. J. Mol. Biol. 144:43–88, 1980.

Structure and refinement of oxymyoglobin at 1.6 Å resolution. S. E. V. Phillips. J. Mol. Biol. 142:531–554, 1980.

Amino acid activation in crystalline tyrosyl–tRNA synthetase from *Bacillus stearothermophilus*. J. Rubin and D. M. Blow. J. Mol. Biol. 145:489–500, 1981.

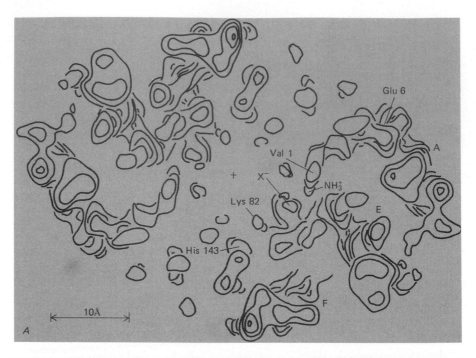

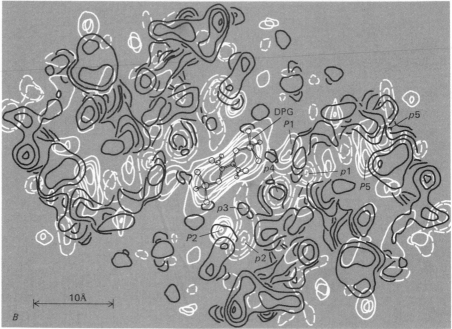

Figure 8–10

*Example of a difference Fourier map, that due to binding of diphosphoglycerate (DPG)
between the two α-amino groups of the β chains of hemoglobin. A, The electron density map
of this portion of normal human deoxyhemoglobin; the "+" indicates the symmetry axis
relating the two β chains. Helices A, E, and F are labeled, as are the residues Val 1, Glu 6*

Chemical Methods of Determining Binding Sites

Because determination of the crystal structure of a complex of protein and ligand often is not possible, other methods of determining where ligands are bound have been developed. These methods usually involve covalent modification of the protein and determination of which part of the covalent structure has reacted. Therefore, binding is related to the one-dimensional primary structure, not the tertiary structure, unless it is known independently. If binding to a single site affects a number of different parts of the covalent structure, most likely the folding of the polypeptide chain has brought them into proximity to make up the binding site.

The classical method has been to examine the reactivity of various groups on a protein in the presence and absence of the ligand. A bound ligand should make inaccessible many groups involved in its binding site and protect them from reaction with an added reagent. Such experiments are simplest if only a single group on the protein is normally modified. Otherwise, protected groups may be determined in a two-step procedure: first reacting the protein with ligand bound, to block all the accessible groups, and then removing the ligand and reacting all the groups made accessible; the second reagent is usually a radioactive form of the first.

In another approach, the modification of various groups on the free protein may be correlated with the ability subsequently to bind the ligand. If modification of a certain group produces a large decrease in affinity, it is likely to be involved directly in binding. However, results with all of these chemical modification approaches can be misleading if conformational changes of the protein follow binding the ligand or modification of the protein.

A technique that is almost certainly free of such limitations is affinity labeling. It uses the specificity of the ligand–protein interaction to label preferentially the binding site by attaching a reactive group to the ligand. Within a protein–ligand complex, the reaction between the two will be essentially intramolecular, when very high effective concentrations are possible with two groups in correct position and proximity (Table 4–7). Such

(the residue changed to Val in sickle cell hemoglobin, p. 101), Lys 82, and His 143; peak X^- is probably a sulfate or phosphate ion.

In B, the difference Fourier map (in white contours) with bound DPG is superimposed on the native map; positive differences are in solid contours, negative in dashed. The large positive peak in the center is the bound DPG molecule, and a symmetry-averaged molecule is superimposed; DPG is not symmetric but is bound on a twofold symmetry axis, so it is probably bound in the two possible orientations with equal probability.

The other positive and negative peaks indicate slight movements of the hemoglobin upon DPG binding. The positive and negative peaks P1 and p1 indicate that the α-amino groups move toward the DPG; peaks P2 and p2 show similar movements of the His 143 side chains. Negative peak p3 indicates a similar movement of Lys 82, but no compensating positive peak is apparent. Negative peak p4 indicates that the inorganic anion X^- has been displaced by the DPG binding. Other positive and negative peaks, such as P5 and p5, indicate the A helix moves slightly toward the center of the molecule.

(Taken by permission from A. Arnone, Nature 237:146–149, 1972, © Macmillan Journals Limited.)

a reaction should be much more rapid than the usual bimolecular reaction of other groups of the protein with an unbound reagent present in solution at rather low concentrations. Unfortunately, the ligand may bind in such a way that no group of the protein capable of reacting is in proximity. Therefore, negative results are not significant.

The constraint of the requirement for close proximity of an appropriate group on the protein may be relaxed by attaching the reactive group to the ligand via a flexible "arm" that may come into contact with a number of groups on the protein within its radius. The price for the convenience of this modification is that the effective concentrations of the protein and ligand groups are correspondingly lower, so that the reaction with the bound ligand is less favored over that with the free ligand.

Bifunctional reagents, with two different reactive groups, are often used in such studies. One group reacts with the ligand, to generate the affinity label; it is then added to the protein, and the second group is encouraged to react with nearby portions of the protein. The reactivities of the two groups must be controlled, so that each reacts only when required. Photoactivated groups are especially useful for the second step, because they are reactive only in the presence of light and are then highly reactive with a wide variety of groups, making reaction with the protein probable. Most commonly used are azide groups, which are totally inert until activated to the nitrene:

$$-N_3 \xrightarrow[N_2]{\text{light}} -\ddot{N}\!\!: \qquad\qquad (8\text{--}23)$$

$$\text{Azide} \qquad\qquad \text{Nitrene}$$

Nitrenes are very reactive, even toward methyl and methylene ($-CH_2-$) groups.

Bifunctional reagents with the same reactive group at each end are also useful if both protein and ligand are reactive when bound to each other. This is the case when the ligand is large or also a protein, as in complicated macromolecular structures, such as ribosomes, viruses, chromatin, membranes, or enzyme complexes. Whether or not the individual components are in close proximity in the complex can often be inferred by whether or not they are cross-linked by bifunctional reagents. When the different components are polypeptide chains, those cross-linked can usually be determined most readily by SDS electrophoresis (Figure 8–11).

This approach relies simply on the tendency of a bifunctional reagent that has reacted at one end with a protein to react readily at the other end with another component of the complex. The two reactive groups of the reagent usually must be inactivated by another competing reaction with some other component of the solvent in order to inactivate those reagent molecules that cannot react readily at both ends. A wide variety of intra- and inter-polypeptide cross-links is usually generated, as well as modifications in which only one end of the reagent has reacted. Therefore, this approach usually identifies not the groups but only the polypeptides that have reacted.

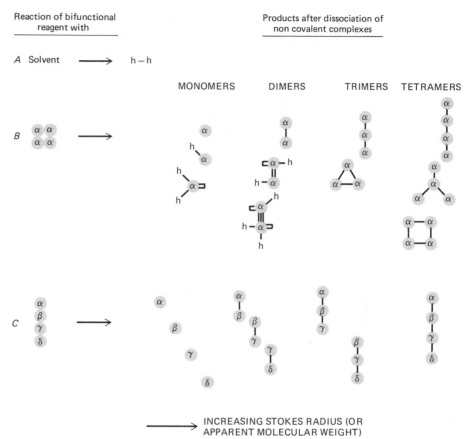

| Reaction of bifunctional reagent with | | Products after dissociation of non covalent complexes | | | |

Figure 8-11

Schematic diagram of the expected products of a limited cross-linking reaction, under dilute conditions where intermolecular contacts are not significant. The competing reaction of the bifunctional reagent with the solvent is shown in A; h is one solvolyzed group. Cross-linking of a tetramer of α chains is shown in B. The possibilities of intrachain cross-links and of monofunctionally linked reagent are shown for the monomer and dimers only; they are present by implication in all of the other complexes. Cross-linking of a linear complex of one molecule of each of α, β, γ, and δ chains is shown in C; note that polypeptide α is not expected to be cross-linked directly to γ or δ, nor β to δ, so only the indicated dimers and trimers would be expected. (Adapted from K. Peters and F. M. Richards, Ann. Rev. Biochem. 46:523–551, 1977.)

The versatility of the approach lies in that the length of the cross-linking group may be varied, in order to serve as a molecular "ruler" for measuring the distances between two molecules in a complex. The nature of the bifunctional reagent may also be varied. For example, membrane proteins that are in proximity at the surface of the membrane should be cross-linked with a polar reagent that is confined to the aqueous solvent; nonpolar reagents that permeate the membrane should cross-link protein molecules that tend to interact there. Cross-linking with bifunctional reagents has been most useful with very large complexes, composed of numerous individual molecules, too large or nonsymmetric to be studied crystallographically.

Covalent labelling of active sites. S. J. Singer. Adv. Protein Chem. 22:1–54, 1967.

Chemical modification by active-site-directed reagents. E. Shaw. In The Enzymes, 3rd ed. P. D. Boyer (ed.). Vol. 1, pp. 91–146. New York, Academic Press, 1970.

Use of dimethyl suberimidate, a cross-linking reagent, in studying the subunit structure of oligomeric proteins. G. E. Davies and G. R. Stark. Proc. Natl. Acad. Sci. U.S.A. 66:651–656, 1970.

Bifunctional reagents. F. Wold. Methods Enzymol. 25:623–651, 1972.

Photogenerated labels for biological receptor-site labelling. J. R. Knowles. Acc. Chem. Res. 5:155–160, 1972.

Chemical cross-linking: reagents and problems in studies of membrane structure. K. Peters and F. M. Richards. Ann. Rev. Biochem. 46:523–551, 1977.

Affinity labelling. W. B. Jakoby and M. Wilchek (eds.). Methods Enzymol. 46:1977.

Photoaffinity labelling of biological systems. V. Chowdhry and R. H. Westheimer. Ann. Rev. Biochem. 48:293–325, 1979.

Chemical cross-linking in biology. M. Das and C. F. Fox. Ann. Rev. Biophys. Bioeng. 8:165–193, 1979.

Mapping of contact areas in protein–nucleic acid and protein–protein complexes by differential chemical modification. H. R. Bosshard. Methods Biochem. Anal. 25:273–301, 1979.

SPECIFICITY OF INTERACTIONS

Can we account for the particular affinity of a protein for any given ligand? Do the structures of the two give any insight into why the affinity is high or low? Only a qualified yes is possible in answer to these questions, in that generally only qualitative conclusions are possible. Yet it should be possible eventually to account for, or even to predict, binding affinities for ligands and to design ligands or proteins with useful binding properties. The implications are immense for chemotherapy and drug design.

The practical difficulties in rationalizing ligand affinities arise from our poor understanding of the energetics of protein structure, as described in Chapter 7, and even of the basic interactions in an aqueous solvent, discussed in Chapter 4. The binding affinity represents the difference in free energy between the complex and the individual protein and ligand, all in aqueous solution. All the factors involved in their interactions with each other and with the solvent are involved, as well as changes in their degrees of freedom, or entropies. Many of the individual factors will compensate each other, and the net, observed effect is probably a small difference remaining after several terms of large and uncertain magnitude are accounted for in some way.

Accounting for Relative Affinities

Interactions between a protein and a ligand always involve a substantial number of individual groups. The general approach to understanding the observed affinity has been to attempt to dissect it into the contributions of each group by measuring the effect of removing them individually. Because this is easiest with the ligand, the method has involved comparison of the affinities of slightly different ligands.

It might be thought the total binding energy would be simply the sum of that of each group, but binding energies are not readily measured from affinity constants (see p. 338). As an example, first given by Jencks, consider a ligand composed of two parts, A and B: one capable of hydrogen bonding, A, and the other hydrophobic, B. The affinity of AB is compared with that of A and B individually:

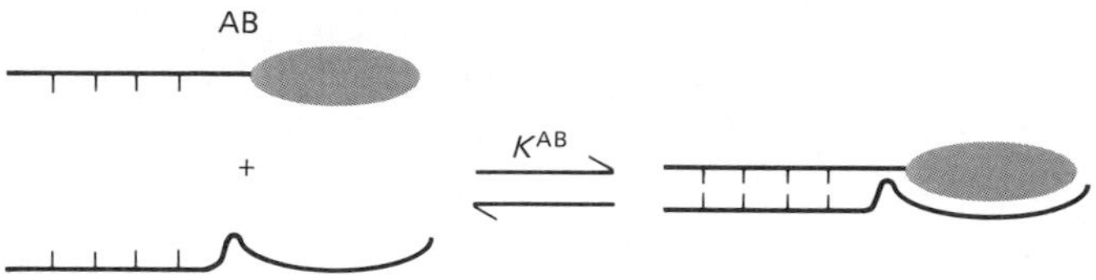

AB

$$K^{AB} \tag{8-24}$$

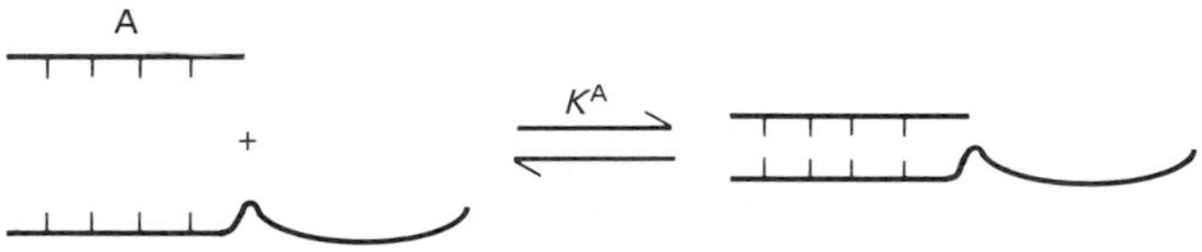

A

$$K^{A} \tag{8-25}$$

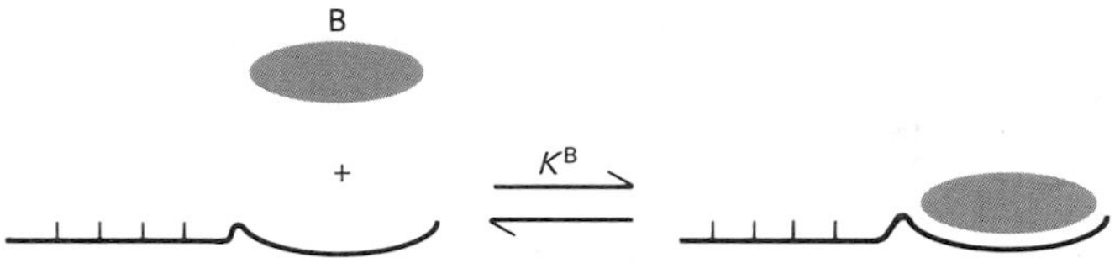

B

$$K^{B} \tag{8-26}$$

where the equilibrium constants K^{AB}, K^{A}, and K^{B} are for either association or dissociation. In general, there is no simple relationship between them, and classical binding energies calculated from them are not additive:

$$-RT \ln K_a^{AB} \neq -RT \ln K_a^{A} - RT \ln K_a^{B} \tag{8-27}$$

even if the standard state is taken as 55 M so that "unitary" binding energies are calculated (see p. 338). This is illustrated by the binding to the protein avidin of biotin and some derivatives in Table 8–2.

The reason for the nonadditive nature of specific binding affinities becomes clear if the binding of AB is dissected into individual steps:

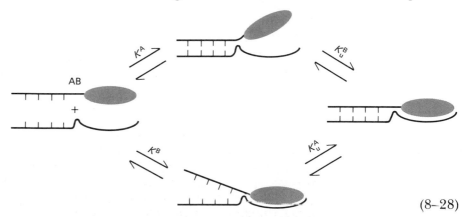

$$\tag{8-28}$$

The initial binding of each part might be analogous to that when it alone is present, and these first steps are assigned the same binding constants.

Table 8–2 Binding of Biotin Derivatives to Avidin

DERIVATIVE	DISSOCIATION CONSTANT (M)	FREE ENERGY CONTRIBUTION TO BINDING (kcal/mole)
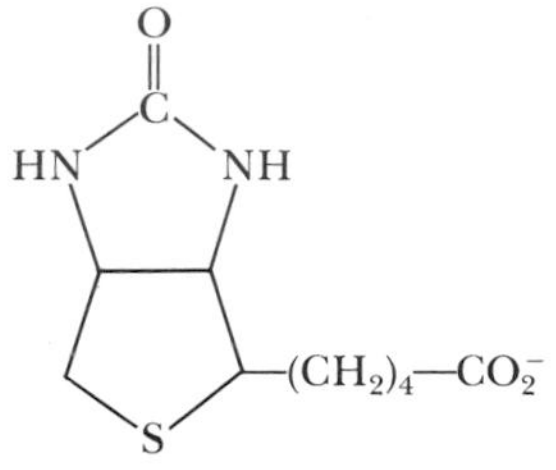 Biotin	1.3×10^{-15}	
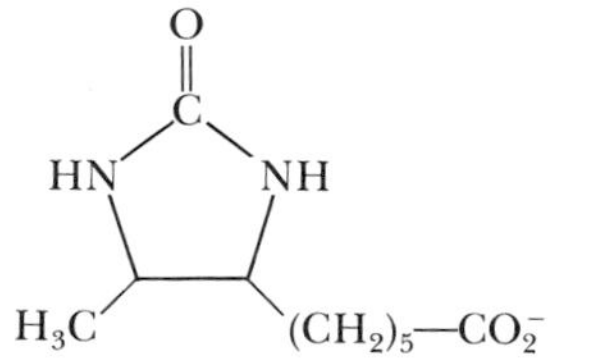 Desthiobiotin	5×10^{-13}	
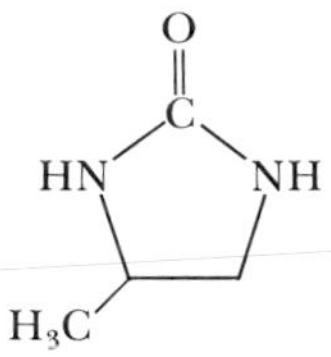	3.4×10^{-5}	-13.3
CH_3—$(CH_2)_4$—CO_2^-	3×10^{-3}	-10.7

Data from N. M. Green, Adv. Protein Chem. 29:85–133, 1975.

However, the second step of binding the other part of the ligand is now a unimolecular step, not a bimolecular step as in binding of that part individually. Consequently, the bimolecular binding constants K^A and K^B do not apply; instead, these steps are assigned the unimolecular equilibrium constants K_u^A and K_u^B. The values of these constants are not independent, since

$$K^{AB} = K^A K_u^B = K^B K_u^A \tag{8–29}$$

The constants K_u^A and K_u^B are often considered the intrinsic binding constants for the A and B parts of the ligand, since they are the ratios of the binding constants of ligands with and without that moiety. The contributions of the free energy of binding of that moiety can then be calculated.

$$\Delta G_i^A = -RT \ln K_u^A = -RT \ln \frac{K_a^{AB}}{K_a^B} \tag{8–30}$$

$$\Delta G_i^B = -RT \ln K_u^B = -RT \ln \frac{K_a^{AB}}{K_a^A} \qquad (8\text{--}31)$$

For example, the free energy contribution to binding of the sulfur atom of biotin (ΔG_i^S) may be estimated from the relative affinities of the first two compounds of Table 8–2 to be -3.5 kcal/mole:

$$\Delta G_i^S = -RT \ln \frac{5 \times 10^{-13} \text{M}}{1.3 \times 10^{-15} \text{ M}} = -3.5 \text{ kcal/mole} \qquad (8\text{--}32)$$

Similarly, the contributions of the remaining five-membered ring and of the acidic hydrocarbon group may be estimated to be -13.3 and -10.7 kcal/mole, respectively.

Such values give a measure of the increased affinity caused by the presence of a group of the ligand, but they need not represent intrinsic binding affinities, for they depend critically upon the relationship between the two parts of the ligand during binding. For example, the two parts A and B could be imagined to bind tightly individually, but their simultaneous binding to require strain and consequently be less favored. The binding affinity of either part would then be decreased by the addition of the other. In this case, the values of K^A and K^B for the individual parts would more accurately reflect their intrinsic affinities.

These binding constants would apply to the equilibrium constant for binding that part of the ligand in the second, intramolecular step, but multiplied by the effective concentration of that part when the other part of the ligand is bound, [A, B].

$$K_u^A = K_a^A [A, B] = \frac{[A, B]}{K_d^A} \qquad (8\text{--}33)$$

$$K_u^B = K_a^B [A, B] = \frac{[A, B]}{K_d^B} \qquad (8\text{--}34)$$

The same effective concentration applies to both parts, since these are linked functions (see Eqs. 8–28 and 8–29).

The effective concentration of either group in an intermediate complex can conceivably be zero, when it is kept away from the binding site and then provides no contribution to binding; be negative, if it interferes with binding; or vary, up to 10^{10} M (Table 4–7) when the two parts A and B of the ligand are always in optimal orientation for simultaneous binding to a perfectly complementary binding site. The large values of effective concentrations in intramolecular reactions result from the entropic effect of covalent linkage of the two parts. A ligand must lose a substantial amount of translational and rotational entropy upon binding; this is one of the factors determining the values of both K^A and K^B. However, in the case of ligand AB, at least some of this entropy is lost in binding the first portion; the remainder of the ligand is now fixed to some extent and need not lose as much entropy upon completing binding as that lost when that part bound by itself. Consequently, the greater the rigidity between the two parts of the ligand, the greater is likely to be the entropic contribution to the effective concentration.

The effective concentration, [A, B], is also affected by any strain in the complex of ligand AB that is not present with either of the two individual parts. For example, if the covalent linkage between parts A and B must be distorted for both to be bound simultaneously, the effective concentration is lowered. This effect is the primary reason for questioning whether values of K_u^A and K_u^B should be regarded as intrinsic binding constants.

Nevertheless, where strain is apparently not involved, very high effective concentrations and free energy contributions to binding may be observed. For example, the data of Table 8–2 for the two halves of desthiobiotin imply that their effective concentrations in the hypothetical intermediate complex are 2×10^5 M, since

$$[A, B] = \frac{K_a^{AB}}{K_a^A K_a^B} = \frac{K_d^A K_d^B}{K_d^{AB}} \qquad (8-35)$$

With such high effective concentrations, ionic and hydrogen bond interactions between ligand and protein may contribute substantially to binding, even in aqueous solution, where they must compete with intermolecular interactions between the solvent and the free protein and free ligand.

Some examples of large incremental contributions to binding, measured by the relative affinities of ligands differing only in the group, are tabulated in Table 8–3. It should be noted that the values for the nonpolar groups are considerably greater than their free energies of transfer from water to nonpolar liquids, which often is considered an analogous process. This discrepancy is further evidence that a protein—at least, its binding site—is not equivalent to an organic liquid. Instead, it has a higher concentration of atoms and probably presents a more rigidly defined cavity for binding nonpolar groups. Consequently, if a part of the ligand normally involved in binding is missing, there might be a void in the interface between protein and ligand; it could be filled by an isolated solvent molecule, or protein and ligand could adapt to attain complementarity. Each possibility seems to be energetically unfavorable.

The data of Table 8–3 demonstrate the origin of protein specificity for ligands. A protein can discriminate very effectively between its proper ligand and one lacking just one small part. Discrimination of ligands containing extra groups can be even more powerful, since the additional groups can interfere sterically with the complementarity between ligand and binding site.

Nevertheless, there are limits to the specificity possible, set by the energetics of the interactions between groups. These limits are exceeded in some instances where extreme specificity is necessary—for example, in the replication, transcription, and translation of the genetic information. DNA replication occurs with an error frequency of 10^{-10}, even though the tautomerization of the bases, which gives incorrect base-pairing, occurs with a frequency of 10^{-5}. Amino acids are also incorporated into proteins

Table 8–3 *Large Contributions to Binding Measured by Affinities to Proteins*

GROUP	FREE ENERGY CONTRIBUTION TO BINDING TO PROTEIN[a] (kcal/mole)	FREE ENERGY OF TRANSFER FROM WATER TO NONPOLAR LIQUID[b] (kcal/mole)
—CH_3	−2.0 to −3.9	−0.5
—CH_2CH_3	−6.5	−1.0
—CH—$(CH_3)_2$	−9.6	−1.5
—CH_2—CH_2—CH_2—CH_3	−7 to −8	−2.6
—SCH_3	−4.9	
—CH_2—CH_2—S—CH_3	−10 to −11	−1.3
—SH	−5.4 to −9.1	
—OH	−8	
—NH_2	−4.5	
—$NH_3{}^+$	−6.7	
—$CO_2{}^-$	−4.3	

[a]Data from W. P. Jencks, Proc. Natl. Acad. Sci. U.S.A. 78:4046–4050, 1981; and A. R. Fersht, Proc. R. Soc. Lond. [Biol.] 212:351–379, 1981.

[b]Data from Y. Nozaki and C. Tanford, J. Biol. Chem. 246:2211–2217, 1971.

with considerably greater fidelity than that expected, even from the data of Table 8–3, much of which comes from binding of amino acids to tRNA synthetases. For example, how does the tRNA synthetase, which carries out the most crucial step of attaching the correct amino acid to the correct tRNA molecule, discriminate effectively against Gly when adding Ala, Val in the case of Ile, and Ser in the case of Thr? Each of these pairs differ only by one —CH_2— group. The answer in this case is that the enzyme seems to check twice, discriminating at the first binding step, then subjecting the selected amino acid to a second check designed to detect the most likely fraudulent amino acid. Any caught at the second step are hydrolyzed from the tRNA and expelled. In such a "double-sieve" editing mechanism, the probability that an incorrect amino acid will be missed by both steps is the product of the two individual probabilities (e.g., $10^{-5} \times 10^{-5} = 10^{-10}$). In this way, biological specificities can be greatly enhanced over those possible with simple physical principles, and comparable multiple checks on specificity appear to be used in DNA replication. However, such methods are used only when absolutely necessary, because there is a "cost" involved in that a certain fraction of correct molecules are also removed at the subsequent recognition steps, owing to the intrinsic limitations on binding specificity.

Binding energy, specificity, and enzymic catalysis. W. P. Jencks. Adv. Enzymol. 43:219–410, 1975.

Entropy, binding energy, and enzymic catalysis. M. I. Page. Angew. Chem. Int. Ed. Engl. 16:449–459, 1977.

On the attribution and additivity of binding energies. W. P. Jencks. Proc. Natl. Acad. Sci. U.S.A. 78:4046–4050, 1981.

Enzymic editing mechanisms and the genetic code. A. R. Fersht. Proc. R. Soc. Lond. [Biol.] 212:351–379, 1981.

Relationship between Protein Conformation and Binding

Binding sites comprise relatively little of the protein structure, often just a small patch on the surface; relatively small, localized alterations of the protein, therefore, can produce large changes in its binding of ligands, without changes in the overall protein conformation. Consequently, similar homologous proteins can bind very different ligands.

A good example is the trypsin protease family (Table 6–1), the members of which all have similar folded conformations. They all bind and hydrolyze peptides and proteins in similar ways, binding about five contiguous residues of the substrates. There is only weak specificity for the side chains of the substrate, except for the residue immediately preceding the peptide bond to be hydrolyzed. The trypsin proteases differ markedly in their specificity for this amino acid side chain; this variation is due to changes of only a few amino acid residues in their otherwise homologous binding sites. Trypsin is quite specific for Arg and Lys residues and has a deep binding site with an Asp residue at the bottom to neutralize the charge of the Arg or Lys residue. Chymotrypsin is relatively specific for aromatic residues (Phe, Tyr, and Trp) and has a large narrow hydrophobic pocket to accommodate their planar aromatic rings. The residue at the bottom of the pocket, which is Asp in trypsin, is Ser in chymotrypsin. Elastase prefers short nonpolar side chains as substrates, and its binding site is less deep owing to the presence of large bulky residues (e.g., Val 216 and Thr 226) at the bottom of the binding pocket, where these two residues are both Gly in trypsin and chymotrypsin (Figure 8–12). The bacterial *Streptomyces griseus* protease A is similar to chymotrypsin in its specificity, preferring large hydrophobic residues, and has a similar large hydrophobic binding site. The highly homologous α-lytic protease prefers small neutral residues and has its binding site shortened by the insertion of five residues into its polypeptide chain at that point, after residue 217.

The very many detailed studies of ligand binding to the trypsin proteases have utilized a wide variety of inhibitors of these destructive enzymes, many of natural origin. One example is bovine pancreatic trypsin inhibitor, BPTI, described in Figures 6–8 to 6–10, 6–12, 6–16, 6–17, and 6–22. Such inhibitors bind tightly but have relatively rigid conformations designed to be closely complementary to the active sites of the proteases they inhibit.

Figure 8–12

Differences in the primary binding sites of α-chymotrypsin and elastase that account for their different specificities. At the top is shown α-chymotrypsin with formyl L-tryptophan bound in the center; the deep binding pocket, which adequately contains the Trp side chain, is due in part to the absence of side chains on Gly 216 and Gly 226. These residues are respectively Val and Thr in elastase (bottom), blocking the pocket so that only small side chains may fit; formyl L-alanine is shown bound. (From B. S. Hartley and D. M. Shotton, in P. D. Boyer (ed.), The Enzymes, 3rd ed. Vol. 3, pp. 323–373. New York, Academic Press, 1971, © Academic Press Inc.)

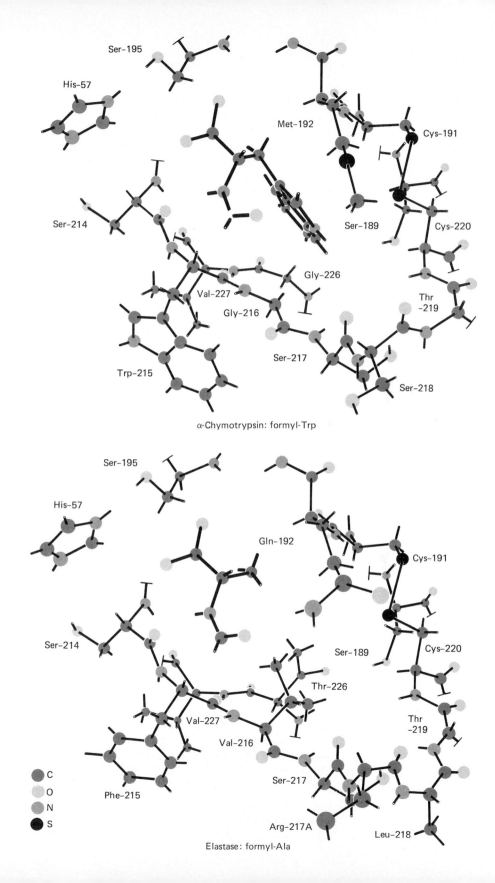

Ser-195

His-57

Met-192

Cys-191

Ser-214

Ser-189

Cys-220

Gly-226

Thr
-219

Val-227

Gly-216

Trp-215

Ser-217

Ser-218

α-Chymotrypsin: formyl-Trp

Ser-195

His-57

Gln-192

Cys-191

Ser-214

Ser-189

Cys-220

Thr-226

Thr
-219

Val-227

Val-216

Phe-215

Ser-217

Arg-217A

Leu-218

C
O
N
S

Elastase: formyl-Ala

Other homologous proteins that bind different ligands are known. The immunoglobulins, described in the following section, provide the most dramatic examples of the variation possible.

Conversely, the same ligand can be bound by unrelated proteins with quite different conformations. This can be illustrated most spectacularly with binding of the heme group (iron-protoporphyrin IX) by a wide variety of proteins, the globins, and various cytochromes. The globins (i.e., hemoglobins, myoglobin, erythrocruorin, leghemoglobin) are all homologous and bind the heme group similarly (Figure 6–27). Very different are the structures of a large number of cytochrome c–like proteins (Figure 6–28), cytochrome b_5, the similar cytochromes c' and b_{562}, and cytochrome c_3, which binds four heme groups in different ways (Figure 8–8).

Thus, a number of different proteins appear to have acquired independently the ability to bind heme groups; the ability to bind the same ligand, therefore, cannot be used to imply that two proteins are related. Yet the structural constraints of binding the same ligand could conceivably produce selective pressures to converge to similar structural features; such proteins might erroneously be thought to be related evolutionarily by very distant divergence. A relatively clearcut example of convergence is the similar geometries of the active sites of the protease subtilisin and the trypsin proteases (Figure 9–8), since the remainders of the structures are clearly not related. However, more intermediate degrees of similarity would be confusing.

Such confusion about evolutionary convergence or divergence has arisen with regard to the nucleotide-binding domains of proteins that bind the dinucleotide coenzymes NAD, NADP, and FAD, or the single nucleotides ATP, ADP, AMP, FMN, and others. These binding sites on proteins generally are formed by the **"Rossmann fold,"** in which each nucleotide is bound by the arrangements of loops between the carboxyl ends of two or three parallel β-strands with their connecting α-helices (Figure 6–30). All dehydrogenases of known structure utilize NAD or NADP as coenzymes, and all have the same dinucleotide binding domain with the Rossmann fold. However, they catalyze the dehydrogenation of different substrates and often have different catalytic domains for binding the substrate; the arrangements of these domains in the primary structure differ considerably (Figure 8–13). In some cases, as with lactate dehydrogenase and malate dehydrogenase, the catalytic domains are also similar; these enzymes probably *did* arise by evolutionary divergence.

The similar nucleotide-binding domains in the other dehydrogenases and in the other nucleotide-binding proteins could conceivably also have arisen by divergence from an ancestral βαβ nucleotide-binding domain, but there often are rather drastic differences in the way in which the nucleotide is bound, and the orders of the strands within the β-sheets are often different. There may also be general geometrical and structural reasons for binding ligands at such positions; in particular, the binding site can be altered without interfering with the central secondary structure, which presumably provides the conformational stability. Furthermore,

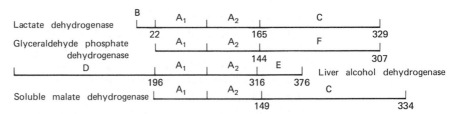

Figure 8–13
Positions within the polypeptide chains of the two nucleotide-binding domains (A$_1$ and A$_2$)
that bind NAD in the dehydrogenases. The different catalytic domains are designated B
through F. The C domains in lactate dehydrogenase and malate dehydrogenase are homolo-
gous. (From W. Eventoff and M. G. Rossmann, Trends Biochem. Sci. 1:227–230, 1976,
© Elsevier Biomedical Press B.V.)

binding acidic nucleotides at the amino ends of the accompanying α-helices
can be favored by their dipoles, which tend to supply net positive charge.
The similar folding patterns may then be fortuitous or the result of evolutionary convergence. This problem is currently an issue of intense debate
and speculation.

Homologies in serine proteinases. B. S. Hartley. Philos. Trans. R. Soc. Lond. [Biol.]
257:77–87, 1970.

Protein proteinase inhibitors. M. Laskowski, Jr., and R. W. Sealock. In The Enzymes, 3rd ed. P. D. Boyer (ed.). Vol. 3, pp. 375–423. New York, Academic
Press, 1971.

Chemical and biological evolution of nucleotide binding proteins. M. G. Rossmann,
et al. Nature 250:194–199, 1974.

Structural and functional similarities within the coenzyme binding domains of dehydrogenases. I. Ohlsson, et al. J. Mol. Biol. 89:339–354, 1974.

The evolution of dehydrogenases and kinases. W. Eventoff and M. G. Rossmann.
Crit. Rev. Biochem. 3:111–140, 1975.

Tertiary structural differences between microbial serine proteases and pancreatic
serine enzymes. L. T. J. Delbaere, et al. Nature 257:758–763, 1975.

Secondary and conformational specificities of trypsin and chymotrypsin. H. T.
Wright. Eur. J. Biochem. 73:567–578, 1977.

Structure of pyruvate kinase and similarities with other enzymes: possible implications for protein taxonomy and evolution. M. Levine, et al. Nature
271:626–630, 1978.

Structural comparisons of heme binding proteins. P. Argos and M. G. Rossmann.
Biochemistry 18:4951–4960, 1979.

Dihydrofolate reductase from *Lactobacillus casei*. Stereochemistry of NADPH binding. D. A. Matthews, et al. J. Biol. Chem. 254:4114–4151, 1979.

Molecular structure of the α-lytic protease from Myxobacter 495 at 2.8 Å resolution. G. D. Brayer, L. T. J. Delbaere, and M. N. G. James. J. Mol. Biol.
131:743–775, 1979.

Structure of the complex formed between the bacterial-produced inhibitor chy-
mostatin and the serine enzyme *Streptomyces griseus* protease A. L. T. J. Delbaere and G. D. Brayer. J. Mol. Biol. 139:45–51, 1980.

Relation between structure and function of α/β proteins. C.-I. Brändén. Quart.
Rev. Biophys. 13:317–338, 1980.

Protein inhibitors of proteinases. M. Laskowski, Jr., and I. Kato. Ann. Rev. Biochem.
49:593–626, 1980.

Immunoglobulins   If no ligands are known for a given protein, they can be "made to order" by preparing antibodies against it. In the normal immunogenic response, a wide variety of antibodies are produced, each one recognizing a portion of the protein surface and having its own affinity for that antigenic determinant. More recently, it has become possible to produce homogenous, monoclonal antibodies. Each antibody appears to recognize the unique spatial distribution of groups in one localized site on the surface of the protein antigen and thus is sensitive to both conformation and amino acid sequence. Antibodies produced against the unfolded protein generally do not bind to the folded protein, and vice versa. Those that do bind to the same groups in both are primarily conformation-specific and may be used to measure the probability that any conformational state will adopt the other (p. 319).

Being proteins, antibody molecules are of great interest in themselves. They can bind specifically a wide variety of antigens, not just other proteins, and thus are capable of overwhelming diversity. Yet this binding diversity is combined with common structural and functional features. The consequences of ligand binding include elimination of the antigen –antibody complex from the bloodstream, complement-induced lysis of cells, histamine release, and stimulation of secretion of antibodies by lymphocytes, depending upon the class of antibody. All immunoglobulin molecules of a given class need to have these common functions, which are then combined with variation of specificity for different antigens. How they are combined is a major question that is only now being answered in molecular terms. The origin of the diversity of the amino acid sequences of the variable regions of the antibody heavy (H) and light (L) chains, especially of the hypervariable sub-regions, was discussed briefly in Chapter 2. Here we shall consider the structural aspects of antibodies; they illustrate very dramatically how binding different ligands does not require different protein conformation.

The basic structure of intact immunoglobulins is a Y-shaped molecule composed of two H and two L chains; in immunoglobulins A and M these molecules are further aggregated. The L chain consists of V_L and C_L domains; the H chains, of V_H, C_{H1}, C_{H2}, and C_{H3} domains. All domains are homologous in primary structure; their arrangement within the Y-shaped molecule is shown schematically in Figure 8–14A. The polypeptide chains between the domains have been shown experimentally to be susceptible to proteases, generating a variety of fragments that led to this schematic picture. Most susceptible is the hinge region linking the two arms to the base of the Y; after cleavage, the two arms are known as the F_{ab} fragments, the base as the F_c fragment. The F_{ab} fragments have the V_L and V_H domains and retain the antigen-binding sites.

A variety of homogenous immunoglobulins and fragments from myelomas have been characterized crystallographically (Table 6–1). They all show that the individual domains have very similar conformations, designated the **immunoglobulin fold,** as expected from their homologous amino acid sequences. The immunoglobulin fold consists of two layers of antiparallel β-sheets (Figure 8–15). The domains interact with each other in a variety of ways in an intact H_2L_2 molecule (Figure 8–14*B*). The two C_{H2} domains interact with each other, as do the two C_{H3} domains, in the F_c portion. Each arm of the molecule is composed of one C_{H1} domain interacting with one C_L domain, plus the V_H and V_L domains. All pairs of C domains associate in a similar manner in which the members of one of their pairs of β-sheets associate isologously, but the V domains use the other β-sheet. Less extensive and presumably less stable interactions also take place between domains adjacent in the primary structure, maintaining the overall shape of the molecule. Nevertheless, the segments of polypeptide chain linking the domains have substantial flexibility, especially the hinge region linking the F_{ab} and F_c portions, which gives the two F_{ab} arms significant flexibility. All of the interdomain interactions involve primarily the packing of amino acid side chains; their variation within the constraints of a common folded conformation, therefore, is possible mainly by varying the amino acid sequence in the appropriate place.

The antigen-binding sites are at the tips of the two F_{ab} arms, situated between the V_L and V_H domains. They are comprised of the residues of the three hypervariable regions of both the L and H chains (Figure 2–2), which are located in irregular loops between the β-strands (Figure 8–16). Consequently, different binding sites are generated with different amino acid side chains in these positions. In addition, deletions and insertions of residues in these positions can alter dramatically their local folding and the topology of the binding site. In contrast, the residues that form the interface between the V_L and V_H chains are highly conserved, so that different chains can be combined to give different specificities.

Unfortunately, the antigens against which these homogenous myeloma immunoglobulins of known structure are directed are not known. Extensive screening using a wide variety of relatively small molecules has identified some for which they have significant affinity, and crystallographic analysis has demonstrated that they bind in reasonable fashion to the presumed antigen-binding sites and that no unique principles appear to be involved (Figure 8–17). However, being small molecules, they only partially fill the sites; therefore, it is currently impossible to assess the details of the specificity and variability of antibodies. Such detailed studies should be possible and of great interest with monoclonal antibodies induced toward known antigens.

A major question to be answered is how binding of antigens to the binding sites at the tips of the F_{ab} arms triggers the effector functions, such as complement activation, which are a property of the F_c portion of the molecule. There is little evidence for conformational changes in immunoglobulins upon antigen binding; indeed, effector functions may well be triggered primarily by the formation of large antigen–immunoglobulin aggregates. In particular, the first protein of the classical complement activation pathway, C1q, is a complex structure resembling a bunch of tulips,

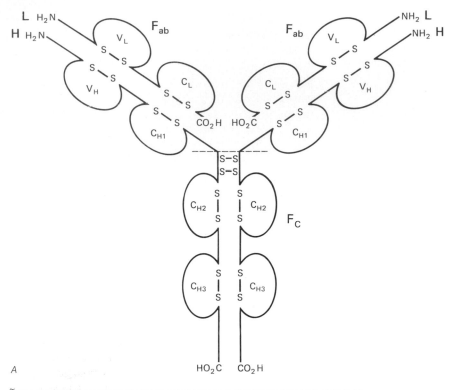

A

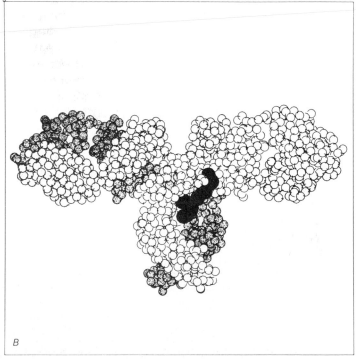

B

Figure 8–14
Schematic representation of a typical immunoglobulin structure (A) *and a space-filling model determined crystallographically* (B).

The schematic in A *shows the intramolecular disulfides linking Cys residues about 60 residues apart in the primary structure that are characteristic of each immunoglobulin domain. The site of cleavage by papain is shown by the dashed line, which yields two F_{ab} fragments and one F_c. If cleavage occurs on the carboxyl side of the disulfide linking the H chains, as occurs with pepsin, the two F_{ab}-like fragments are linked by the disulfide and are generally designated as $F_{ab}{}'$.*

In B, *each sphere represents one amino acid residue. One complete heavy chain is white, the other heavily stippled; both light chains are white. The carbohydrate attached to the C_{H2} domain of each heavy chain is black. The antigen-binding sites are at the tips of the F_{ab} arms, at the far left and right, where the V_H and V_L domains meet. (From E. W. Silverton, et al., Proc. Natl. Acad. Sci. U.S.A. 74:5140–5144, 1977.)*

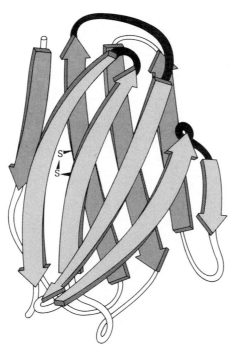

Immunoglobulin V_L domain

Figure 8–15
The immunoglobulin fold. Two layers of antiparallel β-sheet are folded on top of each other to form a sandwich-like structure. Between the two layers are hydrophobic side chains and the indicated disulfide bond linking two Cys residues about 60 residues apart; one Cys residue is in the middle of the second strand from the left of the top sheet, the other is in the middle of the second strand of the bottom sheet. In V_L domains, as illustrated here, the top β-sheet has five strands, the bottom four. The two strands at the right edge of the five-strand sheet are missing in C domains. The loops containing the hypervariable regions in V domains are shaded dark. (Adapted from J. Richardson, Adv. Protein Chem. 34:167–339, 1981.)

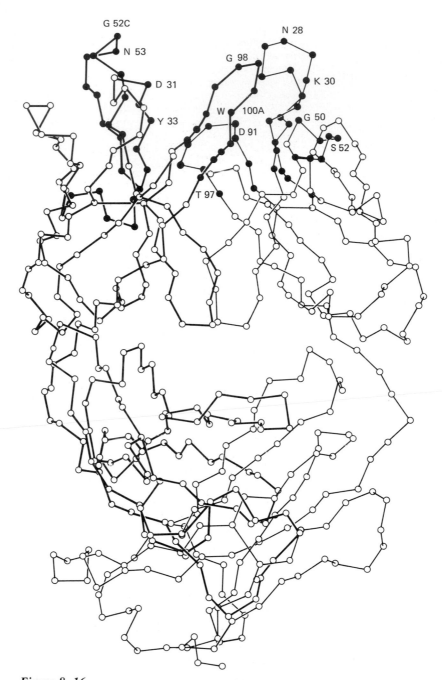

Figure 8–16
The α-carbon backbone of the F_{ab} fragment of immunoglobulin M_{603}, illustrating the hyper-variable residues (●, at top) comprising the antigen-binding site. Two residues of each such segment are identified, using the one-letter code for the amino acids (see Figure 1–1). The heavy chain is shown with thicker lines than the light chain. The top half of the structure comprises the V_H and V_L domains, the bottom half of the C_L and C_{H1} domains. (Figure kindly provided by D. R. Davies and H. Metzger.)

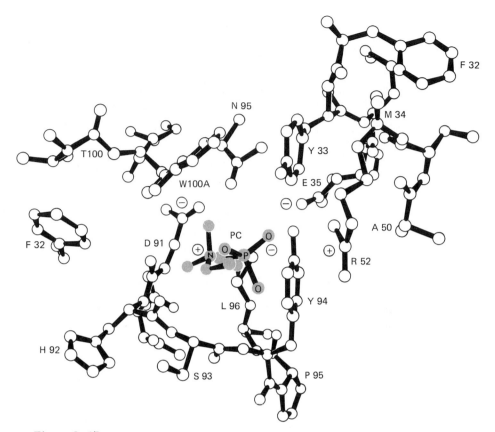

Figure 8–17
Phosphocholine bound to the antigen-combining site of immunoglobulin M_{603}. The view is from the solvent onto the surface of the immunoglobulin, i.e., from the top of Figure 8–16. Phosphocholine (PC) is black, with its phosphate group closest to the solvent; the choline group is inserted into the interior of the combining site. The adjacent residues of the immunoglobulin are shown, identified using the one-letter code for the amino acids (see Figure 1–1). The lower residues (D_{91} to L_{96}) are from the third hypervariable region of the light chain. The other residues are from the three hypervariable regions of the heavy chain. (Figure kindly provided by D. R. Davies and H. Metzger.)

with six globular heads joined by six collagen-like stems that are held together in the lower half. Each head binds to the C_{H2} domain of an immunoglobulin; the simultaneous binding of many heads to aggregated immunoglobulins and antigens may be the trigger that sets off complement activation.

Affinity labelling and topology of the antibody combining site. D. Givol. Essays Biochem. 10:73–103, 1974.
Three-dimensional structure of immunoglobulins. D. R. Davies, et al. Ann. Rev. Biochem. 44:639–667, 1975.
Studies on the three-dimensional structure of immunoglobulins. R. J. Poljak, et al. Prog. Biophys. Mol. Biol. 31:67–93, 1976.

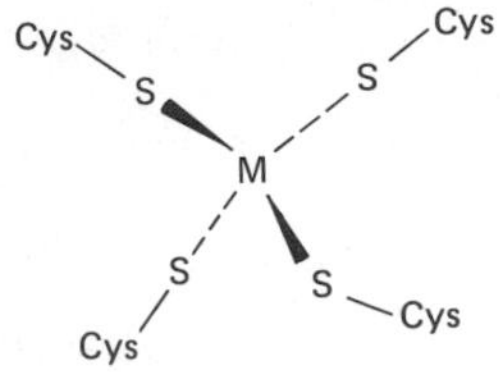

1

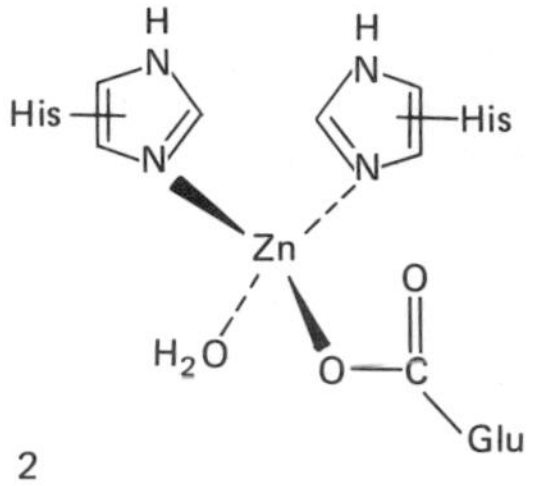

2

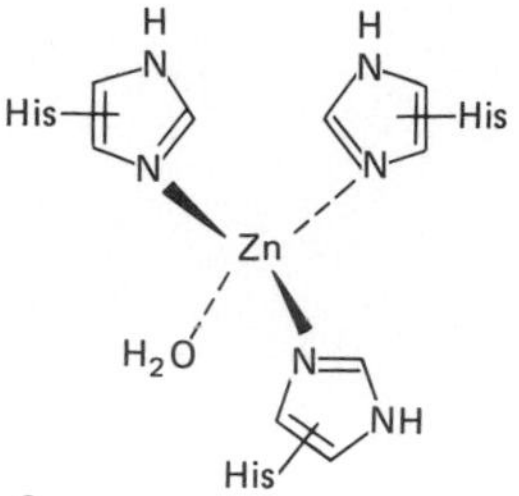

3

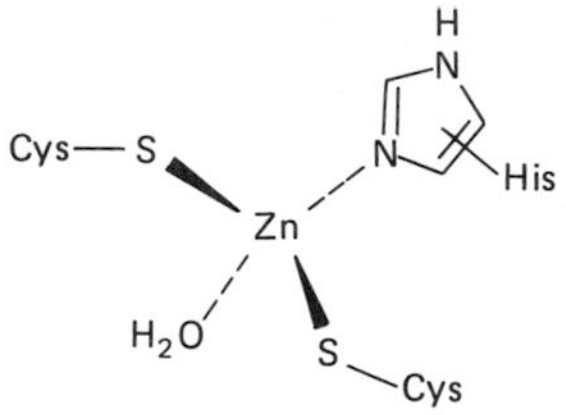

4

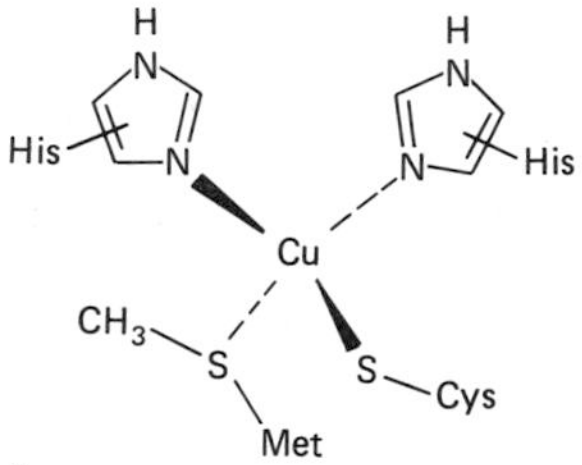

5

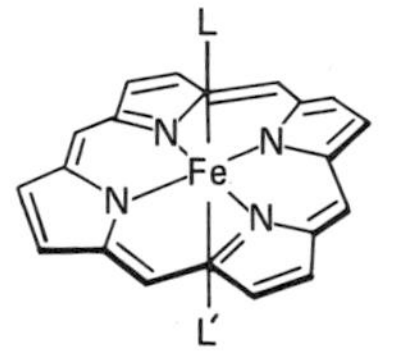

6

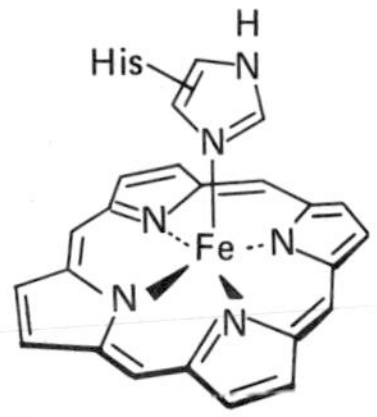

7

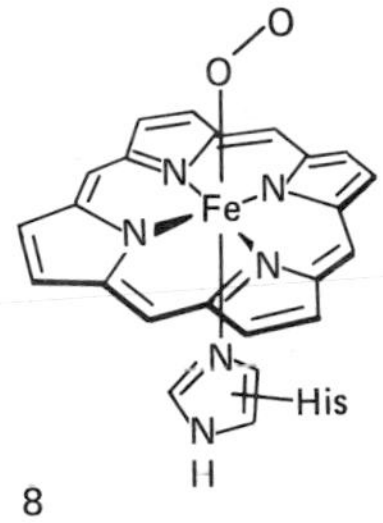

8

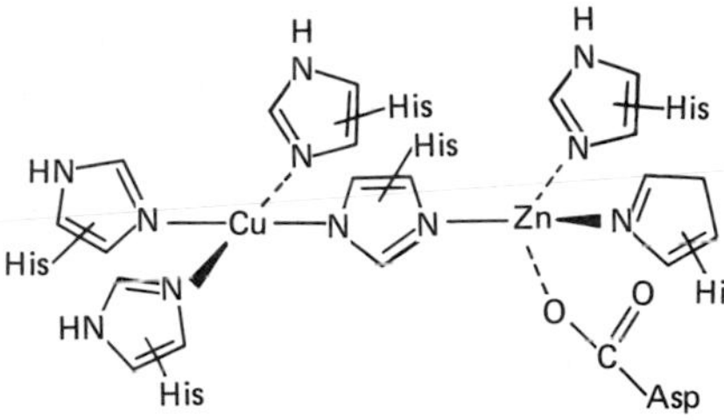

9

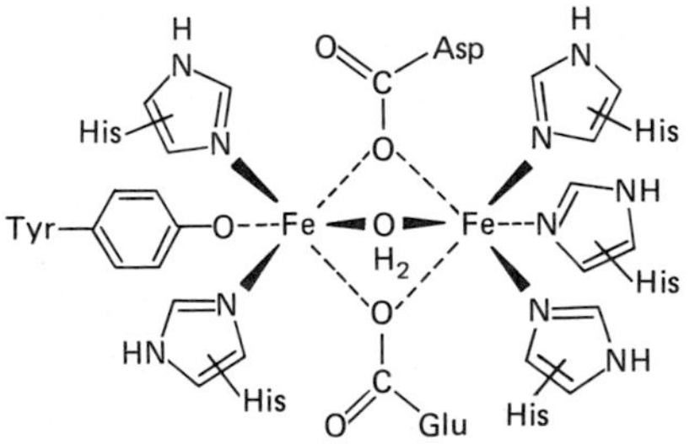

10

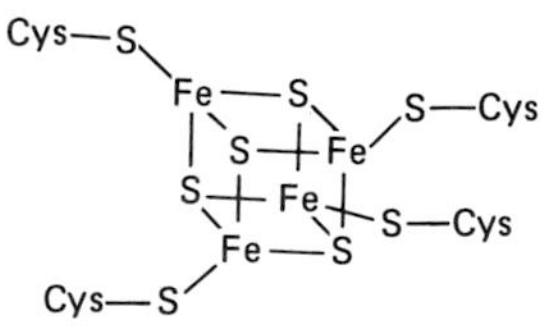

11

Figure 8–18
Examples of some metal ions bound tightly to proteins, with the partial structures of the amino acid side chains to which they are attached. 1, *M is either Fe, as in rubredoxin, or Zn, as in aspartate transcarbamylase and liver alcohol dehydrogenase.* 2, *Carboxypeptidase A.* 3, *Carbonic anhydrase and insulin.* 4, *Liver alcohol dehydrogenase.* 5, *Azurin and plastocyanin.* 6, *Heme group; L is His and L′ is His in cytochrome* b_5 *or Met in cytochrome* c. 7, *Deoxy heme groups in myoglobin and hemoglobin.* 8, *Oxy form of* 7. 9, *Superoxide dismutase.* 10,

Structural basis for the specificity of antibody–antigen reactions and structural mechanisms for the diversification of antigen-binding specificities. E. A. Padlan. Quart. Rev. Biophys. 10:35–65, 1977.

The antibody combining site. J. D. Capra and A. B. Edmundson. Sci. Amer. 236:50–59, 1977.

Correlations between three-dimensional structure and function of immunoglobulins. R. J. Poljak. Crit. Rev. Biochem. 5:45–84, 1978.

Immunoglobulin structure and effector functions. J. L. Winkelhake. Immunochemistry 15:695–714, 1978.

Three-dimensional structure of immunoglobulins. L. A. Amzel and R. J. Poljak. Ann. Rev. Biochem. 48:961–997, 1979.

Activation of the complement system by antibody–antigen complexes: the classical pathway. R. R. Porter and K. B. M. Reid. Adv. Protein Chem. 33:1–71, 1979.

A common mechanism of hapten binding to immunoglobulins and their heterologous chain recombinants. R. Zidovetzki, et al. Biochemistry 19:2790–2795, 1980.

Some properties and applications of monoclonal antibodies. P. A. W. Edwards. Biochem. J. 200:1–10, 1981.

Monoclonal antibodies: a powerful new tool in biology and medicine. D. E. Yelton and M. D. Scharff. Ann. Rev. Biochem. 50:657–680, 1981.

Very Small Ligands

When the ligand is very small, such as an ion or an electron, steric complementarity might seem to be limited in scope, but considerable specificity is still possible. For example, calmodulin, parvalbumin, and other important calcium-binding proteins can bind Ca^{2+} selectively in the presence of 10^3-fold higher concentrations of Mg^{2+}, a rather similar ion. This is accomplished by a specific constellation of six to eight oxygen atoms of Asp and Glu side chains and backbone carbonyl groups of the protein. They are arranged in the correct orientation in a polypeptide loop between two α-helices, a characteristic structure that has come to be known as the **"EF hand"** since its initial discovery in parvalbumin (Figure 8–6A). In general, ion-binding sites are composed of a number of groups of the protein that interact simultaneously with the ion, producing a cooperative effect. The affinities of individual groups are too low.

Nevertheless, more chemical considerations are also involved, because ions tend to bind to protein groups for which they have some intrinsic affinity. For example, Ca^{2+} ions are bound to oxygen atoms, Zn^{2+} to the imidazole rings of His residues, and Fe^{2+} and Fe^{3+} to the sulfur atoms of Cys residues, or to sulfide ions, as in iron–sulfur proteins (see Figure 6–31). Copper ions tend to interact with thiol or imidazole groups; Mg^{2+} ions are also bound along with phosphate groups of ligands. Some metal-binding sites observed in proteins are illustrated in Figure 8–18. All of these ions are bound spontaneously by the apoprotein, even in the case of the seemingly complex iron–sulfur complexes of ferredoxins (Figures 6–31 and 8–18).

Methemerythrin. 11, Some ferredoxins and high-potential iron protein; planar (2Fe–2S) and (3Fe–3S) complexes are also found in other ferredoxins. (Adapted from J. A. Ibers and R. H. Holm, Science 209:223–235, 1980.)

With other small ligands, because sufficient discrimination appears not to be possible with a protein alone, prosthetic groups bound to the protein serve as physical intermediaries. The most numerous examples occur in the physiologically vital binding of O_2 and electrons: Oxygen molecules are bound by either the ferrous heme groups of the globins (Figure 8–23), by two Fe^{2+} atoms held by His, Tyr, Glu, and Asp side chains in hemerythrin (Figure 6–16), or by two Cu^+ ions in hemocyanins of unknown structure. Ferrous heme groups and Fe^{2+} and Cu^+ ions free in solution do not bind O_2 reversibly but instead are oxidized to the Fe^{3+} and Cu^{2+} forms; accordingly, one of the functions of the protein is to provide the appropriate environment for binding without oxidation.

There is a wide variety of proteins that bind electrons reversibly; such proteins are termed **redox** proteins. However, free electrons are not usually present in solution, so they are usually transferred to and from other molecules, often proteins, that are specifically bound as ligands by the redox protein. The specificity in electron binding is not for the electron itself but for its donor and acceptor. How this electron transfer occurs is not known, but the site where the electron is held is usually obvious. Perhaps the simplest redox proteins, in which no prosthetic group is present, are those that use reversible formation of a disulfide bond between two Cys residues, as in thioredoxin. However, such redox proteins seem to be relatively rare, and most use prosthetic groups, such as flavins or NAD. Many use the ability of iron and copper ions to exist in two different redox states, i.e., Fe^{2+}/Fe^{3+} and Cu^+/Cu^{2+}. In some cases, such as the cytochromes, the iron atom is incorporated in a heme group, whereas in others, it is part of an iron–sulfur complex (Figure 8–18). The heme groups of cytochromes bind electrons rather than O_2 molecules, as with the globins, because the sixth coordinating position of the iron atom, where O_2 would be bound, is occupied by a Met side chain (Figure 6–28).

The tightness of binding of very small ligands by proteins via prosthetic groups is determined both by the intrinsic properties of the prosthetic group and by modulation of it by the protein. For example, the redox potentials of heme-containing proteins vary, ranging from -1100 to $+300$ mV—covering some 24 orders of magnitude in affinity for an electron. Understanding how this modulation occurs is still not possible, owing to our insufficient understanding both of protein structure and of the properties of the prosthetic groups. Indeed, these particular groups may be utilized by proteins just because they are sufficiently complex that binding a ligand may produce substantial changes in them that can be transmitted to the rest of the protein, and vice versa. As a very simple example, the Cu^+ ion prefers to be tetrahedrally coordinated, whereas Cu^{2+} prefers planar or octahedral coordination. Because the coordinating groups in proteins are the side chains of some amino acid residues (Figure 8–18), the binding or release of an electron in this case would be expected to produce a tendency to alter the protein structure. Conversely, the orientations of the coordinating side chains should alter the tendency of the copper atom to be in either Cu^+ or Cu^{2+} ligand configurations, that is, the affinity for electrons or redox potential. The effects of ligand binding on the protein and the effect of the protein on ligand affinity are linked functions.

All such perturbations of the intrinsic affinity of a prosthetic group, including the 10^{24}-fold variation in electron affinities of heme groups, imply a suitably rigid protein conformation. Otherwise, the protein would adapt itself to the consequences of ligand binding and would have little power to modulate the affinity, or in the disfavored state it would unfold or the prosthetic group would dissociate. In the case of the redox proteins, very little is known about how the protein modulates the affinity of the prosthetic group, but the covalent linkage between the two probably serves to prevent their dissociation. The known structures of both redox states demonstrate very few, if any, geometrical differences; large energetic effects on binding, therefore, do not require large conformational effects.

In contrast to the above examples, where one or a few ions are bound very specifically by a protein, large amounts of some essential ions, such as iron, need to stored in a benign form to avoid toxicity yet must be available for use when required. Such a function is served by ferritin, one molecule of which can bind about 4500 iron atoms. The protein, apoferritin, provides a spherical coat, consisting of 24 identical polypeptide chains of 163 amino acid residues each, which encloses a microcrystalline inorganic matrix with the approximate composition $(Fe\,OOH)_{8n}$ $(Fe\,O:OPO_3H_2)_n$. The protein does not provide specific binding sites for each stored ferric ion; instead, it probably serves primarily to take up specifically Fe^{2+} ions, oxidize them to Fe^{3+}, and precipitate them in the internal inorganic matrix.

Iron–sulfur proteins: structure and function. W. H. Orme-Johnson. Ann. Rev. Biochem. 42:159–204, 1973.

X-ray structural studies of ferredoxin and related electron carriers. L. H. Jensen. Ann. Rev. Biochem. 43:461–474, 1974.

The cytochrome fold and the evolution of bacterial energy metabolism. R. E. Dickerson, et al. J. Mol. Biol. 100:473–491, 1976.

Calcium-binding proteins. R. H. Kretsinger. Ann. Rev. Biochem. 45:239–266, 1976.

Hemerythrin: alternative oxygen carrier. I. M. Klotz, et al. Science 192:335–344, 1976.

Structure and function of cytochrome *c*. F. R. Salemme. Ann. Rev. Biochem. 46:299–329, 1977.

Haem exposure as the determinate of oxidation–reduction potential of haem proteins. E. W. Stellwagen. Nature 275:73–74, 1978.

Modeling coordination sites in metallobiomolecules. J. A. Ibers and R. H. Holm. Science 209:223–235, 1980.

Structure and evolution of calcium-modulated proteins. R. H. Kretsinger. Crit. Rev. Biochem. 8:119–174, 1980.

Calmodulin—an intracellular calcium receptor. A. R. Means and J. R. Dedman. Nature 285:73–77, 1980.

Calmodulin. C. B. Klee, et al. Ann. Rev. Biochem. 49:489–515, 1980.

Proteins containing 4Fe–4S clusters: an overview. V. W. Sweaney and J. C. Rabinowitz. Ann. Rev. Biochem. 49:139–161, 1980.

b-Type cytochromes. G. von Jagow and W. Sebald. Ann. Rev. Biochem. 49:281–314, 1980.

Iron transport and storage proteins. P. Aisen. Ann. Rev. Biochem. 49:357–393, 1980.

Molecular geometry of cytochrome c and its peroxidase: a model for biological electron transfer. J. Kraut. Biochem. Soc. Trans. 9:197–202, 1981.

Coupling between oxidation state and hydrogen bond conformation in high potential iron–sulfur protein. R. P. Sheridan, et al. J. Biol. Chem. 256:5052–5057, 1981.

Conformation change of cytochrome c: Ferricytochrome c refinement at 1.8 Å and comparison with the ferrocytochrome structure. T. Takano and R. E. Dickerson. J. Mol. Biol. 153:95–115, 1981.

ALLOSTERY: INTERACTIONS BETWEEN BINDING SITES

A major question for the past 20 years has been how the binding of one ligand molecule by a protein can alter the affinity of another site for its ligand, either increasing or decreasing it. The two ligand molecules may be the same, giving positive or negative cooperativity, i.e., *homotropic* allosteric interactions; or they may be different, giving *heterotropic* interactions. The term **allostery** was introduced to distinguish binding at *different* sites by the interacting ligands—which often are structurally unrelated—from competition between structurally related ligands for the *same* binding site, then a more common subject of investigation.

The problem of how different binding sites on a protein interact is very closely related to that of protein flexibility. As discussed in the previous chapter, the flexibility problem is very much one of current debate; consequently, allostery is not fully understood either. Nevertheless, the two related problems have been clarified by the proposal of two opposing, extreme structural models for allostery: the sequential and the concerted.

Allosteric proteins and cellular control systems. J. Monod, et al. J. Mol. Biol. 6:306–329, 1963.

The Sequential Model

One school of thought, led by Koshland and by Weber, holds that a protein is sufficiently flexible that binding of a ligand at one site can readily alter directly its conformation at another, thereby affecting its affinity for a ligand. Such arguments are very plausible on energetic terms, as binding affinities in allosteric systems are usually changed by a few orders of magnitude, at most, which correspond to differences of only a few kcal/mole in binding interactions. Such changes could probably be produced by very small conformational perturbations of the binding site.

In this case, virtually any type of allosteric interaction would be possible, depending only upon what conformational changes are induced upon ligand binding. Each binding site on a protein could be affected independently, its affinity for its particular ligand being either increased, decreased, or unaffected, and binding different ligands at a site could have different effects. Some relatively simple models, proposed by Koshland, Nemethy, and Filner, account for the known allosteric properties of proteins. Their general scheme has come to be known as the **KNF** model or the **sequential model,** the latter because the conformational and affinity effects occur sequentially as a consequence of binding each ligand.

Comparison of experimental binding data and theoretical models in proteins containing subunits. D. E. Koshland, Jr., et al. Biochemistry 5:365–385, 1966.

The molecular basis for enzyme regulation. D. E. Koshland, Jr. In The Enzymes, 3rd ed. P. D. Boyer (ed.). Vol. 1, pp. 341–396. New York, Academic Press, 1970.

Energetics of ligand binding to proteins. G. Weber. Adv. Protein Chem. 29:1–83, 1975.

Role of flexibility in the specificity, control, and evolution of enzymes. D. E. Koshland. FEBS Letters 62:E47–E52, 1976.

Ligand competition curves as a diagnostic tool for delineating the nature of site–site interactions: theory. Y. I. Henis and A. Levitzki. Eur. J. Biochem. 102:449–465, 1979.

Ligand Binding and Protein Flexibility

With the sequential model view of ligand binding, it is almost surprising that so many proteins exhibit normal, nonallosteric binding behavior. The effects of binding a ligand to one site on a protein might be expected invariably to have an effect on another binding site, with significant alterations in its ligand affinity. This does not appear to be the usual situation, although there have been no systematic studies of binding to separate sites on monomeric proteins. Well-characterized monomeric proteins that show interactions between two or more ligands have the binding sites adjacent, and the ligands interact directly. One of the most dramatic examples is concanavalin A: Its binding of saccharides is dependent upon prior binding of two metal ions, usually Ca^{2+} and Mn^{2+}, and binding of Ca^{2+} is dependent on prior binding of Mn^{2+}. The structure of apo-concanavalin A, with no ligands, shows the Mn^{2+} site to be preexisting. Binding Mn^{2+} at this site fixes the position of an Asp side chain with which the Mn^{2+} interacts and of a Glu side chain, which consequently help to form the Ca^{2+} site. The two metal sites are adjacent and share the Asp side chain. The saccharide-binding site is nearby and presumably is fixed in a similar way upon binding the Ca^{2+} ion, but the details of saccharide binding are not yet known. Consequently, such binding interactions are not allosteric but are direct interactions between adjacent sites.

In another case, there are synergistic interactions between the binding of ATP and glucose by hexokinase, caused primarily by the relative movements of the two domains upon binding the two ligands between them (see Figure 9–15). Such interactions are comparable to a quaternary structure change.

It is pertinent to consider what conformational changes occur upon the binding of ligand normally, in the absence of known allosteric effects.

Table 8–4   Examples of Conformational Changes Observed upon Ligand Binding to Nonallosteric Proteins

PROTEIN	LIGAND	AVERAGE RELATIVE MOVEMENT (Å)	
		All Atoms	Main-chain
Trypsin	Bovine pancreatic trypsin inhibitor (BPTI)		0.26 [a]
Trypsinogen	BPTI		0.28 [b]
Lysozyme	Gd^{3+}	0.48 [c]	
Myoglobin	O_2	0.92	0.61 [d]
Concanavalin A	$Ca^{2+} + Mn^{2+}$	1.1	1.0 [e]
Glyceraldehyde 3-phosphate dehydrogenase	NAD	1.2	1.2 [f]
Carboxypeptidase A	Protein inhibitor		0.42 [g]
Streptomyces griseus protease A	Ac-Pro-Ala-Pro-Phe-OH	0.11	0.10 [h]
	Ac-Pro-Ala-Pro-Tyr-OH	0.10	0.09
	Ac-Pro-Ala-Pro-Phe-H	0.14	0.13

[a]From W. Bode and P. Schwyger, J. Mol. Biol. 98:693–717, 1975.

[b]From W. Bode, et al., J. Mol. Biol. 118:99–112, 1978.

[c]From S. J. Perkins, et al., Biochem. J. 173:607–616, 1978.

[d]From S. E. V. Phillips, J. Mol. Biol. 142:531–554, 1980.

[e]From M. Shoham, et al., J. Mol. Biol. 131:137–155, 1979.

[f]From M. R. N. Murthy, et al., J. Mol. Biol. 138:859–872, 1980.

[g]From D. C. Rees and W. N. Lipscomb, Proc. Natl. Acad. Sci. U.S.A. 78:5455–5459, 1981.

[h]M. N. G. James, et al., J. Mol. Biol. 144:43–88, 1980.

Is the protein an inflexible template complementary to the ligand, or how much does it adapt itself to the ligand? The very many crystallographic studies indicate that some alterations usually do take place upon ligand binding. They are greatest near the binding site but may extend elsewhere. Generally, the movements of one part of a protein relative to another are less than 1 Å. One of the largest movements detected is a 12-Å movement of a peptide loop upon binding of the cofactor NAD to lactate dehydrogenase, yet there is no effect on NAD affinity of the other subunits of the tetramer. Some examples of overall conformational changes observed are tabulated in Table 8–4.

That the protein is more like a template, although not totally rigid, should not be unexpected, because this property is necessary for specificity in ligand binding; a very malleable protein would adapt its shape to match that of many ligands and would bind them with similar affinities. The expected relationship between ligand binding and protein flexibility is illustrated schematically in Figure 8–19. A flexible protein could be imagined to bind a suitable ligand to which it was not exactly complementary by becoming so; the extent to which its affinity for this ligand is lower than one exactly complementary would reflect the energy required to distort the normal conformation. In general, ligands with low affinity are not observed to produce large changes in protein conformation upon binding; their low affinities, therefore, reflect primarily their noncomplementarity, and considerably greater energies are required to distort the binding site.

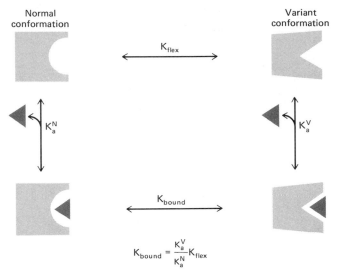

Figure 8–19

Estimation of protein flexibility by the conformational distortions produced upon binding a noncomplementary ligand. The normal conformation of a protein is imagined not to be complementary to a ligand (designated by the triangle), which consequently has poor affinity. However, a variant conformation brought about by flexibility of the protein might be complementary and bind tightly, so $K_a^V \gg K_a^N$, but it would not normally be populated substantially, i.e., $K_{flex} \ll 1$. However, when the ligand is bound, the variant conformation should be more stable by the ratio of the two binding affinities, so it might be detected in the liganded state. For example, if $K_{flex} = 10^{-3}$ and the variant conformation binds the ligand 10_4 times more tightly than the normal conformation, K_{bound} will be 10, so the variant conformation should be present in the complex 90 per cent of the time.

On the other hand, if no substantial changes in the protein are observed upon binding a poor ligand, the protein must be relatively inflexible, so that K_{flex} is very small.

Similar energetic considerations apply to the binding of a flexible ligand; its observed affinity will be lower by the free energy required to pull it into a conformation complementary to the binding site.

Although the conformation may not change greatly, there are usually a substantial increase in stability of the folded state and a decrease in its flexibility, as measured crystallographically or by hydrogen exchange. These alterations are a thermodynamic necessity if the ligand does not bind to the unfolded or perturbed conformations.

Conformation as the determinant of saccharide binding in concanavalin A: Ca^{2+}–concanavalin A complexes. S. H. Koenig, C. F. Brewer, and R. D. Brown. Biochemistry 17:4251–4260, 1978.

Changes in the three-dimensional structure of concanavalin A upon demetallization. G. N. Reeke, Jr., J. W. Becker, and G. M. Edelman. Proc. Natl. Acad. Sci. U.S.A. 75:2286–2290, 1978.

Glucose-induced conformational change in yeast hexokinase. W. S. Bennett and T. A. Steitz. Proc. Natl. Acad. Sci. U.S.A. 75:4848–4852, 1978.

Crystal structure of demetallized concanavalin A: the metal-binding region. M. Shoham, et al. J. Mol. Biol. 131:137–155, 1979.

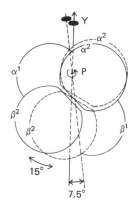

Figure 8–20
The quaternary structures of deoxy (T) and liganded (R) hemoglobins. The $\alpha^1\beta^1$ dimers of both are superimposed and the relative positions of the $\alpha^2\beta^2$ dimer depicted in solid lines for T and dashed lines for R. The $\alpha^2\beta^2$ unit rotates 15 degrees relative to $\alpha^1\beta^1$ about the axis P perpendicular to the plane of the paper and the twofold symmetry axis Y. The only direct subunit contacts are between α^1 and β^1 and between α^1 and β^2, plus the symmetric contacts between α^2 and β^2 and between α^2 and β^1.

The binding site for the organic phosphate allosteric effectors DPG, IPP, and IHP in the T state (Figure 8–10) is at the bottom of the molecule straddling the twofold axis between the two β chains. In the R state, the two chains move closer together, essentially destroying this binding site. (Adapted from J. M. Baldwin, Trends Biochem. Sci. 5:224–228, 1980.)

The transition of bovine trypsinogen to a trypsin-like state upon strong ligand binding. W. Bode. J. Mol. Biol. 127:357–374, 1979.
Substrate stabilization of lysozyme to thermal and guanidine hydrochloride denaturation. C. N. Pace and T. McGrath. J. Biol. Chem. 255:3862–3865, 1980.
Manganese and calcium binding sites of concanavalin A. K. D. Hardman, et al. J. Mol. Biol. 157:69–86, 1982.

The Concerted Model

Proteins demonstrating interactions between nonoverlapping binding sites are invariably oligomeric, consisting of at least several identical polypeptide chains and often also of two different polypeptides. Upon binding an appropriate ligand, they demonstrate very large conformational changes sufficient to disrupt a crystal lattice; in contrast, nonallosteric proteins generally bind ligands within an unaltered crystal lattice unless the lattice contacts are directly perturbed by the ligand. The large conformational changes in allosteric proteins involve rearrangements of relatively unaltered subunits, i.e., primarily quaternary structure changes.

The archetypal allosteric protein is the well-known hemoglobin of vertebrates. Its deoxy and oxy (or other liganded) forms differ by a 15-degree rotation of one pair of α and β subunits relative to the other two in the $\alpha_2\beta_2$ tetramer (Figure 8–20). Owing to this large change, crystals of the two forms of the protein are different, and both crack and dissolve upon converting the protein to the other form.

These properties of allosteric proteins led Monod, Wyman, and Changeux to propose a model for allosteric interactions that envisages the other extreme of protein flexibility to the sequential model. Their scheme imagines the binding of ligand at one site to have no direct effect on the affinity of the other sites, but to alter the conformational equilibrium between two alternative quaternary conformations of the protein: One conformation, having low intrinsic affinity for the ligand at all its sites, is designated "T" (for "tense," since it is imagined to be constrained in some manner); the other, having high affinity for the ligand, is designated "R" (for "relaxed"). These two forms with i ligand molecules bound are often referred to as T_i and R_i. According to the model, these two conformations coexist even in the absence of ligand, with an equilibrium constant L between T_0 and

R_0. The T_0 form is favored, and the protein has relatively low affinity for the first ligand molecule. Since the R conformation has the higher affinity, by a factor c, ligand molecules will be bound to it preferentially. This will pull the conformational equilibrium toward the R state, since the conformational change and ligand binding are linked functions, and the conformational equilibrium between the two conformations with one ligand molecule bound will be cL. The other vacant sites on the R_1 molecule will then be in the high-affinity form; therefore, the average affinity of the vacant sites of the entire population will be increased, giving positive cooperative homotropic interactions.

In this model, called the **concerted model,** heterotropic interactions arise by the preferential binding of other ligands to either of the states. Affinity is decreased by molecules that bind preferentially to the T state, since they pull the conformational equilibrium back toward T. Affinity is increased by any molecule that binds preferentially to the R state. Different ligands control the apparent affinity simply by shifting the equilibrium between the two states.

The concerted model is also known as the **MWC, symmetric,** or **two-state,** allosteric model. It is illustrated in its simplest form and compared with the sequential model in Figure 8–21. The two differ in terms of protein conformation primarily with respect to the conformations of the partially liganded states. Both imply that ligand binding has effects on the protein conformation; in the sequential model, such effects extend directly to the other binding sites, affecting their ligand affinity. In the concerted model, they need extend only to the interface between subunits to alter the conformational equilibrium between the R and T quaternary states, although it would not be surprising if some of these effects were transmitted across the interface to the other subunits. The two models differ in that the concerted model envisages the two conformations to be present even in the absence of ligand, R_0 and T_0, whereas in the sequential model the R conformation is induced only upon ligand binding. The sequential model predicts that the conformational change upon ligand binding should parallel the extent of ligand binding, whereas the two need not coincide with the concerted model, since conformational change should tend to occur at one stage of ligand binding of each molecule: that when Lc^i becomes less than unity (Figure 8–21).

The concerted model is much more restrictive than the sequential. The only parameters that may be varied are L, the conformational equilibrium constant in the absence of any ligand, and the affinities of the two states for each ligand; the relative affinities of the two states specify the allosteric parameter c. Moreover, the two parameters L and c are not observed to be independent: Alteration of one also changes the other. This interdependence probably reflects the linkage between the conformational change and the change in affinity for the ligand in state T.

The greatest restriction of the concerted model is that it does not predict negative cooperativity, since the ligand can pull the conformational

Sequential

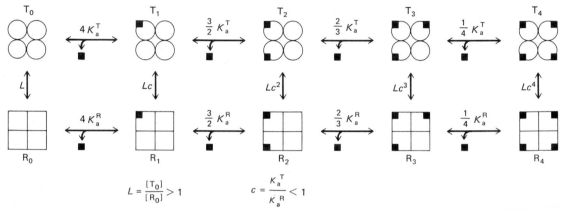

Concerted

$$L = \frac{[T_0]}{[R_0]} > 1 \qquad\qquad c = \frac{K_a^T}{K_a^R} < 1$$

Figure 8–21

Schematic illustration of the sequential and concerted models for allosteric cooperativity of ligand (■) binding, using a hypothetical tetrameric protein. In the sequential model, binding of each ligand induces a conformational change in the subunit to which it is bound and to the other subunits, thereby increasing their affinities for ligand, so that $K_a^1 < K_a^2 < K_a^3 < K_a^4$. The conformations depicted and the sequence of binding by the individual subunits are rather arbitrary. In the concerted model, there are only two conformational states, T and R, with intrinsic affinities for ligand of K_a^T and K_a^R, respectively, which are not altered by ligand binding to other subunits. Instead, binding of each ligand shifts the equilibrium from T to R by the factor c, the ratio of the two intrinsic affinities, K_a^R/K_a^T.

equilibrium only toward the high-affinity form. This phenomenon will be considered further in the following section.

On the nature of allosteric transitions: a plausible model. J. Monod, et al. J. Mol. Biol. 12:88–118, 1965.

On the nature of allosteric transitions: implications of non-exclusive ligand binding. M. M. Rubin and J.-P. Changeux. J. Mol. Biol. 21:265–274, 1966.

The study of allosteric proteins. J. Janin. Prog. Biophys. Mol. Biol. 27:79–120, 1973.

Ligand binding and self-association of proteins. R. F. Steiner. Mol. Cell Biochem. 31:5–24, 1980.

Close correlation between Monod-Wyman-Changeux parameters, *L* and *c*, and its implication for the stereochemical mechanism of haemoglobin allostery. S. Matsukawa, et al. J. Mol. Biol. 150:615–621, 1981.

The Allosteric Properties of Hemoglobin

By far the most extensively studied allosteric protein is vertebrate hemoglobin, particularly that from humans, as its allosteric properties are so obviously of physiological and medical importance in optimizing O_2 transport by erythrocytes. The large number of mutant forms are of medical importance and also provide a vast array of structural alterations with which to correlate effects on the functional properties. Hemoglobin served as the prototype for the concerted allosteric model and still fits most closely that model, although very many details still are not clear.

The fully oxygenated $\alpha_2\beta_2$ tetramer, or that with any other ligand bound to the heme iron, is considered to be in the R state, because the binding properties of its four subunits are considered normal—similar to those of monomeric myoglobin, individual α chains, or $\alpha\beta$ dimers. In contrast, the unique ferrous deoxy tetramer is considered to be in the constrained T state, since the oxygen affinity of its hemes is decreased about 500-fold; the greater is the stability of the T state, the greater are the constraints and the lower is the binding affinity. Although the T state is constrained, it is more stable than the R state when all its hemes are deoxy because, in compensation, its quaternary structure has more extensive interactions between the subunits and extra salt bridges involving certain carboxyl-terminal residues of both chains that are not possible in the R state. The allosteric properties of hemoglobin are completely dependent upon the presence of the tetrameric T quaternary structure.

Besides the positive cooperativity of O_2 binding by the hemes (Figure 8–3), there are heterotropic allosteric interactions between effectors that bind elsewhere on the tetramer, primarily and most avidly to the T structure (Table 8–5); all of these interactions affect each other and are linked

Table 8–5 Allosteric Effectors of O_2 Affinity of Hemoglobin

DECREASE AFFINITY	BINDING SITE ON HUMAN T STATE	
	α-Chain	β-Chain
Organic phosphates (e.g., diphosphoglycerate)		α-NH_3^{+a} His 2 Lys 82 His 143
CO_2	α-NH_2	α-NH_2
Anions (e.g., Cl^-)	α-NH_3^+ Arg 141	Lys 82
H^+ (Bohr effect)	α-NH_2	His 146
INCREASE AFFINITY	BINDING SITE ON R STATE	
	α-Chain	β-Chain
O_2 and other heme ligands	heme Fe	heme Fe

[a]See Figure 8–10

Data from A. Arnone and M. F. Perutz, Nature 249:34–36, 1974; J. V. Kilmartin, Trends Biochem. Sci. 2:247–250, 1977; S. O'Donnell, et al., J. Biol. Chem. 254:12204–12208, 1974; J. V. Kilmartin and L. Rossi-Bernardi, Biochem. J. 124:31–45, 1971; and A. Arnone, Nature 237:146–149, 1972.

functions. Most dramatic are the allosteric effectors 2,3-diphosphoglycerate (DPG) and inositol pentaphosphate and hexaphosphate (IPP and IHP, respectively), which lower the O_2 affinity of hemoglobin in vivo by binding preferentially to the T structure. All bind at the same single site per $\alpha_2\beta_2$ tetramer, straddling the β subunits across the twofold symmetry axis (Figure 8–10). This binding site is destroyed upon the quaternary structure change to the R state, because these parts of the β chains move much closer together (Figure 8–20). DPG is present in most mammalian erythrocytes but is replaced by IPP in birds and by HCO_3^- in crocodiles. Bird hemoglobins have two additional Arg residues on each β chain at positions favorable for participation in binding IPP, presumably to neutralize its additional phosphate groups. Three amino acid replacements in this region of the β chains of crocodile hemoglobins also can explain their preference for HCO_3^- as allosteric effector. These instances of evolutionary replacements of amino acids seem to have occurred for functional reasons; they do not seem to be selectively neutral, as most of the others apparently are.

The oxygen affinity of hemoglobin is also physiologically coupled in two ways to the CO_2 content of the blood, with increased content causing release of O_2 from oxyhemoglobin. Direct interactions occur because CO_2 binds as the carbamate

$$\mathrm{-NH_2 + CO_2 \rightleftharpoons -NH-CO_2H} \qquad (8\text{–}36)$$

to the α-amino groups of both α and β chains in the T state. Binding to the T structure is favored because the ionized carbamate has electrostatic interactions with positively charged groups that are not in the same position in R. Elevated concentrations of CO_2 in the blood also cause the pH to decrease, owing to formation of H^+ and HCO_3^-, which also lowers the O_2 affinity, since more protons are bound by the T than by the R structure. Known as the **Bohr effect,** this phenomenon results primarily from the elevated pK_a values of the His 146 side chain of β chains (8.0 in T_0, 7.1 in R_4) and the α-amino groups of the α chain (about 7.8 and 7.0, respectively) produced by their involvement in salt bridges in the T state, but not in R. Anions such as Cl^- also bind preferentially to the T structure, but the physiological significance is not clear.

The concerted allosteric model is very attractive for characterization of hemoglobin because of the occurrence of only two well-defined quaternary structures in all the many forms of hemoglobin studied, with only very small differences between the subunits. Moreover, the two quaternary structures may be shown to coexist, in the absence of any ligand, by modifying the protein at the $\alpha^1\beta^2$ interface (superscripts refer to the different pairs of subunits within the $\alpha_2\beta_2$ tetramer; see Figure 8–20) or by disrupting the salt bridges that preferentially stabilize the T state. In this way, the value of L is decreased from its normal value of at least 10^4, favoring the T state, to close to unity, so that R becomes populated substantially in the deoxy form. Also, any modification, or the presence of an allosteric effector, that either increases or decreases L sufficiently (up to 10^9 or down to <1) so that the protein stays in either the T or the R state, irrespective of the degree of oxygenation, causes the loss of cooperativity in oxygen binding. If the interconversion between R and T is restored, so is cooperativity. Consequently, this aspect of the concerted model is obeyed.

The greatest difference between the concerted and sequential models lies in the properties of the intermediate, partially liganded states (Figure 8–21). In the concerted model, the quaternary structure of the tetramer changes all at once and thus does not need to parallel ligand binding; in addition, the affinity of individual hemes is not altered without a quaternary structure change. These proposals are difficult to verify experimentally, because the intermediate states are not populated substantially; the cooperativity of oxygen binding results in a partially oxygenated sample of hemoglobin consisting of primarily fully deoxy (T_0) and fully oxy (R_4) molecules (Figure 8–3D). Artificial intermediates may be constructed by altering either the α or the β chains so that they are irreversibly in the liganded states—for example, by oxidizing the iron to the ferric form to give the met derivative, α^+ or β^+, which is similar to oxy heme. The unaltered chains of the $\alpha_2^+\beta_2$ or $\alpha_2\beta_2^+$ hybrids have structural and binding properties similar to those expected of partially oxygenated molecules. The T_0 and R_4 quaternary structures determined crystallographically also suggest that intermediate quaternary structures would not be stable, because their substantial differences at the $\alpha^1\beta^2$ interface involve alternative interdigitations of groups that must be either one way or the other, not in between (Figure 8–22). In T, Thr 41α of the C helix is dovetailed into a groove of the other subunit, whereas in R it is Thr 38α, which is located one turn along the C helix.

A major question is how O_2 binding at the heme affects the R $\rightleftharpoons$ T conversion, and vice versa. There are only very small differences in the tertiary structures of the individual α and β chains between the R and T quaternary structures; hence the answer is not clear. Nevertheless, it appears that ligand binding has complex effects on the iron atom and the heme group, besides interacting directly with other groups in the heme pocket. The spin state and effective radius of the iron atom depend upon its state of ligation; in the ferrous deoxy form, it is somewhat too large to fit within the center of the heme group and is forced out of the plane of the heme. In contrast, liganded hemes tend to have the smaller iron atom very nearly in the plane. These changes in the iron atom tend to distort the heme group, which could affect the many parts of the protein with which it is in contact, but the properties of the heme group are not fully understood. The iron atom is also attached to the proximal His side chain, and it is here that the largest changes in tertiary structure of the individual chains are observed (Figure 8–23). The movement of the iron atom in and out of the plane of the heme group causes movements of the proximal His residue and of the F helix of which it is a part. The low oxygen affinity of the T state is believed to be due to prevention by this part of the protein

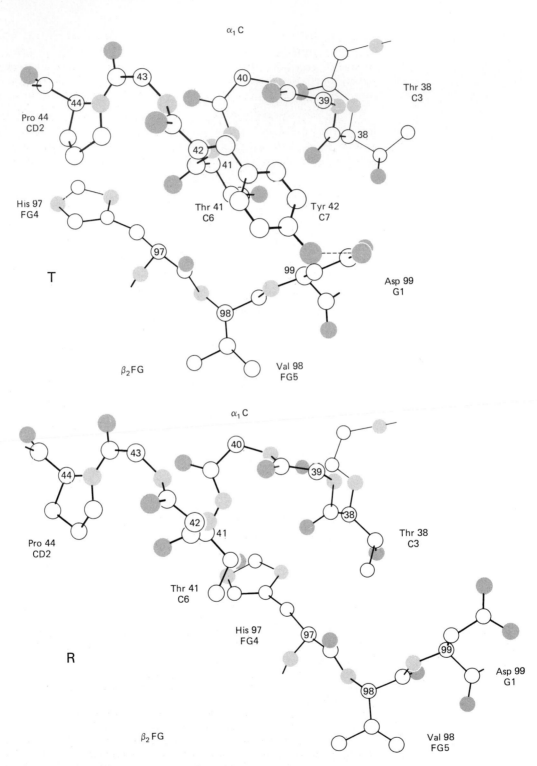

Figure 8–22
Part of the α¹β² interface in the R (oxy) and T (deoxy) states of hemoglobin, illustrating the large differences there. At the top is the C helix of the α¹ chain; at the bottom, the irregular corner linking the F and G helices of the β² chain. The conformations of both are virtually the same in both the R and T structures, but the contacts differ markedly owing to a shift of one subunit relative to the other. His 97 of the β chain is in contact with Thr 41 of the α chain in T, but with Thr 38, one turn back along the C helix, in R. Intermediate positions would be unstable because His 97 and Thr 41 would be too close together. Therefore, molecules must be in either the R or the T quaternary structures, irrespective of the number of heme ligands bound. (Adapted from J. M. Baldwin, Trends Biochem. Sci. 5:224–228, 1980.)

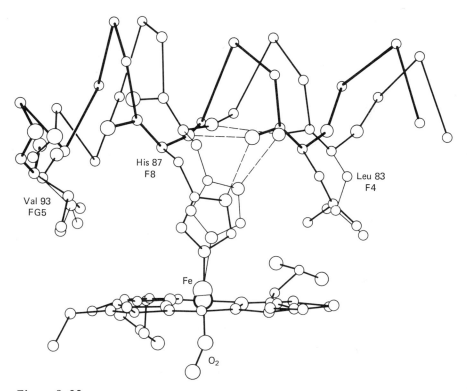

Figure 8–23
Changes at the heme group upon binding of O_2 to the α chains of hemoglobin. Similar changes occur in the β chains. The heme groups of the deoxy (T) and oxy (R) structures were superimposed; some of the neighboring residues of the F helix and the FG corner between the F and G helices are shown in thick lines for oxy, thin lines for deoxy. The two oxygen atoms of bound O_2 are shown below the heme group. Upon binding O_2 (or CO), the iron atom moves about 0.6 Å from above into the plane of the heme group. This pulls with it the proximal His 87 (also known as F8, i.e. the 8th residue of the F helix) and the F helix of which it is a part. This produces changes at the FG corner, on the left, which is in contact with the other chain across the α¹β² interface (see Figure 8–22). It is this interface which changes in the T ⇌ R quaternary structure change, and the above sequence of events describes a plausible, but unproved, mechanism by which O_2 binding can affect the quaternary structure, and vice versa. The dashed lines illustrate hydrogen bonds which are believed to be important for orienting the His F8 side-chain. (Figure kindly provided by B. Shaanan.)

of the iron atom moving toward the heme group, as is favored for O_2 binding. Indeed, with nitrous oxide tightly bound, pulling the tetramer into the T state by a heterotropic effector such as IHP breaks the iron–His bond. This constraint normally just lowers the affinity for O_2 in the T state, but when O_2 is bound, the constraint will be reversed, and the Fe and His groups will be pulled toward the heme, which will also pull the F helix in that direction. If the helix is sufficiently rigid, this constraint will be transmitted to its ends, and the carboxyl end is in contact with the other subunit across the $\alpha^1\beta^2$ interface (Figure 8–22). This is the one that changes markedly in the T $\rightleftharpoons$ R quaternary structure change; this movement of the F helix, therefore, may be a major route by which ligand binding and the quaternary structure change are structurally linked.

The nature of the tertiary and quaternary structure changes appears to make it impossible for one subunit, or one $\alpha^1\beta^1$ dimer, to change significantly its tertiary structure independently of the other subunits. Therefore, changes in oxygen affinity of other subunits within the tetramer, which must arise from changes within their individual tertiary structures, would appear to occur primarily only when there is a quaternary structure change. Hemoglobin is then best described conceptually by the concerted allosteric model. Whether other allosteric proteins will also do so is not yet clear, but all those that have been well studied demonstrate large quaternary structure changes, and they tend to have binding sites for allosteric effectors between subunits, where their binding can readily be coupled with quaternary structure changes (see also pp. 459–466).

Hemoglobin: E. Antonini and M. Brunori. Ann. Rev. Biochem. 39:877–1042, 1970.

Stereochemistry of cooperative effects in haemoglobin. M. F. Perutz. Nature 228:726–739, 1970.

High resolution nuclear magnetic resonance spectra of hemoglobin. III. The half-ligated state and allosteric interactions. S. Ogawa and R. G. Shulman. J. Mol. Biol. 70:315–336, 1972.

The mechanisms of interaction of red cell organic phosphates with hemoglobin. R. E. Benesch and R. Benesch. Adv. Protein Chem. 28:211–237, 1974.

Structure of inositol hexaphosphate–human deoxyhaemoglobin complex. A. Arnone. Nature 249:34–36, 1974.

Cooperative interactions of hemoglobin. S. J. Edelstein. Ann. Rev. Biochem. 44:209–232, 1975.

Structure and function of haemoglobin. J. M. Baldwin. Prog. Biophys. Mol. Biol. 29:225–320, 1975.

Models for cooperative oxygen binding in hemoglobin. J. P. Collman and K. S. Suslick. Pure Appl. Chem. 50:951–961, 1978.

A structural model for the kinetic behaviour of hemoglobin. K. Moffat, et al. Science 206:1035–1042, 1979.

Regulation of oxygen affinity of hemoglobin: influence of structure of the globin on the heme iron. M. F. Perutz. Ann. Rev. Biochem. 48:327–386, 1979.

Hemoglobin: the structural changes related to ligand binding and its allosteric mechanism. J. Baldwin and C. Chothia. J. Mol. Biol. 129:175–220, 1979.

Structure and refinement of oxymyoglobin at 1.6 Å resolution. S. E. V. Phillips. J. Mol. Biol. 142:531–554, 1980.

Regulation of oxygen affinity of mammalian haemoglobins. M. F. Perutz and K. Imai. J. Mol. Biol. 136:183–191, 1980.

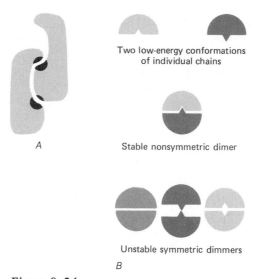

Two low-energy conformations
of individual chains

Stable nonsymmetric dimer

Unstable symmetric dimmers

B

Figure 8–24

A, *Two-dimensional example of a heterologous dimer with nonequivalent binding sites (hatched areas).*

B, *Possible basis for deviations from symmetry in an isologous dimer. The individual chains are assumed to have two alternative low-energy conformations (top) that are complementary so that they can form a stable isologous dimer, with close-packing between the two interfaces, but with deviations from symmetry (middle). Symmetric dimers are not stable (bottom), either because the intermediate conformation that would produce a close-packed interface has high energy (left) or because the two low-energy conformations do not give close-packing when associated (middle and right).*

Interaction of hemoglobin with salts. Effects on the functional properties of human hemoglobin. G. Amiconi, et al. J. Mol. Biol. 152:111–129, 1981.

Allosteric regulation of crocodilian haemoglobin. M. F. Perutz, et al. Nature 291:682–684, 1981.

Negative Cooperativity

Homotropic allosteric interactions in which the binding of one molecule of ligand decreases the affinity of other sites for the same ligand are unlikely with the concerted model; binding of ligand can pull the T $\rightleftharpoons$ R transition only toward the high-affinity state and can give only positive cooperativity. Nevertheless, negative cooperativity is observed with some proteins; this finding is usually taken as evidence against the concerted model. Instead, the sequential model might apply, as there are no restrictions on the ability of one bound ligand to decrease the affinity of another site, or the unliganded protein might be asymmetric, in that covalently identical individual subunits as well as their binding sites have somewhat different conformations.

Nonequivalence of identical polypeptide chains is most likely if they associate heterologously (Figure 6–23), using different binding sets (Figure 8–24A). Oligomeric proteins assembled by isologous association are expected to have exact symmetry, with each subunit being equivalent. However, small departures from exact symmetry are observed in the crystal structures of insulin, α-chymotrypsin, lactate dehydrogenase, and malate dehydrogenase, but these variations could arise from crystal packing interactions since the nonequivalent subunits are in different crystal lattice environments. There are few indications of preexistent asymmetry when the subunits are in comparable environments, as when in solution. Nevertheless, preexistent asymmetry could occur if only two alternative, complementary conformations were energetically most favorable and if the same conformation in all subunits within an oligomer were not possible for steric reasons (Figure 8–24B). In this case, a binding site on one subunit might have one conformation and affinity, while the comparable site on another subunit would have a different conformation and affinity. In the absence of such preexistent asymmetry, binding of ligand at one site could be imagined, for similar reasons, to induce a somewhat different conformation in another subunit, as with the sequential allosteric model. Both preexistent and induced asymmetry could, but need not, disappear upon binding ligand to all the sites.

Other explanations for negative cooperativity are possible: The equivalent binding sites on a symmetric protein oligomer might overlap, or be sufficiently close to interact sterically or electrostatically, because they include, or are near to, symmetry axes of the oligomer. Binding of the first ligand molecule would block, or perhaps just inhibit, binding of a second molecule at the other site. For example, the binding site for organic phosphates on hemoglobin straddles the twofold axis (Figure 8–10), and only one molecule is bound per $\alpha_2\beta_2$ hemoglobin molecule, whereas two would be expected for a symmetric dimer. Effects due to direct interactions between adjacent or overlapping sites are not allosteric in the original sense of the term.

Many proteins show a comparable situation, designated **"half-of-the-sites" reactivity,** in which only half the expected sites or individual groups on an oligomer react with a modifying reagent, and in which the other half reacts slowly or not at all. This type of behavior would be readily explained by preexistent or induced asymmetry or by overlapping sites. Also, this type of behavior is not necessarily incompatible with the concerted model, as reaction with the reagent could pull the R ⇌ T equilibrium toward one state with which it reacts kinetically more slowly.

The one case of negative cooperativity that has been studied extensively is the binding of the coenzyme NAD by glyceraldehyde 3-phosphate dehydrogenase (GPD). With the enzyme from some sources, binding of NAD at one pair of sites on the tetramer appears to induce a decrease in affinity of the other pair, whereas other species of GPD demonstrate positive cooperativity. There are also many instances of half-of-the-sites reactivity of groups in the active site. The pairs of binding sites that probably interact in this way are relatively close in the tetramer, near a twofold axis (Figure 8–25), and it is not unlikely that they interact directly. Binding of NAD and substrates produces only very localized changes in conformation

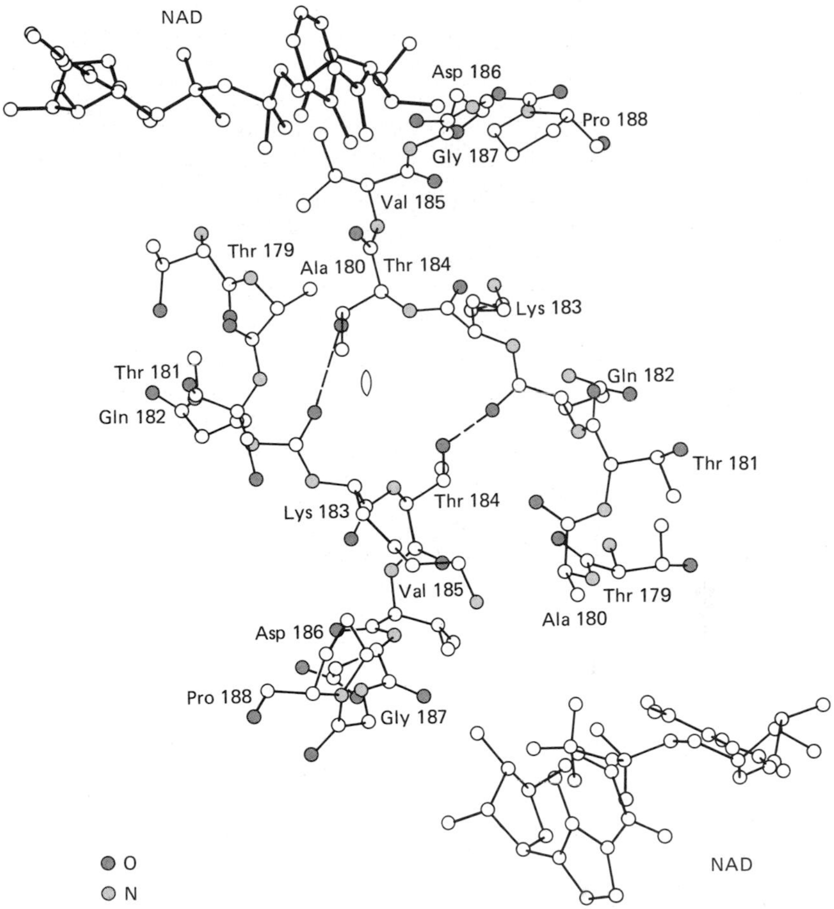

Figure 8–25

The NAD molecules of glyceraldehyde phosphate dehydrogenase that probably bind with negative cooperativity. That bound to one subunit (with thick bonds) is at the top; the other bound to another subunit (with thin bonds) is at the bottom. The twofold axis relating most of the molecule is in the middle, but note the slight departures from symmetry, particularly with Lys 183. (Adapted from D. Moras, et al., J. Biol. Chem. 250:9137–9162, 1975.)

of the protein, quite unlike the R ↝ T transition in hemoglobin. Larger localized changes in conformation upon binding NAD are observed in lactate and alcohol dehydrogenases, where there is no cooperativity, probably because the binding sites are more widely separated.

This example of negative cooperativity is best explained by the sequential model, as it is likely that the pairs of binding sites become different only after binding of the first ligand molecule. The sequential model probably applies in this case because the sites are so close, and the interactions rather direct. Whether induced, sequential changes in conformations can occur between more distantly separated sites or whether they occur primarily by concerted quaternary structure changes remains to be determined. The answer probably reflects the degree of conformational flexibility possible within a folded protein structure.

Negative cooperativity in enzyme action. The binding of diphosphopyridine nucleotide to glyceraldehyde 3-phosphate dehydrogenase. A. Conway and D. E. Koshland, Jr. Biochemistry 7:4011–4022, 1968.

Structure and symmetry of oligomeric enzymes. B. W. Matthews and S. A. Bernhard. Ann. Rev. Biophys. Bioeng. 2:257–317, 1973.

Studies of asymmetry in the three-dimensional structure of lobster D-glyceraldehyde-3-phosphate dehydrogenase. D. Moras, et al. J. Biol. Chem. 250:9137–9162, 1975.

A comparison of the structures of apo dogfish M4 lactate dehydrogenase and its ternary complexes. J. L. White, et al. J. Mol. Biol. 102:759–779, 1976.

Molecular asymmetry in an abortive ternary complex of lobster glyceraldehyde-3-phosphate dehydrogenase. R. M. Garavito, et al. Biochemistry 16:4393–4398, 1977.

Sequence and structure of D-glyceraldehyde-3-phosphate dehydrogenase from *Bacillus stearothermophilus*. G. Biesecker, et al. Nature 266:328–333, 1977.

Structure of lobster apo-D-glyceraldehyde-3-phosphate dehydrogenase at 3.0 Å resolution. M. R. N. Murthy, et al. J. Mol. Biol. 138:859–872, 1980.

Mechanism of negative cooperativity in glyceraldehyde-3-phosphate dehydrogenase deduced from ligand competition experiments. Y. I. Henis and A. Levitzki. Proc. Natl. Acad. Sci. U.S.A. 77:5055–5059, 1980.

Enzymes with asymmetrically arranged subunits. Y. Degani and C. Degani. Trends Biochem. Sci. 5:337–341, 1980.

Molecular symmetry and metastable states of enzymes exhibiting half-of-the-sites reactivity. J. Herzfeld, et al. Biochemistry 20:4936–4941, 1981.

Coenzyme binding in crystals of glyceraldehyde-3-phosphate dehydrogenase. A. G. W. Leslie and A. J. Wonacott. J. Mol. Biol. 165:375–391, 1983.

9

CATALYSIS

The interactions of proteins with other molecules are very often followed by covalent changes in these molecules. An enzyme catalyzes chemical reactions within and between its **substrates,** those molecules that bind in the correct orientation at its active site. In many cases, the chemical reaction can be coupled energetically to other processes, such as mechanical movements in muscle or flagella or the transport of molecules through membranes, often against concentration gradients.

Because enzyme catalysis has been studied extensively for over a century, a thorough description would require much more space than can be provided here. In the classical early studies, much was learned about the kinetics of action of enzymes, but little about the enzymes themselves; emphasis was placed on the substrates and products, and the enzyme was represented either as just "E" or as an ill-defined black box. All this changed in the past two decades, when the structures and properties of the enzymes and of their complexes with ligands were elucidated, as described in the previous chapters of this text.

In this chapter, we will review only briefly the classical kinetics of enzyme action, as this is a very extensive topic that is described adequately in many other books. Instead, we will concentrate on the physical properties of enzymes and the physical principles by which they seem to catalyze reactions at quite extraordinary rates. It is now believed that the kinetic properties of enzymes can be explained, although a wide variety of different terms has been used: for example, "approximation," "orientation," "entropy loss," "propinquity," "rotamer distribution," "anchimeric assistance," "proximity," "orbital steering," "stereopopulation control," "distance distribution function," "togetherness," and "FARCE" (freezing at reactive centers of enzymes). Such jargon, which more often is misunderstood than serves a useful purpose, is avoided in the following description of the general principles involved, which are illustrated with some of the best-understood enzymes.

In addition, the somewhat less certain subject of the regulation of enzyme activity will be described briefly. Finally, the poorly understood and intriguing subject of conversion of chemical energy into mechanical and chemiosmotic forms will be discussed.

The foundations of enzyme action. H. Gutfreund and J. R. Knowles. Essays Biochem. 3:25–72, 1967.

The Enzymes, 3rd ed. P. D. Boyer (ed.). Vols. 1–14. New York, Academic Press, 1970–1981.

Enzymes: Physical Principles. H. Gutfreund. London, J. Wiley & Sons, 1972.

Enzyme Structure and Mechanism. A. Fersht. Reading, England, W. H. Freeman, 1977.

Enzymes, 3rd ed. M. Dixon and E. C. Webb. London, Longmans, Green & Co., 1979.

Structure and catalysis of enzymes. W. N. Lipscomb. Ann. Rev. Biochem. 52:17–34, 1983.

REACTION RATES AND CATALYSIS

A catalyst, such as an enzyme, only increases the rate of a reaction and is not altered itself at the end of the reaction; although it may participate intimately in the reaction, it is regenerated in its original form. Consequently, a catalyst cannot alter the equilibrium between reactants and products that would apply in its absence; instead, it can only increase the rate of attainment of that equilibrium. In the most general terms, it does so by decreasing the energy of the transition state relative to that of the reactants.

A **transition state** is an essentially hypothetical state that cannot be observed directly because it is defined to be the least populated species along the reaction pathway—that with maximum free energy. It decomposes at the vibrational frequency of a covalent bond, v:

$$v = \frac{k_B T}{h} \qquad (9-1)$$

where k_B is Boltzmann's constant and h is Planck's constant. At 25° C, $v = 6.2 \times 10^{12}$ sec^{-1}. As this frequency is assumed to be the same for all transition states, the observed rate constant for a particular reaction is determined by the extent to which the transition state is populated, that is, its free energy relative to the initial reactants, $\Delta G^{\ddagger}$. The superscript $\ddagger$ is used to designate a transition state and thermodynamic quantities applying to it. The value of $\Delta G^{\ddagger}$ is obtained by considering the transition state to be in equilibrium with the ground state of the reactant, A:

$$A \underset{}{\overset{K^{\ddagger}}{\rightleftharpoons}} [A^{\ddagger}] \overset{v}{\longrightarrow} P \qquad (9-2)$$

$$\Delta G^{\ddagger} = -RT \ln K^{\ddagger} = -RT \ln \frac{[A^{\ddagger}]}{[A]} \qquad (9-3)$$

The kinetic equation for the reaction A→P is then

$$\frac{-d[A]}{dt} = \frac{d[P]}{dt} = k_{obs}[A] = \frac{k_B T}{h} [A^\ddagger] = \frac{k_B T\, K^\ddagger}{h} [A]$$

$$= \frac{k_B T}{h} [A] \exp\left(-\Delta G^\ddagger/RT\right) \qquad (9-4)$$

The observed rate constant, k_{obs}, is related to the energy of the transition state by

$$k_{obs} = \frac{k_B T}{h} \exp\left(-\Delta G^\ddagger/RT\right) \qquad (9-5)$$

From the value of k_{obs}, the relative free energy of the hypothetical transition state may be calculated:

$$\Delta G^\ddagger = RT \ln \frac{k_B T}{k_{obs} h} \qquad (9-6)$$

At 25° C with k_{obs} expressed in units of sec^{-1}, this equation has the form

$$\Delta G^\ddagger = (17.4 - 1.36 \log k_{obs}) \text{ kcal/mole} \qquad (9-7)$$

The higher the energy of the transition state, the slower the reaction.

The relative free energies of reactants and transition state are those of the entire system, including their interactions with any components of the solvent and both enthalpic and entropic parameters. Interactions with the solvent are extremely important: Charge separations almost invariably occur in any transition state, because electrons are being redistributed in the process of covalent bond breakage and reformation; polar solvents can greatly stabilize charges. Consequently, reactions involving increased charge separation in the transition state occur much more rapidly in polar solvents. Conversely, reactions with less charge separation occur more rapidly in nonpolar solvents; for example, the reaction

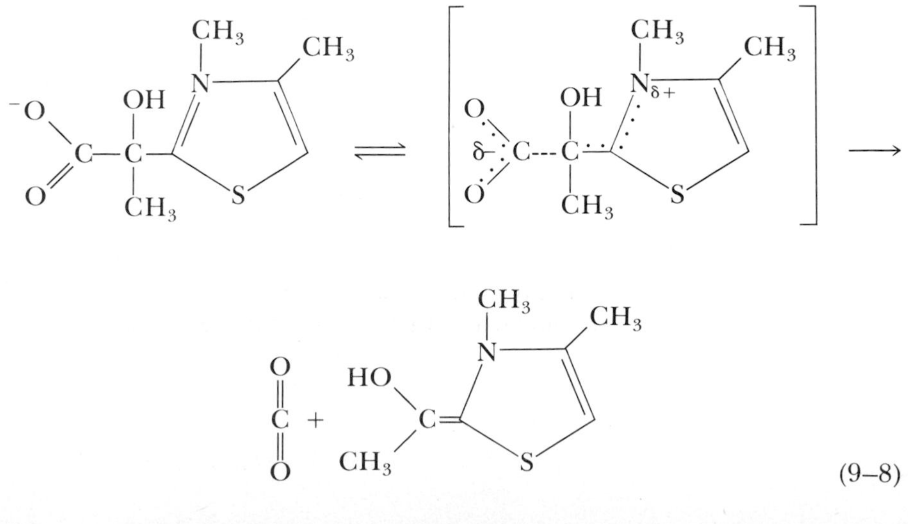

$$(9-8)$$

occurs 10^4 to 10^5 times faster in ethanol than in water.

The relative entropies of the reactants and transition state are also important for determining the rate. Accordingly, unimolecular reactions can occur more rapidly than do those involving multiple reactants, since much more entropy must be lost by independent molecules in order to encounter each other in a transition state complex than by a molecule in which the reactants are already held together by covalent bonds. The rate of a multimolecular reaction with n reactant molecules in the transition state depends upon each of their concentrations, and the rate constant has units of $\sec^{-1}{\rm M}^{-(n-1)}$. In contrast, the half-time of a unimolecular reaction is independent of the reactant concentration. The relative rates of comparable unimolecular and bimolecular reactions give the effective concentration of the two reacting groups within the unimolecular reactant, just as in comparisons of equilibrium constants (see Table 4–7).

The entropy difference between two independent flexible molecules and a single, rigid one can lead to effective concentrations as high as 10^{10} M; the observed value will depend upon how rigidly the reacting groups are normally in position appropriate for the transition state, i.e., how much entropy must be lost in the transition state. This is illustrated in Table 9–1, where the first, relatively flexible unimolecular reactant has an apparent effective concentration of 3×10^4 M evident from its rate of reaction. The addition of other groups, so that the reacting groups have less flexibility for steric reasons and are "locked" into positions appropriate for reaction, leads to an increase in effective concentration to nearly 10^9 M.

In a unimolecular reaction, the relative free energies of transition state and reactant can also be decreased by incorporating strain into the initial reactant, which is relaxed in the transition state. In this case, the rate is increased not by stabilizing the transition state but by destabilizing the initial reactant. Part of the increased rate of the most reactive molecule in Table 9–1 may be due to such strain; however, there are other examples without apparent strain that exhibit comparable kinetic effective concentrations.

Chemical bonds are in the process of being made and broken in a transition state, and its fluctuating nature prevents elucidation of its structure by direct means. Nevertheless, some of its properties may be inferred by the effects of varying the reaction conditions and the nature of the reactants on the rate of the reaction, i.e., the energy of the transition state. The structures of any intermediates, which have intact, normal covalent bonds, along the pathway also give indications of the structure of the transition state, as a reaction is believed to occur primarily by a continuous change of the initial reactants to the final products. The relationships among reactants, intermediates, transition states, and rate constants are illustrated in Figure 9–1.

Enzymes are not the only molecules that can catalyze reactions, although they do so with an efficiency and specificity that is unsurpassed.

Table 9–1 *Relative Rates of the Esterification Reaction*

$$-OH + HO\overset{\overset{\displaystyle O}{\|}}{C}- \longrightarrow -O-\overset{\overset{\displaystyle O}{\|}}{C}- + H_2O$$

REACTANTS	RATE CONSTANT	EFFECTIVE CONCENTRATION OF INTRAMOLECULAR GROUPS
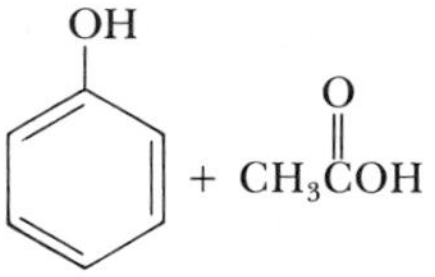	$10^{-10}\ s^{-1}M^{-1}$	—
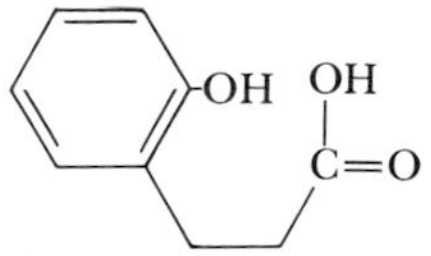	$3.2 \times 10^{-6}\ s^{-1}$	$3.2 \times 10^4\ M$
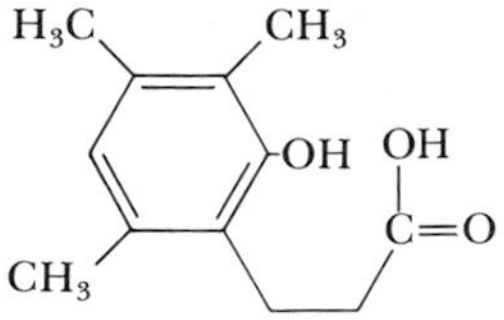	$3.3 \times 10^{-6}\ s^{-1}$	$3.3 \times 10^4\ M$
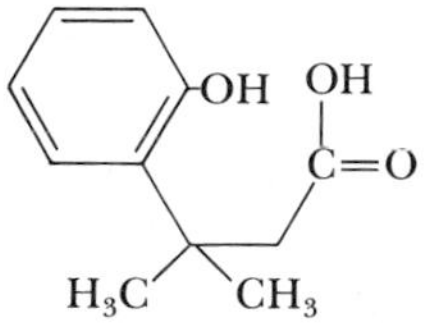	$3.6 \times 10^{-5}\ s^{-1}$	$3.6 \times 10^5\ M$
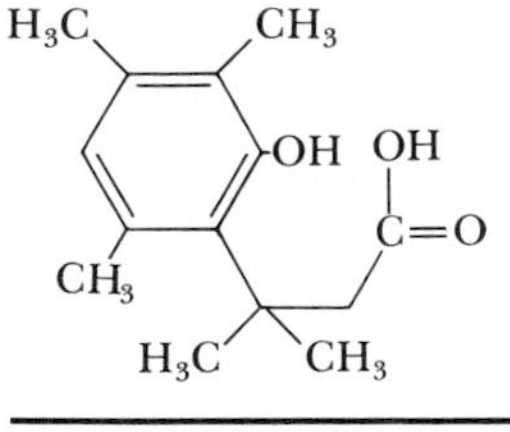	$8.5 \times 10^{-2}\ s^{-1}$	$8.5 \times 10^8\ M$

Rate constants for the uncatalyzed reactions in water from S. Milstien and L. A. Cohen, Proc. Natl. Acad. Sci. U.S.A. 67:1143–1147, 1970.

Micelles of lipid bilayers can catalyze reactions by concentrating nonpolar reactants within the bilayer and by providing a favorable environment. Polymers that strongly bind the reactants can also increase reaction rates, both by concentrating the reactants and by providing a favorable environment.

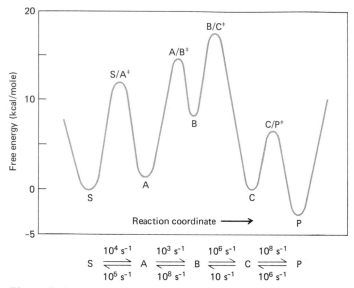

Figure 9–1

Free energy profile for a hypothetical reaction S→P, proceeding sequentially through intermediates A, B, and C. The rate constants for all the interconversions are given on the reaction scheme. The free energy of each transition state, relative to its precursor, was calculated at 25° C as described in the text.

During the reaction from S→P, only intermediate A would accumulate to detectable levels, at most 10 per cent of the molecules present.

More limited rate enhancements by small molecules in solution, generally acids and bases, are often observed. These molecules undoubtedly act by interacting transiently with the transition state and stabilizing it by favoring the necessary charge separation. Some examples are given in Figure 9–2.

Other catalysts are involved more intimately in a chemical reaction, actually forming covalent bonds with the reactants in intermediate stages. An excellent example is seen in the way carbonyl compounds can reversibly form Schiff bases with amines, thereby altering the chemical properties of both:

$$\begin{matrix} R_1 \\ \quad \diagdown \\ \quad\quad C{=}O \ + \ H_2N{-}R_3 \\ \quad \diagup \\ R_2 \end{matrix} \overset{H_2O}{\underset{}{\rightleftharpoons}} \begin{matrix} R_1 \\ \quad \diagdown \\ \quad\quad C{=}N{-}R_3 \\ \quad \diagup \\ R_2 \end{matrix} \tag{9–9}$$

Schiff base

Protonation of the Schiff base occurs under normal conditions, and it then becomes an electron "sink," assisting loss of a proton from groups R_1, R_2, or R_3, thereby making them much more reactive as nucleophiles than they would be in the parent compound; an example follows:

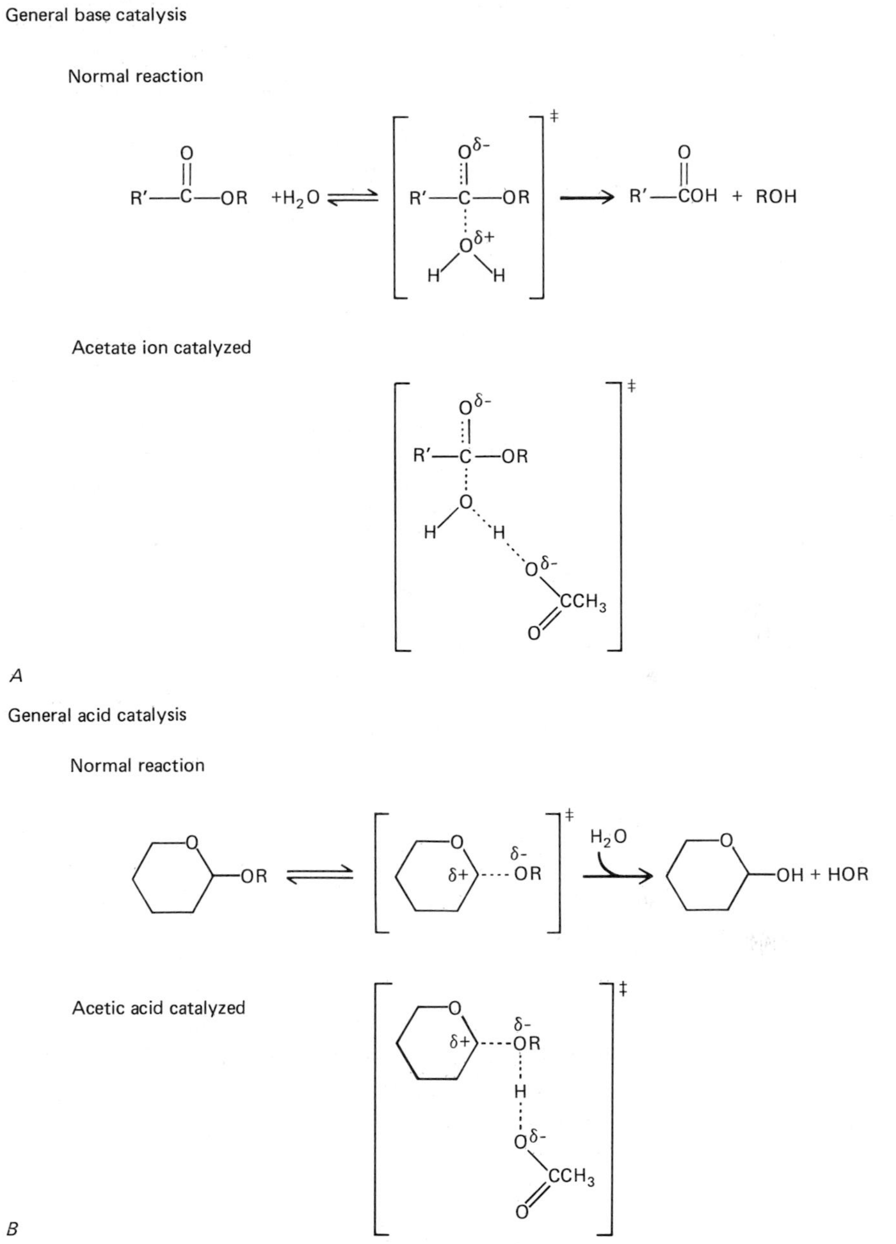

General base catalysis

Normal reaction

Acetate ion catalyzed

A

General acid catalysis

Normal reaction

Acetic acid catalyzed

B

Figure 9–2

Examples of general base (A) and acid (B) catalysis. A, The uncatalyzed hydrolysis of an ester, with the presumed transition state enclosed in brackets. Acetate ion catalyzes this reaction, presumably by participating in the lower energy transition state complex in which the charge separation is more favorable.

B, The uncatalyzed hydrolysis of an acetal, with the presumed transition state in brackets. Acetic acid catalyzes this reaction, presumably by stabilizing the transition state as indicated.

(Adapted from A. Fersht, Enzyme Structure and Mechanism. Reading, W. H. Freeman, 1977.)

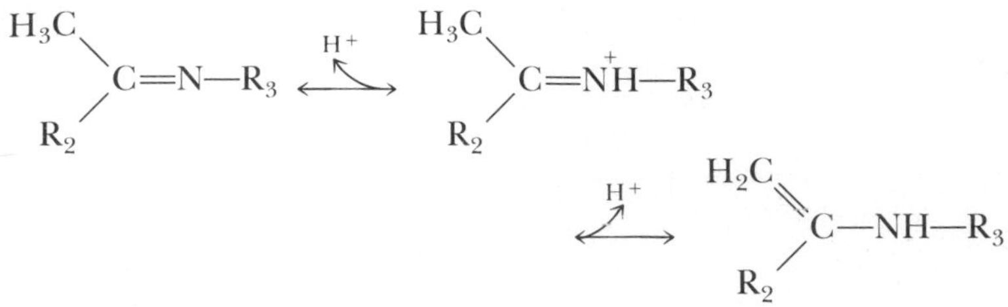

$$\underset{R_2}{\overset{H_3C}{>}}C=N-R_3 \quad \underset{H^+}{\rightleftharpoons} \quad \underset{R_2}{\overset{H_3C}{>}}C=\overset{+}{N}H-R_3$$

$$\underset{H^+}{\rightleftharpoons} \quad \underset{R_2}{\overset{H_2C}{>}}C-NH-R_3 \qquad (9-10)$$

After reaction of one of the activated peripheral groups, the Schiff base can dissociate back to the carbonyl and amine; whichever of the original reactants were not altered would be an apparent catalyst for reaction of the other.

Enzymes engage in such tricks by utilizing suitable groups as cofactors, channeling the catalytic capabilities they exhibit by themselves. For example, many enzymes catalyzing reactions involving amino acids use pyridoxal phosphate as cofactor, which by itself readily forms a Schiff base with amino acids and catalyzes a number of their reactions:

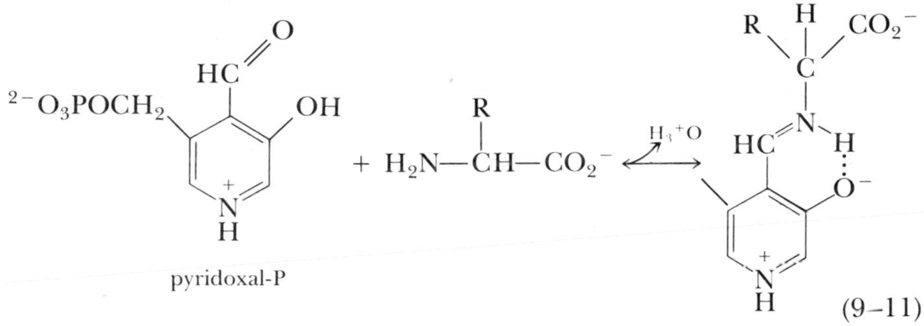

pyridoxal-P

$$(9-11)$$

The pyridine ring of the Schiff base acts as an electron sink, which stabilizes very effectively a negative charge; therefore, each of the groups on the C^α atom of the amino acid may be cleaved off as cations—for example, the hydrogen atom:

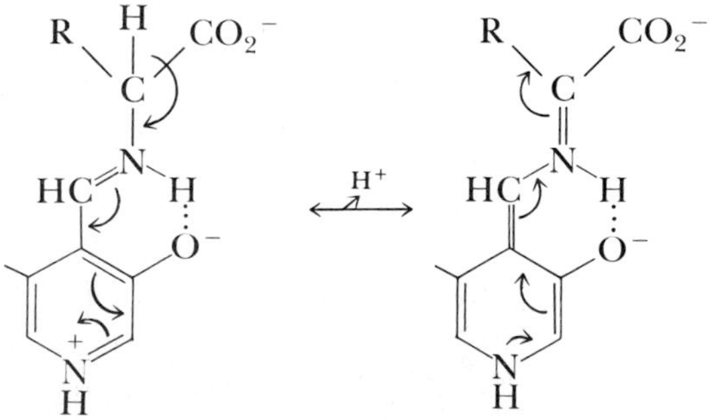

$$(9-12)$$

The consequence of transient removal of the H atom can be racemization of the amino acid, alteration of the amino acid side chain, or loss of the carboxyl group as CO_2. In each case, regeneration of unaltered pyridoxal phosphate occurs; thus it serves as a catalyst. In other reactions, the pyridoxal phosphate is converted from the aldehyde to the amine by cleavage of the C^{α}—N bond.

In summary, catalysts can increase the rates of reactions by stabilizing the normal transition state, or they can generate new reaction pathways, with intermediate steps of faster rate than that of the uncatalyzed reaction.

Bioorganic Mechanisms. T. C. Bruice and S. J. Benkovic. New York, Benjamin, 1966.

Catalysis in Chemistry and Enzymology. W. P. Jencks. New York, McGraw-Hill, 1969.

Schiff base intermediates in enzyme catalysis. E. E. Snell and S. J. DiMari. In The Enzymes, 3rd ed. P. D. Boyer (ed.). Vol. 2, pp. 335–370. New York, Academic Press, 1970.

Rate acceleration by stereopopulation control: models for enzyme action. S. Milstien and L. A. Cohen. Proc. Natl. Acad. Sci. U.S.A. 67:1143–1147, 1970.

A source for the special catalytic power of enzymes: orbital steering. D. R. Storm and D. E. Koshland, Jr. Proc. Natl. Acad. Sci. U.S.A. 66:445–452, 1970.

The sensitivity of intramolecular reactions to the orientation of the reacting atoms. D. A. Dafforn and D. E. Koshland, Jr. Bioorg. Chem. 1:129–139, 1971.

Entropic contributions to rate accelerations in enzymic and intramolecular reactions and the chelate effect. M. I. Page and W. P. Jencks. Proc. Natl. Acad. Sci. U.S.A. 68:1678–1683, 1971.

Intramolecular catalysis. A. J. Kirby and A. R. Fersht. Prog. Bioorg. Chem. 1:1–82, 1971.

Catalytic accelerations of 10^{12}-fold by an enzyme-like synthetic polymer. H. C. Kiefer, et al. Proc. Natl. Acad. Sci. U.S.A. 69:2155–2159, 1972.

The energetics of neighbouring group participation. M. I. Page. Chem. Soc. Rev. 2:295–323, 1973.

Some pertinent aspects of mechanism as determined with small molecules. T. C. Bruice. Ann. Rev. Biochem. 45:331–373, 1976.

Effective molarities for intramolecular reactions. A. J. Kirby. Adv. Phys. Org. Chem. 17:183–278, 1980.

KINETICS OF ENZYME ACTION

In the classical studies of enzymes mentioned earlier, the enzyme itself was used at very low concentrations relative to the substrate, and only the net effect of its presence on the rate of conversion of substrate to product was measured. Under the experimental conditions, the enzyme was generally catalyzing the reaction very rapidly on the time scale of the measurements, so that it was in a steady state in which the concentrations of its various forms were constant and too low to affect the measurements. Because the enzyme was not observed directly, emphasis was placed upon the substrates, products, inhibitors, and so forth. Nevertheless, these studies were pursued rigorously and often ingeniously, permitting the elucidation of complex reaction schemes and some significant insight into enzyme mechanisms. They provided the framework for studying directly the reactions that take place on the enzyme.

Here we will consider briefly the general principles and observations derived from such studies. The following discussion is limited to "simple" enzymes; allosteric enzymes will be described in a subsequent section.

Steady-state Kinetics

Single Substrate and Product

Enzymes as catalysts have distinctive properties: As is well-known, they can catalyze reactions that are negligible in their absence. The velocity of the catalyzed reaction

$$v = \frac{-d[S]}{dt} = \frac{d[P]}{dt} \qquad (9\text{--}13)$$

is proportional to the *substrate concentration* only at low levels, becoming independent at high concentrations (Figure 9–3A); in contrast, at all substrate concentrations the velocity of the reaction is *directly proportional to the enzyme concentration* so long as it is present at catalytic levels. Consequently, two kinetic parameters are required to describe an enzyme's action: the maximum velocity at saturating substrate concentration, V_{max}, and the substrate concentration that gives half V_{max}, K_m. The value of V_{max} is proportional to the total enzyme concentration:

$$V_{max} = k_{cat}[E_T] \qquad (9\text{--}14)$$

whereas that of K_m is independent. The velocity of the enzyme-catalyzed reaction, v, is then expressed by the well-known **Michaelis-Menten equation:**

$$v = \frac{[S]}{K_m + [S]} k_{cat}[E_T] \qquad (9\text{--}15)$$

The dependence of the velocity on the substrate concentration is analogous to a ligand-binding isotherm (see Figure 8–1); such dependence indicates that catalysis occurs upon binding of the substrate to the enzyme:

$$E + S \underset{k_{-1}}{\overset{k_1}{\rightleftharpoons}} ES \xrightarrow{k_2} E + P \qquad (9\text{--}16)$$

The K_m is analogous to the dissociation constant of the Michaelis complex, ES, but the values are the same only if dissociation is more rapid than is conversion to product; i.e., $k_{-1} \gg k_2$. In general,

$$K_m = \frac{k_2 + k_{-1}}{k_1} \qquad (9\text{--}17)$$

The value of K_m is then usually greater than or equal to the equilibrium constant for dissociation of substrate from the enzyme, but it can also be smaller, if intermediate forms of the Michaelis complex along the reaction pathway are present in significant amounts and have smaller dissociation constants. Observed values of K_m for different substrates and different enzymes vary widely; the smaller values are in the region of 10^{-7} M, whereas poor substrates can have very high values. Typical values for physiological substrates are generally in the region of 10^{-3} to 10^{-6} M.

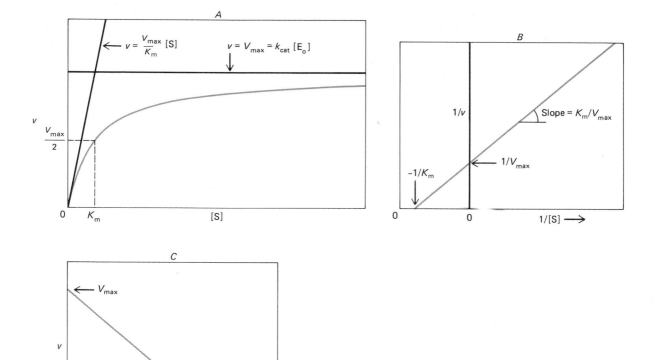

Figure 9–3

The dependence on substrate concentration, [S], of the velocity of an enzyme-catalyzed reaction, v. The normal hyperbolic relationship between v and [S] is shown in A. The double reciprocal Lineweaver-Burke plot shown in B is widely used, as it provides easy estimation of the values of V_{max} and k_m. The Eadie-Hofstee plot in C is the equivalent of the Scatchard plot of binding data (see Figures 8–1 and 8–2). (Adapted from A. Fersht, Enzyme Structure and Mechanism. Reading, W. H. Freeman, 1977.)

    The velocity of an enzyme-catalyzed reaction becomes independent of substrate concentration when the enzyme is fully saturated in the steady state, the concentration of free enzyme becoming negligible. The value of k_{cat} is the rate of breakdown of the ES complex, k_2. It is often referred to as the **turnover number** of the enzyme, representing the number of times each enzyme molecule catalyzes the reaction. Turnover numbers vary widely; the highest values observed are 4×10^7 per second for catalase and 10^6 sec^{-1} for carbonic anhydrase.

    Enzyme kinetic data may be treated graphically in the same manner as that of ligand binding. Eadie-Hofstee plots are exactly comparable to Scatchard plots of binding data (as in Figure 8–1C), plotting v versus $v/[S]$.

Most widely used in enzyme kinetics are the double-reciprocal Lineweaver-Burke plots of $1/v$ versus $1/[S]$ (Figure 9–3B). Inverting the Michaelis-Menten equation (Eq. 9–15) yields

$$\frac{1}{v} = \left(1 + \frac{K_m}{[S]}\right)\frac{1}{V_{max}} \qquad (9\text{–}18)$$

Consequently, the y-intercept gives $1/V_{max}$; the x-intercept is $-1/K_m$; and the slope of the straight line gives K_m/V_{max}.

At very low substrate concentrations, the combination of the substrate with the enzyme becomes rate-limiting. Under these conditions, the Michaelis-Menten equation becomes

$$v = \frac{k_{cat}}{K_m}[E_T][S] \qquad ([S] \ll K_m) \qquad (9\text{–}19)$$

Most of the enzyme is in the free form under these conditions, so Equation 9–19 becomes

$$v = \frac{k_{cat}}{K_m}[E][S] \qquad (9\text{–}20)$$

Expressed in terms of free enzyme, this equation is valid for all substrate concentrations. Consequently, k_{cat}/K_m represents the apparent rate constant for combination of a substrate with the free enzyme and is the most critical parameter in determining the specificity of an enzyme for a substrate. Because an enzyme and substrate cannot combine more rapidly than diffusion permits, there is an upper limit on enzyme catalysis: The value of k_{cat}/K_m cannot be greater than about $10^9\,\text{s}^{-1}\text{M}^{-1}$. It will be lower if the catalytic step is less than k_{-1}, so that some substrate molecules dissociate rather than are converted to product, and if not all encounters between enzyme and substrate result in binding, as with any ligand-binding interaction. Nevertheless, some enzymes have values of k_{cat}/K_m that approach the diffusion limit, indicating extreme efficiency in binding substrate and in converting it to product.

Not altering the equilibrium of a reaction, an enzyme must also catalyze it in the reverse direction. It must then have a Michaelis constant for the product, K_m^P, and a value of k_{cat} for the reverse reaction, k_{cat}^r; also, an enzyme–product complex must exist. The values of K_m^P and k_{cat}^r are not independent of those for the forward reaction; rather, they are related by the normal equilibrium constant for the reaction, since at equilibrium the enzyme must be catalyzing the reaction equally in both directions. The **Haldane relationship** expresses this:

$$\frac{k_{cat}^f/K_m^S}{k_{cat}^r/K_m^P} = K_{eq} = \frac{[P]_{eq}}{[S]_{eq}} \qquad (9\text{–}21)$$

The substrate and product compete for the free enzyme; the outcome depends upon their respective values of k_{cat}/K_m. Consequently, where they have different K_m values, the equilibrium ratio of product to substrate on the enzyme (k_{cat}^f/k_{cat}^r) will be different from that in solution (K_{eq}).

Multiple Substrates and Products

Many reactions catalyzed by enzymes have multiple substrates, products, or both. A number of reaction schemes are possible in each case, and they may usually be distinguished by steady-state kinetic measurements. In brief, the effects on the observed rate of varying independently the concentrations of the substrates and products are measured. Generally, the variation of only one substrate, with all others kept constant, will yield normal Michaelis-Menten kinetics, but different reaction schemes predict various effects of the other variables on the apparent values of K_m and V_{max} for each substrate. The steady-state kinetic behavior expected for any particular reaction scheme can be predicted, most readily using the ingenious graphical procedure of King and Altman. Reaction schemes inconsistent with the kinetic results can be excluded. Remember, however, that kinetic analysis can never prove a mechanism but can only disprove others.

Sequential reactions are those in which all the substrates bind to the enzyme before the first product is formed. The binding of substrates, or release of products, may be essentially random:

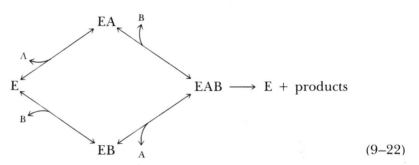

$$(9\text{–}22)$$

although binding of one reactant may alter the affinity for others. An extreme situation in which there are very large differences in affinities for the substrates results in essentially ordered substrate binding, or product release; e.g.,

$$E \overset{A}{\underset{}{\longleftrightarrow}} EA \overset{B}{\underset{}{\longleftrightarrow}} EAB \longrightarrow E + \text{products} \qquad (9\text{–}23)$$

A quite different type of reaction scheme is that in which one (or more) product is formed and released from the enzyme before binding of the other substrates:

$$E \overset{A}{\underset{}{\longleftrightarrow}} EA \overset{P}{\underset{}{\longleftrightarrow}} E' \overset{B}{\underset{}{\longleftrightarrow}} E'B \longrightarrow E + Q \qquad (9\text{–}24)$$

This **substituted-enzyme** type of mechanism, also referred to as **"ping-pong"** or **double-displacement,** generally occurs when a group of substrate A is transferred to substrate B. The intermediate E′ form of the enzyme could merely bind the group very tightly, but we will see that tight binding of ligands is usually detrimental to enzyme catalysis, as is the accumulation of intermediates. Perhaps for this reason, groups to be transferred are held as covalent but unstable acyl, phosphoryl, or Schiff base adducts.

The substituted-enzyme type of reaction often offers the first direct clue to the role of the enzyme in catalysis, as the modified enzyme E′ can usually be prepared by adding substrate A in the absence of the other substrates. In many cases, one of the substrates or products may be water. Because the concentration of water is difficult to vary, other approaches must be used to determine the reaction scheme. For example, an acyl-enzyme intermediate of the serine proteases (described later) was inferred from the observation that substrates with different R′ groups gave the same value of k_{cat}:

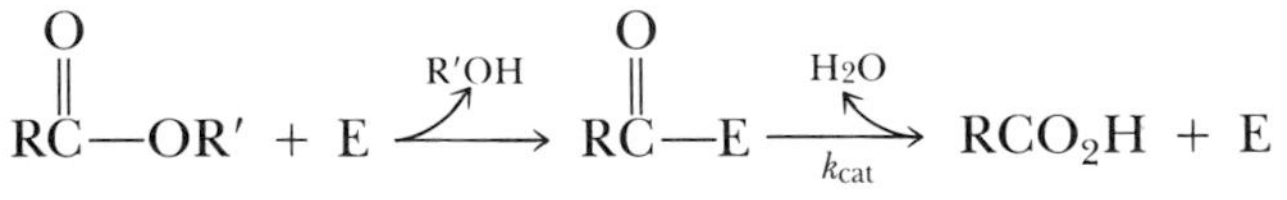

$$\underset{\underset{RC}{\|}{O}}{RC}\!-\!OR' + E \xrightarrow{R'OH} \underset{\underset{RC}{\|}{O}}{RC}\!-\!E \xrightarrow[k_{cat}]{H_2O} RCO_2H + E \qquad (9\text{--}25)$$

Such substrates were hydrolyzed nonenzymatically at different rates; a common enzymic rate, therefore, indicated the rate-limiting breakdown of a common acyl intermediate after release of the varying R′ group. Similarly, the addition of other acceptors that can compete with water

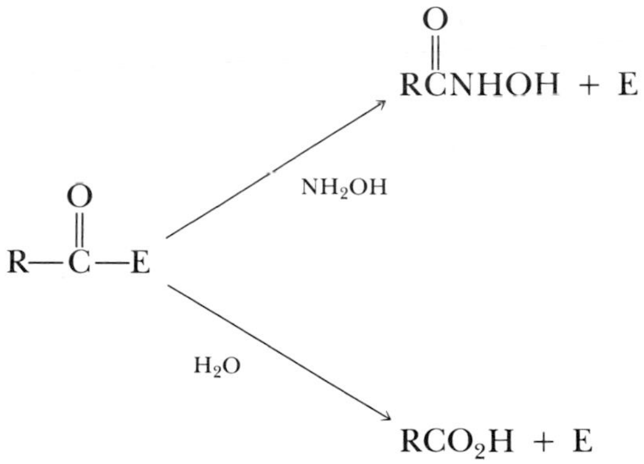

$$
\begin{array}{c}
\overset{\overset{\textstyle O}{\|}}{RCNHOH} + E \\[2mm]
\nearrow \\[1mm]
\underset{\underset{C}{\|}{O}}{R-C-E} \quad \xrightarrow{\ NH_2OH\ } \\[2mm]
\searrow \\[1mm]
\xrightarrow{\ H_2O\ } \\[2mm]
RCO_2H + E \qquad (9\text{--}26)
\end{array}
$$

gave product ratios independent of the R′ group of the substrate, whereas different ratios were obtained with the nonenzymatic reactions.

Inhibitors

Much indirect information about the catalytic activity of an enzyme can be gained from its inhibition, especially if the enzyme has only a limited range of substrates. This discussion excludes inhibitors that react covalently with the enzyme and refers only to those that act by binding reversibly to the enzyme at specific sites. They are usually structural analogues of one of the substrates or products, or bind at some other specific functional site on the enzyme. An inhibitor may be acted upon catalytically, i.e., used as

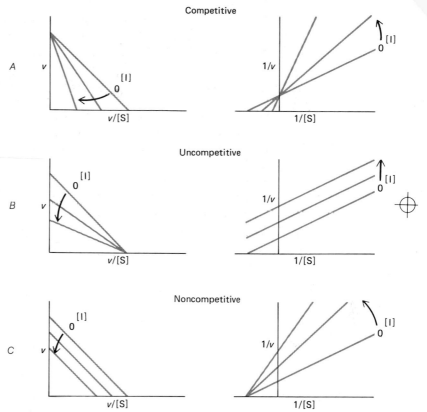

Figure 9–4

Steady-state kinetic analysis of enzyme inhibition. The enzyme velocity, v, is measured at varying concentrations of substrate, S, and of inhibitor, I. The relationship between v and [S] is plotted at different values of [I] in Eadie-Hofstee (left) plots and in Lineweaver-Burke plots (right). (Adapted from A. Fersht, Enzyme Structure and Mechanism, 1977.)

a substrate, but the simplest cases are those in which only reversible binding occurs. Each inhibitor is characterized by its inhibition constant, K_i, analogous to K_m for a substrate; the inhibition constant is usually the same as the dissociation constant if it is not altered by the enzyme.

The modes of action of inhibitors are usually elucidated by their effects, at different concentrations, on the utilization of a substrate, also at different concentrations. The inhibition is generally designated as **competitive, uncompetitive,** or **noncompetitive.** Each classification refers to only one substrate; if there is more than one substrate, the inhibition with respect to each must be determined, and they are unlikely to be the same.

Simplest are inhibitors, such as substrate analogues, that bind at the same site as the substrate and compete with it. This *competitive* inhibition has the effect of increasing the apparent K_m for the substrate, but V_{max} is not altered, since high concentrations of the substrate can compete successfully for the enzyme (Figure 9–4A).

Uncompetitive inhibition is expected if the inhibitor binds only to the enzyme–substrate complex, blocking its catalytic activity. The values of K_m and V_{max} are affected in parallel (Figure 9–4B). This type of inhibition is rare with single-substrate enzymes, but occurs more readily with multisubstrate cases; for example, a competitive inhibitor of one substrate can give uncompetitive inhibition with respect to the other in an ordered, sequential reaction scheme (Eq. 9–23).

An inhibitor that binds to both the free enzyme and the enzyme–substrate complex might affect only V_{max}, giving pure *noncompetitive* inhibition (Figure 9–4C). More commonly, the K_m for the substrate is also affected by the inhibitor, so that reciprocal plots at different inhibitor concentrations intersect elsewhere than on the x-axis. This combination of effects is known as **mixed inhibition.**

Inhibition patterns can be complex and should be predicted analytically from any proposed reaction scheme to show quantitative consistency with the experimental observations. However, they may usually be qualitatively predicted from a few useful rules: An inhibitor affects the slope of a Lineweaver-Burke plot (Figure 9–4) if it and the varied substrate compete directly for the same form of the enzyme or for different forms that are in reversible equilibrium with each other. Inhibitor binding to any other form of the enzyme affects the y-intercept (i.e., V_{max}). These effects can occur separately, producing, respectively, competitive or uncompetitive patterns; jointly, they produce noncompetitive inhibition. If substrate or inhibitor combines with more than one form of the enzyme, the resulting pattern is the sum of the different individual inhibitions.

For example, in the sequential ordered scheme

$$E \xrightarrow{A} EA \xleftrightarrow{B} EAB \longleftrightarrow EPQ \xrightarrow{P} EQ \xleftrightarrow{Q} E \qquad (9\text{–}27)$$

an analogue of B that combines with only EA will be competitive with respect to B but uncompetitive with respect to A. An analogue of A that binds only to E will be competitive with respect to A but noncompetitive to B. Product Q will give the same pattern of inhibition of the forward reaction, since it also competes with A for free enzyme. Product P will be noncompetitive to both A and B.

However, prediction of the inhibitor binding site on an enzyme is dangerous from such measurements alone. Certain noncompetitive inhibitors have been observed to bind at the same site as that for substrate; they can give noncompetitive inhibition, rather than the expected competitive type, if they dissociate very slowly from the enzyme. Analysis of inhibition kinetics, therefore, must include direct studies of inhibitor–enzyme interaction; such studies can be extremely valuable because the complexes formed can be stable, in contrast to those in enzyme–substrate interactions.

Isotopically Labeled Substrates

In order to dissect the kinetic scheme of a complex reaction, the origin of each atom of each product must be known, whether it arises from the solvent or from the substrate, as well as from which atom. This can be determined most readily using specifically labeled substrates and following the fate of each atom. Which groups are transferred in each reaction step can also be inferred from the effect on the rate of substituting a heavier or lighter isotope; lighter isotopes are generally transferred more rapidly than heavier ones. This difference is most dramatic with hydrogen atoms, where the reaction rate for the deuterium and the tritium isotopes may be 1/24 and 1/79, respectively, of that for the normal isotopes. Smaller isotope effects occur with other atoms, but sensitive modern methods of measuring the relative levels of two isotopes can reveal small kinetic differences between the two isotopically labeled forms of the substrate.

In the other techniques using isotopes, the isotope effect on rates is minimized; instead, identification of specific isotopes following reactions permits distinction between atoms of different origin. Most widely used is isotope exchange between different substrates or products. Enzymes with two or more substrates or products often catalyze partial reactions that may be detected by isotope exchange. For example, group-transfer reactions of the type

$$A\!-\!G + B \rightleftharpoons A + B\!-\!G \qquad (9\text{--}28)$$

where group G is transferred from A to B, can occur either by direct exchange on a ternary complex of the two substrates on the enzyme or by a substituted-enzyme scheme in which the group is transiently transferred to the enzyme:

$$E + A\!-\!G \underset{\longleftarrow}{\overset{A}{\longrightarrow}} E\!-\!G \underset{\longleftarrow}{\overset{B}{\longrightarrow}} E + B\!-\!G \qquad (9\text{--}29)$$

In the latter case, the enzyme will catalyze an exchange between A—G and A, in the absence of B so that no net reaction occurs, that can be detected only if either A—G or A is isotopically labeled. A similar reaction between B and B—G will be catalyzed in the absence of A. No such exchange should occur if the group is only transferred directly from one substrate to the other.

This technique is widely used to measure amino acid activation by amino–acyl tRNA synthetases. In the absence of tRNA, to which the correct amino acid becomes attached in the full reaction, the enzyme catalyzes the partial reaction

$$E + ATP + \text{amino acid} \rightleftharpoons \text{aminoacyl adenylate } E + PP_i \quad (9\text{--}30)$$

where the aminoacyl adenylate remains firmly bound to the enzyme. This reaction is readily measured by isotope exchange between ATP and pyrophosphate (PP_i).

Another approach is to follow isotope exchange between a substrate and product when all the substrates are present and the reaction is at equilibrium. No net reaction occurs at equilibrium, but the forward and reverse rates may be followed individually by isotope exchange. For example, the enzyme aldolase was found at equilibrium to catalyze isotope exchange between the substrate fructose 1,6-diphosphate and one of its products, glyceraldehyde-P, much more rapidly than with the other, dihydroxyacetone-P. This was one piece of evidence that glyceraldehyde-P is released first from the enzyme, leaving the other product attached to the enzyme via a Schiff base.

$$E + \text{fructose } 1,6\text{-P}_2 \xleftrightarrow{\quad\text{glyceraldehyde-P}\quad} E \cdot \text{dihydroxyacetone-P}$$

$$\longleftrightarrow E + \text{dihydroxyacetone-P} \qquad (9\text{--}31)$$

A systematic method for deriving the rate laws for enzyme-catalysed reactions. E. L. King and C. Altman. J. Phys. Chem. 60:1375–1378, 1956.

Steady state kinetics. W. W. Cleland. In The Enzymes, 3rd ed. P. D. Boyer (ed.). Vol. 2, pp. 1–65. New York, Academic Press, 1970.

Initial Rate Enzyme Kinetics. H. J. Fromm. Berlin, Springer, 1975.

Kinetics of Enzyme Mechanisms. J. T.-F. Wong. New York, Academic Press, 1975.

Enzyme Kinetics. I. H. Segel. New York, John Wiley & Sons, 1975.

Principles of Enzyme Kinetics. A. Cornish-Bowden. London, Butterworths, 1975.

Kinetic analysis of the properties and reactions of enzymes. H. Gutfreund. Prog. Biophys. Mol. Biol. 29:161–195, 1975.

Determining the chemical mechanisms of enzyme-catalyzed reactions by kinetic studies. W. W. Cleland. Adv. Enzymol. 45:273–387, 1977.

Isotope exchange probes and enzyme mechanisms. P. D. Boyer. Acc. Chem. Res. 11:218–224, 1978.

Enzyme kinetics and mechanism. Part A. Initial rate and inhibitor methods. D. L. Purich (ed.). Methods Enzymol. 63, 1979; Part B. Isotopic probes and complex enzyme systems. D. L. Purich (ed.). Methods Enzymol. 64:1980.

The expression of isotope effects on enzyme-catalyzed reactions. D. B. Northrop. Ann. Rev. Biochem. 50:103–131, 1981.

Use of isotope effects to elucidate enzyme mechanisms. W. W. Cleland. Crit. Rev. Biochem. 13:385–428, 1982.

Reactions on the Enzyme

Steady-state kinetics provide only tantalizing hints of what processes take place on the enzyme to produce catalysis, with postulated complexes of the enzyme with substrates and products, and intermediate, substituted enzymes in "ping-pong" type schemes. To verify the existence of these complexes, and to learn more about what happens between the time when the first substrate is bound and when the last product is released, the enzyme itself must be studied. This requires both substrate-level quantities of enzyme and rapid techniques that permit measurements within the turnover time of the enzyme, which can be as short as 10^{-6} sec.

The existence of enzyme–substrate complexes have been amply demonstrated, using primarily spectral techniques like those used to measure ligand binding in general (Chapter 8), with results comparable to those for binding other ligands. Further changes in the bound substrate can often also be detected, as the rates of these steps are independent of the substrate concentration. However, a basic difficulty with kinetic studies of multistep reactions is that only a few intermediates are likely to accumulate to detectable levels, and such accumulation is energetically unnecessary. Any correlation between the importance of an intermediate and its likelihood of accumulating is likely to be an inverse one: The most important are least likely to accumulate. For example, the transition state is most important for determining the rate of a reaction, but it is the species least likely to accumulate.

Intermediates can be detected most readily if widely varying substrates are available for the enzyme, since they may differ in the rates of individual steps, perhaps leading to accumulation of different intermediates. Some might undergo partial reaction very rapidly, but then not react further at a comparable rate; if a product is released in the first step, it will not be readily reversed and intermediate EX will accumulate:

$$E + S \xrightarrow[P_1]{\text{fast}} EX \xrightarrow{\text{slow}} E + P_2 \qquad (9–32)$$

This kinetic behavior was further evidence for an acyl intermediate of serine proteases (Figure 9–5). Such substrates have very practical uses as active site titrants; upon mixing enzyme with excess substrate, there is an initial "burst" of formation of product P_1—one mole per mole of enzyme active site (Figure 9–5). Subsequent turnovers of the enzyme are limited by the slow second step. In many cases, because the intermediate has the second product covalently attached to the enzyme, the complex can be characterized chemically. In such techniques, intrinsically labile linkages may need to be trapped, while others may be stabilized merely by disruption of the enzyme's structure.

If a substrate normally makes no known covalent intermediate, it may be induced to do so—thus providing the enzymologist with a convenient probe—by altering it so that a very reactive group is generated during the calalytic cycle, which will then react with an appropriate nearby group in the enzyme active site. Such **"suicide substrates"** are similar to affinity labels (pp. 357–358) in that their binding at the active site gives them specificity, but they have the added specificity of not being intrinsically reactive molecules; they become reactive only upon being further subjected to catalysis. Some natural toxins seem to act in just this way, irreversibly inactivating crucial enzymes. Acting irreversibly, they are much more potent than reversible inhibitors, which must compete with the substrate. Penicillin and its analogues are believed to inactivate in this way an enzyme involved in bacterial cell wall synthesis, and analogues that protect penicillin from degradation by the enzyme penicillinase may similarly inactivate this enzyme.

The pH dependence of each step in an enzyme-catalyzed reaction can measure the pK_a values of crucial acids or bases in both the free enzyme and substrate and in the enzyme–substrate complex. The small, relatively simple enzymes that have been studied most extensively usually demonstrate straightforward pH-dependence of substrate binding or of catalysis, in that they appear to depend principally upon the ionization of only one or two groups. However, care must be exercised in assigning such pK_a

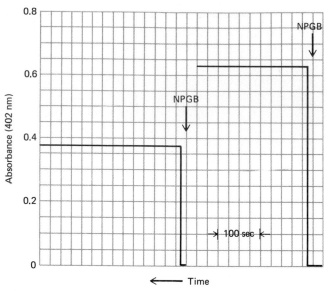

Figure 9–5
Active site titration of 0.05 ml (left) and 0.08 ml (right) of a solution of β-trypsin, using the titrant NPGB:

p-nitrophenyl-p'-guanidinobenzoate
(NPGB)

$HOC_6H_4NO_2$ +
p-nitrophenol

slow

At the times indicated by the arrows, the NPGB was added to a 1.0-ml solution containing the trypsin. The almost instantaneous increase in absorbance is due to the release of p-

Table 9–2   Anomalous pK$_a$ Values of Ionizing Groups in Enzyme Active Sites

ENZYME	IONIZING GROUP	OBSERVED pK$_a$	NORMAL[a] pK$_a$
Acetoacetate decarboxylase[b]	Lys	6.0	10.4
Carboxypeptidase A[c]	Glu 27	7	4.6
Lysozyme[d]	Glu 35	6.5	4.6
Papain[e]	His 159	8.5	6.2
	Cys 25	3.3	9.1 to 9.5
Pepsin[f]	Asp 32	1.5	4.5
Rhodanese[g]	Cys 247	6.5	9.1 to 9.5

[a]From C. Tanford, Adv. Protein Chem. 17:69–165, 1962; see Table 1–1.

[b]From F. C. Kokesh and F. H. Westheimer, J. Am. Chem. Soc. 93:7270–7274, 1971.

[c]From P. H. Petra and H. Neurath, Biochemistry 10:3171–3177, 1971.

[d]From C. C. F. Blake, et al., in Ciba Foundation Symposium. Vol. 60, pp. 137–172. Amsterdam, Excerpta Medica, 1978.

[e]From S. D. Lewis, et al., Biochemistry 20:48–51, 1981.

[f]From J. A. Hartsuck and J. Tang, J. Biol. Chem. 247:2575–2580, 1972.

[g]From J. H. Ploegman, et al., J. Mol. Biol. 127:149–162, 1979.

values to specific groups; those in enzyme active sites are often substantially perturbed from their normal values, and even more perturbed in the enzyme–substrate complex. Some examples are given in Table 9–2. The identities of these and other groups in the active site can be determined by the abolition of substrate binding or catalysis upon their specific covalent modification.

The most direct and detailed information about enzyme action would come from crystallographic determination of complexes with the substrate at various stages of the reaction. The rapid turnover of normal substrates under normal conditions precludes this, but it has been possible to determine the binding of partial sets of the substrates, of substrate analogues that react very slowly, and of inhibitors. Studies yielding such information have provided great insight into how substrates are held in precise orientations at the active site, but they are indirect, and extrapolation of results is necessary to imagine what occurs during catalysis, which could involve significant changes in both the enzyme and the substrate. Inhibitors can also be misleading because they may be inactive as substrates simply because they bind in a different, nonproductive mode. Extensive studies with a number of well-chosen substrate analogues are required to establish the relevance of the observed binding.

nitrophenol upon acylation of trypsin. The subsequent very slow increase is limited by the slow rate of deacylation of the enzyme. Extrapolation to time zero indicates that the concentrations of trypsin active sites in the cuvette were 2.22×10^{-5}M and 3.76×10^{-5}M on the left and right, respectively. Therefore, the stock solution was $4.57 \pm 0.13 \times 10^{-4}$M in active trypsin. (Adapted from T. Chase and E. Shaw, Methods Enzymol. 19:20–27, 1970, © Academic Press Inc.)

$$S \xrightleftharpoons{K_n^{\ddagger}} T^{\ddagger} \longrightarrow P \tag{9-36}$$

$$ES \xrightleftharpoons{K_E^{\ddagger}} ET^{\ddagger} \longrightarrow P \tag{9-37}$$

$$\frac{k_{cat}}{k_n} = \frac{K_E^{\ddagger}}{K_n^{\ddagger}} \tag{9-38}$$

The preceding discussion of effective concentrations in the unimolecular enzyme complex considered essentially only entropic contributions to the relative free energies of the substrate and transition state ($\Delta G^{\ddagger} = -RT \ln K^{\ddagger}$).

In many cases, this free energy difference has been considered to be lowered on the enzyme when the free energy of the substrate bound to the enzyme is increased relative to that in solution, with comparable transition states. This could occur if the bound substrate is strained, either sterically, with distorted geometry, or by placing it in an otherwise unfavorable environment. The free energy "cost" of introducing this strain would be paid for from binding energy, in that the affinity would be lower than for an unstrained substrate. Alternatively, the strain could be present in the enzyme, even before substrate is bound. An example is the perturbed chemical properties of groups often found in the active sites of enzymes (Table 9–2). This strain would be paid for from the energy of the folded conformation, as any such strained properties of the folded conformation necessarily lower its stability. In both cases, the strain must be such as to push the enzyme–substrate complex along the reaction coordinate toward the transition state so that there it is relieved.

Another way of explaining catalysis is to imagine that the energy of the transition state is lowered on the enzyme. This decrease would be achieved by simply having the active site optimally complementary to the transition state. Besides being sterically complementary, such an arrangement is comparable to having the appropriate acid and base catalysts in the appropriate position to stabilize electrostatic charges on the transition state (Figure 9–2).

As a consequence, the enzyme must bind the transition state more tightly than the substrate by the factor by which the enzyme increases the rate of the reaction. This is a result of linkage between binding and the energies of the transition states:

$$
\begin{array}{ccc}
S & \xrightleftharpoons{K_n^{\ddagger}} T^{\ddagger} & \longrightarrow P \\
E \downarrow K_a^S & E \downarrow K_a^T & \\
ES & \xrightleftharpoons{K_E^{\ddagger}} ET^{\ddagger} & \longrightarrow EP
\end{array} \tag{9-39}
$$

It is a thermodynamic requirement that

$$\frac{K_a{}^T}{K_a{}^S} = \frac{K_E{}^{\ddagger}}{K_n{}^{\ddagger}} \tag{9-40}$$

and from Equation 9-38 it follows that

$$\frac{K_a{}^T}{K_a{}^S} = \frac{k_{cat}}{k_n} \tag{9-41}$$

An enzyme optimally complementary to the exact transition state will catalyze both forward and reverse reactions to the same extent. The two reactions may be catalyzed to different extents by an enzyme that is complementary to an intermediate between the transition state and either the substrate or the product; because this substrate or product would be bound more tightly, its utilization by the enzyme would be lower, according to the Haldane relationship (Equation 9-21).

Consequently, enzymes are not expected to have extremely high affinities for their substrates, because they would then not be catalysts. Instead, they would be like immunoglobulins, which generally bind antigens relatively tightly but do not alter them.

Therefore, analogues of a substrate that tend to mimic the transition state will in theory bind more tightly than the normal substrate. Attempts to design transition-state analogues have produced some that do in fact bind substantially more tightly (Table 9-4). It would be unrealistic to imagine that a perfect transition-state analogue could be designed to display the predicted maximum increase in affinity by 6 to 14 orders of magnitudes, since transition states are by definition extremely unstable species, with only partially formed bonds. Also, transition states often have anomalous ionization properties (Table 9-2) that would not be expected to be maintained in solution. Nevertheless, the affinities predicted for transition states are so high that even approximate analogues should bind very tightly. A series of such analogues should provide clues to the nature of the true transition state.

Enzymes may also speed reactions by producing new intermediate states along the reaction pathway. In a single-step reaction the geometry of the transition state will be somewhere between substrate and product and therefore will have a correspondingly higher free energy than either. With an intermediate between reactant and product, the two transition states to it will be closer to either and consequently need not differ so much in geometry or free energy (Figure 9-6). The overall kinetic barrier will not be so high; therefore, reactions entailing a number of small changes can occur faster than those involving one large change. The intermediate need not, and should not, be stable for the overall rate to be increased; however, being closer to the transition state, it is likely to be tightly bound and consequently more stable than if it occurred free in solution during the normal uncatalyzed reaction.

Table 9–4 Examples of Transition-state Analogues

The reaction catalyzed by the enzyme is given, with the presumed structure of the transition state in brackets; below it is given the structure of the transition-state analogue. The dissociation constants used to calculate the increased affinity of the transition-state analogue are given below the ligands.

ENZYME	REACTION		INCREASED AFFINITY OF TRANSITION-STATE ANALOGUE

Triose phosphate isomerase[a]

Reaction:

$$
\begin{matrix} HC{=}O \\ | \\ HCOH \\ | \\ H_2COPO_3^{2-} \end{matrix}
\longleftrightarrow
\left[\begin{matrix} HC{-}OH \\ \| \\ C{-}O^- \\ | \\ H_2COPO_3^{2-} \end{matrix}\right]
\longleftrightarrow
\begin{matrix} H_2C{-}OH \\ | \\ C{=}O \\ | \\ H_2COPO_3^{2-} \end{matrix}
$$

3×10^{-4} M

Transition-state analogue:

$$
\begin{matrix} O \\ \| \\ C{-}O^- \\ | \\ H_2COPO_3^{2-} \end{matrix}
$$

4×10^{-7} M

Increased affinity: 7.5×10^2

Glucose 6-phosphate isomerase[b]

Reaction:

$$
\begin{matrix} HC{=}O \\ | \\ HCOH \\ | \\ HOCH \\ | \\ HCOH \\ | \\ HCOH \\ | \\ H_2COPO_3^{2-} \end{matrix}
\longleftrightarrow
\left[\begin{matrix} HC{-}OH \\ \| \\ C{-}O^- \\ | \\ HOCH \\ | \\ HCOH \\ | \\ HCOH \\ | \\ H_2COPO_3^{2-} \end{matrix}\right]
\longleftrightarrow
\begin{matrix} H_2C{-}OH \\ | \\ C{=}O \\ | \\ HOCH \\ | \\ HCOH \\ | \\ HCOH \\ | \\ H_2COPO_3^{2-} \end{matrix}
$$

3×10^{-4} M

Transition-state analogue:

$$
\begin{matrix} O \\ \| \\ C{-}O \\ | \\ HOCH \\ | \\ HCOH \\ | \\ HCOH \\ | \\ H_2COPO_3^{2-} \end{matrix}
$$

3×10^{-7} M

Increased affinity: 1.0×10^3

Oxaloacetate decarboxylase[c]

Reaction:

$$
{}^-O_2C{-}\overset{\overset{\displaystyle O}{\|}}{C}{-}CH_2{-}CO_2^-
\longleftrightarrow
\left[\,\overset{\displaystyle O}{\overset{\|}{}} \; C{-}C \; \overset{\displaystyle CH_2}{}\;\right]
\longleftrightarrow
{}^-O_2C{-}\overset{\overset{\displaystyle O}{\|}}{C}{-}CH_3
$$

1×10^{-3} M

Transition-state analogue:

$$
\overset{O}{\underset{{}^-O}{}}C{-}C\overset{O}{\underset{O^-}{}}
$$

3.5×10^{-6} M

Increased affinity: 3.5×10^3

[a]From R. Wolfenden, Nature 223:704–705, 1969; and F. C. Hartman, et al., Biochemistry 14:5274–5279, 1975.

[b]From J. M. Chirgwin and E. A. Noltmann, J. Biol. Chem. 250:7272–7276, 1975; and P. J. Shaw and H. Muirhead, FEBS Letters 65:50–55, 1976.

[c]From A. Schmitt, et al., Hoppe-Seylers Z. Physiol. Chem. 347:18–34, 1966.

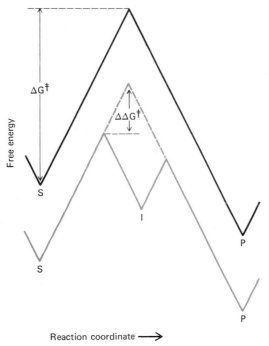

Figure 9–6
Naïve illustration of how the occurrence of an intermediate can increase the rate of a reaction. The free energy of the molecule is postulated to be directly proportional to the distance along the reaction coordinate from either the reactant, S, the product, P, or any intermediate, I. At the top is illustrated such a free energy profile for a reaction occurring in a single step, showing the rate-determining free energy of the single transition state. Below is shown the same reaction that has an unstable intermediate midway along the reaction coordinate. The rate-determining transition state is then on the left; its free energy is lower than that in the absence of the intermediate by $\Delta \Delta G^{\ddagger}$.

Intramolecular models depicting the kinetic importance of "fit" in enzymatic catalysis. T. C. Bruice and U. K. Pandit. Proc. Natl. Acad. Sci. U.S.A. 46:402–404, 1960.

Proximity effects and enzyme catalysis. T. C. Bruice. In The Enzymes, 3rd ed. P. D. Boyer (ed.). Vol. 2, pp. 217–279. New York, Academic Press, 1970.

Analog approaches to the structure of the transition state in enzyme reactions. R. Wolfenden. Acc. Chem. Res. 5:10–18, 1972.

Enzymatic catalysis and transition-state theory. G. E. Lienhard. Science 180:149–154, 1973.

Binding energy, specificity, and enzymic catalysis: the Circe effect. W. P. Jencks. Adv. Enzymol. 43:219–410, 1975.

Transition state analog inhibitors and enzyme catalysis. R. Wolfenden. Ann. Rev. Biophys. Bioeng. 5:271–306, 1976.

Transition states, standard states and enzymic catalysis. M. I. Page. Int. J. Biochem. 11:331–335, 1980.

Electrostatic basis of structure–function correlation in proteins. A. Warshel. Acc. Chem. Res. 14:284–290, 1981.

Substrate Specificity: Induced Fit

The specificity of an enzyme for only particular substrates could be imagined to be due simply to its specificity in binding ligands at its active site. An additional factor could be a requirement to bind substrates in a very specific orientation for catalysis, thereby discriminating further against ligands that bind nonproductively in other modes. The specificity exhibited by an enzyme for its substrate, however, is apparently even greater than that in ligand binding. Koshland, therefore, proposed that binding of a substrate must induce the active conformation of the enzyme, which would not exist in its absence. As an example of this **"induced fit,"** hexokinase catalyzes the phosphorylation of glucose

$$\text{glucose} + \text{ATP} \rightleftharpoons \text{glucose 6-P} + \text{ADP} \tag{9-42}$$

4×10^4 times more rapidly than the phosphorylation of water

$$\text{H}_2\text{O} + \text{ATP} \rightleftharpoons \text{ADP} + \text{P}_i \tag{9-43}$$

The enzyme is not particularly specific, readily phosphorylating other sugars also; yet water is used as substrate only very poorly, even though it is present in the solution at a concentration of 55 M. Other phosphoryl-transferring enzymes are also very specific; e.g., phosphoglucomutase discriminates against H_2O by a factor of 3×10^{10} in favor of its normal substrate, glucose 6-P. Because the hydrolysis of ATP would clearly be detrimental to metabolism, it is imperative that these enzymes discriminate effectively against water.

Modest degrees of specificity would not seem to require an induced fit; that observed in hexokinase conceivably could result from binding of glucose and ATP so that their effective concentration is $4 \times 10^4 \times 55$ M ($= 2 \times 10^6$ M) in the hexokinase active site to explain the discrimination against bulk water. Possibly, water is excluded from the active site, so that its concentration there is less than 55 M. On the other hand, the more extreme examples, such as 3×10^{10} in the case of phosphoglucomutase, may not be possible on this basis alone. Induced fit need not contribute to increase the rate of the reaction, and it produces a greater K_m for the substrate than otherwise would be necessary, since some of the binding energy of the substrate must be used to pull the enzyme into the active conformation.

Nevertheless, there is kinetic evidence of an induced fit in hexokinase: A sugar that binds to its active site, but that cannot accept a phosphate group, is known to increase 18-fold the hydrolysis of ATP by water and to decrease 40-fold its K_m. Consequently, the binding of this substrate analogue has activated its catalytic capability. The structural basis of this phenomenon is described in the following section.

The concept of induced fit can account for another model of substrate specificity: The enzyme may need to embrace the substrate intimately, bringing reactive groups into its proximity from all directions and preventing the transition state from dissociating. Yet the enzyme also needs to bind substrates and to release products into the solvent; induced fit would be a mechanism for giving access to an otherwise inaccessible active site. Indeed, the release of products is often observed to be the rate-determining step in an enzyme-catalyzed reaction.

Evidence for conformation changes induced by substrates of phosphoglucomutase. J. A. Yankeelov, Jr. and D. E. Koshland, Jr. J. Biol. Chem. 240:1593–1602, 1965.

Induced fit in yeast hexokinase. G. Dela Fuente, et al. Eur. J. Biochem. 16:226–233, 1970.

What limits the rate of an enzyme-catalyzed reaction? W. W. Cleland. Acc. Chem. Res. 8:145–151, 1975.

EXAMPLES OF ENZYME MECHANISMS

In this section, the possible mechanisms of enzyme function are elaborated for a few particular enzymes, chosen for the amount of data available, for the degree to which they appear to be understood, and for their illustration of important principles.

Serine Proteases

The serine proteases are the most extensively studied and best-understood enzymes; in fact, the term *enzyme* was first used to describe one of them, trypsin. Extensive kinetic studies of their catalytic activity, especially of α-chymotrypsin, made good use of their broad substrate specificities to investigate in detail the effects upon k_{cat} and K_m of varying the substrate and inhibitors. Some of the enzymes were also characterized chemically to a high degree, with direct determination of some of the residues involved in their activity. Consequently, when determination of their crystal structures was first accomplished about 15 years ago, there was a vast amount of enzymological data to explain. Many aspects of how they probably function have now become clear, although the process has not been straightforward. Only the historical background considered necessary to read the literature is given in the following description, which presents the current consensus on serine protease function.

This class of proteases is characterized by the conspicuous presence of a unique Ser residue of apparently exceptional reactivity that forms covalent adducts with some substrates and inhibitors. Two major evolutionary families that have very similar mechanisms of action, even though not otherwise detectably related, are represented in this class; this similarity is one of the most striking examples at the molecular level of apparent evolutionary convergence. One family is represented by the bacterial protease **subtilisin**. The other, more extensive and more thoroughly studied, is the **trypsin** family, which includes chymotrypsin, trypsin, and elastase from mammals, as well as several bacterial proteases of known structure (Table 6–1). It also includes the well-characterized but physiologically very important enzymes thrombin, plasmin, kallikrein, acrosin, and others, involved in a diverse range of cellular functions including blood clotting, complement activation, hormone production, and fertilization. This discussion focuses on the more familiar trypsin family, designating the homologous residues at the active site by their residue numbers in chymo-trypsinogen.

The normal, presumably physiological, reaction catalyzed by these enzymes is the hydrolysis of peptide bonds in proteins and peptides:

$$
\begin{array}{c}
\overset{R_n}{\underset{|}{\ }}\ \overset{O}{\underset{\|}{\ }}\ \overset{R_{n+1}}{\underset{|}{\ }} \\
-NH-CH-C-NH-CH- \xrightarrow{\ H_2O\ }
\end{array}
$$

$$
-NH-\underset{\underset{R_n}{|}}{CH}-CO_2H\ +\ H_2N-\underset{\underset{R_{n+1}}{|}}{CH}- \qquad (9\text{--}44)
$$

They catalyze this reaction in very similar ways, but differ most strikingly in their preference for amino acid side chains at position R_n. Trypsin cleaves bonds only after Lys and Arg residues, chymotrypsin after large hydrophobic residues; the others have less distinct preferences at position n and also depend to varying extents on the residues at neighboring positions (Figure 8–12). The usual terminology for specificity of proteases designates residues relative to the peptide bond that is cleaved:

$$
-P_4-P_3-P_2-P_1-P_1{}'-P_2{}'- \xrightarrow{\ H_2O\ }
$$

$$
-P_4-P_3-P_2-P_1-OH\ +\ H-P_1{}'-P_2{}'- \qquad (9\text{--}45)
$$

These proteases usually also catalyze related reactions of small amides and esters, especially their hydrolysis:

$$
R_1-\overset{O}{\overset{\|}{C}}-NH-R_2 \xrightarrow{\ H_2O\ } R_1-\overset{O}{\overset{\|}{C}}OH\ +\ H_2N-R_2
$$

$$
R_1-\overset{O}{\overset{\|}{C}}-OR_2 \xrightarrow{\ H_2O\ } R_1-\overset{O}{\overset{\|}{C}}OH\ +\ HOR_2 \qquad (9\text{--}46)
$$

Exchange of the oxygen atoms of the carboxylic acids with those of solvent can also be catalyzed:

$$
R_1-\overset{O}{\overset{\|}{C}}OH\ +\ H_2{}^{18}O \rightleftharpoons R-\overset{O}{\overset{\|}{C}}{}^{18}OH\ +\ H_2O \qquad (9\text{--}47)
$$

Because acyl acceptors other than water can also be utilized, the most general representation of the reaction catalyzed by these proteases is

$$
R_1-\overset{O}{\overset{\|}{C}}-X\ +\ HY \rightleftharpoons R_1-\overset{O}{\overset{\|}{C}}-Y\ +\ HX \qquad (9\text{--}48)
$$

The primary specificity of each enzyme is important for determining which substrates are utilized, depending primarily upon the nature of the group R_1.

The nonenzymic cleavage of amides and esters is believed to proceed through a tetrahedral intermediate, or a transition state approximating such a structure, generated by nucleophilic attack by group Y on the acyl carbon:

$$R_1-\underset{\underset{YH}{\uparrow}}{\overset{\overset{O}{\parallel}}{C}}-X \rightleftharpoons \left[\underset{Y}{\overset{R_1 \quad O^- \quad X}{C}} \right] + H^+ \longrightarrow R_1-\overset{\overset{O}{\parallel}}{C}-Y + HX \tag{9-49}$$

It is designated a **tetrahedral** intermediate because of the configuration about the acyl carbon atom, which contrasts with the trigonal geometry in the original amide or ester. Consequently, groups R and X must move relative to C and O to attain the tetrahedral configuration. The tetrahedral species is unstable and breaks down by expelling either group X or Y.

Much effort has been made to detect this tetrahedral species during protease catalysis, but the evidence for its occurrence is still only indirect. However, the enzymic reaction was found to proceed through an **acyl-enzyme** intermediate in which the $R_1-\overset{\overset{O}{\parallel}}{C}-$ group is transferred first to the enzyme via the uniquely reactive Ser residue that gives this class of proteases its name:

$$R_1-\overset{\overset{O}{\parallel}}{C}-X + E-CH_2OH \longrightarrow ECH_2O-\overset{\overset{O}{\parallel}}{C}-R_1 + HX \tag{9-50}$$

E—CH$_2$OH represents the enzyme and this Ser side chain, which is at residue 195 in the trypsin family and at 221 in subtilisin. The acyl enzyme then reacts with the second substrate to complete the reaction:

$$E-CH_2O-\overset{\overset{O}{\parallel}}{C}-R_1 + HY \longrightarrow E-CH_2OH + R_1\overset{\overset{O}{\parallel}}{C}-Y \tag{9-51}$$

The acyl enzyme can accumulate with reactive ester substrates and can lead to a burst of the first product (Figure 9–5), because it can be formed more rapidly than it is hydrolyzed. However, it does not accumulate with the less reactive amides. That the rate-limiting step in ester hydrolysis does not depend upon the nature of the group X was an initial indication that it has been released as product prior to the rate-determining step, hydrolysis of the acyl enzyme (p. 410). Subsequently, acyl-enzyme intermediates have been trapped, isolated, and characterized.

The enzymic reaction, then, is of the "ping-pong" or substituted-enzyme type (Eq. 9–24); this was not determined, however, by the usual steady-state kinetic analysis, since the second substrate is H$_2$O, whose concentration is difficult to vary independently. Beside dividing up the reac-

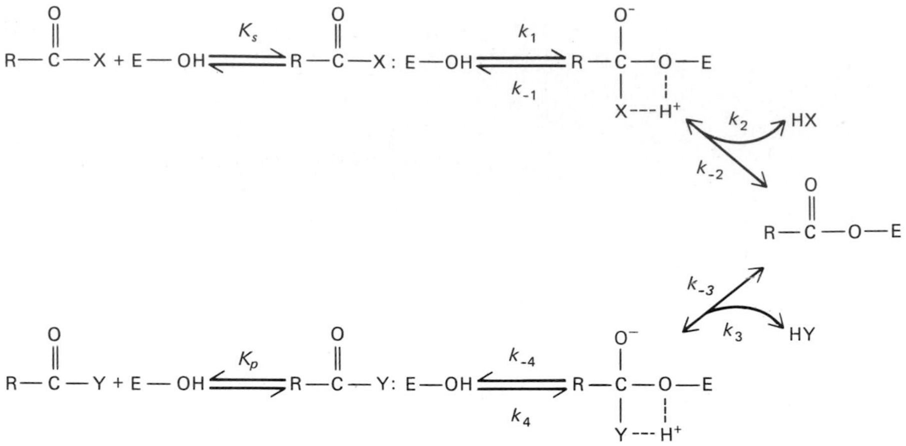

Figure 9–7
General reaction sequence for serine proteases (E—OH, the OH group being that of the
$$O$$
reactive Ser residue). The enzyme and substrate (R—C—X) are illustrated forming a Michaelis complex, with dissociation constant K_s. This then forms the tetrahedral intermediate or transition state, in which the proton is transferred from the Ser side chain to the leaving group, X; the acyl enzyme results. It is hydrolyzed (if Y is OH) by the exact reverse of the first half of the reaction, shown at the bottom. Consequently, the overall reaction is symmetric. (By permission, from J. Kraul, Ann. Rev. Biochem. 46:331–358, 1977, © Annual Reviews Inc.)

tion into two steps, this type of mechanism results in a symmetric reaction sequence in which hydrolysis of the acyl enzyme can occur by the precise reverse of its formation (Figure 9–7). There are then two acyl transfers, first from the substrate to the enzyme, forming an ester with the Ser side chain, and then to H_2O or other acceptor HY. In both transfers, it seems likely that a tetrahedral intermediate or transition state would be involved. As can be seen in Figure 9–7, one major question is how the hydrogen atom of the Ser residue is effectively transferred to the leaving group, X, and then replaced from the second substrate, HY.

The unique reactivity of this Ser side chain with substrates and inhibitors such as diisopropyl fluorophosphate led to its being considered an extraordinarily potent nucleophile. A nonionized His residue was also found to be involved in both catalysis and the reactivity of this Ser side chain; it reacted with affinity labels, and any modification, including protonation with a pK_a of about 7, greatly diminished both properties. The obvious chemical use of a His side chain would be to interact as a base with the Ser hydroxyl group and to promote its nucleophilicity.

Accordingly, the crystal structure of the first serine protease to be determined, α-chymotrypsin, showed the reactive Ser 195 to be in the active site, with the important His 57 nearby. Moreover, on the other side of His 57 was the buried side chain of Asp 102. Buried carboxyl groups are very rare in proteins, presumably because such conformations are energetically unfavorable for folding; its occurrence here, therefore, suggested an important function. The unfavorable negative charge was imagined to be transferred to the Ser 195 side chain via His 57 in what has come to be known as the **charge-transfer relay system:**

Asp 102   His 57   Ser 195

$$(9-52)$$

Exceedingly similar constellations of these residues were found subsequently in the other serine proteases, including subtilisin, where there is no other homology detectable in its primary or tertiary structures; here the three residues are Asp 32, His 64, and Ser 221. That the same catalytic triad arose independently in subtilisin by convergent evolution is also indicative of its catalytic importance.

However, careful analysis of the refined crystal structures of these enzymes indicated that there was little, if any, direct interaction between the His and Ser side chains; thus the latter was unlikely to be significantly abnormal in the free enzyme, without enhanced nucleophilicity. It was also realized that a very nucleophilic Ser side chain would promote formation of the acyl enzyme but would then equally inhibit its deacylation in the second half of the reaction.

The steps that do occur during catalysis have been inferred from the crystal structures of the enzymes with various acyl groups, products, and inhibitors bound, but with few good substrates, owing to their rapid turnover. The only structure of an enzyme–substrate Michaelis complex at low temperatures shows the substrate bound at the active site, moving the His 57 side chain toward Ser 195, perhaps inducing a hydrogen bond interaction between the two. The most detailed models of enzyme–substrate complexes come from the complexes of trypsin with bovine pancreatic trypsin inhibitor (BPTI; see Figures 6–9, 6–10, and 6–15), trypsin with soybean trypsin inhibitor, and subtilisin with its protein inhibitor. These **protease inhibitors** bind tightly and specifically to the active sites of the respective proteases and thereby inhibit them. They are substrates of the proteases, being hydrolyzed at one peptide bond that is in the appropriate position within the enzyme active site. However, they are tightly folded protein molecules that remain folded upon cleavage of this peptide bond. Consequently, the newly generated α-amino group of the P_1' residue remains approximately in its original position, inhibiting access of water to the cleavage site, and can readily reverse the hydrolysis step, resynthesizing the peptide bond:

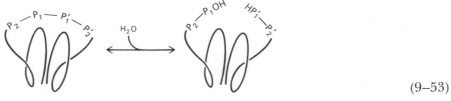

$$(9\text{--}53)$$

Equilibrium is reached in which roughly half the molecules are cleaved, the precise value depending upon the inhibitor and the conditions. The hydrolysis reaction is readily reversed because it is a unimolecular reaction, in contrast to the normal cleavage of a peptide into two products (Eq. 9–44). The latter reaction is normally irreversible with small concentrations of products, but high concentrations do lead to reversal and demonstration of an equilibrium constant for Equation 9–44 of approximately 1 M. Therefore, the reversal of this reaction with the protease inhibitors requires only an effective concentration of about 1 M between the newly generated amino and carboxyl groups of residues P_1 and P_1' in the cleaved, but folded, inhibitor. On the enzyme, equilibrium greatly favors the original form with the intact peptide bond ($k_{cat}^r > k_{cat}^f$), and the cleaved form has the higher K_m as expected from the Haldane relationship (Eq. 9–21).

Even though they are not homologous, the three protease inhibitors of known structure bind very similarly to the serine proteases. The P_1 amino acid side chain is bound in the primary specificity pocket of the enzyme, designated subsite S_1. The molecular basis of the substrate specificities at this site of some of the serine proteases was discussed earlier (see Figure 8–12). There are additional subsites S_2, S_3, and S_4, which interact to varying extents with residues P_2, P_3, and P_4, respectively, of the inhibitor; this part of the polypeptide backbone of the inhibitor is extended and forms an antiparallel β-sheet interaction with the enzyme. Further but less extensive interactions occur between residues P_1' and P_2', on the other side of the scissile bond, and subsites S_1' and S_2'.

The interaction of residue P_1 of a substrate with the primary subsite S_1 is crucial for orienting it properly for catalysis (Figure 9–8). Where this interaction is extensive, as with the large Lys and Arg side chains in the case of trypsin and with Phe, Tyr, and Trp in chymotrypsin, this interaction is of primary importance for catalysis, and the other sites have little effect on the rate of catalysis. However, where the S_1 site is small and makes limited interactions with the P_1 residue, as with elastase, interactions at subsites S_2, S_3, and S_4 are much more important and substantially affect catalysis (Table 9–5). Strong interactions at these sites are exhibited primarily as increases in k_{cat}, rather than increasing binding affinity, so they are used to enhance catalysis, probably by assisting the productive orientation of the substrate at the bond to be cleaved. Interactions at the other side of the scissile bond presumably are much less extensive because this portion is to be expelled first.

Two other interactions within the catalytic center serve to fix the scissile bond of the substrate in a fixed orientation. One is a hydrogen bond

Table 9–5 *Kinetic Parameters for the Hydrolysis of Peptide Substrates by Two Relatively Nonspecific Serine Proteases*

SUBSTRATE	ELASTASE			α-LYTIC PROTEASE		
	k_{cat} (s^{-1})	K_m (mM)	k_{cat}/K_m (s^{-1}M^{-1})	k_{cat} (s^{-1})	K_m (mM)	k_{cat}/K_m (s^{-1}M^{-1})
P$_5$ -P$_4$ -P$_3$ -P$_2$ -P$_1$ -P$_1'$ -P$_2'$ $\downarrow$						
Ac-Ala-NH$_2$	<0.0008	160	<0.005	<0.0004	300	<0.0013
Ac-Pro-Ala-NH$_2$	0.007	100	0.07	0.07	300	0.23
Ac-Ala-Pro-Ala-NH$_2$	0.09	4.2	21	0.70	110	6.3
Ac-Pro-Ala-Pro-Ala-NH$_2$	8.5	3.9	2200	1.1	23	48
Ac-Ala-Pro-Ala-Pro-Ala-NH$_2$	5.3	3.9	1360	0.97	15	64
Ac-Pro-Ala-Pro-Gly -NH$_2$	0.1	22	5	0.12	42	2.9
Ac-Pro-Ala-Pro-Val -NH$_2$	6.0	35	208	0.42	28	15
Ac-Pro-Ala-Pro-Leu-NH$_2$	3.0	11	270	<0.023	>9.9	<3.2
Ac-Pro-Ala-Pro-Ala-Gly -NH$_2$	26	4.0	6500	3.0	12.7	236
Ac-Pro-Ala-Pro-Ala-Ala -NH$_2$	37	1.5	24700	2.67	5.5	500
Ac-Pro-Ala-Pro-Ala-Phe-NH$_2$	18	0.64	28800	1.52	5.2	288
Ac-Pro-Ala-Pro-Ala-Ala -Ala-NH$_2$	—	—	—	17.5	8.3	2100

Ac- is the $CH_3\overset{\displaystyle O}{\overset{\|}{C}}$— group blocking the α-amino group of the next residue; —NH$_2$ is the amide group on the α-carboxyl group.

These data demonstrate that k_{cat}/K_m increases over 10^6-fold when the size of the substrate increases from a single residue to a hexapeptide, primarily due to increases in k_{cat}. Interactions with substrate residue P$_2$ are most important for α-lytic protease, but with elastase it is residue P$_4$. Elastase prefers an Ala residue at P$_1$, but α-lytic protease tolerates both smaller and slightly larger side chains. Interactions with residues P$_1'$ and P$_2'$ can increase k_{cat}/K_m, but there is little specificity for the amino acid side chain.
Data from R. C. Thompson and E. R. Blout, Biochemistry 12:57–65, 1973; from C. A. Bauer, et al., Biochemistry 15:1296–1299, 1976; and C. A. Bauer, et al., Eur. J. Biochem. 120:289–294, 1981.

between the backbone —NH— group of the P$_1$ residue to the backbone carbonyl group of residue 214 of the enzyme. The other is two hydrogen bonds from the carboxyl oxygen of the P$_1$ residue to the two —NH— groups of residues 213 and 215 of the enzyme, known as the **oxyanion binding site.** With these three points of attachment, the position of the peptide bond to be cleaved is now fixed, within the limits imposed by the flexibility of the enzyme; however, note that the enzyme uses primarily groups of the more rigid backbone, rather than side chains. Any potential substrate that cannot participate simultaneously in these interactions tends to bind with a different geometry, nonproductively.

However, optimum interactions between substrate and enzyme at these sites in the catalytic center and at the other subsites occur only if the carbonyl carbon of the scissile bond is distorted toward the tetrahedral geometry expected in the transition state (Eq. 9–49). In particular, the carbonyl oxygen can only then form geometrically favorable hydrogen bonds to the donors of the oxyanion binding site (Figure 9–8B). At the same time, the dipole interaction with these residues can be imagined to induce polarization of the carbonyl group, producing accumulation of

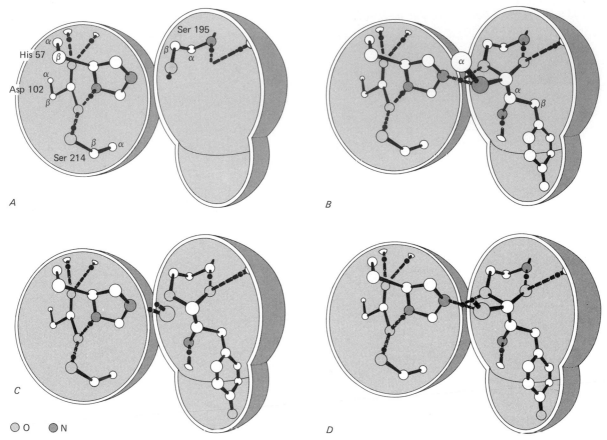

○ O ● N

Figure 9–8

Schematic and stylized views of the active site of a serine protease during catalysis, based on the structure of Streptomyces griseus *protease A. Atoms other than carbon are labeled; important hydrogen atoms are indicated by black dots; hydrogen bonds are broken cylinders. The superscript designations of some carbon atoms are indicated.*

A, The native enzyme before substrate is bound. The hollow disk on the left contains the side chains of His 57 and Asp 102 of the catalytic triad and Ser 214, which serves to orient Asp 102, as do two hydrogen bonds to its other oxygen atom from the backbone NH groups of residues 56 and 57.

The disk on the right shows the side chain of Ser 195, along with potential hydrogen bonds, from the backbone NH groups of residues 193 and 195, which make up the oxyanion binding site.

B, Part of a bound peptide substrate, including residue P_1 (Tyr) and atoms N and C^α of residue P_1', approximating the presumed tetrahedral intermediate. The side chain of residue P_1 is bound in the S_1 subsite of primary specificity, at bottom right; its NH group is hydrogen-bonded to the backbone carbonyl oxygen of residue 214. The carbonyl oxygen is bound in the oxyanion hole; the C' atom is in the tetrahedral configuration and bonded, at least partially, to Ser 195. The proton originally on Ser 195 is in the process of being transferred by His 57 $N^{\epsilon 2}$ to the NH group of the leaving residue, P_1'.

C, The P_1' residue has left, and a water molecule has taken its place, ready to hydrogen bond with His 57 $N^{\epsilon 2}$. The remaining part of the substrate is present as the ester with Ser 195; the C' carbon is shown as still being tetrahedrally distorted owing to the complementarity of the active site.

negative charge on the oxygen atom and of positive charge on the carbonyl carbon. This carbon atom is close to the hydroxyl of Ser 195, and its developing positive charge will favor nucleophilic attack by Ser 195. The proton from the Ser hydroxyl would be transferred, at least transiently, to His 57.

Quite simply, the active site is more complementary, both sterically and electrostatically, to the tetrahedral state, than to the normal geometry of a substrate. Although the tetrahedral intermediate has not been seen directly, that it is close to the transition state is supported by the analogy with many of the inhibitors of serine proteases. They produce tetrahedral adducts with the Ser residue and bind with the expected geometry (Figure 9–9) and thus may be considered transition-state analogues. This includes the classical diisopropyl fluorophosphate (DFP), which after reacting at the Ser residue can be hydrolyzed to the monoisopropyl phosphate derivative:

$$
\begin{array}{c}
\text{Ser } 195 \\
| \\
CH_2 \\
| \\
O \\
| \\
(CH_3)_2CH - O - P - O - CH(CH_3)_2 \\
\| \\
O
\end{array}
\xrightarrow[\ ]{H_2O \quad H^+}
$$

$$
\begin{array}{c}
CH_2 \\
| \\
O \\
| \\
{}^-O - P - O - CH(CH_3)_2 \\
\| \\
O
\end{array}
\quad + \ HOCH(CH_3)_2
$$

$$(9\text{--}54)$$

The sulfonyl portion of chloromethyl ketone affinity labels forms a tetrahedral adduct with the Ser residue, and the reactive chloromethyl group then reacts further with the nearby His residue. The protein inhibitors also may be considered transition-state analogues, because they appear to be strained in their normal conformation somewhat toward tetrahedral geometry.

During investigations of the functional significance of the charge-relay system, much controversy arose about the fate of the proton abstracted from the Ser residue by the His. Was the positive charge retained by the His, or was it transferred to the buried Asp?

D, *The presumed tetrahedral intermediate in deacylation, with the proton of the water molecule being transferred by His 57 to Ser 195. The ester is then hydrolyzed and the carboxylic acid product dissociates, regenerating the native enzyme* (A). *(Adapted from M. N. G. James, et al., J. Mol. Biol. 144:43–88, 1980.)*

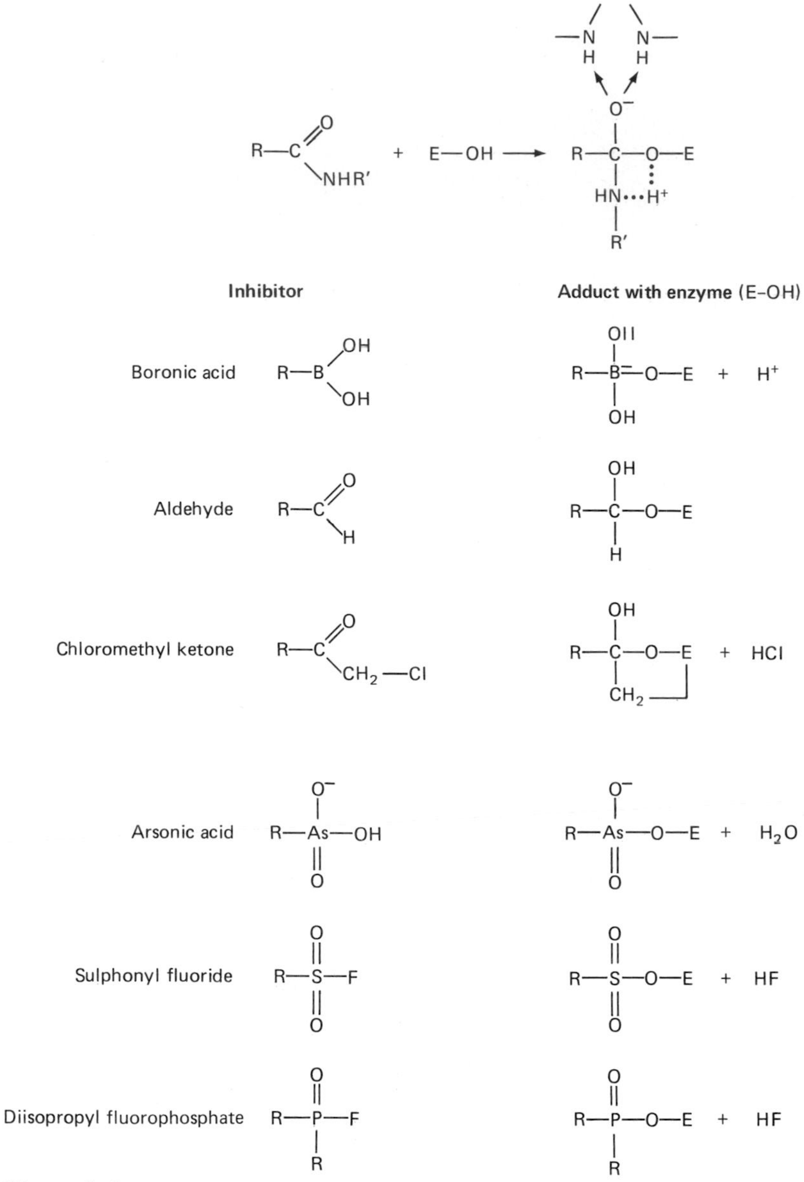

Figure 9–9

Tetrahedral transition-state analogues of serine proteases. At the top is shown a normal amide substrate, the enzyme (E—OH, the OH group being that of the active-site Ser residue), and the presumed tetrahedral intermediate. The carbonyl oxygen is shown interacting with the NH groups of the enzyme "oxanion hole," and the proton is being transferred from the Ser to the substrate leaving group.

Below are the inhibitors known to form covalent bonds with the active site Ser residue. Note that, in each case, the adducts are tetrahedral, with an oxygen atom or hydroxyl group in position to interact with the oxanion binding site.

With the chloromethyl ketone inhibitors, the ClCH$_2$— group reacts covalently with the adjacent His residue of the charge-relay group. (Adapted from J. Kraut, Ann. Rev. Biochem. 46:331–358, 1977.)

Asp 102   His 57   Ser 195

$$
\begin{array}{ccc}
-\text{C}\!\!\begin{array}{c}\nwarrow \text{O}\\ \text{O}_- \end{array} \cdots \text{HN} \overset{+}{\cdots} \text{NH} & & \begin{array}{c}\text{CH}_2\\ |\\ \text{O}\\ |\\ \text{C}=\text{O}\\ |\\ \text{R}\end{array}
\end{array}
\tag{9--55}
$$

$$
\begin{array}{ccc}
-\text{C}\!\!\begin{array}{c}\nwarrow \text{O}\\ \text{OH} \end{array} \cdots \text{N} \diagup \text{NH} & & \begin{array}{c}\text{CH}_2\\ |\\ \text{O}\\ |\\ \text{C}=\text{O}\\ |\\ \text{R}\end{array}
\end{array}
\tag{9--56}
$$

Experimental evidence was found to support both possibilities. However, a detailed neutron diffraction study, a technique that can readily position hydrogen atoms (pp. 220, 280), of trypsin modified by DFP, a presumed transition-state analogue, shows it to be present on the His residue, as in Equation 9–55. Other experimental observations also seem now to be consistent with this finding.

Within the tetrahedral intermediate, the His 57 side chain is in good proximity to transfer the proton to the leaving group of the substrate. This results in the group's expulsion from the tetrahedral intermediate and from the enzyme active site, since there are few interactions between it and the enzyme at subsites S_1', S_2', and so forth. The remainder of the substrate remains as the acyl intermediate, esterified to Ser 195.

Half the reaction is now completed, and there remains only to hydrolyze the acyl intermediate, presumably by the reverse of the process just described. The acyl intermediate is observed to be less stable and more reactive than a normal ester. These properties may also be due to its being constrained toward a tetrahedral species by the nature of the catalytic site; consequently, it is activated toward attack by nucleophiles, such as water. A water molecule is observed crystallographically to occupy the site vacated by the leaving group, hydrogen-bonded via one of its protons to the $N^{\epsilon 2}$ atom of His 57 and possibly to another group on the protein. Its lone-pair electrons are oriented for nucleophilic attack on the acyl carbon. In the process, the water proton can be transferred transiently to His 57, to form more fully the tetrahedral intermediate. This proton can then be transferred to Ser 195, producing breakdown of the tetrahedral intermediate to generate the free acid product and the intact enzyme. With dissociation of the product, the catalytic cycle is complete.

In summary, serine proteases appear to catalyze amide and ester hydrolysis by means of an active site more complementary to the tetrahedral transition state than either the substrates, products, or acyl-enzyme intermediate. Energetically important interactions occur between enzyme and substrate throughout the binding subsites, and in the oxyanion site to polarize the acyl group toward tetrahedral geometry and nature. The Ser 195 residue participates primarily because it is in the correct proximity for interacting with the tetrahedral distorted carbon atom of the substrate, and because His 57 is nearby to accept its proton transiently. It is apparently not especially potent as a nucleophile but is especially reactive because bound ligands are activated toward the tetrahedral state. His 57 serves as both a base in transiently accepting a proton and as an acid in transferring it again, a role for which imidazole groups are admirably suited. Its role is undoubtedly augmented by its interaction with Asp 102, although the detailed energetics of this interaction are not clear. Asp 102 also hydrogen-bonds through its other oxygen atom to another group on the protein; its orientation is therefore fixed. It then probably serves to keep the His 57 side chain in the appropriate orientation and to keep it in the correct tautomeric state, with the hydrogen atom on atom $N^{\delta1}$ rather than $N^{\epsilon2}$, which must be nonprotonated initially to participate in catalysis. The effects of its charge on the ionization properties of His 57 are not clear, since the pK_a value of about 7 is apparently normal for a His side chain.

Finally, note that no movements of the enzyme need occur according to the mechanism of Figure 9–8, although His 57 could no doubt more readily shuttle the proton between Ser 195 and the leaving group if it rocked between two conformations. Transient fluctuations of the enzyme and substrate may be important but have not been established. The primary catalytic function of the serine proteases appears to be stabilization of the tetrahedral transition state believed to participate in the nonenzymic reaction, both by favorable electrostatic interactions and by steric complementarity. The enzyme also provides a novel reaction intermediate, the acyl enzyme, thereby permitting the reaction to occur in two symmetric half-steps. A possible advantage of this aspect is that the leaving and attacking groups (HX and HY, respectively, of Figure 9–7) can share the same position, so that the same catalytic groups can be involved in the two half-reactions.

Role of a buried acid group in the mechanism of action of chymotrypsin. D. M. Blow, J. J. Birktoft, and B. S. Hartley. Nature 221:337–340, 1969.

Crystallographic studies of substrates and inhibitors bound to the active site of α-chymotrypsin. T. A. Steitz, et al. J. Mol. Biol. 46:337–348, 1969.

The structure of indoleacryloyl-α-chymotrypsin and its relevance to the hydrolytic mechanism of the enzyme. R. Henderson. J. Mol. Biol. 54:341–354, 1970.

Protein inhibitors of proteolytic enzymes. G. E. Means, D. S. Ryan, and R. E. Feeney. Acc. Chem. Res. 7:315–320, 1974.

X-ray crystallographic study of boronic acid adducts with subtilisin BPN′ (Novo). A model for the catalytic transition state. D. A. Matthews, et al. J. Biol. Chem. 250:7120–7126, 1975.

Polypeptide halomethyl ketones bind to serine proteases as analogs of the tetrahedral intermediate. T. L. Poulos, et al. J. Biol. Chem. 251:1097–1103, 1976.

Structure and mechanism of chymotrypsin. D. M. Blow. Acc. Chem. Res. 9:145–152, 1976.

Serine proteases: structure and mechanism of catalysis. J. Kraut. Ann. Rev. Biochem. 46:331–358, 1977.

Re-examination of the charge relay system in subtilisin and comparison with other serine proteases. D. A. Matthews, et al. J. Biol. Chem. 252:8875–8883, 1977.

The state of histidine in the catalytic triad of α-lytic protease. Implications for the charge-relay mechanism of peptide bond cleavage by serine proteases. W. Bachovchin and J. D. Roberts. J. Am. Chem. Soc. 100:8041–8047, 1978.

Crystallographic and kinetic investigations of the covalent complex formed by a specific tetrapeptide aldehyde and the serine protease from *Streptomyces griseus*. G. D. Brayer, et al. Proc. Natl. Acad. Sci. U.S.A. 76:96–100, 1979.

Crystal structures of Streptomyces subtilisin inhibitor and its complex with subtilisin BPN′. Y. Mitsui, et al. Nature 277:447–452, 1979.

Protein inhibitors of proteinases. M. Laskowski, Jr., and I. Kato. Ann. Rev. Biochem. 49:593–626, 1980.

Direct determination of the protonation states of aspartic acid-102 and histidine-57 in the tetrahedral intermediate of the serine proteases: neutron structure of trypsin. A. A. Kossiakoff and S. A. Spencer. Biochemistry 20:6462–6474, 1981.

Catalytic mechanism of serine proteases: reexamination of the pH dependence of the histidyl $^1J_{13C2-H}$ coupling constant in the catalytic triad of α-lytic protease. W. W. Bachovchin, et al. Proc. Natl. Acad. Sci. U.S.A. 78:7323–7326, 1981.

Lack of evidence for a tetrahedral intermediate in the hydrolysis of nitroanilide substrates by serine proteases. J. L. Markley, et al. Eur. J. Biochem. 120:477–485, 1981.

Refined crystal structure of the molecular complex of *Streptomyces griseus* protease B, a serine protease, with the third domain of the ovomucoid inhibitor from turkey. M. Fujinaga, et al. Proc. Natl. Acad. Sci. U.S.A. 79:4868–4872, 1982.

Crystallographic and NMR studies of the serine proteases. T. A. Steitz and R. G. Shulman. Ann. Rev. Biophys. Bioeng. 11:419–444, 1982.

Zymogen Activation

In common with many proteases, mammalian serine proteases are synthesized as **zymogen** precursors, which are essentially inactive proteolytically. They have detectable activity toward small substrates, but no more than that of a comparable solution of imidazole. Synthesis as the inactive precursor is undoubtedly of physiological importance, permitting its export to the appropriate location before the destructive catalytic powers of the enzyme are unleashed. In the mammalian gut, the process is initiated by a proteolytic enzyme, enterokinase, attached to the external side of the brush border membrane. It is extremely specific for activating trypsinogen; the product trypsin then activates other molecules of trypsinogen and also proelastase and chymotrypsinogens A, B, and C.

Activation of these serine protease zymogens occurs by proteolytic cleavage of the peptide bond between residues 15 and 16 in chymotrypsinogen and the homologous bond in the others. Further cleavages may then take place, as after residues 13, 146, and 148 in chymotrypsinogen A, releasing the dipeptides 14–15 and 147–148, to give the final disulfide-cross-linked, three-polypeptide-chain α-chymotrypsin (see Figure 2–7). However, only the initial cleavage is necessary for generation of catalytic activity.

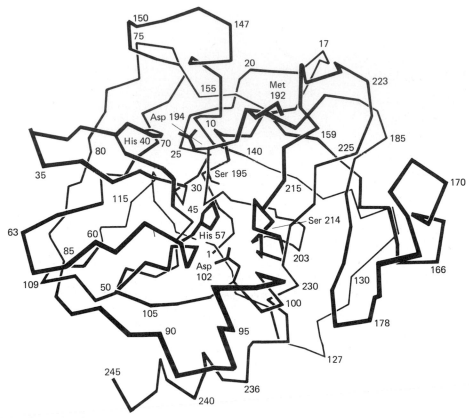

Figure 9–10
Comparison of the polypeptide backbone conformation of chymotrypsinogen A (left) and α-chymotrypsin (right). The latter has had residues 14, 15, 147, and 148 removed proteolytically upon activation. The side chains of some residues important for catalysis and for activation are also shown. (By permission, from S. T. Freer, et al., Biochemistry 9:1997–2009, 1970, © American Chemical Society.)

The structural basis of this activation has been investigated by comparing the structures of chymotrypsinogen A and α-chymotrypsin and of trypsinogen and trypsin. The initially surprising finding was that the structures are so similar (Figure 9–10). The two pairs of zymogen and of active enzyme may each be superimposed as closely as expected with the experimental error in two independent determinations of the same structure. In particular, the catalytic sites of the zymogens and enzymes are similar, with the same orientations of the Asp 102, His 57, Ser 195 catalytic triads. The greatest difference observed has been that one of two structure determinations of trypsinogen has found four segments of residues 16–19, 142–152, 184–193, and 216–223 to be flexible, and consequently invisible in the electron density map; these segments border on the active site. They become fixed in a trypsin-like conformation if the protein inhibitor BPTI is bound at the active site of trypsinogen; its affinity for the zymogen is

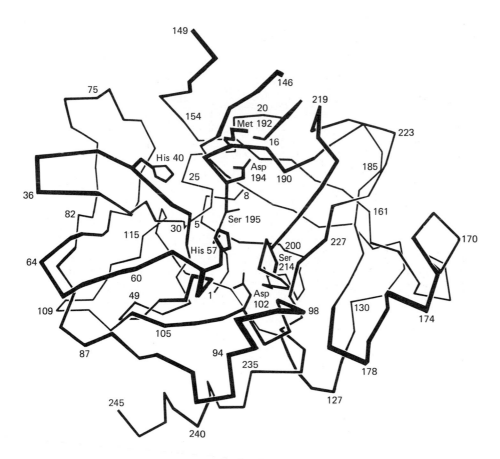

only 10^{-7} that for trypsin. However, the discrepancy with the other structure of trypsinogen, which shows these segments fixed as in trypsin, has not yet been resolved.

The activation of these zymogens does not result from a large conformational rearrangement to generate the enzyme active site, as might have been imagined, but results from rather subtle molecular changes. Closer comparison of these structures has revealed small structural changes in the S_1 primary binding site and in the orientation of the hydrogen bond donors of the oxyanion binding site; the latter point in the wrong direction in chymotrypsinogen and may be moved by 1.9 Å in trypsinogen. These changes are the most probable reason for the decreased catalytic powers and affinities for ligands of the zymogens, as the groups altered seem to be especially important for stabilizing the tetrahedral-like transition-state and inhibitor complexes. Disrupting the orientation of only one group of a hydrogen bond interaction seems to have large energetic consequences.

The trigger for these small conformational changes upon activation is the generation of an α-amino group on residue 16 by cleavage of that peptide bond. In the active enzymes, this —NH$_3^+$ group forms an internal salt bridge with Asp 194, a second buried, and therefore probably signif-

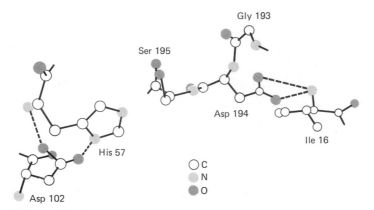

Figure 9–11
The position of the Ile 16 α-amino salt bridge to Asp 194 within the interior of α-chymo-
trypsin, relative to the catalytic triad of Asp 102, His 57 and Ser 195 at the active site. These
residues virtually span the interior of the globular folded structure, with the entrance to the
active site at the left, front; the rear surface is at the right, rear (see Figure 9–10). (Adapted
from D. M. Blow, in The Enzymes, 3rd ed. P. D. Boyer (ed.). Vol. 3, pp. 185–212. New
York, Academic Press, 1971.)

icant, carboxyl group in these proteins. Asp 194 is adjacent to Ser 195
(Figure 9–11) and therefore is directly linked to the catalytic center of the
enzymes, even though the cleaved 15–16 peptide bond was originally on
the surface of the protein opposite the active site (Figure 9–10). Catalytic
activity requires that the amino group of residue 16 be ionized; if it loses
a proton, with a pK_a value of about 9, chymotrypsin is inactive.

The residue 16 amino group does not exist in the zymogen, so Asp 194
then finds an alternative, more polar environment, interacting with His 40.
Consequently, the side chains of residues 15 and 16 are exposed in the
zymogen but become buried upon activation; the peptide backbone around
residue 16 is rotated about 180 degrees in the transition. The bacterial
serine proteases have no known zymogen precursors, but they also have a
buried Asp 194 side chain; it is salt-bridged to the Arg side chain of residue
138. Its occurrence in these proteins also is indicative of the importance
of this ionic interaction for maintaining the active site in the appropriate
orientation.

The very similar conformations of the zymogens and active enzymes
point out that rather small conformational changes in folded proteins can
have substantial energetic consequences—in this case, in the bound transition states of the enzymatic reaction. Our understanding of this phenomenon is very poor, as can be further illustrated by α- and γ-chymotrypsins.
These two forms of chymotrypsin are covalently identical but differ in the
way they crystallize, unfold in urea, and tend to dimerize. The two forms
are interconvertible in a pH-dependent equilibrium, but the process is
extremely slow, requiring days in some instances. Yet the two proteins
are almost indistinguishable in their folded conformations, and there cur-

rently is no explanation for the slowness of their interconversion. Clearly, there is much to be learned about the energetics of protein folded conformations.

Chymotrypsinogen: 2.5 Å crystal structure, comparison with α-chymotrypsin, and implications for zymogen activation. S. T. Freer, et al. Biochemistry 9:1997–2009, 1970.

A detailed structural comparison between the charge relay system in chymotrypsinogen and in α-chymotrypsin. J. J. Birktoft, et al. Biochemistry 15:4481–4485, 1976.

Mechanism of zymogen activation. R. M. Stroud, et al. Ann. Rev. Biophys. Bioeng. 6:177–193, 1977.

Probes of the mechanism of zymogen catalysis. J. D. Lonsdale-Eccles, et al. Biochemistry 17:2805–2809, 1978.

Structural basis of the activation and action of trypsin. R. Huber and W. Bode. Acc. Chem. Res. 11:114–122, 1978.

Crystal structure analysis and refinement of two variants of trigonal trypsinogen. W. Bode and R. Huber. FEBS Letters 90:265–269, 1978.

The transition of bovine trypsinogen to a trypsin-like state upon strong ligand binding. W. Bode. J. Mol. Biol. 127:357–374, 1979.

Recent developments in the activation process of bovine chymotrypsinogen A. S. K. Sharma and T. R. Hopkins. Bioorg. Chem. 10:357–374, 1981.

Refined crystal structure of γ-chymotrypsin at 1.9 Å resolution. G. H. Cohen, et al. J. Mol. Biol. 148:449–479, 1981.

Other Proteases

In addition to the serine proteases, three other classes of proteases are known. These also are designated by the principal functional group in their active sites, namely, thiol, carboxyl, and metallo. Since we have just examined the serine proteases in detail, it is worthwhile to compare and contrast examples of these other classes, even though they are not as fully understood.

Thiol Proteases

Thiol proteases are the class most comparable to the serine proteases, with a Cys side chain playing the role of the Ser in the serine proteases. The known thiol proteases are not detectably homologous to the serine proteases, and there is no known evolutionary variant in which the Ser and Cys residues at the active site are interchanged. However, the active site Ser side chain of subtilisin has been converted chemically to Cys, but the enzyme then is no longer active proteolytically, only being able to hydrolyze active esters. Nevertheless, the thiol proteases are catalytically similar to the serine proteases, catalyzing similar reactions involving tetrahedral intermediates and an acyl-enzyme intermediate in which the acyl group is bonded to the Cys sulfur atom.

The structures of two homologous thiol proteases, papain and actinidin, are known in great detail. Both are composed of two structural domains comprising respectively the two halves of the polypeptide chain of about 212 amino acid residues. The active site lies in a deep cleft between these two domains, and consists of seven subsites: S_4, S_3, S_2, S_1, S_1', S_2', and S_3'. The substrate specificity arises primarily from interactions at subsites S_2 and S_1', which are specific for hydrophobic side chains of polypeptide substrates. The catalytic site is defined by residues from both domains. In particular, the Cys 25 residue that forms the acyl enzyme interacts closely with His 159 of the other domain, which is believed to be analogous to His 57 of the serine protease catalytic triad. However, there is no obvious apparent analogue of Asp 102; the nearest acidic group is 7.5 Å away, although it has been proposed to interact transiently with the His side chain during catalysis by fluctuations of the structure. There is an equivalent of the oxanion binding site. The catalytic site is probably sterically and electrostatically complementary to the tetrahedral-like transition state, just as in the serine proteases.

With a thiol group required for activity, the enzymes are extremely susceptible to inhibition by the many reagents that react with such groups. The enzymes are usually isolated to a great extent in inactive form; mixed disulfides with other compounds and some other adducts may be removed by treatment with thiol reagents, and these enzymes activated. Other derivatives are irreversibly inhibited, such as the oxidation product $-SO_2^-$, which appears to be formed readily owing to the adjacent oxanion binding site, in which one of the oxygen atoms binds.

Ionization of the Cys 25 thiol group and His 159 side chain account for the pH-dependence of the catalytic activity. Their ionization is complex, because close interaction between them results in marked interdependence among their ionization states. At the present time, ionization of one group is believed to affect the pK_a of the other by 4.2 pH units:

$$(9\text{--}57)$$

The active form of papain is believed to be that at the bottom, where His 159 and Cys 25 are both ionized and interacting electrostatically. The difference from the serine proteases, where the His and the Ser are not ionized, is simply due to the different tendencies of the sulfur and oxygen atoms to ionize.

Mapping the active site of papain; specific peptide inhibitors of papain. I. Schechter and A. Berger. Biochem. Biophys. Res. Commun. 32:898–902, 1968.

Simulated mutation at the active site of biologically active proteases. L. Polgar and M. L. Bender. Adv. Enzymol. 33:381–400, 1970.

The structure of papain. J. Drenth, et al. Adv. Protein Chem. 25:79–115, 1971.

Papain and other plant sulfhydryl proteolytic enzymes. A. N. Glazer and E. L. Smith. In The Enzymes, 3rd ed. P. D. Boyer (ed.). Vol. 3, pp. 401–546. New York, Academic Press, 1971.

On the reactivity of the thiol group of thiolsubtilisin. L. Polgar, et al. Eur. J. Biochem. 39:421–429, 1973.

Binding of chloromethyl ketone substrate analogues to crystalline papain. J. Drenth, et al. Biochemistry 15:3731–3738, 1976.

Mechanism of action of papain with a specific anilide substrate. Detection and stabilization of a tetrahedral intermediate in papain catalysis. K. Y. Angelides and A. L. Fink. Biochemistry 18:2355–2369, 1979.

Rates of thiol–disulfide interchange reactions involving proteins and kinetic measurements of thiol pK_a values. Z. Shaked, et al. Biochemistry 19:4156–4166, 1980.

Effect of cysteine-25 on the ionization of histidine-159 in papain as determined by proton nuclear magnetic resonance spectroscopy. Evidence for a His-159–Cys-25 ion pair and its possible role in catalysis. S. D. Lewis, et al. Biochemistry 20:48–51, 1981.

Carboxyl Proteases

Carboxyl proteases have long intrigued enzymologists because the best-known representative of this enzyme class, pepsin, is active only in the extreme pH region of between 1 and 5 (no doubt because it must function in the acidic environment of the stomach). It is also a very acidic protein, having about 39 carboxyl side chains and only 2 Arg, 1 Lys, and 1 His. The enzymes of this class are also known as the acidic proteases, although not all require low pH for activity.

Pepsin best hydrolyzes peptides with hydrophobic residues on either side of the scissile bond, positions P_1 and $P_1{}'$. It can catalyze the transfer of not only the acyl portion of the peptide but also the amino half, thereby differing from the serine and thiol proteases, which can transfer only the acyl portion (Figure 9–7). Consequently, either or both portions of the substrate must be retained by the enzyme in such a manner that they can be transferred to other acceptors, but the mechanism is not known.

The structures of several carboxyl proteases have only recently been solved. Their interactions with substrates, products, and inhibitors at the active site are only now being studied, therefore, but their structures are sufficiently intriguing to give a brief description here.

The known structures are clearly homologous both in amino acid sequences and in folded conformations and therefore are related evolutionarily. They are also homologous internally; the two domains comprising the two halves of the polypeptide chain are very similar and probably arose by gene duplication. The active site is located within the cleft between the domains, as with the thiol proteases. Two Asp residues, 32 and 215, within the active site are the groups for which this class is named. They have equivalent positions within the two homologous domains but have rather different environments and, consequently, quite different pK_a values. They also interact closely, sharing one proton in the active form; thus, one has an anomalously low pK_a value (about 1.5), and the other, a somewhat high value (4.7). Asp 32 has been assigned the lower value because it is modified specifically by epoxides, which react with ionized carboxyls, and Asp 215 reacts with diazo compounds specific for nonionized carboxyl groups within the appropriate pH region.

$$\text{(Asp 32)}-CO_2H \qquad\qquad\qquad -CO_2^-$$

$$\text{(Asp 215)}-CO_2H \qquad\qquad\qquad -CO_2^- \qquad (9\text{--}58)$$

The ionization of these two groups accounts for the pH-dependence of pepsin action if only the middle species is catalytically active.

The positions of these residues within the active site are consistent with the use of Asp 32 as base to attack the acyl carbon of the substrate, comparable to Ser and Cys of the serine and thiol proteases, and with transfer by Asp 215 of a proton to the NH leaving group as an acid, comparable to the His residue of the other proteases (Figure 9–12). Both may form covalent adducts with the two portions of the substrate, which could explain both the acyl and amino group transfer reactions.

In summary, carboxyl proteases clearly share similarities with the serine and thiol proteases; they differ primarily in the use of different functional groups to serve as acid and base, depending upon the pH at which they are required to operate (Figure 9–12).

Pepsin is synthesized as the inactive zymogen precursor, pepsinogen, with 44 additional residues at the amino end of the polypeptide chain. They may be cleaved, and the zymogen activated, by other pepsin molecules, but the zymogen also appears to have the potential of cleaving its own amino terminus at acid pH. This novel feature may serve to ensure that the enzyme is activated immediately upon entering the acidic environment of the stomach.

The mechanism of the catalytic action of pepsin and related acid proteases. J. S. Fruton. Adv. Enzymol. 44:1–36, 1976.

Pepsin and pepsinogen: models for carboxyl (acid) proteases and their zymogens. J. Tang. Trends Biochem. Sci. 1:205–208, 1976.

Mechanism of acid protease catalysis based on the crystal structure of penicillopepsin. M. N. G. James, et al. Nature 267:808–813, 1977.

Acid Proteases: Structure, Function and Biology. J. Tang (ed.). New York, Plenum Press, 1977.

An X-ray crystallographic approach to enzyme structure and function. M. N. G. James. Can. J. Biochem. 58:251–271, 1980.

Gastric proteinases—structure, function, evolution and mechanism of action. B. Foltmann. Essays Biochem. 17:52–84, 1981.

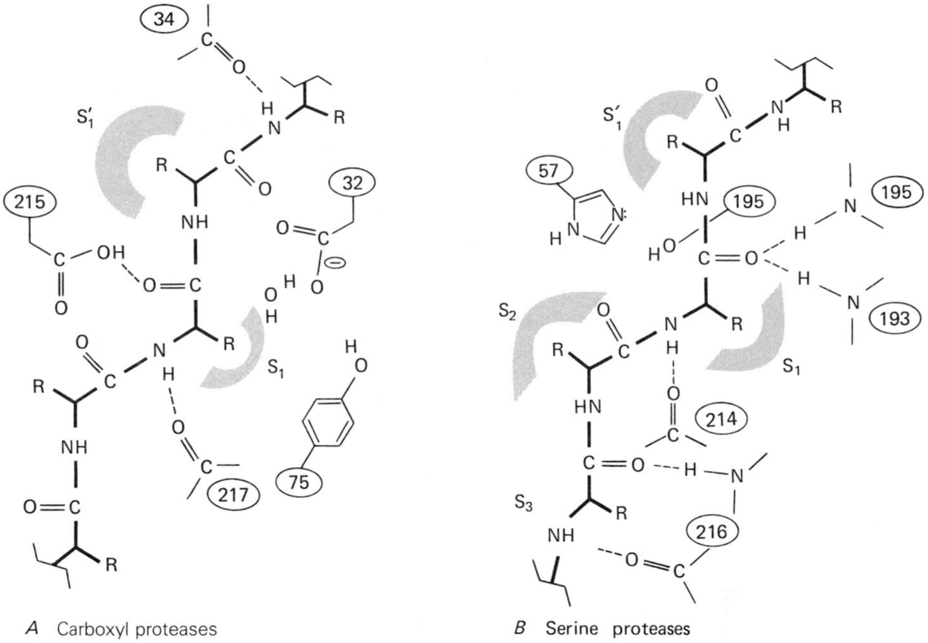

A Carboxyl proteases   *B* Serine proteases

Figure 9–12

Schematic comparison of the active sites of (A) *carboxyl and* (B) *serine proteases. The poly-peptide substrates are indicated by the thick lines; functional groups of the enzymes are also included. Hydrogen bond interactions are indicated by dotted lines; nonpolar interactions by the shaded areas. (From M. N. G. James, Can. J. Biochem. 58:251–271, 1980.)*

Metalloproteases

Metalloproteases use bound metals at their active sites, usually Zn^{2+}, and include carboxypeptidases A and B and thermolysin as their most thoroughly studied representatives. The carboxypeptidases catalyze the step-wise removal of single residues from the carboxyl-terminus of a polypeptide substrate and are specific for the nature of that P_1' residue. Carboxypeptidase A is specific for large hydrophobic side chains; B, for basic residues. The two enzymes are homologous and have very similar conformations, but A has a hydrophobic S_1' site, while that of B has an Asp residue to interact with basic P_1' side chains. Therefore, these two **exopeptidases** are undoubtedly related evolutionarily in the same way as the chymotrypsin and trypsin variants of the serine proteases. Thermolysin is not homologous with the carboxypeptidases in either primary or tertiary structures. It is an **endopeptidase,** catalyzing hydrolysis of nonterminal peptide bonds, especially those with hydrophobic P_1' residues. Neverthe-less, its active site shows many similarities to those of the carboxypeptidases, presumably as a result of convergent evolution. This relationship is anal-ogous to that of subtilisin to the other serine proteases.

The mechanisms of action of these metalloproteases are not known in detail, but in each case the Zn^{2+} ion interacts with the carbonyl oxygen of the scissile peptide bond (Figure 9–13), presumably serving the role of the

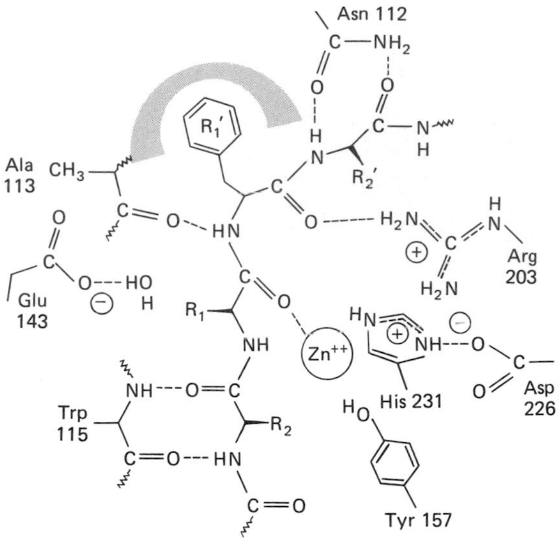

Figure 9–13
Schematic drawing of the binding to thermolysin of an extended substrate of residues P_2-P_1-P_1'-P_2', as inferred from the binding of inhibitors. The side chain of a Phe P_1' residue is labeled R_1', bound in subsite S_1'. The peptide bond to be hydrolyzed is that between residues P_1 and P_1'. (By permission, from W. R. Kester and B. W. Matthews, Biochemistry 16:2506–2516, 1977, © American Chemical Society.)

oxyanion binding sites of the serine and thiol proteases. The Zn^{2+} ion is five-coordinated in the native enzyme to two His side chains, the two carboxylate oxygen atoms of a Glu residue, and a water molecule. Upon binding of a substrate, its carbonyl oxygen replaces the water molecule. The nucleophiles that interact with the carbonyl carbon of the scissile bond are either the carboxylate oxygen atoms of Glu residues or water molecules hydrogen-bonded to them. With one particular substrate, a mixed anhydride between it and the Glu residue has been trapped, comparable with the acyl enzyme of the serine and thiol proteases:

$$E-CO_2^- + R_1-\overset{\overset{\displaystyle O}{\|}}{C}-X \longrightarrow E-\overset{\overset{\displaystyle O}{\|}}{C}-O-\overset{\overset{\displaystyle O}{\|}}{C}-R_1 + X^- \quad (9-59)$$

However, it is not yet clear whether this intermediate is involved with other substrates. The acidic group that protonates the leaving amino group may be Tyr 248 in carboxypeptidase A or His 231 in thermolysin (Figure 9–13). The orientation of the Tyr 248 side chain in carboxypeptidase A changes markedly upon substrate binding, suggestive of an induced-fit type of conformational change. However, it has relatively high mobility within the native enzyme; this finding has produced much controversy, and the significance for catalysis of such movement is not clear.

Carboxypeptidase A: a protein and an enzyme. F. A. Quiocho and W. N. Lipscomb. Adv. Protein Chem. 25:1–78, 1971.

Carboxypeptidase A: a mechanistic analysis. E. T. Kaiser and B. L. Kaiser. Acc. Chem. Res. 5:219–224, 1972.

Unified picture of mechanisms of catalysis by carboxypeptidase A. R. Breslow and D. L. Wernick. Proc. Natl. Acad. Sci. U.S.A. 74:1303–1307, 1977.

Comparison of the structures of carboxypeptidase A and thermolysin. W. R. Kester and B. W. Matthews. J. Biol. Chem. 252:7704–7710, 1977.

Crystallographic study of the binding of dipeptide inhibitors to thermolysin: implications for the mechanism of catalysis. W. R. Kester and B. W. Matthews. Biochemistry 16:2506–2516, 1977.

A crystallographic study of the complex of phosphoramidon with thermolysin. A model for the presumed catalytic transition state and for the binding of extended substrates. L. H. Weaver, et al. J. Mol. Biol. 114:119–132, 1977.

Carboxypeptidase A mechanisms. W. N. Lipscomb. Proc. Natl. Acad. Sci. U.S.A. 77:3875–3878, 1980.

Zinc environment and *cis* peptide bonds in carboxypeptidase A at 1.75 Å resolution. D. C. Rees, et al. Proc. Natl. Acad. Sci. U.S.A. 78:3408–3412, 1981.

Binding of ligands to the active site of carboxypeptidase A. D. C. Rees and W. N. Lipscomb. Proc. Natl. Acad. Sci. U.S.A. 78:5455–5459, 1981.

Structure of a mercaptan–thermolysin complex illustrates mode of inhibition of zinc proteases by substrate–analogue mercaptans. A. F. Monzingo and B. W. Matthews. Biochemistry 21:3390–3394, 1982.

Catalytic conformation of carboxypeptidase A: The structure of a true reaction intermediate stabilized at subzero temperatures. L. C. Kuo, et al. J. Mol. Biol. 163:63–105, 1983.

Lysozyme

Hen egg-white lysozyme was the first enzyme to have its structure determined crystallographically, but up until that time very little was known about its catalytic properties. Lysozymes hydrolyze various oligosaccharides, especially those of bacterial cell walls. The complex structures and chemical properties of their substrates have not encouraged large numbers of detailed enzymatic studies of these enzymes. Nevertheless, the structure of hen lysozyme and of its complexes with several saccharides provided Phillips and coworkers sufficient clues to propose a plausible model of how catalysis might take place; this model has become the archetypal example used in most biochemistry textbooks. Subsequent investigations have largely supported that model, although there have emerged some unforeseen complexities and a change in emphasis, and many questions remain unanswered.

Briefly, the enzyme's active site is a cleft capable of binding simultaneously six sugar residues of an appropriate polysaccharide at subsites designated A through F (Figures 8–6 and 9–14). Hydrolysis occurs between the sugar residues at sites D and E; thus, the six sites are comparable to subsites P_4 through P_2' of the proteases. Binding at all six subsites is important for catalysis, since the enzyme hydrolyzes a hexasaccharide 10^7 times more rapidly than a disaccharide. Of its substrates, *N*-acetyl glucosamine (GlcNAc, often abbreviated NAG) residues can be accommodated at all six subsites, whereas the additional lactyl group of an *N*-acetyl muramic acid (MurNAc, or NAM) residue interferes with the binding to subsites A, C, and E, limiting it to bind to the others. Consequently, the

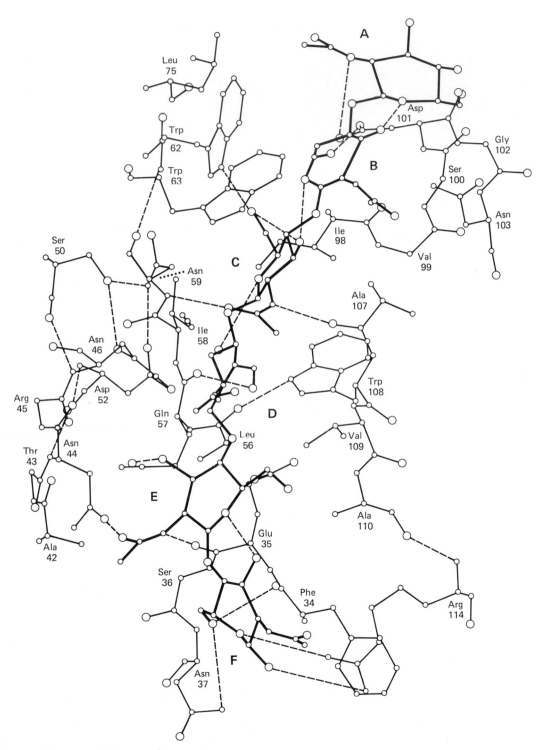

A

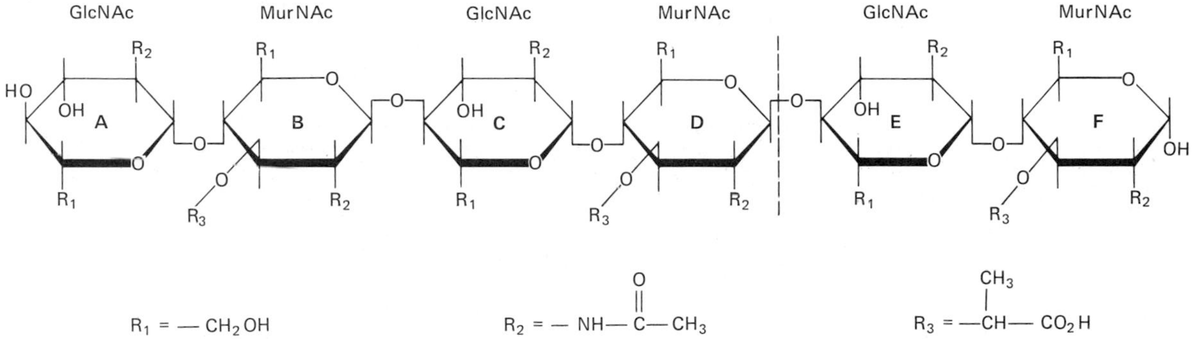

B

$R_1 = -CH_2OH$

$R_2 = -NH-\overset{\overset{\displaystyle O}{\|}}{C}-CH_3$

$R_3 = \overset{\overset{\displaystyle CH_3}{|}}{-CH}-CO_2H$

Figure 9–14

A, *Proposed binding of a hexasaccharide to the active site of lysozyme, showing the six subsites A through F, each binding one saccharide unit. B, The structure of six saccharide residues of the normal substrate of lysozyme, an alternating polymer of N-acetyl-D-glucosamine (GlcNAc) and N-acetyl muramic acid (MurNAc). This substrate is normally hydrolyzed by lysozyme at the dashed line, when bound between subsites D and E. (Adapted from S. J. Perkins, et al., Biochem. J. 193:553–572, 1981.)*

alternating oligosaccharide (-GlcNAc-MurNAc-) of bacterial cell walls is hydrolyzed only adjacent to the MurNAc residue bound to site D (Figure 9–14).

Model-building a substrate into the active site of the enzyme suggested that a normal, chair-shaped sugar residue could not be accommodated in site D unless it were sterically distorted into the "half-chair," or "sofa," conformation believed to occur in the transition state:

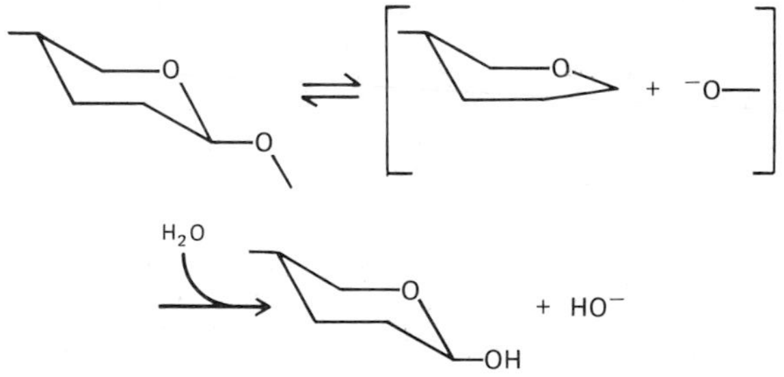

$$(9\text{--}60)$$

The oxocarbonium ion form of the sugar residue, which would be in subsite D of the enzyme, is most stable with a planar conformation due to resonance, spreading the positive charge onto the oxygen atom. In this way, lysozyme could catalyze hydrolysis simply by being sterically complementary to the presumed sofa transition state. However, more detailed model-building analysis has not supported the suggestion that binding of the half-chair conformation would be more favorable, and one substrate bound to subsite D has been found crystallographically not to be strained in this way, although it was not bound closely to the site.

Nevertheless, studies of binding of saccharides to the active site have indicated that binding to subsite D is significantly less favorable than to the others (Table 9–6). Interactions to subsite C are the most extensive,

Table 9–6   Energetics of Binding Saccharides to the Active Site of Hen Lysozyme

A	B	C	D	K_d (M)	INCREMENTAL BINDING CONTRIBUTION OF GROUP IN ITALICS (kcal/mole)
		GlcNAc		3.2×10^{-2}	
	GlcNAc	- GlcNAc		3.1×10^{-4}	-2.7
	GlcNAc	- GlcNAc	- *MurNAc*	7×10^{-4}	$+0.5$
GlcNAc	- GlcNAc	- GlcNAc		8.6×10^{-6}	-2.1
GlcNAc	- GlcNAc	- GlcNAc	- *(GlcNAc lactone)*	7.5×10^{-8}	-2.8
GlcNAc	- GlcNAc	- GlcNAc	- *XylNAc*	4.3×10^{-6}	-0.4
	MurNAc	- GlcNAc		1.0×10^{-4}	3.4
GlcNAc	- MurNAc			9.0×10^{-5}	
GlcNAc	- MurNAc	- GlcNAc		3.2×10^{-6}	-2.0
GlcNAc	- MurNAc	- *GlcNAc*			-2.0
GlcNAc	- MurNAc	- GlcNAc	- *XylNAc*	7.1×10^{-7}	-0.9
GlcNAc	- MurNAc	- GlcNAc	- *GlcNAc*	1.1×10^{-6}	-0.6
GlcNAc	- MurNAc	- GlcNAc	- *MurNAc*	4.1×10^{-4}	$+2.9$
GlcNAc	- MurNAc	- GlcNAc	- $\Delta^{2,3}GlcNAc$	5.1×10^{-6}	$+0.3$

Abbreviations: GlcNAc: *N*-acetyl-D-glucosamine
MurNAc: *N*-acetyl muramic acid   } See Figure 9–14.

XylNAc:

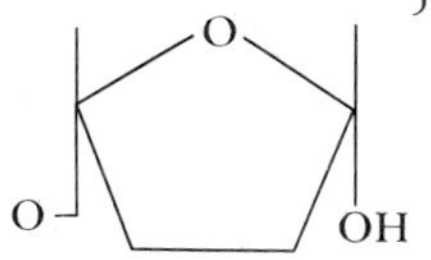

GlcNAc lactone:

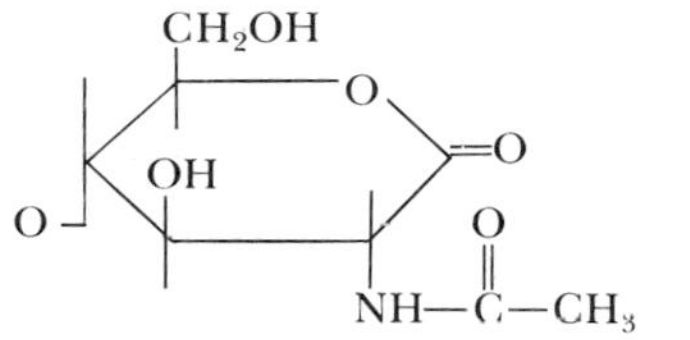

$\Delta^{2,3}$GlcNAc:

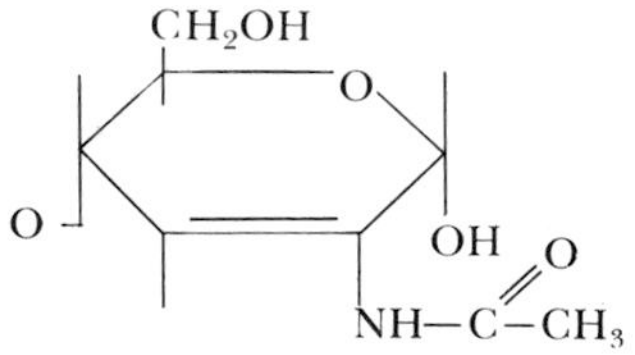

Many discussions of the energetics of binding to the subsites of lysozyme use "unitary" free energies of binding, calculated from association constants and a saccharide concentration of 55 M. As discussed on page 338, these values are not considered to have any special significance and should not be compared with the incremental free energies used here, which are calculated from the change in overall binding affinity produced by adding a single residue (i.e., $-RT \ln (K_d^A/K_d^{AB})$ for group B).
Data from M. Schindler, et al., Biochemistry 16:423–431, 1977.

and sugars are bound most readily to it. The change in overall binding affinity when adjacent residues are added to the saccharide gives the incremental binding energies of the additional group (Eq. 8–30). As can be seen in Table 9–6, GlcNAc residues in subsite A, B, or C increase the affinity about 10^2-fold, but only 3-fold in subsite D. Binding of the alternative residue MurNAc actually decreases the affinity 10^2-fold, indicating unfavorable interactions involving the lactyl side chain. Lactone and de-hydro derivatives of GlcNAc, designed to mimic the transition state, bind only slightly more tightly in one case, and less tightly in the other (Table 9–6).

Detailed interpretations of such data are complicated by crystallo-graphic observations that the positions of the bound saccharides can vary considerably as a result of small conformational changes in the protein. The position adopted depends upon what residues are in the adjacent sites. In particular, binding closely within subsite D is avoided, unless very fa-vorable interactions also occur between sugar residues in subsites E and F, which, together with binding to subsites A, B and C, would be expected to pull a residue into subsite D. Unfortunately, binding to subsites E and F have not been observed crystallographically because they are blocked by the crystal lattice, and there are no quantitative data for their incremental binding energies, although these are believed to be substantial.

Clearly, binding of a normal substrate to subsite D is energetically unfavorable, but this is now believed to be primarily for electrostatic rather than steric reasons. Within the active site cleft are two acidic groups, Glu 35 and Asp 52, which should be in very close proximity to the bond to be cleaved when a substrate is inserted fully into subsite D (Figure 9–14). In the free enzyme, Asp 52 is in a very polar environment and has a nearly normal pK_a value of 3.5 ± 0.5. In contrast, Glu 35 is within a hydrophobic environment and, probably as a consequence, has a substantially elevated pK_a value of 6.3 ± 0.2. Hen lysozyme is maximally active at pH 5, and its catalytic activity decreases in a manner consistent with catalytic activity resulting from only the enzyme with Asp 52 ionized, not Glu 35—com-parable with the situation in the carboxyl proteases. Removing the ionized Asp 52 from contact with the solvent may be the energetically unfavorable aspect of binding to subsite D. On the other hand, this unfavorable inter-action would be expected to be used to stabilize the developing positive charge on the oxocarbonium ion of the transition state (Eq. 9–60), thereby stabilizing the transition state. At the same time, the nonionized side chain of Glu 35 should be ideally placed to transfer its proton to the oxygen atom of the leaving residue in subsite E, thereby cleaving the bond. The saccharide bound to subsites E and F may then diffuse away from the enzyme. A water molecule may then protonate Glu 35 once again and provide a hydroxyl ion to react with the oxocarbonium-ion form of the remainder of the substrate. With its release, the catalytic cycle is complete.

Subsites E and F have sufficient affinity for oligosaccharides that they may compete with water in this last step and may then be joined to the residual saccharide in subsite D, regenerating a glycosyl bond. This trans-glycosylation reaction, plus the many possible modes of substrate binding to the six subsites, complicates the kinetic study of lysozyme catalysis.

The lysozyme from phage T4 exhibits similar structural and binding properties; therefore, it may catalyze hydrolysis of the same substrates in a similar manner. This enzyme has no detectable amino acid sequence homology with the mammalian lysozymes, but it has barely sufficient structural similarities to suggest that they all have diverged from an ancient common ancestor.

The three-dimensional structure of an enzyme molecule. D. C. Phillips. Sci. Amer. 215:78–90, 1966.

Crystallographic studies of the activity of hen egg-white lysozyme. C. C. F. Blake, et al. Proc. R. Soc. Lond. [Biol.] 167:378–388, 1967.

The binding and cleavage by lysozyme of N-acetylglucosamine oligosaccharides. J. A. Rupley. Proc. R. Soc. Lond. [Biol.] 167:416–428, 1967.

Vertebrate lysozymes. T. Imoto, et al. In The Enzymes, 3rd ed. P. D. Boyer (ed.). Vol. 7, pp. 665–868. New York, Academic Press, 1972.

Physical organic models for the mechanism of lysozyme action. B. M. Dunn and T. C. Bruice. Adv. Enzymol. 37:1–60, 1973.

Crystal structure of a lysozyme–tetrasaccharide lactone complex. L. O. Ford, et al. J. Mol. Biol. 88:349–371, 1974.

Reaction of N-acetyl glucosamine oligosaccharides with lysozyme. Temperature, pH and solvent deuterium isotope equilibrium, steady-state, and pre-steady state measurements. S. K. Banarjee, et al. J. Biol. Chem. 250:4355–4367, 1975.

Theoretical studies of enzymic reactions: dielectric, electrostatic and steric stabilization of the carbonium ion in the reaction of lysozyme. A. Warshel and M. Levitt. J. Mol. Biol. 103:227–249, 1976.

Crystallographic study of turkey egg-white lysozyme and its complex with a disaccharide. R. Sarma and R. Bott. J. Mol. Biol. 113:555–565, 1977.

Mechanism of lysozyme catalysis: role of ground-state strain in subsite D in hen egg-white and human lysozymes. M. Schindler, et al. Biochemistry 16:423–431, 1977.

X-ray crystallography of the binding of the bacterial cell wall trisaccharide NAM-NAG-NAM to lysozyme. J. A. Kelly, et al. Nature 282:875–878, 1979.

Relation between hen egg-white lysozyme and bacteriophage T4 lysozyme: evolutionary implications. B. W. Matthews, et al. J. Mol. Biol. 147:545–558, 1981.

Crystallographic determination of the mode of binding of oligosaccharides to T4 bacteriophage lysozyme: implications for the mechanism of catalysis. W. F. Anderson, et al. J. Mol. Biol. 147:523–543, 1981.

Theoretical calculations on enzyme–substrate complexes: the basis of molecular recognition and catalysis. M. R. Pincus and H. A. Scheraga. Acc. Chem. Res. 14:299–306, 1981.

Amino acid substitutions far from the active site of bacteriophage T4 lysozyme reduce catalytic activity and suggest that the C-terminal lobe of the enzyme participates in substrate binding. M. G. Grütter and B. W. Matthews. J. Mol. Biol. 154:525–535, 1982.

Goose lysozyme structure: an evolutionary link between hen and bacteriophage lysozymes? M. G. Grütter, et al. Nature 303:828–830, 1983.

Kinases

Kinases are the class of enzymes that provide the most dramatic example of the phenomenon of induced fit, in contrast to the relatively small changes induced by substrate binding in the other enzymes described earlier. The kinases catalyze the transfer of phosphoryl groups from ATP to other molecules, especially sugars, and it is extremely important physiologically that they transfer these groups very specifically to the appropriate substrate and not to the much more abundant water. Otherwise, they would function as ATPases and would drain a cell of its supply of ATP. The structures of five different kinases are known at the present time (Table 6–1). None show significant homology in either primary or tertiary structures, other than having similar nucleotide-binding domains related to their common use of ATP (see pp. 260–262), but all have very similar bilobed structures of two domains, with a marked cleft between them containing the active site. This common structural feature probably reflects the need for conformational changes upon substrate binding and catalysis, as has been shown quite dramatically in the case of hexokinase.

Upon binding its substrate, glucose, the two lobes of hexokinase rotate, essentially as rigid bodies, by 12 degrees relative to each other, resulting in relative movements of as much as 8 Å (Figure 9–15). Ligands that bind but do not produce this conformational change are not substrates; such movement, therefore, is probably essential for catalysis. Following rotation of the lobes, the glucose molecule is almost entirely engulfed by the enzyme, so that it cannot enter or leave the active site; dissociation of product from the enzyme is quite slow and rate-limiting, probably being limited by the opening of the cleft.

As a consequence of glucose binding and the conformational change, the apparent affinity for ATP increases 50-fold. It also binds within the cleft, interacting with both domains, in a way that is favored by the conformational change. Consequently, the conformational change induced by glucose, or presumably by analogues that enhance the ATPase activity, appear to be important for activating the enzyme as a catalyst toward ATP hydrolysis. If sugars that cannot accept the phosphoryl group are added to the enzyme with ATP, the phosphoryl group tends to be slowly transferred to an adjacent Ser residue in the enzyme. Water does not cause this change, so it is a poor acceptor of the phosphoryl group of ATP. That the conformational change involves rotation of otherwise unchanged domains, rather than rearrangements within them, is suggestive of a relatively rigid architecture for each.

Another function of this conformational change may be to permit the enzyme to embrace totally the substrate in the transition state, placing functional groups in the proper orientation all around the substrate and providing them with the appropriate environment. For example, a hydrophobic environment would be expected to facilitate the electrostatic interactions between enzyme and substrate believed to be of prime importance for catalysis. A conformational change would be necessary simply to permit products to dissociate and substrates to bind. Such an embracing action seems to inhibit dissociation of the substrate; thus, it may contribute directly to increasing the rate of the reaction (Eq. 9–41) only if it is more effective with the transition state.

Further conformational changes probably take place in hexokinase after binding the second substrate, ATP, since the glucose 6-hydroxyl and ATP γ-phosphoryl groups are observed to be 6 Å apart, too far for the reaction to occur in one step. The nature of these changes is not yet known.

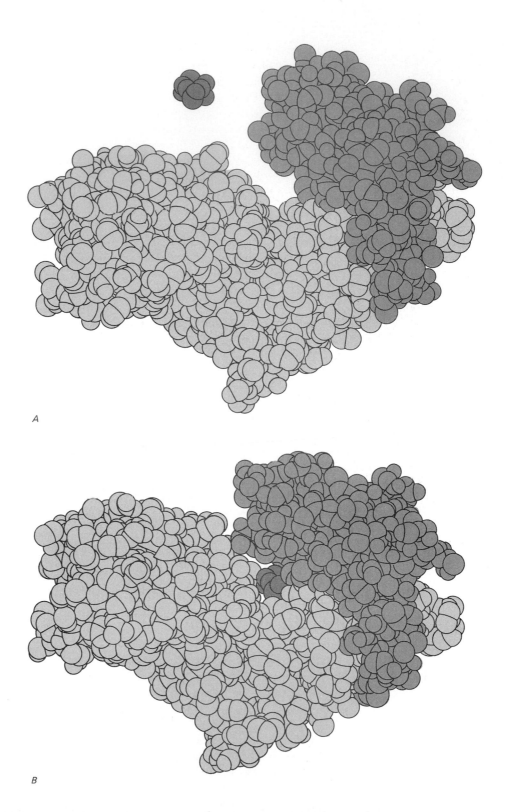

A

B

Other complexities have been observed with hexokinase, such as negative cooperativity in ATP binding and slow conformational transitions. However, these are now known to result from Al^{3+} contamination of ATP preparations. Complex results obtained in binding studies with all nucleoside triphosphates must be suspected to have arisen from such contamination.

A refined model of the sugar binding site of yeast hexokinase B. C. M. Anderson, et al. J. Mol. Biol. 123:207–219, 1978.

Glucose-induced conformational change in yeast hexokinase. W. S. Bennett and T. A. Steitz. Proc. Natl. Acad. Sci. U.S.A. 75:4848–4852, 1978.

Space-filling models of kinase clefts and conformation changes. C. M. Anderson, et al. Science 204:375–380, 1979.

Proton-dependent inhibition of yeast and brain hexokinases by aluminium in ATP preparations. F. C. Womack and S. P. Colowick. Proc. Natl. Acad. Sci. U.S.A. 76:5080–5084, 1979.

Crystallographic studies of model building of ATP at the active site of hexokinase. M. Shoham and T. A. Steitz. J. Mol. Biol. 140:1–14, 1980.

Structure of a complex between yeast hexokinase A and glucose. W. S. Bennett and T. A. Steitz. J. Mol. Biol. 140:183–230, 1980.

Activation of yeast hexokinase by chelators and the enzymic slow transition due to metal–nucleotide interactions. K. E. Neet, T. C. Furman, and W. J. Hueston. Arch. Biochem. Biophys. 213:14–25, 1982.

REGULATION OF ENZYME ACTIVITY

In vivo, enzymes do not normally act independently and in isolation, in the manner described thus far. Instead, their function is generally just one of many steps in metabolism, which consists of a maze of converging degradative pathways, on the one hand, by which food and endogenous reserves are converted to energy and a few simple products, and, on the other hand, of a maze of diverging and interlocking biosynthetic pathways by which the degradative products are converted to proteins, nucleic acids, lipids, and so forth. In order for this complex metabolic system to work, each enzyme must have the appropriate catalytic properties. Most enzymes have existed since the early stages of evolution and, during this time, have been subjected to constant selective pressure to maintain and improve these properties. Consequently, their catalytic powers are now observed to be highly refined to suit their roles; a variety of molecular devices are used in regulation of function.

Biosynthesis of an enzyme requires energy; consequently, enzymes

Figure 9–15

The conformational change that occurs upon glucose binding to hexokinase. A, A space-filling model of the free enzyme, with a glucose molecule approaching the active site cleft. B, A model of the complex, showing the bound glucose within the closed cleft. The small lobe of the enzyme (darkly shaded) has rotated counterclockwise, relative to the large lobe (lightly shaded). (From W. S. Bennett and T. A. Steitz, J. Mol. Biol. 140:211–230, 1980, © Academic Press Inc., London.)

would be expected to have evolved toward maximum catalytic efficiency, so that the minimum number of molecules needed to provide the required catalytic power can be synthesized. The limit to this efficiency is that the ratio of k_{cat}/K_m cannot exceed the limit by which substrate and enzyme encounter each other by diffusion (p. 408). Some enzymes appear to have reached this limit; triose phosphate isomerase is believed to be 60 per cent maximally efficient. Other considerations may prevent attainment of this perfection, such as the need to sacrifice catalytic efficiency for specificity based on induced fit, which effectively increases K_m above the limiting value. Some enzymes are also synthesized coordinately with others of the same operon, so there may be only limited selective pressure to increase their individual catalytic efficiencies.

The catalytic capability required for each enzyme will depend upon its metabolic role. Accumulation of any intermediate metabolite to high levels is probably undesirable; each enzyme, therefore, is presumably able to cope with the maximum rate of formation of its substrate. Consequently, the first enzyme in any single pathway usually has the lowest V_{max} and is rate-determining for that pathway. The intermediates do not accumulate to high levels in the steady state, the actual value of each depending primarily upon its K_m for the enzyme that utilizes it; the K_m value is usually higher than the observed steady-state level of the substrate. In some cases, the enzymes of a pathway are physically associated in a complex; additional aggregates that have not been detected yet may exist in vivo.

In addition to these conditions, the next most crucial requirement is that the catalytic capabilities of the enzymes respond to changes in metabolism, providing an orderly supply of energy, catabolites, and end-products in the required proportions. This control of enzyme catalysis is accomplished by varying both the quantity of enzyme present and its catalytic activity. Control by synthesis (Chapter 2) occurs most often in rapidly growing organisms, such as bacteria; control via the rate of enzyme degradation is less common (Chapter 10). In most organisms, where the level of an enzyme is relatively constant, control is achieved primarily by varying the catalytic activity. The dependence of velocity on substrate concentration is one regulating factor, but the activity is usually modulated further by the levels of metabolites other than its substrates and products. Such further modulation could be imagined to occur by classical enzyme inhibition (Figure 9–4), e.g., a regulating metabolite competing directly with substrate for the active site; but this is not a common phenomenon. Where such inhibition occurs, the inhibiting ligand is usually another protein molecule, the activity of which is regulated by other metabolites; an example is the cyclic AMP–dependent kinases to be described.

The two predominant ways in which enzyme activity is controlled are by allosteric interactions, with the regulatory metabolites binding to regulatory sites on the enzyme, and by covalent modification. For example, in most pathways the initial, rate-determining enzyme is inhibited allosterically by the end-product of the pathway. Covalent alterations of an enzyme, with catalytic consequences, are produced by another enzyme, whose activity is controlled by appropriate metabolites. Such regulatory effects can be indirect and complex, operating by amplifying enzyme cascades in which a metabolite activates one modifying enzyme, which modifies and activates a second modifying enzyme, and so on for several steps, until the catalytic enzyme is modified and regulated.

Where an enzyme central to metabolism is responsible for numerous end-products, its regulation can be quite complex. For example, glutamine synthetase plays an essential role in bacterial nitrogen metabolism; its synthesis is repressed by ammonia, and its catalytic activity is inhibited cumulatively by alanine, glycine, histidine, tryptophan, cytidine triophosphate (CTP), AMP, carbamyl phosphate, and glucosamine 6-phosphate, and also by reversible adenylylation of a specific Tyr residue catalyzed by an enzyme system that is regulated by other metabolites. In the following section we will consider some simpler systems that are more fully understood at the molecular level.

The molecular basis for enzyme regulation. D. E. Koshland, Jr. Mechanisms of enzyme regulation in metabolism. E. R. Stadtman; Enzymes as control elements in metabolic regulation. D. E. Atkinson. In The Enzymes, 3rd ed. P. D. Boyer (ed.). Vol. 1, pp. 341–396, 398–459, and 461–489. New York, Academic Press, 1970.

Multienzyme systems. A. Ginsburg and E. R. Stadtman. Ann. Rev. Biochem. 39:429–472, 1970.

The regulation of enzyme activity and allosteric transition. E. Whitehead. Prog. Biophys. Mol. Biol. 21:321–397, 1970.

Regulation of enzyme activity. G. G. Hammes and G.-W. Wu. Science 172:1205–1211, 1971.

Glutamine synthetase of *Escherichia coli:* some physical and chemical properties. A. Ginsburg. Adv. Protein Chem. 26:1–79, 1972.

Amino acid biosynthesis and its regulation. H. E. Umbarger. Ann. Rev. Biochem. 47:533–606, 1978.

Regulation of tryptophan biosynthesis. I. P. Crawford and G. V. Stauffer. Ann. Rev. Biochem. 49:163–197, 1980.

Allosteric Regulation

Metabolites may regulate the catalytic activity of an enzyme by binding at sites other than the active site. This is exactly analogous to allosteric effects on binding affinity for ligands (pp. 380–396), with the additional property that k_{cat} may be affected also. Allosteric interactions may be analyzed by their effect on enzyme activity in the same way as in ligand binding (Figure 8–3), but the additional complexity—that both affinity and catalytic activity may be affected—makes mechanistic deductions from enzyme activity alone very tenuous. Physical and structural analysis is also required, and in no case is an allosteric enzyme fully understood.

The aspartate transcarbamylase (ACTase) of *E. coli* is the most thoroughly studied allosteric enzyme. It catalyzes the formation of carbamyl aspartate from carbamyl phosphate and aspartate:

$$NH_2-\overset{O}{\overset{\|}{C}}-P_i + H_2N-\overset{\overset{\displaystyle CO_2^-}{|}}{\underset{}{\overset{\displaystyle CH_2}{|}}}CH-CO_2^- \longrightarrow$$

$$H_2N-\overset{O}{\overset{\|}{C}}-NH-\overset{\overset{\displaystyle CO_2^-}{|}}{\underset{}{\overset{\displaystyle CH_2}{|}}}CH-CO_2^- + P_i \qquad (9\text{--}61)$$

The carbamyl group is transferred directly from carbamyl-P to aspartate on the enzyme. Binding of the substrates is ordered, with carbamyl-P binding first. A bisubstrate analogue

$$P_i-CH_2-\overset{O}{\overset{\|}{C}}-NH-\overset{\overset{\displaystyle CO_2^-}{|}}{\underset{}{\overset{\displaystyle CH_2}{|}}}CH-CO_2^- \qquad (9\text{--}62)$$

N-(phosphonacetyl)-L-aspartate (PALA)

which may also be a transition-state analogue, binds at least 10^3 times more tightly than either single substrate.

Carbamyl aspartate is the first metabolite that is unique to the biosynthetic pathway for pyrimidine nucleotides. Consequently, the synthesis of ATCase is repressed by one product of this pathway, uracil, and its enzymic activity is inhibited by the end-product, CTP. It is activated by ATP, presumably to coordinate pyrimidine and purine nucleotide synthesis. These allosteric regulators affect only K_m, not V_{max} (Figure 9–16A). However, the enzyme does not obey normal Michaelis-Menten kinetics, in that the dependence of velocity on concentration of either substrate is sigmoidal, rather than hyperbolic; therefore, heterotropic interactions are accompanied by homotropic interactions, as occurs so often in allosteric systems. CTP and ATP are mutually competitive, but their binding at the same site produces opposite effects on homotropic cooperativity and on apparent K_m for the substrates (Figure 9–16A). A most striking consequence of the homotropic interactions is that inhibitors that are competitive with the substrate aspartate, and that undoubtedly bind at the active site, activate the enzyme at low concentrations (Figure 9–16C). Presumably, the homotropic effect on the other active sites of the enzyme of inhibitor binding at one site more than compensates for the inhibiting effect on the site to which it is bound.

The enzyme consists of six copies of each of two polypeptide chains, noted here as c and r, because one is primarily catalytic, the other regulatory. The c_6r_6 complex can be dissociated reversibly by mild treatments with heat, mercurials, or urea into two c_3 trimers (often designated C) and three r_2 dimers (or R):

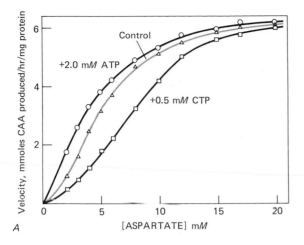

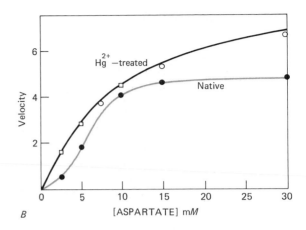

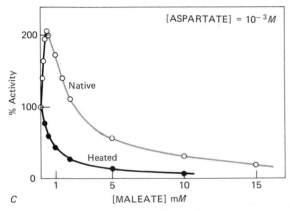

Figure 9–16

Steady-state kinetic behavior of aspartate transcarbamylase. A, The sigmoidal dependence of enzyme velocity on aspartate concentrations, at a fixed concentration of the other substrate, carbamyl-P (3.6 mM). CTP lowers the apparent affinity for aspartate and increases the cooperativity, whereas ATP has the opposite effect.

B, The kinetic behavior of native and of mercuric ion–treated enzyme. The treated enzyme is dissociated into catalytic trimers and regulatory dimers; the kinetic response to aspartate concentration is now hyperbolic, i.e., normal Michaelis-Menten, and the value of V_{max} is increased.

C, The effect of the inhibitor, maleate, which competes with aspartate, on the enzymatic activity of normal and heat-dissociated aspartate transcarbamylase. With the dissociated enzyme, maleate acts as a normal competitive inhibitor, but it activates the native enzyme at low concentrations of both maleate and aspartate. Its inhibitory effect of binding at one or a few of the six active sites on the native enzyme must be more than compensated by an allosteric activating effect on the remaining active sites, increasing their affinity for aspartate. (Adapted from J. C. Gerhart, Curr. Top. Cell Reg. 2:275–325, 1970; and J. C. Gerhart and A. B. Pardee, Cold Spring Harbor Symp. Quant. Biol. 28:491–496, 1963.)

$$c_6r_6 \rightleftharpoons 2\ c_3 + 3\ r_2$$

or

$$C_2R_3 \rightleftharpoons 2\ C + 3\ R \qquad\qquad (9\text{--}63)$$

The dissociation products are somewhat more active than the original enzyme and have lost all allosteric properties (Figure 9–16B and C). The r_2 dimers have no catalytic activity but bind the allosteric effectors. Many studies with modified c and r chains and with different aggregates have demonstrated that both the homotropic and heterotropic allosteric interactions depend upon the presence of the two different polypeptide chains—one catalytic, the other regulatory—assembled into the large quaternary structure.

The nature of this quaternary structure is shown in Figure 9–17. The c_3 and r_2 structural units into which it may be dissociated are demonstrated vividly. The regulatory binding sites on the r chains are at the very periphery of the structure, at least 60 Å away from the catalytic sites, which face the large interior cavity of this structure. Each of the six catalytic sites may be shared between pairs of catalytic chains adjacent within the same trimer, involving residues from both neighboring chains.

A substantial change in the overall shape of the molecule, and in many other of its physical properties, occurs when substrates or ATP molecules are bound: this change tends to be inhibited by CTP. The extent of the change is not proportional to the degree of saturation of the enzyme with ligands; instead, it occurs over a narrow range. This is expected with a concerted quaternary structure change of the T↔R type in the Monod-Wyman-Changeux allosteric model (see pp. 384–386). The T state would predominate with unliganded ATCase and would bind CTP more strongly than R, which would bind preferentially the substrates and ATP. For

---

Figure 9–17

A schematic model of aspartate transcarbamylase. A, Two views of the overall quaternary structure. Left, A view down the threefold axis, showing a trimer of three catalytic subunits (large spheres) lying above a second trimer just barely visible below it. The two trimers are linked on the outside by three dimers of regulatory subunits (small spheres). The arrow indicates the direction in which the top trimer is rotated upon binding PALA. Three twofold axes relating the top and bottom halves are indicated. Right, A view down one such twofold axis, showing the two catalytic subunits (c) linked structurally via two regulatory subunits (r). The arrow indicates the direction in which the top trimer rotates upon binding PALA; the trimers also move apart along the threefold axis.

*B, A more detailed drawing of the course of the polypeptide chains of one half of the molecule. The regulatory subunits are drawn with dashed lines. Each is composed of two structural domains; the one furthest from the center binds CTP or ATP at the effector site (**E**); the other binds one Zn^{2+} atom (black sphere) coordinated to four Cys residues. The Zn^{2+} atom is believed to have only a structural role. The catalytic subunits are drawn with solid lines; the probable locations of the active sites are designated by **A**.*

Based on a picture produced by a computer program of A. M. Lesk and K. D. Hardman and kindly provided by A. M. Lesk.

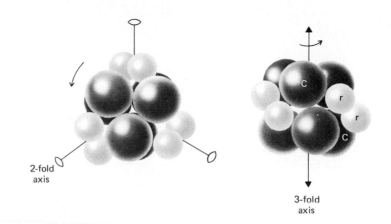

2-fold
axis

3-fold
axis

A

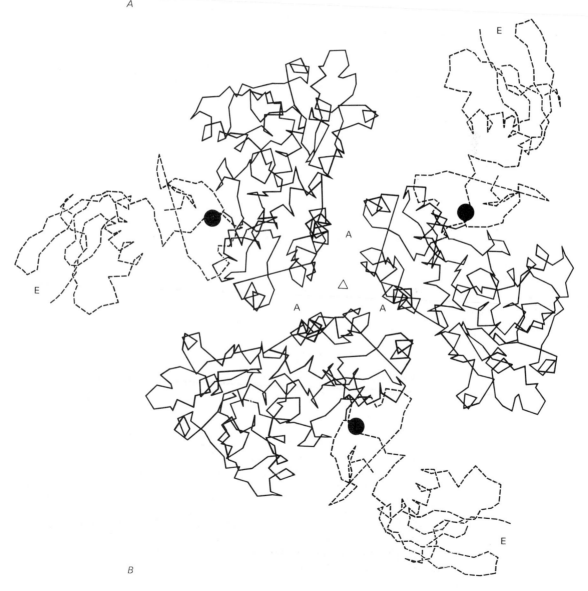

E

E

A

A

A

E

B

ATCase, the greatest apparent exception to this scheme is the negative cooperativity observed in the binding of CTP and ATP; some results have indicated that only half the six regulatory sites readily bind the nucleotides. However, the numerous studies have not given consistent results, and the observation that prevalent metal contamination of ATP preparations was responsible for many anomalous kinetic properties of hexokinase (see p. 457), suggests that most binding studies utilizing nucleoside triphosphates must be suspect.

The crystal structure of ATCase just described is believed to correspond to the T state; only recently has a tentative structure been determined for the presumed R state, induced by the bisubstrate analogue PALA (Eq. 9–62). In this structure, the two catalytic trimers have separated from each other along the threefold axis by 11 to 12 Å and have rotated relative to each other around that axis by 8 to 9 degrees (see Figure 9–17). This rotation places the trimers in a more nearly eclipsed configuration. The two regulatory chains of each dimeric unit have also rotated 14 to 15 degrees around the twofold axis and have become more nearly perpendicular. Consequently, the anticipated quaternary structure changes upon binding an allosteric effector have been observed, but many questions about the allosteric mechanism remain to be answered. Unfortunately, the enzyme crystallizes at a relatively low pH, where it exhibits little enzymic activity or allosteric control.

ATCase is somewhat unusual in having separate polypeptide chains for catalysis and regulation. In contrast, glutamine synthetase is composed of a single type of 50,000-molecular-weight polypeptide chain, even though it presumably has at least eight different regulatory binding sites (p. 459). The structure of this complex is known only at very low resolution; it consists of two rings of six subunits each associated about the same sixfold symmetry axis, comparable to the two trimers of ATCase (Figure 9–17).

The other allosteric enzyme with known atomic structure is bacterial phosphofructokinase. It catalyzes the key control step of glycolysis:

$$\text{fructose 6-P} + \text{ATP} \longrightarrow \text{fructose 1,6-P}_2 + \text{ADP} \qquad (9\text{--}64)$$

The enzymes from eukaryotic sources are usually inhibited by ATP and by citrate and are activated by AMP, ADP, and cyclic AMP. The bacterial enzymes are simpler but rather diverse; that of known structure is activated by ADP, thereby responding to the energy level of the cell. It is inhibited by phosphoenolpyruvate, an end-product of glycolysis, and gives sigmoidal rate dependence on the concentration of fructose 6-P. The regulatory effects are due to apparent changes in K_m for substrates, not to changes in k_{cat}. The tetrameric structure determined crystallographically (Figure 9–18) is the equivalent of the allosteric R state, with high affinity for substrates. Crystallographic binding studies have demonstrated distinct sites for binding the substrates fructose 6-P and ATP and an allosteric site that binds either the activator ADP or the inhibitor phosphoenolpyruvate. The active site is in an extended cleft between the two domains of each subunit.

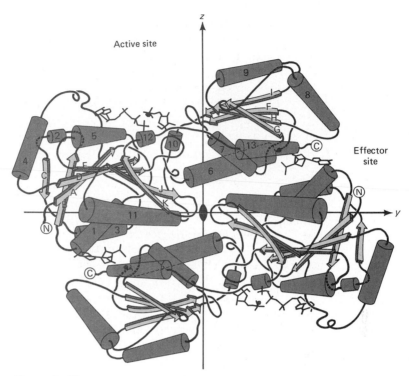

Figure 9–18
Schematic view of two subunits of the phosphofructokinase from Bacillus stearothermophilus. *Two active sites, with bound substrates, ATP and fructose 6-P, and two regulatory
effector sites, with bound activator ADP, are shown. The other two subunits of the tetramer
would be below, related by the* y *and* z *twofold symmetry axes. The helices and β-strands of
one subunit are labeled. (From P. R. Evans, et al., Philos. Trans. R. Soc. Lond. [Biol.]
293:53–62, 1981.)*

The effector sites lie between, and bridge, different subunits of the tetramer, as does that for the phosphate group of the substrate fructose 6-P.
These findings suggested that the allosteric effects could involve primarily
changes in quaternary structure upon ligand binding; also, crystals of the
T structure crack upon adding fructose 6-P. However, a recently determined crystal structure of the T state has demonstrated that substantial
changes in conformational states are also involved.

The enzymology of control by feedback inhibition. J. C. Gerhart and A. B. Pardee.
J. Biol. Chem. 237:891–896, 1962.
The effect of the feedback inhibitor, CTP, on subunit interactions in aspartate
transcarbamylase. J. C. Gerhart and A. B. Pardee. Cold Spring Harbor Symp.
Quant. Biol. 28:491–496, 1963.
Distinct subunits for the regulation and catalytic activity of aspartate transcarbam-
ylase. J. C. Gerhart and H. K. Schachman. Biochemistry 4:1054–1062, 1965.
Aspartate transcarbamylases. G. R. Jacobson and G. R. Stark. In The Enzymes, 3rd
ed. P. D. Boyer (ed.). Vol. 9, pp. 225–308. New York, Academic Press, 1973.

Anatomy and physiology of a regulatory enzyme–aspartate transcarbamylase. H. K. Schachman. Harvey Lect. 68:67–113, 1974.

Allosteric regulation of aspartate transcarbamylase. Analysis of the structural and functional behavior in terms of a two-state model. G. J. Howlett, et al. Biochemistry 16:5091–5099, 1977.

Asymmetry of binding and physical assignments of CTP and ATP sites in aspartate transcarbamylase. P. Suter and J. P. Rosenbusch. J. Biol. Chem. 252:8136–8141, 1977.

Structure and control of phosphofructokinase from *Bacillus stearothermophilus*. P. R. Evans and P. J. Hudson. Nature 279:500–504, 1979.

E. coli aspartate transcarbamylase. E. R. Kantrowitz, et al. Trends Biochem. Sci. 5:124–128, 150–153, 1980.

On the detection of homotropic effects in enzymes of low cooperativity. Application to modified aspartate transcarbamylase. P. Hensley, Y. R. Yang, and H. K. Schachman. J. Mol. Biol. 152:131–152, 1981.

Communication between subunits in aspartate transcarbamoylase. C. M. Wang, et al. J. Biol. Chem. 256:7028–7034, 1981.

Coupling of homotropic and heterotropic interactions in *Escherichia coli* aspartate transcarbamylase. P. Tauc, et al. J. Mol. Biol. 155:155–168, 1982.

Crystal and molecular structures of native and CTP-liganded aspartate carbamoyltransferase from *Escherichia coli*. R. B. Honzatko, et al. J. Mol. Biol. 160:219–263, 1982.

Gross quaternary changes in aspartate carbamoyltransferase are induced by the binding of *N*-(phosphonacetyl)-L-aspartate: a 3.5 Å resolution study. J. E. Ladner, et al. Proc. Natl. Acad. Sci. U.S.A. 79:3125–3128, 1982.

A fully nonequilibrium concerted model for enzymes. J. S. Shiner. Biopolymers 21:2241–2252, 1982.

Reversible Covalent Modification

Another level of control, which can respond to different signals, utilizes reversible covalent modification of the enzyme. Modification is catalyzed by one enzyme; removal of the modification, by a second. These two reactions are not simply the reverse of each other; if this were the case, both enzymes would have to catalyze the modification reaction in both directions, with kinetic parameters dictated by the Haldane relationship (Eq. 9–21). Instead, the sum of the two reactions involves a favorable net reaction, so that both are energetically favorable, with insignificant reversibility. For example, phosphorylation of an enzyme generally uses a phosphate group of ATP, whereas its removal is by simple hydrolysis:

$$
\begin{aligned}
\text{phosphorylation:} \quad & \text{E} + \text{ATP} \longrightarrow \text{E—P} + \text{ADP} \\
\text{dephosphorylation:} \quad & \text{E—P} + \text{H}_2\text{O} \longrightarrow \text{E} + \text{P}_i \\
\hline
\text{net:} \quad & \text{ATP} + \text{H}_2\text{O} \longrightarrow \text{ADP} + \text{P}_i \quad (9\text{–}65)
\end{aligned}
$$

Because the net result is hydrolysis of ATP, the two reactions are both favorable, but they must also be regulated to prevent wasteful cycling. The potential for regulating the two reactions independently is undoubtedly the reason that this principle is used throughout metabolism.

Besides phosphorylation, which normally occurs to specific Ser, Thr, and Tyr residues, several other reversible, regulatory modifications have been observed: nucleotidylation of Tyr residues; ADP ribosylation, primarily of Arg residues; methylation of Glu or Asp carboxyl groups; acetylation of Lys amino groups; and removal and replacement of Tyr residues from the carboxyl end of the polypeptide chain. The chemical natures of these reactions are described in Chapter 2 (pp. 70–87).

This type of regulation has been known for a long time in a few specific cases, but only recently has the prevalence and generality of the phenomenon come to be recognized. With this new awareness, many more examples are being uncovered, and the implications for understanding the control of metabolism are immense. Here only one system will be described, that most thoroughly characterized at the molecular level, but it is also the archetypal example and illustrates well the general principles.

Glycogen Phosphorylase

Glycogen phosphorylase has the vital role of controlling the metabolic utilization of the storage polysaccharide, glycogen, catalyzing the first step in its degradation:

$$(\alpha\text{-1,4-glucoside})_n + P_i \rightleftharpoons (\alpha\text{-1,4-glucoside})_{n-1} + \text{glucose 1-P} \quad (9\text{–}66)$$

The glucose 1-P produced is utilized in the muscle primarily for the energy to produce contraction and in liver for the maintenance of blood sugar levels.

Phosphorylase normally occurs in two forms, designated a and b, which differ covalently only in that the former has Ser 14 of the 841-residue polypeptide chain phosphorylated. Yet the two forms differ markedly in their allosteric regulatory properties: The b form is inhibited by ATP, ADP, glucose, and glucose 6-P, thereby responding to the energy requirements of the cell. The a form escapes the controls by ATP, ADP, AMP, and glucose 6-P, but it is still inhibited by glucose; consequently, this form degrades glycogen even if the cell is not energy-deficient.

The structures of the two forms have been determined independently in different laboratories. They exhibit few substantial differences, the greatest involving about 17 amino-terminal residues, including Ser 14 at which they differ covalently. In the nonphosphorylated form, b, this segment is flexible and not visible on the electron density map. When Ser 14 is phosphorylated, these residues adopt a fixed conformation, extended along the subunit contact and interacting with the other subunit (Figure 9–19). Both a and b forms are probably dimeric in their functional state, although b is known to become tetrameric at high concentrations.

The allosteric properties of phosphorylase are considered to be consistent, at least to a first approximation, with a two-state concerted transition between R and T states (see pp. 384–386) (Table 9–7). Both crystal structures are believed to be in the T state, which is essentially inactive. Substantial conformational changes occur upon conversion to the R state; a preliminary structural analysis indicates that most of the changes are confined to the areas close to the subunit contacts (Figure 9–19). Nevertheless, they interconnect all the regulatory binding sites; therefore direct, induced interactions between sites may also be involved.

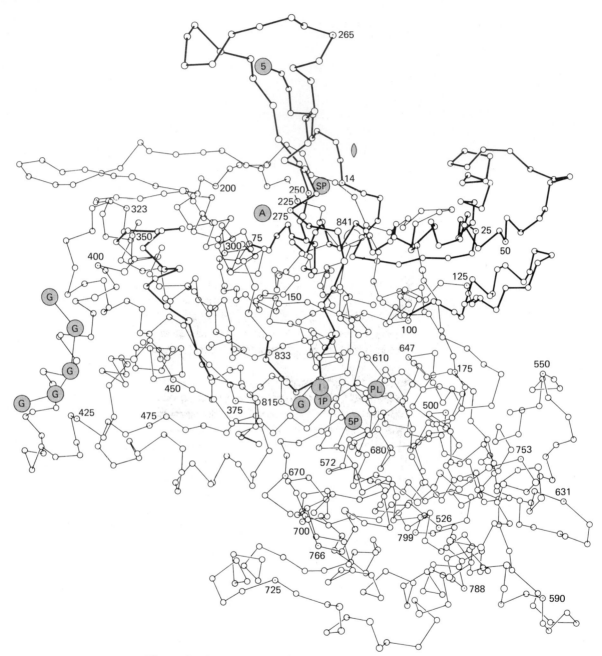

Figure 9–19
Conformation of the backbone of one subunit of phosphorylase a, *with the positions of the binding sites. The twofold axis relating this subunit to the other one of the dimer is shown approximately by* ◖ *. The portions of the polypeptide backbone believed to move upon the transition to the more active R state are drawn with thick lines.*

 The known binding sites of Table 9–7 are labeled as follows: G *and* 1P, *the active site where glucose 1-P binds;* I, *the inhibitor allosteric site;* A, *the activating site;* SP, *the phos-*

Table 9–7 **Ligand Interactions with Phosphorylases a and b**

LIGAND	BINDING SITE	CONFORMER PROMOTED	FUNCTION
Glycogen	Storage	R	Allosteric activation and
	Active	—	binding substrate
Glucose 1-P	Active	R	Substrate
	Activator	—	?
P_i	Active	R	Substrate
	Activator	—	?
Glucose	Active	T	Inactivation of *a*
AMP	Activator	R	Activates *b*, *a*
ATP	Activator	T	Inhibits *b*
Glucose 6-P	Activator	T	Inhibits *b*
Purines and analogues	Inhibitor	T	Inactivates *a*
UDPG	Active	R	Inhibits phosphorylase
Phosphoryl group	Ser 14	R	Hormonal activation

From R. J. Fletterick and N. B. Madsen, Ann. Rev. Biochem. 49:31–61, 1980.

The active site is in a very narrow cleft between two domains of each subunit (Figure 9–19). There are two regulatory sites: One binds inhibitors in such a way that they directly block access to the active site. The other binds activators and is quite distant from the active site but is part of the subunit contact area. There is also a glycogen-binding site, presumably for anchoring the enzyme to its substrate. Pyridoxal phosphate is bound tightly to the protein, even though it has never been found to have its usual functional role in catalysis; it appears simply to provide a phosphate group for the catalytic site.

The difference between regulatory properties of the *a* and *b* forms is believed to result simply from a shift in the allosteric conformational equilibrium constant, *L*, from at least 10^3 in *b*, favoring the T state, to a value approaching unity in *a*. Consequently, *a* is almost in the active R state even in the absence of activators and is in fact active in their absence unless an inhibitor such as glucose is present to shift the equilibrium back toward T. This molecular basis for the effect of phosphorylation of Ser 14 on the regulatory properties of phosphorylase is plausible in view of the presence of this residue within a part of the protein apparently involved in allosteric conformational changes, its interactions with the other subunit in the quaternary structure, and the effect of phosphorylation on these interactions. However, the details are not yet clear.

phate group on Ser 14; G-G-G-G-G-, the binding site for 5 glucose units of glycogen; PL and 5P, the sites of pyridoxyl and phosphate moieties of pyridoxal-P (its role apparently is only to provide a phosphate group to the catalytic site). (Adapted from N. B. Madsen, P. J. Kasvinsky, and R. J. Fletterick, J. Biol. Chem. 253:9097–9101, 1978; the revised coordinates at 2.5 Å resolution from S. Sprang and R. J. Fletterick, J. Mol. Biol. 131:523–551, 1979.)

α-Glucan phosphorylases—chemical and physical basis of catalysis and regulation. D. J. Graves and J. H. Wang. In The Enzymes, 3rd ed. P. D. Boyer (ed.). Vol. 7, pp. 435–482. New York, Academic Press, 1970.

Complete amino acid sequence of rabbit muscle glycogen phosphorylase. K. Titani, et al. Proc. Natl. Acad. Sci. U.S.A. 74:4762–4766, 1977.

Allosteric transitions of phosphorylase *a* and the regulation of glycogen metabolism. N. B. Madsen, et al. J. Biol. Chem. 253:9097–9101, 1978.

Crystallographic studies on the activity of glycogen phosphorylase *b*. I. T. Weber, et al. Nature 274:433–437, 1978.

The structure of glycogen phosphorylase *a* at 2.5 Å resolution. S. Sprang and R. J. Fletterick. J. Mol. Biol. 131:523–551, 1979.

Nucleotide binding to glycogen phosphorylase *b* in the crystal. L. N. Johnson, et al. J. Mol. Biol. 134:639–653, 1979.

The structure and related functions of phosphorylase *a*. R. J. Fletterick and N. B. Madsen. Ann. Rev. Biochem. 49:31–61, 1980.

Site-site interactions in glycogen phosphorylase *b* probed by ligands specific for each site. N. B. Madsen, et al. Biochemistry 22:4460–4465, 1983.

Interconversion of Phosphorylase *a* and *b*

Elucidation of the factors involved in this regulation of phosphorylase has been one of the major advances in biochemistry, with widespread repercussions for the mechanism of hormone action and the control of metabolism. A description of all of these aspects is beyond the scope of this text but is available in modern biochemistry textbooks; we will consider here only those most pertinent to phosphorylase.

Within muscle, phosphorylase is converted from the *b* to the *a* form upon muscle contraction, or upon stimulation by the hormone adrenaline (also known as epinephrine). The peptide hormone glucagon plays a comparable role in the liver. These stimuli are signals that energy or glucose is going to be required in the near future and that phosphorylase should be activated, irrespective of the current energy state of the cell, which regulates phosphorylase *b* activity allosterically.

These and other hormones act by stimulating cyclic AMP production, the ubiquitous "second messenger." Cyclic AMP is formed from ATP by adenylate cyclase, a membrane-bound complex of at least three different proteins: One is the receptor for the specific hormone, on the outer surface of the membrane. Another is the catalytic protein, on the inner side of the membrane, which produces cyclic AMP from ATP within the cell, but only when hormone is bound to the receptor. Communication between the two components is mediated by the third component, but only when it has GTP bound. The interactions between these components are intricate and only partly understood.

The increased level of cyclic AMP caused by hormone binding to the cell surface activates two different cyclic AMP–dependent kinases. In the absence of cyclic AMP, both proteins are catalytically inactive. Both are composed of two regulatory and two catalytic subunits, R_2C_2; they differ only in the R chains. Cyclic AMP dissociates this complex by binding more tightly to the free regulatory dimers, R_2, and thereby releases active catalytic subunits:

$$R_2C_2 + 4 \text{ cAMP} \rightleftharpoons (R(\text{cAMP})_2)_2 + 2 \text{ C} \qquad (9\text{–}67)$$

$$\underset{\text{inactive}}{} \qquad \qquad \underset{\text{active}}{\phantom{(R(\text{cAMP})_2)_2}}$$

This process illustrates the way in which enzymes may be regulated by their combination with inhibiting or modifying proteins.

Activation of the cAMP-dependent protein kinase is also regulated by covalent modification and allosteric regulation. One type of R subunit can be phosphorylated by the catalytic subunit, thereby increasing its affinity for cyclic AMP and its tendency to dissociate. With the other type of regulatory subunit, ATP binds tightly to the R_2C_2 form, thereby inhibiting its tendency to be dissociated and activated by cyclic AMP.

The active cyclic AMP–dependent kinases phosphorylate at least ten different enzymes, each on accessible Ser (or Thr) residues occurring in the sequence -Lys-Arg-X-X-Ser or Arg-Arg-X-Ser. X can be almost any amino acid, and Thr can be present rather than Ser. As a consequence of these phosphorylations, degradative enzymes such as glycogen phosphorylase are activated, whereas biosynthetic enzymes such as glycogen synthetase are inhibited. However, these metabolic effects often are not direct but occur via intermediary enzymes in cascades. For example, cyclic AMP–dependent kinases directly phosphorylate phosphorylase kinases, not glycogen phosphorylase. The phosphorylated phosphorylase kinase then phosphorylates phosphorylase *b*.

Phosphorylase kinase is a complex protein, consisting of four copies of each of four different polypeptide chains. Two of these chains can each be phosphorylated on a specific Ser side chain. Phosphorylation of one chain increases the rate of phosphorylation of the other. This kinase is active only if the first subunit is phosphorylated and if another of the subunits has bound Ca^{2+} ion. This latter subunit is calmodulin, the ubiquitous Ca^{2+}-binding regulator protein. It is primarily this protein that makes activation of glycogen phosphorylase sensitive to muscle contraction, which is triggered by Ca^{2+} ion release.

The activated phosphorylase kinase then specifically phosphorylates Ser 14 of phosphorylase *b*. The details of this process are not clear, but Ser 14 is on a flexible portion of polypeptide chain in this enzyme form, so that it is probably relatively accessible to the kinase.

All of these activation steps are balanced by reverse steps, and under normal conditions there is probably a steady-state condition, even though ATP is hydrolyzed in the process. The adenylate cyclase middle component hydrolyzes its bound GTP and thereby temporarily stops the hormone from causing cyclic AMP formation. Existent cyclic AMP is hydrolyzed to AMP by a specific phosphodiesterase, thereby favoring inhibition of the cyclic AMP–dependent kinase by aggregation with its regulatory subunits (the reverse of Eq. 9–67). The two phosphoryl groups on phosphorylase kinase are removed by two different phosphatases; the phosphoryl group that was added first is also removed first. The responsible phosphatase also removes the phosphoryl group on Ser 14 of phosphorylase *a* and on glycogen synthetase. The rate at which it acts on Ser 14 probably depends upon the conformational state of phosphorylase *a*. The phosphatase appears to bind tightly to phosphorylase *a*, but to cleave the phosphoryl group only if the protein is pulled into the T state, when the phosphoryl group may become accessible. Therefore, this covalent modification is also susceptible to allosteric control within the substrate.

These two phosphatases are also regulated directly in their activity, one by calmodulin and Ca^{2+} binding, and the other by two protein inhibitors. One of these inhibitors is active only if phosphorylated by cyclic AMP–dependent protein kinase. This amplifies the effect of the hormone, since the inhibitor is activated, which then inhibits the phosphatase; consequently, the rate of inactivation of phosphorylase *a* is decreased, which tends to increase the rate of glycogen breakdown.

At least some of these components have additional roles in other metabolic pathways, thereby serving to integrate many aspects of metabolism. Virtually nothing is known about how these various components act at the molecular level, but the regulatory properties of those allosteric proteins described in detail here, such as hemoglobin (pp. 387–393), aspartate transcarbamylase, and glycogen phosphorylase, provide plausible models of how they could occur.

Studies on the mechanism of hormone action. E. W. Sutherland. Science 177:401–408, 1972.

Calmodulin: an intracellular calcium receptor. A. R. Means and J. R. Dedham. Nature 285:73–77, 1980.

The control of glycogen metabolism in the liver. H. G. Hers. Ann. Rev. Biochem. 45:167–190, 1976.

Phosphorylated proteins as physiological effectors. P. Greengard. Science 199:146–152, 1978.

The hormonal control of glycogen metabolism in mammalian muscle by multivalent phosphorylation. P. Cohen. Biochem. Soc. Trans. 7:459–480, 1979.

Phosphorylation–dephosphorylation of enzymes. E. G. Krebs and J. A. Beavo. Ann. Rev. Biochem. 48:923–959, 1979.

Chemical and regulatory properties of phosphorylase kinase and cyclic AMP–dependent protein kinase. G. M. Carlson, et al. Adv. Enzymol. 50:41–115, 1979.

The role of hormone receptors and GTP-regulatory proteins in membrane transduction. M. Rodbell. Nature 284:17–22, 1980.

Biochemical properties of hormone-sensitive adenylate cyclase. E. M. Ross and A. G. Gilmore. Ann. Rev. Biochem. 49:533–564, 1980.

Interconvertible enzyme cascades in cellular regulation. P. B. Chock, et al. Ann. Rev. Biochem. 49:813–844, 1980.

Studies of the mechanism of action and regulation of cAMP-dependent protein kinase. J. Granot, et al. Arch. Biochem. Biophys. 205:1–17, 1980.

The role of calcium ions, calmodulin and troponin in the regulation of phosphorylase kinase from rabbit skeletal muscle. P. Cohen. Eur. J. Biochem. 111:563–574, 1980.

Complete amino acid sequence of the catalytic subunit of bovine cardiac muscle cyclic AMP–dependent protein kinase. S. Shoji, et al. Proc. Natl. Acad. Sci. U.S.A. 78:848–851, 1981.

Bimodal regulation of adenylate cyclase. D. M. F. Cooper. FEBS Letters 138:157–163, 1982.

The role of protein phosphorylation in neural and hormonal control of cellular activity. P. Cohen. Nature 296:613–620, 1982.

Regulatory mechanisms in the control of protein kinases. D. A. Flockhart and J. D. Corbin. Crit. Rev. Biochem. 12:133–186, 1982.

ENERGY TRANSDUCTION

Proteins catalyze many processes other than chemical reactions in solution; in some of these processes, one form of energy can be considered to be converted to another form. The ubiquitous ATP is usually the currency of chemical energy and is involved in most energy interconversions, being either utilized or generated. For example, the chemical energy of ATP is converted to mechanical energy to move myosin thick filaments relative to actin thin filaments in muscle and to move microfilaments within cells. Many other cellular functions rely, either directly or indirectly, on the transport of ions across membranes, either storing or utilizing free energy in the form of concentration gradients. These include oxidative and photosynthetic phosphorylation, active transport of metabolites, and generation of nerve impulses. Much of the free energy of metabolic breakdown of food is converted to ATP by oxidative phosphorylation indirectly: This free energy is first converted to chemiosmotic energy in the form of proton gradients across membranes. Conversely, a large portion of this chemical energy is used to pump ions against concentration gradients across membranes.

A thorough discussion of all these topics is beyond the scope of this text, but they are outlined in modern textbooks on biochemistry. The brief discussion here concentrates on the principles of energy interconversion and assumes a familiarity with the basic properties of muscle structures and contraction, of oxidative phosphorylation and metabolism, and of membrane structure and transport. We will consider only how the energy transduction might occur at the atomic level, focusing on the proteins involved in this step. As virtually nothing is known about the detailed structures of these proteins, the following discussion is brief and general. Nevertheless, it seems worthwhile to attempt to spell out the general principles that must apply and the questions to be answered, because these topics are being studied intensively at the present time and will undoubtedly supply in the near future some of the greatest excitement in biochemistry.

Reaction Coupling

How can the chemical energy of ATP be used to drive an otherwise unfavorable reaction, and how can some other form of energy be used to drive the otherwise unfavorable formation of ATP from ADP and P_i? We are concerned primarily with reaction equilibria and assume that if a reaction is energetically favorable, some enzyme will be capable of catalyzing it at an adequate rate. The simplest way of coupling two reactions is for both processes to occur on the same enzyme molecule and only at the same time, since then the energy may be transferred directly. A simple example is encountered with the kinases, where ATP is used to phosphorylate small molecules (pp. 454–457) and proteins (pp. 466–473); such reactions would be very unfavorable with P_i as phosphate donor. The phosphoryl group is generally transferred directly from ATP to the substrate, and the only "trick" is to inhibit the hydrolysis of ATP in the absence of that substrate. This is a rather simple problem of enzyme specificity that can be solved, at least theoretically, by invoking induced fit (pp. 426–427).

An alternative method of coupling two reactions would be to have them occur independently, but to be linked energetically via some common intermediary. For example, the hydrolysis of ATP might produce a modified factor, F*, which is required for the second, energetically unfavorable step, A→B:

$$F + ATP \rightleftharpoons F^* + ADP + P_i$$
$$\underline{F^* + A \rightleftharpoons F + B}$$

$$\text{net:} \quad A + ATP \rightleftharpoons B + ADP + P_i \qquad (9\text{--}68)$$

This is intuitively most obvious if we describe F* as a "high-energy" species to which the free energy of ATP has been transferred, but it is not necessary thermodynamically. The relative free energies of F and F* are immaterial to the equilibrium constant of the net reaction. On the other hand, it may be important kinetically for both F and F* to be present at comparable concentrations; thus, the equilibrium constant for the more rapid reaction is often observed to be near unity. F and F* may be different forms of the enzyme catalyzing the reaction or many other things, such as an ion on the two sides of a membrane.

Energy coupling will occur to the extent that the two steps require the appropriate form of the factor F, so that ATP is hydrolyzed by this system only after one molecule of A has been converted to B. It also is imperative that the two forms of F not be interconvertible in any other way.

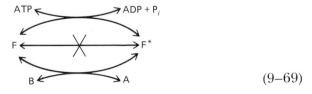

$$(9\text{--}69)$$

Otherwise, the two reactions would occur independently and in the direction dictated by their individual equilibria; i.e., ATP would be hydrolyzed, and B converted back to A.

These simple examples illustrate the important principle that coupling simply requires that certain steps be forbidden. Otherwise, the cycle of free energy transduction would be short-circuited, and the two reactions uncoupled. The process of free energy transduction then becomes simply one of specificity of the proteins catalyzing the reactions.

In the few examples that follow, energy transduction will be shown to occur by a cyclic sequence of chemical reactions occurring on the protein catalyst. Transduction requires only that the cycle not be short-circuited by additional interconversions. All such cycles are fully reversible, and the direction of cycling depends upon the input of energy. In many cases, catalysis of the system is usually inhibited when the energy situation is unfavorable, when the cycle could tend to run backwards. The importance of this regulation, which usually is produced by protein inhibitors, can be appreciated if it is realized that otherwise plants would luminesce at night by the reverse of photosynthesis and at the expense of the energy captured during the day. Although biological catalysts cannot affect reaction equilibria, their catalytic potency can be inhibited when they would otherwise be required to catalyze a reaction in the wrong direction.

The energetics of the individual steps in such cycles are not important, as only the net reaction matters; consequently, it is futile to assign energy transduction to a specific step or to look for "high-energy" states, as is often done. Nevertheless, the energies of the various states are often found to be relatively comparable, presumably so that all accumulate to similar levels and so that each step is kinetically feasible.

Biochemical cycles and free energy transduction. T. L. Hill. Trends Biochem. Sci. 2:204–207, 1977.

Free Energy Transduction in Biology. T. L. Hill. New York, Academic Press, 1977.

Non-equilibrium thermodynamics of energy conversion in bioenergetics. H. Rottenberg. Biochim. Biophys. Acta 549:225–253, 1979.

The utilization of binding energy in coupled vectorial processes. W. P. Jencks. Adv. Enzymol. 51:75–106, 1980.

Can free energy transduction be localized at some crucial part of the enzymatic cycle? T. L. Hill and E. Eisenberg. Q. Rev. Biophys. 14:463–511, 1981.

Mechanical Energy

Chemical energy can be converted to mechanical energy in muscle, cilia, flagella, and other motility systems. This conversion involves movements of proteins relative to each other, undoubtedly produced by changes in their conformations and by changes in their binding to each other. The most thoroughly studied system is that of muscle; it is complex both structurally and catalytically, owing in part to the obvious need to control closely muscle contraction. It is generally triggered by the release of Ca^{2+} ions from either the sarcoplasmic reticulum or the cell membrane. Depending upon the muscle, the released Ca^{2+} ions bind to either the troponin part of the actin thin filament, to one of the light chains of myosin, or to a kinase, which then phosphorylates the light chains. In the last instance, the myosin kinase can also be activated by cyclic AMP–dependent phosphorylation. These regulatory effects are analogous to those described earlier for the regulation of enzyme activity and will not be discussed further.

The result of triggering the release of Ca^{2+} ions is initiation of a physical interaction between the actin thin filaments and the globular heads of myosin, which serve as cross-bridges between the thick and thin fila-

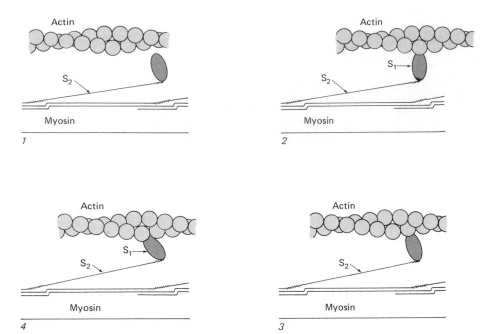

Figure 9–20

A schematic representation of the cross-bridge model of muscle contraction. Each diagram illustrates (top) *one thin filament of polymerized actin molecules and* (bottom) *one thick filament comprising many myosin molecules aggregated via the amino-terminal portion of their coiled-coil regions. One myosin cross-bridge is illustrated, consisting of one S_2, carboxyl-terminal portion of the coiled coil plus the globular head, S_1, which has the essential ATPase activity. Myosin molecules are dimeric, but only one of the two heads is shown.*

One cycle of ATP hydrolysis and filament movement probably involves the sequence $1\rightarrow2\rightarrow3\rightarrow4\rightarrow1$. The detached cross-bridges in one conformation attach to the actin filament at the closest appropriate binding site (2) and then change conformation (3 and 4) in a process coupled to ATP hydrolysis, moving the attached actin filament relative to the rest of the thick filament. The cross-bridge then detaches (1), changes conformation, and binds to a different site on the actin filament, some 100 Å away from the first.

Many different cross-bridges are operating in this way between adjacent thick and thin filaments, to produce a smooth movement of the two relative to each other.

ments. The myosin heads have the ability to hydrolyze ATP and are believed to provide the catalytic power for muscle contraction. In this process, the thick and thin filaments move past each other, propelled by a series of cyclic events in which each cross-bridge dissociates from one actin molecule of a thin filament, changes conformation to interact with a different actin molecule about 100 Å further along the thin filament, and then pulls the thin filament so that this second actin molecule is now in the same relative position as the first (Figure 9–20). Consequently, mechanical work is performed, at the expense of one molecule of ATP hydrolyzed in each cycle.

The enzymic hydrolysis of ATP by the globular heads of myosin has been extensively studied in solution, including the presence of soluble, nonaggregated actin. No mechanical work is performed; only the hydrolysis of ATP takes place. These conditions differ substantially from those in muscle, where the components are part of a complex supramolecular lattice. Extrapolation of these observations to muscle is dangerous, because the effective concentrations there of the actin and myosin components are not known. Physical and catalytic interactions may also occur between the two globular heads of a myosin dimer, between those of other myosin molecules of the same or adjacent thick filaments, and between different actin molecules of thin filaments. Nevertheless, the observed enzymic properties fit in sufficiently well with how muscle is believed to function (Figure 9–20) to permit some explanations of how the energy transduction takes place.

We consider here only a single myosin head, designated M, and two actin monomers, designated A_x, where x is the linear position along the thin filament, relative to the thick filament. Including the substrate ATP and the products ADP and P_i, one complete cycle can be represented most simply as

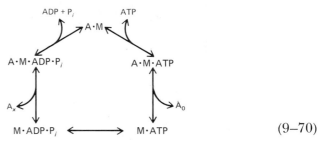

$$(9\text{–}70)$$

Although the cycle is fully reversible, the high, far-from-equilibrium concentration of ATP in vivo drives it in a clockwise direction.

ATP–myosin binding and actin–myosin binding are both strong but are mutually antagonistic: Binding of ATP causes a 10^5-fold decrease in affinity for actin, causing it to dissociate. Only then does myosin catalyze the hydrolysis of ATP. This requirement that actin dissociate is crucial for energy transduction; otherwise, ATP would simply be hydrolyzed unproductively by $A \cdot M$:

$$(9\text{–}71)$$

Hydrolysis of ATP restores the affinity of myosin for actin, so that binding occurs again. Energy transduction requires that in muscle the $M \cdot ADP \cdot P_i$ state not be able to rebind the original actin molecule, even though this would be energetically the most favorable; without this restriction the entire cycle of Equation 9–70 would result in only net ATP hydrolysis. In solution, there is no such restriction, and no work is performed. Mechanical work requires that the second actin molecule bound be about 100 Å further along the thin filament (i.e., $x = 100$ Å in Eq. 9–70). This restriction on which actin molecule can bind to $M \cdot ADP \cdot P_i$ must be due to steric restrictions resulting from a conformational change in the myosin molecule upon dissociation from the first actin molecule and hydrolysis of ATP (Figure 9–20), but its nature is not known.

Upon binding of the second actin molecule, the cycle is completed only when ADP and P_i are dissociated and when the second actin moves into the same position as the original molecule. Dissociation of ADP and P_i is usually very slow but is catalyzed by the binding of the actin. This also is probably crucial for energy transduction; otherwise, the free energy of ATP would be wasted, in that the free myosin head might bind another molecule of ATP and repeat the hydrolysis step. After binding of the second actin molecule, ADP and P_i dissociate and the actin thin filament and the attached myosin head move relative to the thick filament. The details are not at all clear, and only minor conformational changes have been observed in soluble myosin heads. The scheme of Equation 9–70 is not complete, in that it does not include these various conformational states, because the detailed interpretations are not yet certain. The major problem is to relate this scheme to the structural properties of muscle (Figure 9–20).

We also have not considered the detailed energetics of the various steps, because these are not of fundamental importance; only the energetics of the entire cycle are important:

$$ATP + A_X \rightleftharpoons ADP + P_i + A_0 \qquad (9\text{--}72)$$

This illustrates that the net equilibrium constant in muscle is the difference between the chemical energy of hydrolyzing one molecule of ATP and the mechanical energy required to move an actin thin filament by the distance x. The crucial requirement for energy transduction is that the cycle not be short-circuited by the conversion of one state to any state other than the two on either side of it in the cycle. This necessary limitation depends ultimately on enzyme specificity—which provides appropriate affinities for ligands and permits the hydrolysis of ATP or the release of products only at the appropriate stage—and on sufficient change in the conformation of myosin that it binds to different actin molecules during each cycle.

Under these restrictions, chemical energy will be converted to mechanical energy, but the rate is not specified. Each step must be catalyzed by the proteins to permit one cycle to occur in about 10^{-3} sec. How these processes occur at the molecular level is yet to be determined.

The mechanism of muscular contraction. H. E. Huxley. Science 164:1356–1366, 1969.

Proposed mechanism of force generation in striated muscle. A. F. Huxley and R. M. Simmons. Nature 233:533–538, 1971.

Molecular mechanism of contraction. Y. Tonomura and F. Oosawa. Ann. Rev. Biophys. Bioeng. 1:159–190, 1972.

Muscular contraction and cell motility. H. E. Huxley. Nature 243:445–449, 1973.

Molecular control mechanism in muscle contraction. A. Weber and J. M. Murray. Physiol. Rev. 53:612–673, 1973.

Theoretical formalism for the sliding filament model of contraction of striated muscle. Parts I and II. T. L. Hill. Prog. Biophys. Mol. Biol. 28:267–340, 1974; 29:105–159, 1975.

Muscle filament structure and muscle contraction. J. M. Squire. Ann. Rev. Biophys. Bioeng. 4:137–163, 1975.

Proteins of contractile systems. H. G. Mannherz and R. S. Goody. Ann. Rev. Biochem. 45:427–465, 1976.

Kinetic analysis of ATPase mechanisms. D. R. Trentham, et al. Q. Rev. Biophys. 9:217–281, 1976.

Non-muscle contractile proteins: the role of actin and myosin in cell motility and shape determination. M. Clarke and J. A. Spudich. Ann. Rev. Biochem. 46:797–823, 1977.

Biochemistry of actomyosin-dependent cell motility. E. D. Korn. Proc. Natl. Acad. Sci. U.S.A. 75:588–599, 1978.

Mechanism of actomyosin ATPase and the problem of muscle contraction. E. W. Taylor. Crit. Rev. Biochem. 6:103–164, 1979.

Regulation and kinetics of the actin–myosin–ATP interaction. R. S. Adelstein and E. Eisenberg. Ann. Rev. Biochem. 49:921–956, 1980.

Millisecond time-resolved changes in X-ray reflections from contracting muscle during rapid mechanical transients, recorded using synchrotron radiation. H. E. Huxley, et al. Proc. Natl. Acad. Sci. U.S.A. 78:2297–2301, 1981.

The Structural Basis of Muscular Contraction. J. Squire. New York, Plenum Press, 1981.

Time-resolved x-ray diffraction studies on vertebrate striated muscle. H. E. Huxley and A. R. Faruqi. Ann. Rev. Biophys. Bioeng. 12:381–417, 1983.

Active Transport

Pumping of ions and other metabolites against concentration gradients across membranes is a major metabolic activity of cells. ATP hydrolysis is used directly to pump H^+, Ca^{2+}, K^+, and Na^+ ions; other metabolites use these concentration gradients to drive energetically their active transport. Consequently, the transport of the ions H^+, Ca^{2+}, K^+, and Na^+ are the primary processes in transduction of chemical energy into concentration gradients.

Three such ion pumps are now known; they catalyze the following translocations, where the subscript i refers to the inside of the membrane and o to the outside:

$$2\ K_o^+ + 3\ Na_i^+ + ATP_i \rightleftharpoons ADP_i + P_i + 3\ Na_o^+ + 2\ K_i^+ \quad (9\text{–}73)$$

$$2\ Ca_i^{2+} + ATP_i \rightleftharpoons ADP_i + P_i + 2\ Ca_o^{2+} \quad (9\text{–}74)$$

$$n\ K_o^+ + n\ H_i^+ + ATP_i \rightleftharpoons ADP_i + P_i + n\ H_o^+ + n\ K_i^+ \quad (9\text{–}75)$$

These three ion pumps show similarities both in their mechanisms of action and in their composition—each is composed of polypeptide chains having molecular weights of about 100,000. In addition, the $(Na^+ + K^+)$ ATPase (Eq. 9–73) has equal numbers of a smaller subunit. Ion pumps do not rotate across the membrane and do not simply carry an ion from one side to another; consequently, they are believed to use channels through which the ions are transported. Along with most other membrane transport proteins, they were believed to exist as oligomers in the membrane and to have a channel between the individual subunits. The subunits were then imagined to cooperate and give complex kinetic behavior. However, the physical state of membrane proteins has been found to be notoriously difficult to establish directly and unambiguously. In particular, the $(Na^+ + K^+)$ ATPase has recently been demonstrated convincingly to function as a monomer; oligomers were found to be inactive. Therefore, ions are likely to be transported through a channel within a single asymmetric protein unit, as also appears to be the case with the proton-pump bacteriorhodopsin, as described in the following section.

The presumed channels must be carefully controlled so that ions do not pass freely and thereby short-circuit the membrane. At the present time, these three ion pumps are believed to function similarly, although no mechanisms are yet well-established. The current proposed reaction scheme for the $(Na^+ + K^+)$ ATPase is illustrated by

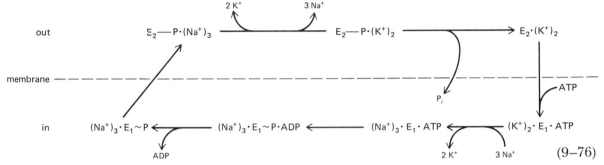

$$(9-76)$$

All the steps are reversible, but the cycle is indicated as operating in the clockwise direction—that usually dictated by the high intracellular concentration of ATP. The ion pump enzyme is shown binding simultaneously $2 \, Na^+$ or $3 \, K^+$ ions, only because this is the observed net stoichiometry. The net reaction is Equation 9–73.

Four forms of the protein are required, E_1 and E_2, each with and without a phosphate group esterified to a particular Asp side chain. The E_1 form of the protein would have its channel open to the inside, E_2 to the outside. During the interconversion between the two forms, the ions somehow pass from one end of the channel to the other.

It might be thought that pumping of ions against their concentration gradients would require that the $E_1 \cdot ATP$ form of the enzyme have greater affinity for Na^+ than K^+ and the E_2—P form have greater affinity for K^+, and that the E_2 state of the enzyme be preferred over E_1 when it is phosphorylated and vice versa for the nonphosphorylated form, thereby pushing the steps in a clockwise direction. However, these considerations are relevant only to the relative energetics of individual transitions, not to the energetics of the overall cycle. It is the latter that determines whether ions will be pumped against a concentration gradient, and the suggested considerations are not necessary.

Efficient energy transduction between ATP hydrolysis and ion pumping requires only that the cycle just described not be short-circuited by other interconversions, which is primarily a matter of enzyme specificity. For example, it is important that only the E_1 form of the protein, not E_2, catalyze hydrolysis of ATP and that only the E_2 form, not E_1, be able to catalyze removal of the phosphoryl group. Otherwise, ATP hydrolysis would occur without transport. The restriction on transport is that the phosphorylated form of the enzyme interconvert between the E_1 and E_2 states, i.e., permit transport across the membrane, only when Na^+ is bound, not K^+. Similarly, the nonphosphorylated form should open its channel only when K^+ is bound, not Na^+. Without these restrictions, K^+ and Na^+ would leak back through the membrane. With only these two sets of restrictions, ion transport and ATP hydrolysis are coupled and energy transduction occurs.

The principles involved are seemingly no more rigorous than we have already encountered in enzyme catalysis in terms of specificity and regulation by covalent modification, especially phosphorylation (see pp. 466–473). Nevertheless, the molecular mechanisms utilized will be of great interest. At the present time, a major question concerns the interactions between different polypeptides of any oligomeric pumps. There are numerous experimental indications of half-of-the-sites reactivity, i.e., extreme negative cooperativity; whether this reflects experimental artifacts, physical coupling of action of the monomers, or sharing by two or more polypeptides of a single ion channel remains to be determined.

Simple allosteric model for membrane pumps. O. Jardetsky. Nature 211:969–970, 1966.

The molecular organization of membranes. S. J. Singer. Ann. Rev. Biochem. 43:815–833, 1974.

Molecular mechanism of active calcium transport by sarcoplasmic reticulum. M. Tada, et al. Physiol. Rev. 58:1–79, 1978.

Biochemical mechanism of the sodium pump. E. T. Wallick, et al. Ann. Rev. Physiol. 41:397–411, 1979.

Transport of ions. E. Racker. Acc. Chem. Res. 12:338–344, 1979.

The ($Na^+ + K^+$)-activated ATPase. Enzymatic and transport properties. J. D. Robinson and M. S. Flashner. Biochim. Biophys. Acta 549:145–176, 1979.

Energy interconversion by the Ca^{2+}-dependent ATPases of the sarcoplasmic reticulum. L. de Meis and A. L. Vianna. Ann. Rev. Biochem. 48:275–292, 1979.

Passive and active calcium fluxes across plasma membranes. P. V. Sulakhe and P. J. St. Louis. Prog. Biophys. Mol. Biol. 35:135–195, 1980.

The structure of proteins involved in active membrane transport. A. S. Hobbs and R. W. Albers. Ann. Rev. Biophys. Bioeng. 9:259–291, 1980.

Membrane protein oligomeric structure and transport function. M. Klingenberg. Nature 290:449–454, 1981.

Molecular considerations relevant to the mechanism of active transport. J. Kyte. Nature 292:201–204, 1981.

Structural basis of proton-translocating protein function. Y. A. Ovchinnikov, et al. Ann. Rev. Biophys. Bioeng. 11:445–463, 1982.

Monomer of sodium and potassium ion activated adenosine triphosphatase displays complete enzymatic function. W. S. Craig. Biochemistry 21:5707–5717, 1982.

Mechanism of free energy coupling in active transport. C. Tanford. Ann. Rev. Biochem. 52:379–409, 1983.

Proton ATPases: structure and mechanism. L. M. Amzel and P. L. Pedersen. Ann. Rev. Biochem. 52:801–824, 1983.

Proton Pumps and Light Energy

Much of the energy from oxidative metabolism and that from photosynthesis is converted to ATP only indirectly, via the generation of chemiosmotic proton gradients across membranes. The systems that generate these proton gradients, and those that then use the proton gradient to form ATP, are large, complex supramolecular structures not yet known at the molecular level. They are being studied intensively and will undoubtedly yield their secrets in the near future, but as yet there is too little known about their molecular aspects for them to be described here.

Nevertheless, there is one relatively simple proton pump driven by light that, although not yet fully understood, is likely in the very near future to give a detailed picture of how it works. Consequently, it seems worthwhile to present here a brief description of the system so that readers will be able to appreciate the undoubtedly elegant story soon to unfold.

This simple system is the bacteriorhodopsin of the purple membrane of *Halobacterium halobium*. It uses light energy to pump protons out of the bacterial cell under anaerobic conditions; this proton gradient is then used to generate ATP. For each photon absorbed, approximately one proton is pumped across the membrane. Bacteriorhodopsin is a membrane-bound single polypeptide chain of 248 residues, containing one molecule of retinal attached via a protonated Schiff base to the amino group of Lys 216:

$$(9\text{--}77)$$

The same retinal chromophore is used in the vision system of all animals.

The three-dimensional structure of bacteriorhodopsin, within the regular two-dimensional lattice it forms within the purple membrane, has been determined to 7 Å resolution, using a novel combination of electron diffraction and microscopy (Figure 9–21). Seven presumably helical segments of polypeptide chain are visible traversing the membrane bilayer, but neither the position of the retinal nor the pathway of the polypeptide chain protruding into the aqueous phase at both membrane surfaces, and connecting the seven helical segments, can be discerned. However, one particular arrangement of the polypeptide chain seems most probable from physical principles (Figure 9–22).

The physical principles of membrane protein architecture have not been discussed here, because so few facts are available; this model of bacteriorhodopsin is the most detailed structural information available at present. Nevertheless, it would be expected that a membrane protein would have a higher proportion of nonpolar amino acid residues than that found in a protein soluble in aqueous solution. This appears to be the case, but the differences are not drastic. As can be seen in the bacteriorhodopsin sequence (Figure 9–22), there are seven predominantly nonpolar segments

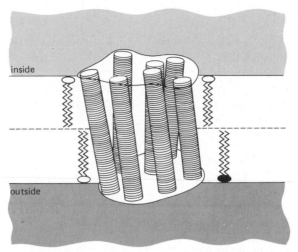

Figure 9–21

Low-resolution model of one molecule of bacteriorhodopsin within a lipid bilayer. The seven continuous rods through the membrane are believed to be α-helices. The polypeptide segments, which must protrude into the aqueous phase on both sides of the membrane to link these seven helices of the single polypeptide chain, are not visible in the electron density map, nor is the retinal chromophore. The proton channel is believed to lie in the middle of the molecule, between the helices; there is no aqueous channel, only amino acid side chains. (From R. Henderson, Ann. Rev. Biophys. Bioeng. 6:87–109, 1977.)

that are likely to be the membrane-submersed segments, but they are not devoid of polar and normally ionized side chains, as might have been expected for the nonpolar membrane interior. Of course, nonpolar side chains alone are unlikely to be sufficient to maintain a folded conformation within a membrane. The hydrophobic effect will not be operative; thus, purely nonpolar segments of the polypeptide chain would have no intrinsic tendency to associate with each other, as they do in bacteriorhodopsin. The polar, protruding segments could limit the mobility of the membrane-submersed segments, but more stabilizing interactions between them could be imagined to come from hydrogen bonds and salt bridges within the membrane. The polar amide groups might be used, as in the β-barrels of proteins like triose phosphate isomerase (see Figure 6–16), but bacteriorhodopsin and many other membrane proteins apparently pair these hydrogen bonds within α-helices. Consequently, polar interactions between side chains must be used and may be a reason for the occurrence of a few polar amino acids within the membrane. The amino acid sequence of bacteriorhodopsin possibly is arranged so that each of the seven helical segments has all of its basic and acidic residues along one side of the helical cylinder, where they would be expected to face the interior, pair in salt bridges, and hold the seven helical segments together. Under such an arrangement, membrane-submersed proteins are thus inside-out in comparison to water-soluble proteins, with nonpolar side chains on the protein

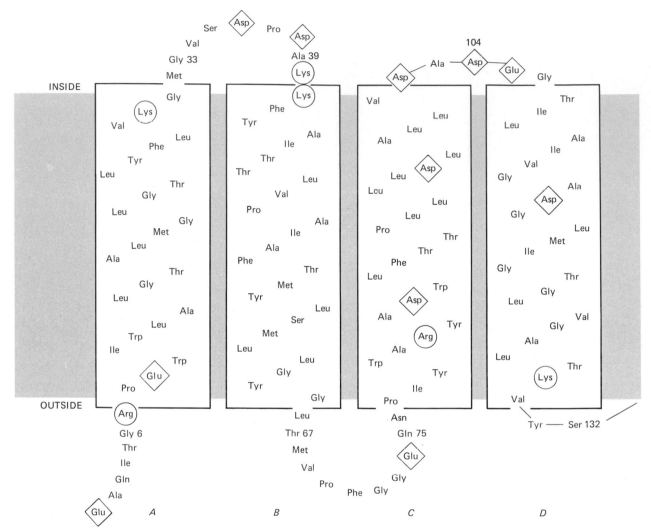

Figure 9–22
Tentative arrangement of the polypeptide chain of bacteriorhodopsin in the seven helices, labeled A to G, traversing the lipid membrane region, indicated by the cross-hatching. Each α-helix cylinder is opened out and laid flat, so that residues aligned vertically would be on the same side of the cylinder. Basic residues are circled; acidic residues, enclosed by diamonds. Note that most of the charged residues are in the connecting segments and presumably are exposed to the aqueous solvent. The basic and acidic residues within the buried helices tend to occur on just one side of each helix, presumably facing inward, where they would pair and stabilize the three-dimensional arrangement of the helices (see Figure 9–21). (Adapted from D. Engelman, et al., Proc. Natl. Acad. Sci. U.S.A. 77:2023–2027, 1980.)

exterior within the nonpolar membrane interior, and polar residues internal and paired in hydrogen bonds and salt bridges.

Another reason for bacteriorhodopsin to have polar residues on the interior, between the seven helices, would be to provide a channel through

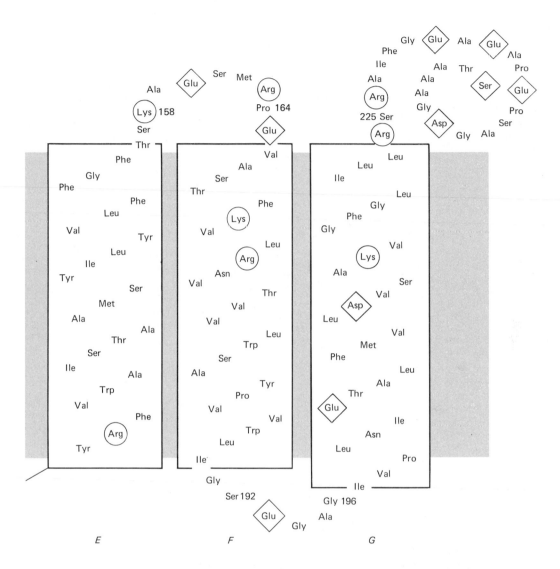

which protons can be pumped. Upon absorbing light, a proton is believed to be released on the outside of the membrane, and then one taken up on the inside. This proton must then be effectively transported through the membrane to occupy the position of the proton first released, so that one cycle is completed and the process may be repeated. The most obvious way for a protein to release or take up protons is to have ionizable groups with pK_a values that are altered in its different states. The energetics of the bacteriorhodopsin cycle require that such cyclic pK_a changes be of the order of 5 pH units. How a proton might be effectively transported through the presumed interior channel of bacteriorhodopsin is not known, but transport could be imagined to occur via chains of hydrogen bonds through the channel:

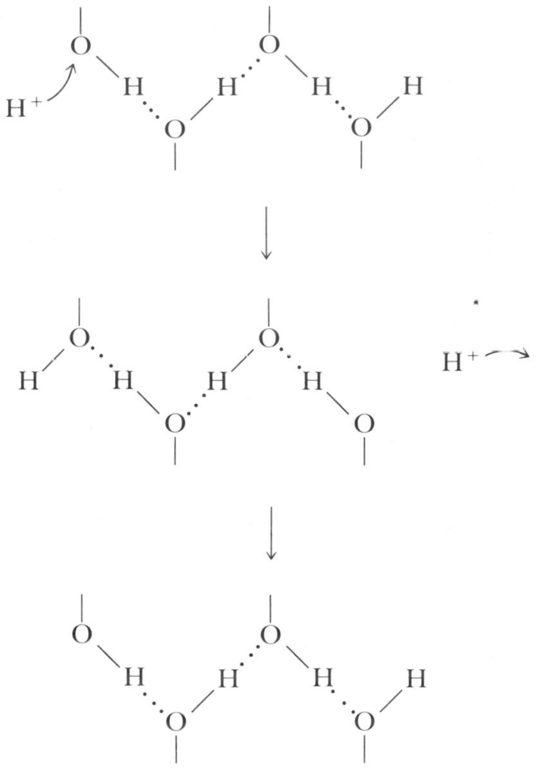

$$(9-78)$$

as occurs in ice. However, this process must be regulated to occur in essentially only one direction, so that net transport occurs and the membrane is not short-circuited. The relative energies of the two tautomeric states in Equation 9–78 need not have the same free energies and could vary with relatively small changes in geometry of the interacting groups. Other hydrogen-bonding groups, such as carboxyl, amino, guanidinium, imidazole, and even amide and peptide groups, could be imagined to participate transiently in high-energy states of the proton induced by light absorption.

Because one cycle of light absorption and protein pumping by bacteriorhodopsin occurs within a few milliseconds, spectral probes are required for study. The current model of the sequence of events is illustrated in Figure 9–23. As in any sequential kinetic reaction, only those intermediates that are formed more rapidly than they disappear will accumulate to detectable levels; therefore, those that have been detected after a light flash are involved in increasingly slower transitions around the cycle. The initial, very rapid step is the only one to require light; thus, all of the light energy required for proton translocation must be stored immediately within the protein. This is believed to occur by an isomerization of the excited retinal, from that in which all the double bonds are of the *trans* configuration (Eq.

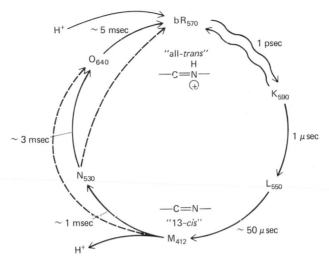

Figure 9–23

*Tentative photoreaction cycle of bacteriorhodopsin. Normal light-adapted bacteriorhodopsin has all-*trans*, protonated retinal and absorbs light maximally of 570-nm wavelength; hence it is designated* bR_{570}. *Upon absorption of light, it undergoes a clockwise cycle of transitions, with the indicated half-times, through intermediates L, M, N, and O, with their absorbance maxima indicated by the subscripts. The wavy transitions require light absorption. The kinetic roles of intermediates N and O are not certain. The indicated release and uptake of protons occur on opposite sides of the membrane. (Adapted from W. Stoeckenius, Acc. Chem. Res. 13:337–344, 1980.)*

9–77) to that where the double bond at carbon 13 is *cis*. An analogous isomerization occurs in rhodopsin of the eye, but the change is from 11-*cis* to *trans*. The relative energetics of the various isomers of retinal clearly depend upon the protein in which it is embedded. Rotation of 180 degrees about either double bond produces a substantial change in the geometry of the chromophore, which must have some effect on the protein.

As a consequence of light absorption, a proton is released from bacteriorhodopsin on the outside of the membrane; at about this stage, the Schiff base between the retinal and Lys residue loses its proton. The steps beyond this intermediate, M_{412} (Figure 9–23), are not well-established; presumably, however, a proton is taken up by bacteriorhodopsin from the inside of the membrane, a proton is transferred through the protein to occupy the site of the original proton released, the retinal resumes its normal all-*trans* isomer, and the Schiff base becomes protonated.

As in the previous cases of energy transduction, this cycle must not be short-circuited by transitions avoiding proton release or uptake; the protein must be capable of storing at least part of the light energy, and protons must be released, taken up, and transferred by specific groups within the protein. Some insight into how these processes take place should be gained from a more detailed structure of this relatively simple but proficient protein.

Three-dimensional model of purple membrane obtained by electron microscopy. R. Henderson and P. N. T. Unwin. Nature 257:28–32, 1975.

The purple membrane from *Halobacterium halobium.* R. Henderson. Ann. Rev. Biophys. Bioeng. 6:87–109, 1977.

Molecular mechanisms for proton transport in membranes. J. Nagle and H. J. Morowitz. Proc. Natl. Acad. Sci. U.S.A. 75:298–302, 1978.

Bacteriorhodopsin and the purple membrane of halobacteria. W. Stoeckenius, et al. Biochim. Biophys. Acta 505:215–278, 1979.

Purple membrane of halobacteria: a new light-energy converter. W. Stoeckenius. Acc. Chem. Res. 13:337–344, 1980.

Path of the polypeptide in bacteriorhodopsin. D. Engelman, et al. Proc. Natl. Acad. Sci. U.S.A. 77:2023–2027, 1980.

Bacteriorhodopsin is an inside-out protein. D. M. Engelman and G. Zaccai. Proc. Natl. Acad. Sci. U.S.A. 77:5894–5898, 1980.

Site of attachment of retinal in bacteriorhodopsin. H. Bayley, et al. Proc. Natl. Acad. Sci. U.S.A. 78:2225–2229, 1981.

Attachment site(s) of retinal in bacteriorhodopsin. N. V. Katre, et al. Proc. Natl. Acad. Sci. U.S.A. 78:4068–4072, 1981.

Bacteriorhodopsin and related pigments of halobacteria. W. Stoeckenius and R. A. Bogomolni. Ann. Rev. Biochem. 52:587–616, 1982.

Rhodopsin and bacteriorhodopsin: structure–function relationships. Y. A. Ovchinnikov. FEBS Letters 148:179–191, 1982.

10

DEGRADATION

Proteins are not immortal, even though they have so many other remarkable properties. They succumb to old age: Their sulfur atoms become oxidized, their Gln and Asn residues become deamidated, and they racemize about their C^α atoms to the D isomer and epimerize about any additional asymmetric centers of side chains (i.e., Thr, Ile). Of the many other chemical reactions possible (see Chapter 1), one of particular significance to aging is the nonenzymatic reaction between protein amino groups and reducing sugars, such as glucose and glucose 6-P, to form a Schiff base, which then undergoes an Amadori-type rearrangement to the more stable ketoamine:

$$
-NH_2 + \quad
\begin{array}{c}
HC{=}O \\
| \\
HCOH \\
| \\
HOCH \\
| \\
HCOH \\
| \\
HCOH \\
| \\
CH_2OH
\end{array}
\quad \rightleftharpoons \quad
\begin{array}{c}
HC{=}N- \\
| \\
HCOH \\
| \\
HOCH \\
| \\
HCOH \\
| \\
HCOH \\
| \\
CH_2OH
\end{array}
\quad \xrightarrow[\text{rearrangement}]{\text{Amadori}} \quad
\begin{array}{c}
CH_2{-}NH- \\
| \\
C{=}O \\
| \\
HOCH \\
| \\
HCOH \\
| \\
HCOH \\
| \\
CH_2OH
\end{array}
$$

glucose   aldimine (Schiff base)   ketoamine   (10–1)

This product then undergoes a variety of dehydrations to form yellow-brown fluorescent products and cross-links between protein molecules.

These aging processes have been especially evident in eye lens proteins, where they result in decreased solubility and pigmentation, with obvious deleterious effects on vision and formation of cataracts. Besides the extreme importance of the physical properties of lens proteins for their function, they are also excellent subjects for protein aging studies because lens cells do not die and there is little degradation of their proteins. Therefore, proteins synthesized at the embryonic stage are still present in the human eye lens 70 or more years later. Many structural proteins are also long-lived; others exist for the life of the cell they are in, such as hemoglobin, which is stable throughout the three-month average life span of an erythrocyte.

Relatively few proteins are so long-lived. There is extensive turnover of proteins within most cells; the proteins are constantly being degraded proteolytically to their constituent amino acids. This process is costly energetically but appears to be of physiological significance. Intracellular protein levels often need to change during development or in response to changes in the environment. They may be increased by accelerated synthesis but can decrease only if there is some mechanism for degrading proteins; the greater the rate of degradation, the greater the degree of control possible of the protein level. Cells also need mechanisms for surviving temporary conditions of starvation, so that they can degrade "luxury" proteins in order to utilize their amino acids for biosynthesis of those more essential for survival. There is also a need for a mechanism for degrading abnormal proteins, either those that have succumbed to old age or those that were synthesized in an incorrect form, owing to mistakes in the biosynthetic machinery for assembly and folding of the polypeptide chain.

Although these reasons for protein degradation are all plausible and consistent with many of the observations, there are very many unexplained observations, and the phenomenon of protein turnover is not understood. Only in the past decade has the importance and intriguing nature of protein degradation been widely appreciated; consequently, far less is known about it than about protein biosynthesis. However, the current activity and excitement in the field indicates that the situation will soon be rectified.

Control of enzyme levels in animal tissues. R. T. Schimke and D. Doyle. Ann. Rev. Biochem. 39:929–976, 1970.

Control of enzyme levels in mammalian tissues. R. T. Schimke. Adv. Enzymol. 37:135–187, 1973.

Evolution and the distribution of glutaminyl and asparaginyl residues in proteins. A. B. Robinson. Proc. Natl. Acad. Sci. U.S.A. 71:885–888, 1974.

Intracellular protein degradation in mammalian and bacterial cells. Part I. A. L. Goldberg and J. F. Dice. Ann. Rev. Biochem. 43:835–869, 1974; Part 2. A. L. Goldberg and A. C. St. John. Ann. Rev. Biochem. 45:747–803, 1976.

Amino acid racemization reactions and their geochemical implications. J. L. Bada and R. A. Schroader. Naturwissenschaften 62:71–79, 1975.

Intracellular protein degradation. F. J. Ballard. Essays Biochem. 13:1–37, 1977.

Intracellular Protein Catabolism. V. Turk and N. Marks (eds.). New York, Plenum Press, 1977.

Intracellular protein degradation. J. Kay. Biochem. Soc. Trans. 6:789–797, 1978.

Protein Turnover in Mammalian Tissues and in the Whole Body. J. C. Waterlow, et al. North Holland, Amsterdam, 1978.

Aspartic acid racemization in heavy molecular weight crystalline and water-insoluble protein from normal human lenses and cataracts. P. M. Masters, et al. Proc. Natl. Acad. Sci. U.S.A. 95:1204–1208, 1978.

The glyosylation of hemoglobin: relevance to diabetes mellitus. H. F. Bunn, et al. Science 200:21–27, 1978.

Protein degradation in health and disease. Ciba Found. Symp., Vol. 75, 1980.

Non-enzymatic browning in vivo: possible process for aging of long-lived proteins. V. M. Monnier and A. Cerami. Science 211:491–493, 1981.

Aging of protein molecules: lens crystallins as a model system. J. S. Zigler and J. Goosey. Trends Biochem. Sci. 6:133–136, 1981.

Lens proteins. H. Bloemendal. Crit. Rev. Biochem. 12:1–38, 1982.

RATES OF DEGRADATION IN VIVO

Physiologically significant measurements of rates of protein degradation need to be made in vivo, which introduces significant experimental complexities. The most prevalent methods utilize radioactive tracers; either a labeled protein is injected into the appropriate site, or it is synthesized in situ by giving a pulse of radioactive amino acids or precursors. The disappearance of the radioactivity in the relevant protein is then followed as a function of time. It is assumed that the radioactive protein is indistinguishable from the normal protein, but this may not be the case if it was prepared by covalent modification. Pulse synthesis in situ of a protein from ^{3}H, ^{14}C, or ^{35}S isotopic forms of the normal amino acids is preferable on this basis but suffers from the tendency of the radioactive amino acids released from the degraded protein to be utilized for further rounds of protein synthesis. This tendency can be minimized by adding a large excess of unlabeled amino acid after the initial radioactive pulse, but this may alter the rate of degradation. With all such manipulations there is uncertainty about the size and nature of the intracellular pools of amino acids used for protein synthesis, as well as only limited ability to manipulate them. Nevertheless, this method has been used extensively and further refined by using at different times two pulses with different isotopes, e.g., first ^{14}C, than ^{3}H at a later time. All other factors being constant or normalized, the subsequent ratio of the two isotopes present within a protein will depend upon the degree of degradation of the protein labeled by the first isotope that occurred during the time interval between the two pulses.

Nonradioactive methods are also useful. For example, the rate constant for degradation (k_D) of a protein (P) may be calculated from the difference between its measured rate of synthesis (k_s) and the rate of increase in its cellular level ($d[P]/dt$)

$$\text{amino acids} \xrightarrow{k_s} \text{P} \xrightarrow{k_D} \text{degradation} \qquad (10\text{--}2)$$

$$\frac{d[P]}{dt} = k_s - k_D[P] \qquad (10\text{--}3)$$

Some proteins have unique amino acids, generated by posttranslational covalent modification, that are not utilized metabolically after degradation. For example, certain His residues in the muscle proteins actin and myosin are quantitatively methylated. This methyl-His amino acid does not occur in other proteins; it is not utilized but is excreted in the urine. Consequently, accurate measurements of the degradation of these two proteins can be made by measuring the rate of excretion, the total protein pool, and any dietary intake of this unique amino acid. Similarly, hydroxyproline in the urine has long been used clinically as a measure of collagen degradation.

When the conditions are kept constant, protein degradation is found to follow first-order kinetics, with the rate determined by a single apparent rate constant. Consequently, degradation of protein molecules is random, and newly synthesized proteins are just as likely to be degraded as old proteins. Moreover, a single event must be sufficient to cause degradation. The nature of this event is not specified, and the observed rate may be the sum of the rates of a number of different events. However, first-order kinetics rule out the possibility that a protein molecule is degraded only after suffering two or more covalent modifications or proteolytic cleavages—two likely examples of degradative events.

A very wide range in the rates at which proteins are degraded has been observed under normal physiological conditions (Table 10–1), even within the same cell or tissue. The proteins that turn over most rapidly in normal cells tend to be those catalyzing rate-determining metabolic reactions. For example, the most rapidly degraded protein known in rat liver, ornithine decarboxylase, has a half-life of only 11 minutes and catalyzes the rate-determining step in polyamine biosynthesis. It is not subject to any known regulation by allosteric control or covalent modification. Instead, its synthesis may be varied 1000-fold, which, with its rapid rate of degradation, indicates that its activity is regulated via its levels in the liver.

The molecular basis for the varying rates of protein degradation has been sought by looking for correlations with specific molecular properties of the proteins. Seemingly significant correlations have been reported with susceptibility to thermal unfolding, dissociation of stabilizing ligand, and susceptibility to protease digestion in vitro. Such correlations would be expected if the rate-limiting step in degradation of a protein were its initial cleavage by a protease, which is more likely to occur in a transiently unfolded state. A further correlation has been reported with the rate of deamidation of a protein's Gln and Asn residues measured in vitro within short peptides of appropriate sequence. It could then be imagined that the covalent and folded structures of proteins have been selected evolutionarily to "self-destruct" at the appropriate rate.

However, additional correlations that are not so readily explicable have been observed. Analysis by isoelectric focusing shows acidic proteins to be degraded more rapidly than neutral or basic ones. SDS electrophoresis demonstrates that those composed of large polypeptide chains are de-

Table 10–1   Degradation Rates of Various Enzymes and Proteins

PROTEIN	TISSUE	HALF-LIFE
Ornithine decarboxylase	Rat liver	11 min
RNA polymerase I	Rat liver	1.3 h
Tyrosine aminotransferase	Rat liver	1.5 h
Tryptophan oxygenase	Rat liver	2 h
Phosphoenolpyruvate carboxykinase	Rat liver	6 h
Hexokinase	Rat liver	1 day
Acetyl CoA carboxylase	Rat liver	2 days
Glyceraldehyde phosphate dehydrogenase	Rat liver	3–4 days
Arginase	Rat liver	4–5 days
α-Actinin	Rat cardiac muscle	5–6 days
Myosin heavy chain	Rat cardiac muscle	5–6 days
Myosin light chain	Rat cardiac muscle	9 days
Actin	Rat cardiac muscle	7–8 days
Troponin	Rabbit skeletal muscle	10–15 days
α-Actinin	Rabbit skeletal muscle	20–25 days
Tropomyosin	Rabbit skeletal muscle	20–25 days
Myosin	Rabbit skeletal muscle	30 days
Actin	Rabbit skeletal muscle	>50 days
Hemoglobin	Reticulocytes/erythrocytes	≈∞

From J. Kay, Biochem. Soc. Trans. 6:789–797, 1978.

graded most rapidly. The rate of degradation is also proportional to the amount of exposed apolar surface area on the folded protein, as measured by its affinity for liver fat, organic phases, membranes, and hydrophobic chromatographic resins. Glycoproteins are also degraded more rapidly than those without attached carbohydrates. These correlations are not readily explained by simple proteolysis. In addition, intracellular protein degradation is almost invariably found to be energy-dependent, which is not expected for simple, energetically favorable proteolysis.

However, any such correlation can hold only to the extent that the same mechanisms are used for the degradation of all the proteins. That this is probably not the case is indicated by many observations that the rates of degradation of proteins change individually under different conditions. For example, a number of enzymes not required when glucose is present in the growth medium are "catabolite-inactivated" by specific degradation in yeast. Maturation of reticulocytes into erythrocytes is accompanied by the selective degradation of mitochondria, ribosomes, and many proteins no longer required once the cell has synthesized its complement of hemoglobin. In diabetes or starvation, there are many proteins whose degradation rates are markedly enhanced and those that are little affected.

Nevertheless, it is clear that the folded conformation of a protein is important for at least minimizing its rate of degradation. Abnormal proteins, such as those resulting from chemical modification, mutation, chain termination, or incorporation of amino acid analogues, are almost invariably found to be rapidly degraded proteolytically in cells. Excess, but normal, hemoglobin α chains synthesized as a result of a block in formation of β chains are usually also degraded, presumably because they cannot adopt their normal, more stable, conformation within the $\alpha_2\beta_2$ hemoglobin tetramer. Consequently, abnormal proteins are readily recognized by most cells and are rapidly degraded, but again, the process is not simple proteolysis, since it is energy-dependent.

Studies on the synthesis and degradation of proteins of the endoplasmic reticulum of rat liver. I. M. Arias, et al. J. Biol. Chem. 244:3303–3315, 1969.

β-Galactosidase: rates of synthesis and degradation of incomplete chains. S. Lin and I. Zabin. J. Biol. Chem. 247:2205–2211, 1972.

Studies on the correlation between size and relative degradation rate of soluble proteins. J. F. Dice, et al. J. Biol. Chem. 248:4220–4228, 1973.

Relationship between degradation rates of proteins *in vivo* and their susceptibility to lysosomal proteases. H. I. Segal, et al. J. Biol. Chem. 249:6364–6365, 1974.

Imbalanced globin chain synthesis in heterozygous β-thalassemic bone marrow. G. Chalevelakis, et al. Proc. Natl. Acad. Sci. U.S.A. 72:3853–3857, 1975.

Relationship between *in vivo* degradative rates and isoelectric points of proteins. J. F. Dice and A. L. Goldberg. Proc. Natl. Acad. Sci. U.S.A. 72:3893–3897, 1975.

Is protein turnover thermodynamically controlled? G. McLendon and E. Radany. J. Biol. Chem. 253:6335–6337, 1978.

Control of proteolysis. H. Holzer and P. C. Heinrich. Ann. Rev. Biochem. 49:63–91, 1980.

Properties of abnormal proteins degraded rapidly in reticulocytes. Intracellular aggregation of the globin molecules prior to hydrolysis. Y. Klemes, et al. J. Biol. Chem. 256:8436–8444, 1981.

Early steps in the processing of large premature termination fragments of β-galactosidase in *Escherichia coli*. J. L. McKnight and V. A. Fried. J. Biol. Chem. 256:9652–9661, 1981.

Role for the adenosine triphosphate–dependent proteolytic cleavage in reticulocyte maturation. F. S. Boches and A. L. Goldberg. Science 215:978–980, 1982.

Oxidized proteins in erythrocytes are rapidly degraded by the adenosine triphosphate–dependent proteolytic system. A. L. Goldberg and F. S. Boches. Science 215:1107–1109, 1982.

MECHANISMS OF PROTEIN DEGRADATION

Most studies of protein degradation have concentrated on proteolytic enzymes, since hydrolysis to amino acids seems to be a universal aspect of this process. However, cells and organisms have a variety of processes other than degradation that must be catalyzed by proteases, such as postbiosynthetic processing of "pre" and "pro" polypeptide chains (Chapter 2) activation of zymogen precursors of many proteins (Chapter 9), and digestion of exogenous proteins for their utilization as foodstuffs. Consequently the identification of a proteolytic enzyme does not imply that it is involved in protein turnover. Also, the initial, rate-determining event in degradation need not be proteolytic cleavage of the protein. Nevertheless, understanding degradation will require information about the relevant proteases; thus, this is a logical place to start studying the process.

In studies of protein degradation, attention naturally focused on **lysosomes,** those intracellular vesicles of potential hydrolytic destruction known to be involved in the uptake and digestion of exogenous proteins. They contain a variety of proteases, known as the cathepsins, which are related to the serine, carboxyl, and thiol proteases (pp. 427–449). They may be distinguished by their sensitivity to the unusual inhibitors from microorganisms, such as pepstatin and leupeptin:

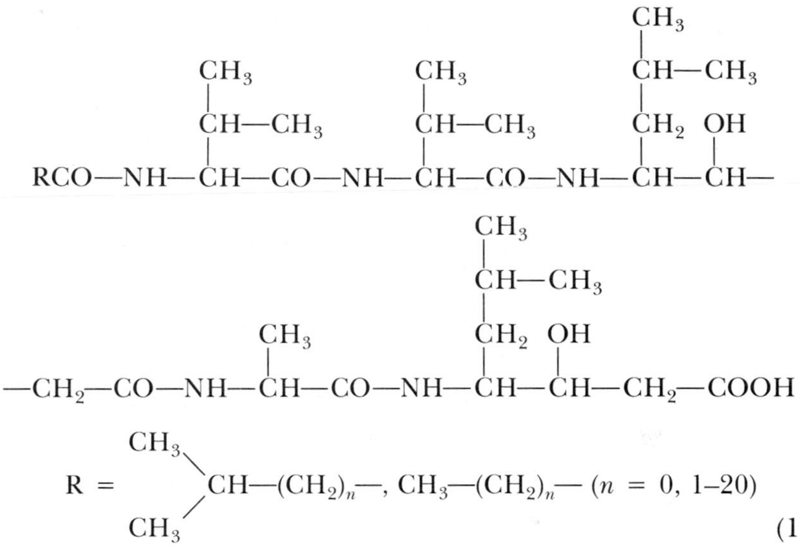

$$
\begin{array}{c}
\text{CH}_3 \\
|\\
\text{CH}_3 \qquad\qquad \text{CH}_3 \qquad\qquad \text{CH—CH}_3 \\
|\qquad\qquad\quad |\qquad\qquad\quad\;\; |\\
\text{CH—CH}_3 \quad\;\; \text{CH—CH}_3 \quad\;\; \text{CH}_2 \;\; \text{OH} \\
|\qquad\qquad\quad |\qquad\qquad\quad |\quad\;\; |\\
\text{RCO—NH—CH—CO—NH—CH—CO—NH—CH—CH—}
\end{array}
$$

$$
\begin{array}{c}
\text{CH}_3 \\
|\\
\text{CH—CH}_3 \\
|\\
\text{CH}_3 \qquad\qquad \text{CH}_2 \;\; \text{OH} \\
|\qquad\qquad\quad\; |\quad\;\; |\\
\text{—CH}_2\text{—CO—NH—CH—CO—NH—CH—CH—CH}_2\text{—COOH}
\end{array}
$$

$$
\text{R} = \quad {\begin{array}{c}\text{CH}_3 \\ \diagdown \\ \diagup \\ \text{CH}_3\end{array}}\text{CH—(CH}_2)_n\text{—, CH}_3\text{—(CH}_2)_n\text{— } (n = 0, 1\text{–}20)
$$

$$(10\text{–}4)$$

pepstatin

$$
\begin{array}{c}
\text{HO} \qquad\quad \text{OH} \\
\diagdown\quad\;\diagup \\
\text{CH} \qquad\qquad\qquad\qquad\qquad\qquad\;\; \text{NH}_2 \\
|\qquad\qquad\qquad\qquad\qquad\quad\; \diagup \\
\text{RCO—Leu—Leu—NH—CH—CH}_2\text{—CH}_2\text{—CH}_2\text{—NH—C} \\
\diagdown \\
\text{NH}
\end{array}
$$

$$\updownarrow$$

$$
\begin{array}{c}
\text{CHO} \qquad\qquad\qquad\qquad\qquad\qquad \text{NH}_2 \\
|\qquad\qquad\qquad\qquad\qquad\quad\;\; \diagup \\
\text{RCO—Leu—Leu—NH—CH—CH}_2\text{—CH}_2\text{—CH}_2\text{—NH—C} \\
\text{(L)}\qquad\qquad\qquad\qquad\qquad\qquad \diagdown \\
\text{NH}
\end{array}
$$

$$\updownarrow$$

$$
\begin{array}{c}
\qquad\qquad\qquad\qquad\qquad\quad \text{NH}_2 \\
\qquad\qquad\qquad\qquad\quad\;\; \diagup \\
\text{HO} \qquad\quad\; \text{C} \\
\diagdown\qquad\quad \diagdown \\
\text{RCO—Leu—Leu—NH—} \qquad \text{N} \qquad \text{NH}
\end{array}
$$

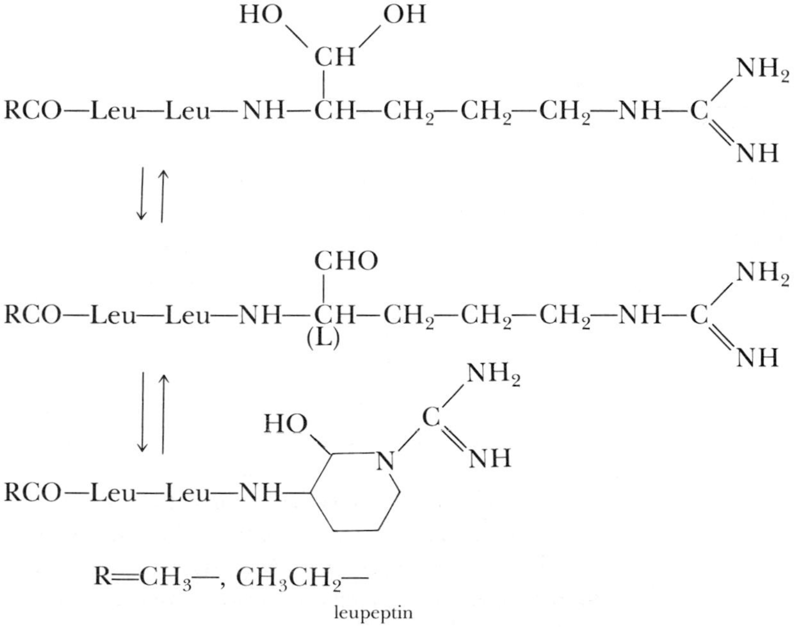

$$\text{R}{=}\text{CH}_3\text{—, CH}_3\text{CH}_2\text{—}$$

$$(10\text{–}5)$$

leupeptin

and by their requirement for relatively acidic conditions, which are maintained in the lysosome by active transport. Weakly basic reagents, which can penetrate membranes, such as ammonia and chloroquine,

$$\text{Cl} \quad \begin{array}{c} \overset{\displaystyle \bigcirc\!\!\!\!\bigcirc}{} \\ \text{HN—CH—(CH}_2)_3\text{—N—(C}_2\text{H}_5)_2 \\ | \\ \text{CH}_3 \end{array}$$

(10–6)

raise the internal pH of the lysosome and inhibit the cathepsins. The partial sensitivity of protein degradation in vivo to these inhibitors suggests that some, but not all, is catalyzed by the lysosomal proteases.

Direct uptake of proteins by lysosomes has not been observed; most likely, therefore, they act on intracellular proteins by the process of autophagy, in which a discrete volume of cytoplasm is sequestered by the cytomembrane and then fuses with a lysosome. All the sequestered proteins are believed to be hydrolyzed completely, so different rates of degradation of different proteins would require specificity in the uptake mechanism of autophagy. Such a membrane-related mechanism could explain the correlation of degradation rate with hydrophobic surface area of the protein. It may also account for the increased rates of degradation caused by starvation, deprivation of insulin in diabetes, and administration of glucagon, since these conditions affect the activities of lysosomes. The usual correlations of rate of degradation with acidity, large polypeptide chains, and glycosylation do not then apply. The energy requirement for degradation could also be explained by the need for the process of autophagy and for lysosomes to maintain a low pH by active transport.

On the other hand, not all cells have lysosomes or equivalent hydrolytic vacuoles, and there are numerous indications that much protein degradation does not occur via lysosomes. Other proteases must be involved, and numerous ones may be detected in cell-free extracts. However, the most active proteases are often found to arise from specific cells within tissues, such as mast cells, where they are not in contact with the proteins that are degraded. Mutational studies in microorganisms are most useful for assessing the in vivo roles of proteases; mutant strains lacking several active proteases have been found to have unaltered rates of in vivo degradation, indicating that those proteases are not required.

A variety of proteases of varying specificity appear to be involved in degradation of some proteins and in some disease states, such as muscular dystrophy, arthritis, degenerative skin disorders, respiratory and gastrointestinal diseases, and malignancy. Two such prominent proteases are collagenase and elastase, which degrade the structural proteins collagen and elastin. Muscle cells have a Ca^{2+}-activated protease that degrades selectively some of the muscle proteins. There are also reported to be proteases specific for pyridoxal phosphate-containing and NAD-containing proteins, but only the apoproteins are hydrolyzed; therefore, the proteases must be specific not for the cofactor but for its empty binding site. It can be imagined that a protease disguised as a cofactor binds to the cofactor binding site and then suddenly unleashes its proteolytic activity on the unsuspecting protein. With the possibility of a large number of different proteases, with varying specificities for different proteins, it is easy to envisage different rates of degradation, with different regulatory features.

However, the nature of the initial event in degradation is not known, for there are few observations of partially degraded proteins, and this event need not be proteolysis. It could be chemical aging of the protein or specific, enzymic modification that labels the protein for degradation. Alternatively, it could be attachment to certain receptors on membranes for autophagy, or transient unfolding, perhaps upon dissociation of a stabilizing ligand. The rate of degradation could be determined either by the protein or by the apparatus that produces the initiating event.

One of the most striking aspects of protein degradation in vivo is the requirement for energy within the cell. Because protein hydrolysis per se does not have this energy requirement, it was presumed to be due to an indirect effect, such as autophagy or maintenance of a low pH within lysosomes, as suggested earlier. However, energy is now known to be required for the initiation of proteolytic degradation of at least some proteins, and an ATP-dependent proteolytic system has recently been discovered in cell-free extracts. It seems likely that this system, which in *E. coli* has been assigned to a single gene product, may be the prime agent of normal protein turnover. Such turnover of at least a few proteins is apparently of physiological importance; *E. coli* mutants that lack this protease exhibit a wide variety of physiological defects, including overproduction of capsular polysaccharide, enhanced sensitivity to DNA-damaging agents, abnormal control of cell division, and lysogenic response to bacteriophage infection.

The responsible protease requires ATP for protein degradation and hydrolyzes it in the process, suggesting a complex mechanism of action. The initial observations with this system have been surprising, intriguing, but also contradictory. Unfortunately, at present it is clear that a description of these observations would be premature; we can only follow this intriguing story as it unfolds.

Proteases and Biological Control. E. Reich et al., (eds.). New York, Cold Spring Harbor Laboratories, 1975.

Lysosomes. R. T. Dean and A. J. Barrett. Essays Biochem. 12:1–40, 1976.

Proteinases in Mammalian Cells and Tissues. A. J. Barrett (ed.). Amsterdam, North Holland, 1977.

New intracellular proteases and their role in intracellular enzyme degradation. N. Katunuma. Trends Biochem. Sci. 2:122–125, 1977.

The many forms and functions of cellular proteinases. A. J. Barrett. Fed. Proc. 39:9–14, 1980.

The genetics of protein degradation in bacteria. D. W. Mount. Ann. Rev. Genetics 14:279–319, 1980.

Enzyme inactivation via disulphide–thiol exchange as catalysed by a rat liver membrane protein. G. L. Francis and F. J. Ballard. Biochem. J. 186:581–590, 1980.

Control of protein degradation in reticulocytes and reticulocyte extracts by hemin. J. D. Etlinger and A. L. Goldberg. J. Biol. Chem. 255:4563–4568, 1980.

Structure, specificity and localization of the serine proteases of connective tissue. R. G. Woodbury and H. Neurath. FEBS Letters 114:189–196, 1980.

ATP-dependent conjugation of reticulocyte proteins with the polypeptide required for protein degradation. A. Ciechanover, et al. Proc. Natl. Acad. Sci. U.S.A. 77:1365–1368, 1980.

Proposed role of ATP in protein breakdown: conjugation of proteins with multiple chains of the polypeptide of ATP-dependent proteolysis. A. Herschko, et al. Proc. Natl. Acad. Sci. U.S.A. 77:1783–1786, 1980.

Stimulation of ATP-dependent proteolysis requires ubiquitin with the COOH-terminal sequence Arg-Gly-Gly. K. D. Wilkinson and T. K. Audhya. J. Biol. Chem. 256:9235–9241, 1981.

E. coli contains eight soluble proteolytic activities, one being ATP dependent. K. H. S. Swamy and A. L. Goldberg. Nature 292:652–654, 1981.

Selective depletion of small basic non-glycosylated proteins in diabetes. F. C. Samaniago, F. Berry, and J. F. Dice. Biochem. J. 198:149–157, 1981.

Proteinase action *in vitro* versus proteinase function *in vivo:* mutants shed light on intracellular proteolysis in yeast. D. H. Wolf. Trends Biochem. Sci. 7:35–37, 1982.

Protease La from *Escherichia coli* hydrolyzes ATP and proteins in a linked fashion. L. Waxman and A. L. Goldberg. Proc. Natl. Acad. Sci. U.S.A. 79:4883–4887, 1982.

Mechanisms of intracellular protein breakdown. A. Hershko and A. Ciechanover. Ann. Rev. Biochem. 51:335–364, 1982.

Mutations in the *lon* gene of *E. coli* K12 phenotypically suppress a mutation in the sigma subunit of RNA polymerase. A. D. Grossman, et al. Cell 32:151–159, 1983.

Ubiquitin: roles in protein modification and breakdown. A. Hershko. Cell 34:11–12, 1983.

INDEX